Clusters of Galaxies
Probes of Cosmological Structure and Galaxy Evolution

Clusters of galaxies are the largest and most massive collapsed systems in the universe, and as such they are valuable probes of cosmological structure and galaxy evolution. The advent of extensive galaxy surveys, large ground-based facilities, space-based missions such as *HST, Chandra*, and *XMM-Newton* and detailed numerical simulations makes now a particularly exciting time to be involved in this field. The review papers in this volume span the full range of current research in this area, including theoretical expectations for the growth of structure, survey techniques to identify clusters, metal production and the intracluster medium, galaxy evolution in the cluster environment and group–cluster connections. With contributions from leading authorities in the field, this volume is appropriate both as an introduction to this topic for physics and astronomy graduate students, and as a reference source for professional research astronomers.

JOHN S. MULCHAEY's research has focused on groups of galaxies. In 1993, he provided some of the strongest evidence to date that galaxy groups are dominated by dark matter. More recently he has played an important role in the discovery and study of "fossil groups," massive systems that contain very few galaxies.

ALAN DRESSLER has made many fundamental contributions to the study of large-scale structure in the Universe over the last 30 years. Recently, he participated in the MORPHS project, using *Hubble Space Telescope* images to show that bursts of star formation were much more common in galaxies 5 billion years ago than they are today.

AUQUSTUS OEMLER has devoted much of his research career to understanding how galaxies have evolved to their present form. In collaboration with H. Butcher, he showed that clusters at intermediate redshifts contain a large excess of blue galaxies (now known as the Butcher–Oemler effect). He recently finished a seven-year term as director of Carnegie Observatories.

All three editors are staff astronomers at the Carnegie Observatories.

This series of four books celebrates the Centennial of the Carnegie Institution of Washington, and is based on a set of four special symposia held by the Observatories in Pasadena. Each symposium explored an astronomical topic of major historical and current interest at the Observatories, and each resulting book contains a set of comprehensive, authoritative review articles by leading experts in the field.

Series Editor: Luis C. Ho.
Luis Ho received his undergraduate education at Harvard University and his Ph.D. in astronomy from the University of California at Berkeley. He is currently a staff astronomer at the Carnegie Observatories, where he conducts research on black holes, accretion physics in galactic nuclei, and star formation processes.

Carnegie Observatories Astrophysics Series
Volume 3

CLUSTERS OF GALAXIES

Probes of Cosmological Structure and Galaxy Evolution

Edited by

JOHN S. MULCHAEY

ALAN DRESSLER

and

AUGUSTUS OEMLER

CAMBRIDGE
UNIVERSITY PRESS

CAMBRIDGE UNIVERSITY PRESS
Cambridge, New York, Melbourne, Madrid, Cape Town, Singapore,
São Paulo, Delhi, Dubai, Tokyo, Mexico City

Cambridge University Press
The Edinburgh Building, Cambridge CB2 8RU, UK

Published in the United States of America by Cambridge University Press, New York

www.cambridge.org
Information on this title: www.cambridge.org/9780521143523

First published 2004
First paperback printing 2010

A catalogue record for this publication is available from the British Library

ISBN 978-0-521-75577-1 Hardback
ISBN 978-0-521-14352-3 Paperback

Contents

Introduction

While the term "cluster of galaxies" dates back to at least the 18th century (and the work of Charles Messier), it was the discovery of Cepheids in M31 in the 1920's by Carnegie astronomer Edwin Hubble that established the extragalactic nature of these objects. With this realization, the study of clusters of galaxies was born. In the decades that followed, many Carnegie astronomers made important contributions to this growing field. Milton Humason, Allan Sandage and others measured redshifts for many clusters including Virgo and Coma, at the Mount Wilson and Palomar Observatories. These observations provided an essential database for the early study of clusters and helped establish that few galaxies occur in isolation. These early studies also included much poorer systems (i.e. groups). Edwin Hubble was responsible for noting that we live in such a system, which he named the Local Group. In 1958, Abell published a paper that included his famous cluster catalog. This work extended far beyond the catalog, however. In his paper, Abell showed there was a large variation in cluster richness, and the richness scale he defined is still widely used today. Walter Baade, for which one of the Magellan 6.5 meter telescopes is named, suggested in a 1951 paper with Spitzer that collisions between galaxies in clusters could transform a spiral into an early-type galaxy. This was the first paper on galaxy evolution in clusters.

Given Carnegie's extensive role in the study of clusters of galaxies, it was an easy decision to dedicate one of the Centennial Symposia to this topic. For me it was a great pleasure to organize this event with two of my colleagues who themselves have made fundamental contributions to this field. In fact, it is fair to say that one could not write a fair history of cluster research in the last three decades without mentioning the work of Alan Dressler and Gus Oemler.

The third Carnegie Observatories Centennial Symposium, "Clusters of Galaxies: Probes of Cosmological Structure and Galaxy Evolution" was held January 27–31, 2003 in Pasadena, California. Although we formally limited the attendance to 140 people, well over 160 people attended the meeting. Over the course of four days, there were 28 invited review talks, 21 contributed talks and over 60 poster presentations. Some of the highlights of the meeting included detailed numerical simulations of clusters, new results from the Sloan and 2dF redshift surveys, and exciting new results from *Chandra*, *XMM-Newton*, *HST* and Sunyaev-Zel'dovich surveys. I had many people tell me during the course of the meeting and in the months since that this was one of the most exciting meetings on clusters of galaxies in recent memory. I believe the quality of the scientific presentations is reflected in the review articles that appear in this volume.

The Symposium was made possible with the help of many people. Steve Wilson, Silvia

Hutchison and Becky Lynn were responsible for handling the meeting logistics. Scott Rubel was the Symposium photographer. Many people helped referee the invited review papers in this volume and I believe their participation greatly improved the quality of these articles. Finally, I'd like to extend my personal gratitude to Luis Ho for the tremendous amount of effort he put into the entire Centennial series and the resulting volumes. It was Luis' idea to hold these Symposia, and he was involved with every aspect of these events from the very beginning. I'm particularly indebted to him for his work on this volume. His guidance and insight have assured the content of these books will have a long-lasting impact on astronomy.

John Mulchaey
Carnegie Observatories
October 2003

List of Participants

Barrientos, Felipe	Universidad Catolica, Chile
Bautz, Mark	MIT, USA
Birkinshaw, Mark	University of Bristol, UK
Biviano, Andrea	Osservatorio Astronomico di Trieste, Italy
Blanton, Elizabeth	University of Virginia, USA
Blindert, Kris	University of Toronto, Canada
Borys, Colin	Caltech, USA
Bower, Richard	University of Durham, UK
Brown, Michael	National Optical Astronomy Observatory, USA
Burns, Jack	University of Colorado, USA
Carlberg, Ray	University of Toronto, Canada
Carrasco, Rodrigo	Gemini Observatory, USA
Cavaliere, Alfonso	University of Rome, Italy
Chapman, Scott	Caltech, USA
Choi, Philip	Caltech, USA
Christlein, Daniel	University of Arizona, USA
Cohen, Judith	Caltech, USA
Couch, Warrick	University of New South Wales, Australia
Crawford, Carolin	Institute of Astronomy, UK
Dave, Romeel	University of Arizona, USA
Davies, Roger	University of Durham, UK
Demarco, Ricardo	European Southern Observatory, Germany
Desai, Vandana	University of Washington, USA
Diaferio, Antonaldo	Universita' di Torino, Italy
Dickinson, Mark	Space Telescope Science Institute, USA
Donahue, Megan	Space Telescope Science Institute, USA
Dressler, Alan	Carnegie Observatories, USA
Duc, Pierre-Alain	CEA-Saclay, France
Dwarakanath, K. S.	NRAO, USA
Edge, Alastair	University of Durham, UK
Eisenhardt, Peter	JPL, USA
Ellingson, Erica	University of Colorado, USA

Ellis, Richard   Caltech, USA
Ellis, Simon   University of Birmingham, UK
Ettori, Stefano   European Southern Observatory, Germany
Evrard, Augustus   University of Michigan, USA
Feldmeier, John   Case Western Reserve University, USA
Ferrari, Chiara   Observatoire de la Cote d'Azur, France
Finn, Rose   University of Arizona, USA
Franx, Marijn   Leiden University, Netherlands
Fritz, Alexander   Universitots-Sternwarte Gottingen, Germany
Fujita, Yutaka   National Astronomical Observatory, Japan
Gal, Roy   Johns Hopkins University, USA
Gioia, Isabella   IRA-CNR, Italy
Gladders, Michael   Carnegie Observatories, USA
Gonzalez, Anthony   University of Florida, USA
Goto, Tomo   Carnegie Mellon University, USA
Helsdon, Stephen   Carnegie Observatories,USA
Hinz, Joannah   University of Arizona, USA
Ho, Luis   Carnegie Observatories, USA
Holden, Bradford   University of California at Davis, USA
Huchra, John   Harvard-Smithsonian Center for Astrophysics, USA
Im, Myungshin   SIRTF Science Center, USA
Jeltema, Tesla   MIT, USA
Jones, Laurence   University of Birmingham, UK
Kauffmann, Guinevere   MPA-Garching, Germany
Kelson, Dan   Carnegie Observatories, USA
Kocevski, Dale   University of Hawaii, USA
Koo, David   UCO/Lick Observatory, USA
Krick, Jessica   University of Michigan, USA
Lapi, Andrea   Univ. di Roma, Italy
Lin, Kai-Yang   National Taiwan University, Taiwan
Lin, Lihwai   National Taiwan University, Taiwan
Lopez-Cruz, Omar   INAOE-Tonatzintla, Mexico
Madau, Piero   U. C. Santa Cruz, USA
Malkan, Matt   UCLA, USA
Margoniner, Vera   Bell Laboratories, USA
Marmo, Chiara   University of Padova, Italy
Martini, Paul   Carnegie Observatories, USA
Maurogordato, Sophie   Observatoire de la Cote d'Azur, France
McNamara, Brian   Ohio University, USA
Medvedev, Mikhail   University of Kansas, USA
Melott, Adrian   University of Kansas, USA
Metevier, Anne   U. C. Santa Cruz, USA
Mihos, Chris   Case Western University, USA
Miles, Trevor   University of Birmingham, UK
Miller, Christopher   Carnegie Mellon University, USA
Miller, Neal   NASA/GSFC, USA
Moore, Ben   University of Durham, UK
Morrison, Glenn   Caltech-IPAC, USA

Motl, Patrick	University of Colorado, USA
Mulchaey, John	Carnegie Observatories, USA
Mullis, Christopher	European Southern Observatory, Germany
Mushotzky, Richard	NASA/GSFC, USA
Nagai, Daisuke	University of Chicago, USA
Nanduri, Vidyardhi	Cosmology Research Center, India
Nichol, Robert	Carnegie Mellon University, USA
Norman, Michael	University of California at San Diego, USA
Oegerle, William	NASA/GSFC, USA
Oemler, Augustus	Carnegie Observatories, USA
Olowin, Ronald	Saint Mary's College, USA
O'Shea, Brian	U. C. San Diego, USA
O'Sullivan, Ewan	Harvard-Smithsonian CfA, USA
Ostriker, Jeremiah	Institute of Astronomy, UK
Ouchi, Masami	University of Tokyo, Japan
Perlman, Eric	University of Maryland at Baltimore County, USA
Poggianti, Bianca	Osservatorio Astronomico di Padova, Italy
Ponman, Trevor	University of Birmingham, UK
Postman, Marc	Space Telescope Science Institute, USA
Pracy, Michael	University of New South Wales, Australia
Rakos, Karl	Institute for Astronomy, Austria
Renzini, Alvio	European Southern Observatory, Germany
Roeser, Hermann-Josef	MPI fuer Astronomie, Germany
Romer, Kathy	Carnegie Mellon University, USA
Rosati, Piero	European Southern Observatory, Germany
Rudnick, Gregory	Max-Planck-Institut fuer Astrophysik, Germany
Sakai, Shoko	UCLA, USA
Sanderson, Alastair	University of Illinois at Urbana-Champaign, USA
Schade, David	National Research Council of Canada, Canada
Scharf, Caleb	Columbia Astrophysics Laboratory, USA
Schombert, James	University of Oregon, USA
Shapley, Alice	Caltech, USA
Smail, Ian	University of Durham, UK
Smith, Graham	Caltech, USA
Steidel, Charles	Caltech, USA
Tran, Kim-Vy	U. C. Santa Cruz, USA
Treu, Tommaso	Caltech, USA
Tucker, Douglas	Fermilab, USA
van Breugel, Wil	LLNL, USA
van Gorkom, Jacqueline	Columbia University, USA
Webb, Tracy	Leiden Observatory, Netherlands
Wechsler, Risa	University of Michigan, USA
White, Simon	MPA-Garching, Germany
Willman, Beth	University of Washington, USA
Yamada, Toru	National Astronomical Observatory of Japan, Japan
Yee, Howard	University of Toronto, Canada

1

Galaxy clusters as probes of cosmology and astrophysics

AUGUST E. EVRARD
*Departments of Physics and Astronomy, Michigan Center for Theoretical Physics,
University of Michigan*

Abstract
Clusters of galaxies emerge as nodes in the gravitationally evolving cosmic web of dark matter and baryons that defines the large-scale structure of the Universe. X-ray and optical observations offer plentiful evidence of clusters' dynamical youth, yet bulk measures derived from these observations are tightly correlated, indicating a high degree of structural regularity that makes the population an attractive probe of cosmology. Accurate constraints on cosmological parameters require a precise and unbiased model relating observables to total mass, as well as a statistical characterization of the massive halo population within a given cosmology. In this contribution, I focus on the latter by providing evidence from simulations for $\mathcal{O}(10\%)$ calibration of the space density as a function of mass and for $\mathcal{O}(1\%)$ calibration of the dark matter virial relation. Matching the observed space density as a function of X-ray temperature for a ΛCDM world model is presented as an example of astrophysical/cosmological confusion. The resulting constraint $\beta \sigma_8^{-5/3} = (1.10 \pm 0.07)$ combines β, the ratio of specific energies in dark matter and intracluster gas, with σ_8, the normalization of the mass fluctuation spectrum. Disentangling astrophysical and cosmological factors for upcoming large statistical surveys is the main challenge in the quest to use galaxy clusters as sensitive probes of dark matter and dark energy.

1.1   Introduction
Cosmology is now a data-rich subject, with empirical support from at least four independent channels: the cosmic microwave background (CMB) radiation, light element nucleosynthesis, Type Ia supernovae, and large-scale cosmic structure. The latest CMB observations from the *Wilkinson Microwave Anisotropy Probe* (*WMAP*; Bennett et al. 2003) and ground-based experiments have revealed the series of acoustic peaks expected in an inflationary, hot Big Bang picture (Scott, Silk, & White 1995) and spotlight a fundamentally unexpected model containing roughly two-thirds vacuum or dark energy, one-third dark matter, and about four percent ordinary (baryonic) matter (Spergel et al. 2003). The nature of the principal dark components remains a mystery.

The ornery energy of Fritz Zwicky, though unlikely to dominate the Universe, is nonetheless involved in all this. Zwicky's revelation of dark matter in the Coma cluster (Zwicky 1933, 1937) established the roots of the modern era, and it is fitting to recognize his seminal contributions at this centennial event of the Carnegie Observatories. The problem of *missing mass*, as it was then known, derived from the humble virial theorem, helped to spark the

Table 1.1. *Mass Hierarchy in the Coma Cluster*

Component	$M(< 1.5\,h^{-1}\ \mathrm{Mpc})$ $(M_\odot)$	M/M_{vis}
Total[a]	$1.3 \pm 0.3 \times 10^{15}\ h_{70}^{-1}$	9.0 ± 2.5
Intracluster gas	$1.3 \pm 0.2 \times 10^{14}\ h_{70}^{-5/2}$	0.90 ± 0.02
Galaxies	$1.4 \pm 0.3 \times 10^{13}\ h_{70}^{-1}$	0.10 ± 0.03

[a]Estimated from gas dynamic simulations.

now vast and complex hunt at large collider experiments and underground laboratories for the particle constituent of dark matter.

A half-century after Zwicky's study, advances in X-ray astronomy revealed the full mass hierarchy within clusters: galaxies are outweighed by an encompassing hot intracluster medium (ICM), and the combined visible mass remains a minority of the total. High-throughput spectroscopy of the ICM established its nearly isothermal nature, and high-resolution imaging from the *ROSAT* mission (Briel, Henry, & Böhringer 1992) enabled accurate estimates of ICM masses as well as estimates of total masses, under a hydrostatic equilibrium assumption, which confirmed, within $\sim 30\%$ statistical errors, the optical virial estimates of Dressler (1978) and many others (see Girardi et al. 2000). Estimates of the mass hierarchy for the Coma cluster (White et al. 1993) are listed in Table 1.1 for a Hubble constant $H_0 = 70\,h_{70}$ km s^{-1} Mpc^{-1}.

Why was galaxy formation in Coma so inefficient? To what extent is the environment within Coma representative of the state of clusters in general? Or of the Universe as a whole? Answers to many of these questions are now becoming available from increasingly large statistical surveys of galaxies and clusters, along with detailed investigation of individual clusters using 8 m-class optical telescopes. Papers presented at this meeting offer many excellent examples of both such approaches.

On the theory side, modeling the development of galaxies and the ICM within a cosmological framework of hierarchical clustering poses a formidable task. Although the physical processes—gravity, heating by shocks, cooling and (especially in low-density regions) heating by radiation, magnetic fields, conduction, turbulent mixing, etc.—are now firmly in hand, the complex, nonlinear interactions that govern their time evolution are analytically intractable, even for highly simplified geometries. Computational solutions are progressing (Kauffman et al. 1999; Somerville & Primack 1999; Cole et al. 2000), but the accuracy of solutions is limited by uncertainties associated with the "mesoscopic" processes involved in stellar birth, evolution, and death. The role of central black holes/active galactic nuclei on the galaxy/ICM interaction is only beginning to be explored (cf. Brüggen & Kaiser 2002).

Such modeling uncertainties are a cause for concern to cosmologists, since biased parameter estimates will result from application of an incorrect astrophysical model. Let $\mathcal{C}$ represent the vector of cosmological parameters (clustered mass density Ω_{m}, baryon mass density Ω_{b}, vacuum energy density Ω_Λ or dark energy Ω_{DE}, spectrum normalization σ_8, etc.) and let $\mathcal{R}$ represent a set of observations from a cluster survey. Then Bayes' theorem makes clear that identifying the most likely cosmology is dependent on knowing precisely how likely are the observations within that world model:

$$p(\mathcal{C} \mid \mathcal{R}) \propto p(\mathcal{R} \mid \mathcal{C}) \, p_{\text{prior}}(\mathcal{C}). \tag{1.1}$$

For the CMB, the relevant likelihoods $p(\mathcal{R} \mid \mathcal{C})$ are calculable to high accuracy with codes such as CMBFAST (Seljak & Zaldarriaga 1996). For Type Ia supernovae, only the luminosity distance $d_L(z)$ is needed to compute the apparent brightness of standard candles as a function of redshift (although proving the standard candle nature is challenging). For galaxy clusters and most other large-scale structure signatures, nonlinear dynamics and astrophysical uncertainties complicate the computation of the observable likelihood $p(\mathcal{R} \mid \mathcal{C})$.

There are computational and empirical reasons to suspect that the problem, though complex, is still tractable. Tight correlations, such as that between ICM mass M_{ICM} and X-ray temperature T_X (Mohr, Mathiesen, & Evrard 1999), are observed among intrinsic properties of local clusters. Simulated clusters follow narrow scaling relations (Evrard 1989; Navarro, Frenk, & White 1995, hereafter NFW; Evrard, Metzler, & Navarro 1996; Bryan & Norman 1998). These findings motivate a basic framework in which clusters are essentially a one-parameter family ordered by a size parameter, typically taken to be the total mass M. (Exactly how mass is defined is a detail discussed below.) Tight scaling relations between M and T_X reflect virial equilibrium, while retention of the cosmic baryon fraction and inefficient cooling/star formation within cluster environments lead to a strong correlation $M_{\text{ICM}} \sim (\Omega_b/\Omega_m) f_{\text{hot}} M$, with f_{hot} the fraction of baryons not processed into cold gas or stars.

In this paper, I take the perspective of separating the problem into two, quasi-independent pieces. For a given cosmology, the question of computing the likelihood of an observable, say the expected number of clusters with temperate T at redshift z, can be split into two parts, namely

- How many clusters of mass M exist in this cosmology at redshift z?
- What is the likelihood that a cluster of mass M at redshift z will have temperature T?

In general, the answer to the second question will require an astrophysical model defined by some set of parameters $\mathcal{A}$. For example, varying the efficiency of supernova or active galactic nuclei heating will affect the detailed form of the joint likelihood $p(M,T)$ at a particular epoch.

Under this separable assumption, the likelihood of forming a cluster of temperature T at redshift z will be a convolution

$$p(T,z \mid \mathcal{C}, \mathcal{A}) = \frac{\int dM \, p(M,z \mid \mathcal{C}) \, p(T \mid M,z,\mathcal{A})}{\int dM \, p(M,z \mid \mathcal{C})}, \tag{1.2}$$

where $p(M,z \mid \mathcal{C})$ gives the likelihood that a cluster of mass M exists at redshift z in cosmology $\mathcal{C}$ and $p(T \mid M,z,\mathcal{A})$ gives the likelihood that such a cluster has temperature T for the particular astrophysical model $\mathcal{A}$.

In § 1.2, after briefly reviewing the framework of nonlinear structure formation, I present recent calibrations of the cluster space density by large simulations and show that this problem, modulo some inherent arbitrariness in assigning mass to halos, is now on quite firm footing. In § 1.3, the virial relation linking mass to ICM temperature is discussed and discrepancies between computational and empirical approaches are noted. This motivates a "back-door" approach to calibrating the mass scale of the cluster population that stems from very precise determination of the dark matter virial relation discussed in § 1.4. Implications emerge in § 1.5, where the observed cluster space density as a function of temperature forges a link between σ_8 (a cosmological parameter) and the ratio of cluster component energies

β (a purely astrophysical parameter). Unless sources of substantial systematic error can be identified, the low normalization $\sigma_8 \approx 0.7-0.8$ inferred from *WMAP* analysis will require significantly more heating of the ICM plasma than is provided by gravitational collapse alone. A concluding discussion is provided in § 1.6.

1.2   Clusters as Dark Matter Potential Wells

An early inflationary period in the history of the Universe is thought to seed the Universe with fluctuations in energy density (Kolb & Turner 1990). Depending on the chosen scenario, the spectrum of primordial density fluctuations can differ from a power law, and a general expression is a form with a running spectral index:

$$P_{prim}(k) = P(k_0) \left(\frac{k}{k_0} \right)^{n_s(k)} , \tag{1.3}$$

where $n_s(k) = n_s(k_0) + dn_s/d\ln k \ln(k/k_0)$. Recent analysis of *WMAP* thermal fluctuations combined with 2dF galaxy and Lyα forest power spectrum estimates suggest, at the 2σ level, a nonzero value for the spectral index derivative $dn_s/d\ln k = -0.031^{+0.016}_{-0.017}$ (Spergel et al. 2003).

During the pre-recombination era, when fluctuation amplitudes are small and a linear treatment of independent wavemodes is valid, the primordial matter perturbations undergo stagnant growth on small scales and (for warm/hot components) strong damping due to free-streaming. The net result of such physics, which is particularly sensitive to the matter density Ω_m and the baryon fraction Ω_b/Ω_m, is summarized by a transfer function $T(k)$ that defines the post-recombination power spectrum $P_{rec}(k) \equiv T^2(k)P_{prim}(k)$ (Bond & Efstathiou 1984).

As discussed below, the mass dependence of the cluster space density is sensitive to the logarithmic slope of the linear, post-recombination spectrum $n_{eff} = d\ln P_{rec}(k)/d\ln k$ on $\sim 10h^{-1}$ Mpc scales. A statistically precise measurement of $n_{eff}(k)$ can be sought using upcoming large cluster samples, after systematic selection and projection effects are addressed and understood.

1.2.1   *Nonlinear Gravitational Condensation*

In the linear regime, the growth rate of the fluctuation amplitude at any wavenumber is controlled by a function $D(a)$, the form of which is determined by the mix of matter and energy components in the Universe (Peebles 1980; Carroll, Press, & Turner 1992). As linearized wave amplitudes δ approach unity, mode coupling becomes important, and the linear treatment must be extended to second and higher order (Bernadeau et al. 2002). Comparison with *N*-body simulations confirm the validity of calculations to tenth order and higher (Szapudi et al. 2000), but precise solution of the deeply nonlinear ($\delta \gtrsim 10^2$) evolution of the matter density and velocity fields is exclusively achieved by numerical simulation (Bertschinger 1998).

Simulations of large comoving volumes show that, as mode-coupling strengthens, the density field develops the texture of a "cosmic web" (Bond, Kofman, & Pogosian 1996), an example of which is shown in Figure 1.1. Filaments and walls surround lower density voids and bound ellipsoidal structures—the halos/clusters housing astrophysical objects—develop through gravitational collapse, with the largest and rarest clusters forming at nodes defined by major filament intersections. As discussed below, the spectrum of halos sizes can be derived analytically from $P_{rec}(k)$ under spherical or ellipsoidal evolution approximations. For

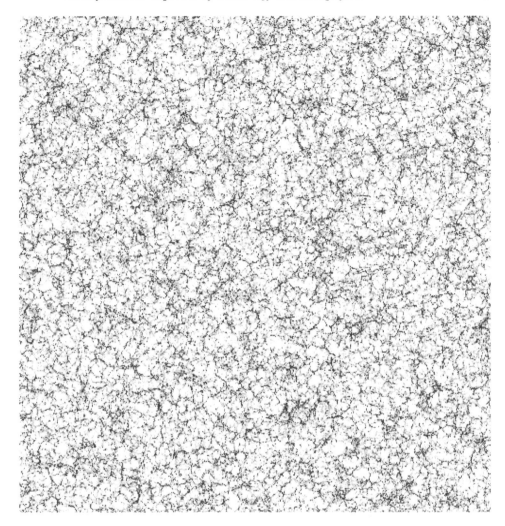

Fig. 1.1. Large-scale structure of cold dark matter in a Hubble Length ($3000\,h^{-1}$ Mpc) slice through a simulated ΛCDM Universe at $z = 0$ (Evrard et al. 2002). The slice thickness with $30\,h^{-1}$ Mpc and the greyscale show regions with above-average density smoothed on a $10^{13}\,h^{-1}M_\odot$ scale.

effectively power-law spectra, the characteristic mass scale M_* of the distribution evolves in redshift according to $M_*(a) \propto [D(a)]^{6/(n_{\mathrm{eff}}+3)}$ (Kaiser 1986), and this feature is the basis for deducing a low matter density Universe from the sky surface density of massive clusters at $z \approx 0.5 - 1$ (Bahcall & Fan 1998; Donahue et al. 1998; Borgani et al. 2001).

Since both the Gaussian random initial conditions and the process of gravitational amplification have no sharp intrinsic scales, clusters do not develop obvious physical boundaries. Instead, a roughly hydrostatic and dynamically older core connects seamlessly to an infall region fed by material drawn mainly from the embedding filaments. Accretion of small halos occurs nearly continuously up to the present, and major mergers that cause significant phase space rearrangement happen stochastically every few dynamical times. Due to this complexity, a unique definition of cluster mass is hard to justify. Instead, a few, quite similar measures are in common use (Lacey & Cole 1994; White 2001, and references therein), each defined by a threshold density that attempts to separate the quasi-equilibrium, or "virialized", cluster interior from its surrounding, infalling streams.

One way to address the mass ambiguity is to consider the relative merits of different measures. What "added value" does each bring to descriptive analysis of the cluster population? As shown below, it appears that masses set by thresholds defined relative to the *mean* matter density $\rho_m(a)$ are convenient for counting clusters, while those defined relative to the critical density $\rho_c(a)$ are advantageous for addressing internal structure issues such as the virial relation linking mass to dark matter velocity dispersion or ICM thermal temperature.

In spherical models of perturbation evolution (Gunn & Gott 1972; Bertschinger 1985; Lokas & Hoffmann 2000), a cluster develops from a local peak in the linear density field. Peaks in Gaussian random fields exist on a wide range of scales, so to select those at a particular scale $M \propto R^3$ it is useful to smooth the continuous density field $\rho(x)$ with a spatial filter $W(|x'-x|/R)$. The variance in the smoothed density field is

$$\sigma^2(M) = \frac{1}{(2\pi)^3} \int d^3k \, P_{\rm rec}(k) \, \hat{W}^2(kR), \tag{1.4}$$

where $\hat{W}(kR)$ is the Fourier transform of the spatial filter function. The use of a spherical Heaviside, or "top-hat," function [$W(r/R) \equiv 1$ for $r/R \leq 1$ and 0 otherwise] with comoving scale $8\,h^{-1}$ Mpc defines a conventional measure of the present, linearly evolved power spectrum amplitude $\sigma_8 \equiv \sigma(M_8)$, with $M_8 = 1.785(\Omega_m/0.3) \times 10^{14}\,h^{-1}M_\odot$. The probability density function (PDF) of the filtered density field is Gaussian normal in the variable $\nu \equiv \delta/\sigma(M)$.

1.2.2   The Cluster Space Density

Counting the number of clusters as a function of size is currently an inexact exercise for observers and theorists alike. In the sky, optical/infrared catalogs can be searched for galaxy concentrations in redshift/color space (Bahcall et al. 2003; Nichol 2004). Detection of the ICM via its X-ray emission (Rosati, Borgani, & Norman 2002) or its spectral distortion of the microwave background radiation (Carlstrom, Holder, & Reese 2002) is another means of cluster identification. Although the observable signatures are strongly correlated (to first order, via the ever-useful "bigger is bigger" maxim), the scaling relations linking pairs of observables display typically tens of percent intrinsic scatter (Borgani et al. 1999; Mohr et al. 1999; Sanderson et al. 2003), and each measure is subject to different sources of systematic and random errors.

The upshot is that an X-ray temperature-limited sample of clusters will differ somewhat from an optical richness-limited sample, and both will differ to some degree with an Sunyaev-Zel'dovich-limited set of clusters. Given enough signal-to-noise ratio, the most massive clusters will be identified by any method, but differences in how projected signals add, along with intrinsic scatter among observables, will, at the minimum reorder, and more likely reorder and blend a set of objects detected at even high signal-to-noise ratio. Under realistic conditions, confusion will become more severe as one pushes to smaller systems near the sample detection threshold (Bahcall et al. 2003).

With the advantage of full spatial information, theorists working with simulated volumes are afforded higher precision. For a point set of simulation particles, percolation methods, such as the "friends-of-friends" algorithm of Davis et al. (1985), have been a popular way to define halos. Complex geometries typically bound the objects defined with this method (see Fig. 4 of White 2001), and this aspect can complicate attempts to connect this mass measure to observations.

A geometrically simpler approach is to identify a cluster as material lying within a sphere, centered on a local density maximum or potential minimum, whose radial extent r_Δ is defined by an enclosed isodensity condition $M(< r_\Delta)/(4\pi r_\Delta{}^3/3) = \rho_t(z)$. Such spherical overdensity (SO) masses require a choice of threshold density $\rho_t(z)$ that is typically written as a multiple Δ of either the mean mass density $\rho_m(z)$ or the critical density $\rho_c(z)$. These measures are referred to as "mean" and "critical" masses below. Although many treatments employ a time-varying Δ in cosmologies with $\Omega_m \neq 1$ (Eke, Cole, & Frenk 1996), evidence below suggests that this complication is unnecessary, and Δ here is assumed constant.

Given a mass measure M, the space density $n(M,z|\mathcal{C})$, or *mass function*, describes the probability of finding a cluster at redshift z with total mass in the interval M to $Me^{d\ln M}$ within a suitably small comoving volume element dV:

$$p(M,z|\mathcal{C}) = n(M,z|\mathcal{C})dV. \qquad (1.5)$$

An analytic form for the mass function, based on spherical dynamics and a Gaussian random density field, was first developed by Press & Schechter (1974, hereafter PS) and rederived using a rigorous excursion set approach by Bond et al. (1991). The resulting shape of $n(M,z)$ is dictated by the linear power spectrum $\sigma(M)$ and its logarithmic derivative. The mass fraction in halos of mass M at redshift z can be expressed in terms of a single function $f(\sigma^{-1})$:*

$$f(\sigma^{-1},z) \equiv \frac{M}{\rho_m(z)}\, n(M,z)\, \frac{d\ln M}{d\ln \sigma^{-1}}. \qquad (1.6)$$

The dependence on cosmology is implicit, determined by the fluctuation spectrum and its rate of linear evolution $D(a)$.

The PS treatment employs a spherical collapse model that assumes equilibrium is reached when the linearly evolved interior density reaches a critical threshold δ_c (equal to 1.686 in an Einstein-de Sitter cosmology). This leads to a mass function of the form

$$f(\sigma^{-1}) = \sqrt{2/\pi}\,(\delta_c\sigma^{-1})\,\exp[-(\delta_c\sigma^{-1})^2/2]. \qquad (1.7)$$

Initial comparison of the model with N-body simulations yielded good agreement (Efstathiou et al. 1988), but as computational dynamic range and fidelity improved, disagreement in the detailed shape of the mass function emerged. Introduction of ellipsoidal, rather than spherical, perturbation evolution greatly improved agreement with simulations (Sheth & Tormen 1999; Sheth, Mo, & Tormen 2001, hereafter ST). The mass function in this case takes the form

$$f(\sigma^{-1}) = A_{ST}\sqrt{2a/\pi}\,[1+(a\delta_c\sigma^{-1})^{-p}]\,(\delta_c\sigma^{-1})\,\exp[-(a\delta_c\sigma^{-1})^2/2], \qquad (1.8)$$

with parameters $A_{ST} = 0.3222$, $a = 0.707$, and $p = 0.3$.

Massively parallel computers, with aggregate memory in excess of one terabyte, have enabled production of very large statistical samples of virtual clusters. Calibrations of the mass function from such samples are reported in a number of recent papers (Governato et al.

---

* Note that, since $\sigma(M)$ is a monotonic decreasing function in cold dark matter models, its inverse $\sigma^{-1}(M)$ has the same sense as mass; high σ^{-1} implies high mass and *vice versa*.

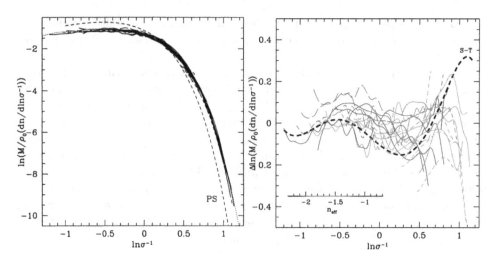

Fig. 1.2. Calibration of the mass function from a set of 29 halo samples extracted from Virgo Consortium simulations (Jenkins et al. 2001). *Left:* Mass fraction in halos as a function of the similarity variable $\sigma^{-1}(M)$. The dotted curve shows the fit to the Jenkins mass function (Eq. 1.9), while the dashed curve shows that the PS expectation (Eq. 1.7) has the wrong shape. *Right:* Residuals in number density between the binned simulated samples and the Jenkins mass function fit are shown as thin lines. The dashed line shows the difference between the Jenkins mass function and ST prediction (Eq. 1.8), with parameters given in the text.

1999; Bode et al. 2001; Jenkins et al. 2001; Evrard et al. 2002; Hu & Kratsov 2003; Reed et al. 2003). The results from analysis of a suite of simulations performed by the Virgo Consortium (Jenkins et al. 2001) are shown in Figure 1.2. The mass fraction as a function of $\sigma^{-1}(M)$ is shown for 29 halo samples identified in 13 simulations and covering epochs $z \approx 0-5$ in four different cosmological models (see Table 2 of Jenkins et al. 2001).

All of the models are well fit by a function (hereafter the Jenkins mass function) of the form

$$f(\ln \sigma^{-1}) = A \, \exp[-|\ln \sigma^{-1} + B|^{\epsilon}]. \tag{1.9}$$

The parameter B controls the location of the peak in the collapsed mass fraction (e^B plays the role of δ_c in the PS and ST forms) while A controls the overall mass fraction in halos and ϵ stretches the function to fit the overall shape of the simulation results. Values $A=0.315$, $B=0.61$, and $\epsilon=3.8$ fit the data in Figure 1.2, which uses a friends-of-friends mass measure with a linking length of 0.2 times the mean interparticle spacing. Note that these parameters are independent of cosmological model and epoch. In this sense, it may be said that Nature prefers to do *accounting* relative to the mean mass density.

We will see below that Nature appears to prefer doing *dynamics* relative to the critical density. For a critical SO(200) mass measure, Evrard et al. (2002) show that Equation (1.9) provides a good fit to the mass function with Ω_m-dependent fit parameters $A(\Omega_m)=0.27 - 0.07\Omega_m$ and $B(\Omega_m)=0.65+0.11\Omega_m$ (and $\epsilon=3.8$). Hu & Kravtsov (2003) provide independent confirmation of this fit for the case $\Omega_m=0.15$.

The right panel of Figure 1.2 shows that this form predicts the space density to an accu-

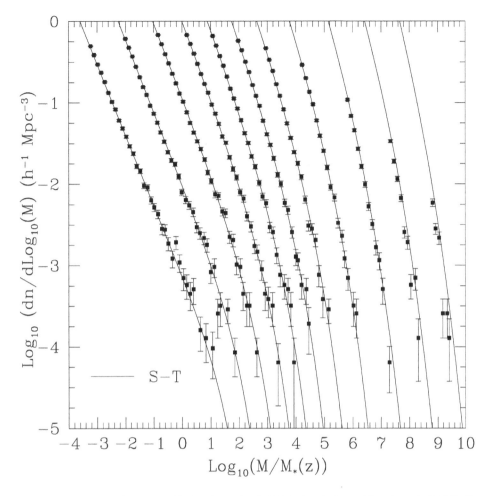

Fig. 1.3. The mass function of halos from the ΛCDM simulation of Reed et al. (2003). Data points with Poisson error bars show the simulation results, while curves are ST predictions at redshifts (from left to right) 0, 1, 2, 3, 4, 5, 6.2, 7.8, 10, 12.1, and 14.5.

racy of $\sim 20\%$. This panel also shows deviations with respect to the ST form with recalibrated parameters $A_{ST} = 0.353$, $a = 0.73$, and $p = 0.175$. With the exception of the rarest objects ($\ln \sigma^{-1} \gtrsim 0.7$), this model gives an equally good fit. The PS formula for the rarest systems underpredicts their abundance by more than a factor 10. The use of an ellipsoidal collapse model (with two added parameters calibrated by simulation) is clearly superior to the original PS spherical treatment.

Reed et al. (2003) reach a similar conclusion on the ST predictions, and offer a correction factor to apply at high masses. As shown in Figure 1.3, their 432^3 particle simulation of a $50 h^{-1}$ Mpc region, which probes a more shallow region of the power spectrum compared to cluster scales, shows stunning agreement with the ST predictions over a wide dynamic

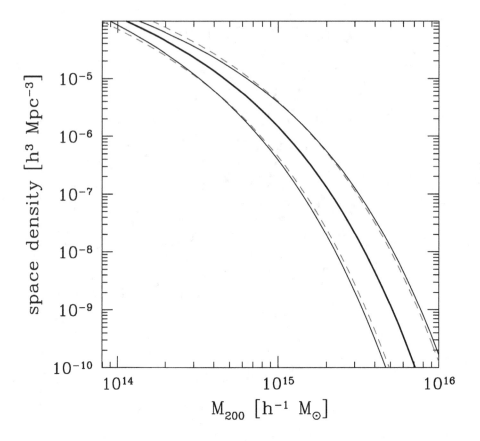

Fig. 1.4. The bold line shows the Jenkins mass function expectations for a ΛCDM cosmology with a default value $\sigma_8 = 0.9$. Solid lines to the right and left of the default give the Jenkins mass function for modified normalizations $\sigma_8 = 0.9\,e^{\pm 0.148}$. The dashed lines show the default Jenkins mass function with masses displaced by amounts $e^{\pm 0.371}$. The degeneracy between mass scale and σ_8 is apparent at space densities below $10^{-5}h^3$ Mpc^{-3}.

range in number density and epoch. The finding that the fit works well over 10 decades in normalized mass $M/M_*(z)$ is particularly impressive.

1.2.3   *Mass-scale Uncertainty and σ_8*

The space density of the most massive clusters at a given epoch is exponentially sensitive to σ_8, but extracting σ_8 from observations of massive clusters requires accurate knowledge of cluster masses. From the form of the Jenkins mass function, Evrard et al. (2002) derive $d\ln\sigma_8/d\ln M \simeq 0.4$ for massive clusters in a ΛCDM cosmology, meaning that, at fixed spatial abundance, systematic errors of, say, 25% in the mass scale of clusters translate into a systematic uncertainty of 10% in σ_8. A simple demonstration of this effect is given in Figure 1.4. The shapes of the functions, scaled separately in mass or σ_8, are nearly identical over the space density range 10^{-5} to $10^{-10}h^3$ Mpc^{-3}.

In summary, computational modeling now provides $\sim 10\%$-level accurate calibration of the space density $n(M,z)$ for halos spanning the entire observable range of galaxy cluster masses. In this sense, the theoretical situation is in good shape. The hard task remains of relating mass to readily observable properties of clusters.

1.3   Connecting Mass to Observables: Virial Scaling Relations

Since ICM temperatures and galaxy velocity dispersions are directly observable measures of cluster size, the virial theorem, as Zwicky had long ago appreciated, is a natural starting point for establishing an observable-mass connection.

1.3.1   What is the "Virial Mass" of a Cluster?

Since clusters are dynamically evolving within the large-scale cosmic web, some deviation from hydrostatic and virial equilibrium must be expected. Simulations of colliding pairs of clusters show significant departures from hydrostatic equilibrium during major mergers (Roettiger, Burns, & Loken 1996; Ricker & Sarazin 2002), yet simulations of clusters formed from Gaussian random density fields show that the population as a whole adheres to a surprisingly tight virial relation (Evrard 1989; Navarro et al. 1995; Bryan & Norman 1998; Mathiesen & Evrard 2001). These seemingly inconsistent findings can both be true. Major mergers are relatively rare (Bower 1991; Lacey & Cole 1994; Sheth & Tormen 1999), so small departures from equilibrium are statistically far more likely than order unity variations. Furthermore, dynamical relaxation occurs rapidly, essentially on a single crossing time (Peebles 1970), and the net effect is to drive clusters very nearly along the equilibrium relation (Navarro et al. 1995), even during phases of major mergers (see Fig. 14 of Evrard & Gioia 2002).

The virial relation connects the dark matter velocity dispersion σ_{DM} and ICM temperature T, both mass-weighted within a sphere of radius r_Δ, to the corresponding SO mass M_Δ via the form

$$\sigma_{DM}^2 \propto (kT/\mu m_p) = \varepsilon \, GM_\Delta/r_\Delta, \qquad (1.10)$$

where μ is the mean molecular weight of the ICM, m_p the proton mass, and ε a dimensionless constant that depends primarily on the internal density profile of clusters. For mass defined against the critical density, the right-hand side scales as $[h(z)M_\Delta]^{2/3}$, where $H(z) = 100\,h(z)$ km s^{-1} Mpc^{-1}* is the Hubble parameter at redshift z. If ε is independent of mass and redshift (i.e., under a self-similar assumption), or is at least approximately so over some finite range, then we expect scalings of the form

$$\sigma_{DM} = \sigma_{DM,15} \, [h(z)M_\Delta/10^{15}M_\odot]^\alpha, \qquad (1.11)$$

$$kT = kT_{15} \, [h(z)M_\Delta/10^{15}M_\odot]^{\alpha_T} \qquad (1.12)$$

with slope $\alpha = \alpha_T/2 = 1/3$ and where the values of the intercepts will depend on the choice of threshold scale Δ. Note that if a mean mass measure is used, then $h(z)M_\Delta$ is replaced by $(1+z)^{3/2}M_\Delta$ (Bryan & Norman 1998).

1.3.2   Estimates of ICM Virial Scaling

Tension has developed between observational and computational estimates of the scaling between mass and ICM temperature. From X-ray observations of surface brightness

---

* This term is often written with the notation $E(z) = H(z)/H_0$ used by Peebles (1993).

Table 1.2. *Empirical Mass-temperature Relations*

Reference	M_{500} (6 keV) $(10^{15} h^{-1} M_\odot)$	$d \ln M / d \ln T$
Horner, Mushotzky, & Scharf (1999)[a]	0.40	1.78
Mohr, Mathiesen, & Evrard (1999)	0.47	1.64
Nevalainen, Markevitch, & Forman (1999)	0.17	1.79
Finoguenov, Reiprich, & Böhringer (2001)	0.30	1.85
Allen, Schmidt, & Fabian (2002)	0.39	1.54
Shimizu et al. (2003)	0.40	1.90
Sanderson et al. (2003)	0.44	1.84

[a]Isothermal model estimates.

and spectral temperature, one can construct the radial pressure profile of the ICM and use the assumption of strict hydrostatic equilibrium, $GM(< r)/r^2 = -\nabla P/\rho$, to derive mass estimates. This exercise has been done for samples of moderate size (many tens), and results are summarized in Table 1.2. Because T is the independent observable, fits to the inverse of Equation (1.12) are quoted, pivoted about a typical cluster temperature of 6 keV. Values of the mass intercept are defined at a critical contrast of 500, a scale accessible at moderate signal-to-noise ratio for many X-ray cluster images. In cases where a different scale is quoted, I assume an NFW profile with concentration $c = 5$ to rescale to $\Delta = 500$ (convenient expressions for this are given in an Appendix of Hu & Kravtsov 2003). Statistical errors quoted in the references are typically $\sim 10\%$ in each parameter.

The fits listed in Table 1.2 consistently indicate a slope that is steeper than the self-similar expectation $1/\alpha_T = 3/2$. With the exception of Nevalainen et al. (1999), the intercepts are in good agreement with M_{500} (6 keV) $= (0.40 \pm 0.05) \times 10^{15} h^{-1} M_\odot$.

The fact that observed scaling relations differ from self-similar predictions motivates the addition of ICM physics beyond gravitational shock heating. Voit et al. (2003) discuss the roles of radiative cooling and heating associated with galactic feedback, and focus on the gas entropy $K = kT/\rho_{ICM}^{2/3}$ as the controlling parameter.

The tight color-magnitude relation of early-type cluster galaxies observed in low- to moderate-redshift clusters (Bower, Lucey, & Ellis 1992; Stanford, Eisenhart, & Dickinson 1998) implies that most of the stars in galaxies were formed at high redshifts $z \gtrsim 4$. If these galaxies hosted vigorous outflows during their rapidly star-forming phase, as appears to be the case for $z \simeq 3$ Lyman-break galaxies (e.g., Adelberger et al. 2003), then the thermal energy deposited by the outflows could raise the entropy of the surrounding intergalactic/protocluster environment. This "preheating" of the gas offers a relatively simple mechanism capable of breaking self-similarity in the ICM (Evrard & Henry 1991; Kaiser 1991).

Bialek, Evrard, & Mohr (2001) show that simulated clusters that employ an elevated initial ICM entropy at a level $K = 55 - 150$ keV cm^{-2} offer an acceptable match to the local observed scalings of ICM mass, bolometric luminosity, and isophotal size with X-ray temperature. Figure 1.5 shows the relation between virial mass and temperature for a set of 68 clusters, simulated assuming initial $K = 106$ keV cm^{-2}, at 12 epochs ranging from $z \simeq 2$ to

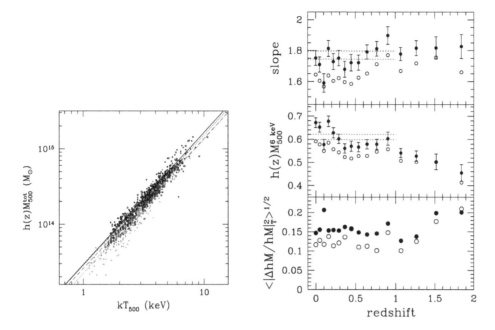

Fig. 1.5. The virial relation from a set of 68 preheated cluster simulations that match the local, observed ICM scaling relations (Bialek et al. 2001). *Left:* Virial mass $h(z)M_{500}$ against spectral temperature for redshift ranges $z < 0.2$ (filled circles), $0.2 < z < 0.5$ (open circles), $0.5 < z < 1$ (filled triangles), and $1 < z < 2$ (open triangles). *Right:* Parameters of power-law fits to these data using either spectral (filled points with error bars) or mass-weighted (open) temperatures: slope (upper panel), mass intercept $M_{500}(6 \text{ keV})$ in units of $10^{15}M_\odot$ (middle panel), and rms dispersion in mass about the best-fit relation (lower panel). Dashed lines in the upper two panels show the 90% confidence range of the slope and intercept derived from a simultaneous fit to all $z < 1$ outputs.

the present. The left panel employs spectral temperatures derived from single-temperature `mekal` fits to synthetic X-ray spectra created by summing the `mekal` contributions from all gas lying within r_{500} of the cluster center. For comparing to observations, this measure of temperature is more appropriate to use than a mass-weighted measure. Low-entropy gas in the cores of infalling satellites has a larger effect when emission weighted [$\propto \rho^2 \Lambda(T)dV$] rather than mass weighted ($\propto \rho dV$). Depending on the energy passband, temperature differences at the ten percent level are expected between these two measures (Mathiesen & Evrard 2001).

The simulated cluster sample is well fit by a power law with $\sim 17\%$ scatter in mass at fixed temperature. At redshifts $z \lesssim 1$, the virial scaling under preheating is slowly evolving. A simultaneous fit to all data from these epochs, treating all outputs as independent, leads to a mean relation

$$h(z)M_{500} = (0.61 \pm 0.01) \left(\frac{kT}{6\,\text{keV}} \right)^{1.77 \pm 0.02} \times 10^{15} M_\odot. \tag{1.13}$$

The combined effects of preheating, which affects the energetics of shallower potential

Table 1.3. *Computational Mass-temperature Intercepts*

Reference	M_{500} (6 keV) $(10^{15} h^{-1} M_\odot)$
Evrard, Metzler, & Navarro (1996)	0.52
Bryan & Norman (1998)	0.78
Mathiesen & Evrard (2001)	0.83
Thomas et al. (2002)	0.40
Borgani et al. (2004)	0.43
Bialek et al. (2004)	0.61

wells more so than deep ones, and use of spectral instead of mass-weighted temperatures produce a slope $d\ln M/d\ln T$ that is $\sim 20\%$ steeper than the self-similar expectation and more consistent with the observational estimates. However, the intercept of Equation (1.13) is $\sim 50\%$ high compared to the empirical estimates.

This tension between computational and empirical determinations of the cluster mass scale appears generically true for simulations that ignore radiative cooling of the ICM. Table 1.3 lists estimates from simulated samples. The models of Thomas et al. (2002) and Borgani et al. (2004) employ radiative cooling and provide values consistent with the observed mean. The remaining models, which ignore radiative cooling, paradoxically tend to produce an ICM that is *cooler* than the radiative solution for a given mass halo. The PdV work done in compressing gas into regions where efficient cooling has removed gas is the mechanism that explains why "cooling causes heating" (Bryan 2000; Pearce et al. 2000).

The cooling models may relieve the tension in the ICM virial relation, but there may be a cost to this cure. The reason that the recent 2×480^3 particle GADGET-2 simulation of Borgani et al. (2004) produces a lower mass intercept is that the gas-phase structure is inverted relative to that displayed by most non-cooling simulations: the dense core gas is hotter, not cooler, than the surrounding lower density material. The inverted phase structure helps drive up the emission-weighted temperature at fixed mass, hence down the mass at 6 keV, but Borgani et al. show that the price paid is that radially averaged temperature profiles of the model clusters fall too steeply compared to observations (Markevitch 1998; DeGrandi & Molendi 2001; Pratt & Arnaud 2002). Adding transport processes such as turbulent thermal conduction (Narayan & Medvedev 2001) may help smooth the gradients, but perhaps this inclusion would then also damage agreement with observed ICM scalings.

In short, direct simulation methods are making headway and will continue to improve as the effects of coupled physical processes are sorted out. In the meantime, it is worthwhile considering a simpler approach to calibrating the ICM virial relation.

1.3.3   A "Back Door" Approach: The ICM Virial Relation from $n(T)$

Another way to constrain the mass-temperature relation is by a simple counting argument—identify the mass at which the theoretical space density of halos $n(M, z)$ matches the observed space density of clusters as a function of temperature $n(T, z)$. In practice, one needs to account for scatter by convolving

$$n(T,z)\mathrm{d}\ln T \;=\; \int \mathrm{d}\ln M\, n(M,z)\, p(T|M,z), \tag{1.14}$$

where the form of the conditional PDF $p(T|M,z)$ provides the critical link between theory and observation.

The detailed form of $p(T|M,z)$ will be set by the physical processes governing the ICM thermodynamical evolution, by the chosen mass measure, and to some degree by the assumed cosmology. Assuming for now that gravitational heating is the dominant process, then the ICM temperature should follow the depth of the dark matter potential as measured by the (one-dimensional) dark matter velocity dispersion σ_{DM}. Here $\sigma_{DM}^2 \equiv \int dV \rho_{DM} |\mathbf{v} - \bar{\mathbf{v}}|^2 / 3 M_\Delta$ is the second moment of the fluid velocity $\mathbf{v}$ relative to the cluster mass-weighted mean $\bar{\mathbf{v}}$, and the integral is performed over the same spherical volume used to define the mass. Following the lead of Cavaliere & Fusco-Femiano (1976), introduce the parameter

$$\beta \;=\; \sigma_{DM}^2 / (kT/\mu m_p) \tag{1.15}$$

that measures the ratio of specific energies in dark matter and ICM gas.

By wrapping our ignorance of ICM physics into the parameter β, we can proceed to pose the question of how well the virial relation for the dark matter (Eq. 1.11) is understood from simulations.

1.4   The Dark Matter Virial Relation

Calibrating the virial scaling of dark matter is a far more tractable exercise compared to that for the ICM, especially if one limits the analysis to the standard assumption of weakly interacting cold dark matter (CDM). Collisionless N-body simulations are a sufficient and mature methodology for this exercise. Still, there remains the question of whether different N-body codes will produce consistent answers in the deeply nonlinear regime.

The answer to this exercise is presented in Figures 1.6 and 1.7 and Table 1.4. The panels of Figure 1.6 show results from numerical cluster samples produced by different investigators employing five independent simulation codes (see caption for references). Clusters are identified using a critical SO(200) condition. Some methods allow the r_{200} spheres of clusters to overlap, but the center of one cannot lie within the boundary of another. In the overlap case, label the larger member of an overlapping pair the *primary* and the smaller member the *satellite*. The latter account for $5\% - 10\%$ of the population between 10^{14} and $10^{15} h^{-1} M_\odot$. As shown in panel (a) of Figure 1.6, the satellite population follows a slightly different virial relation from the primaries, with a distribution of σ_{DM} skewed to larger velocities. This population contains objects in the throes of ongoing mergers, and it will be ignored in the statistical analysis to follow. Their omission changes the virial relation fits at a level below one percent.

The simulation samples and the unbinned, least-squares fit parameters to Equation (1.11) are described in Table 1.4. There is a wide range in sample size and mass resolution, from Hubble Volume samples with roughly one-half million objects but only 450 particles in a

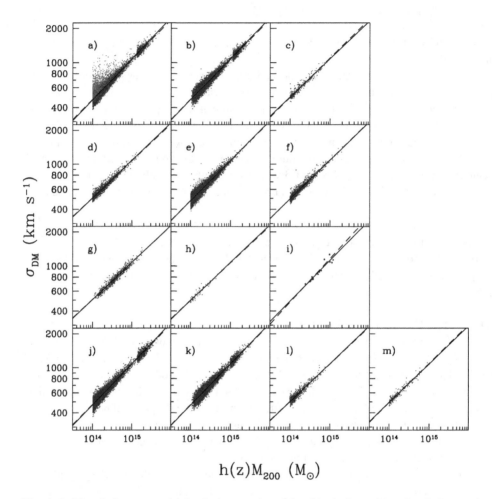

$$\text{h(z)M}_{200} \ (\text{M}_\odot)$$

Fig. 1.6. The dark matter virial relation produced by five independent simulation codes. All panels show $z=0$ data except (*b*) and (*k*), which show sky survey data extending to $z \simeq 1.5$. For large samples, the most massive 1,000 clusters are shown along with a random sampling of 9,000 above a minimum mass of $h(z)M_{200} = 10^{14} M_\odot$. Panels (*a*)–(*i*) use a ΛCDM cosmology with $\Omega_m = 0.3$, $\Omega_\Lambda = 0.7$ with spectrum normalizations in the range $\sigma_8 \simeq 0.8 - 1$. Panels (*a*)–(*f*) and (*j*)–(*m*) model only dark matter, while (*g*)–(*i*) are two-fluid simulations that include a $\sim 10\%$ baryonic mass component. Specific references are as follows: (*a*)–(*b*) Hubble Volume $z=0$ and combined sky survey samples, respectively, from Evrard et al. (2002); (*c*) Virgo model of Jenkins et al. (1998); (*d*) combined lcdm300A, 300B, 256A, 128A runs of White (2002); (*e*)–(*f*) 256^3 particle simulations of size 768 and 384 h^{-1} Mpc, respectively, from M. Warren (priv. comm.); (*g*) P3MSPH preheated ICM models of Bialek et al. (2004); (*h*) combined concordance150 SPH models from M. White (priv. comm.); (*i*) GADGET SPH models of Rasia, Tormen & Moscardini (2004); (*j*)–(*k*) τCDM ($\Omega_m = 1$) Hubble Volume $z=0$ and combined sky survey samples, respectively, from Evrard et al. (2002); (*l*) τCDM Virgo model of Jenkins et al. (1998); (*m*) open CDM ($\Omega_m = 0.3$) from M. White (priv. comm.). Points in each panel show primary cluster data (except for panel *a*, which shows both primary and satellites; see text for definition), while the solid and dashed lines are fits using no binning and binning primary data in intervals of 0.1 in decimal $\log[h(z)M_{200}]$, respectively. Parameter values of the former are listed in Table 1.4.

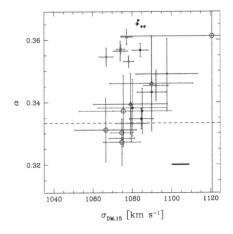 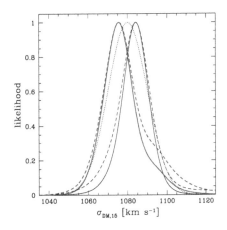

Fig. 1.7. *Left:* Fit parameters of the dark matter virial relation (Eq. 1.11) derived from log-linear, least-squares fits using equal weight per data point (thin lines) or equal weight for points binned in intervals of 0.1 in decimal $\log[h(z)M_{200}]$ (bold lines). The dashed line marks a self-similar slope of $1/3$, and the bar denotes a systematic uncertainty of 0.5% in σ_{DM}, a conservative estimate of the energy conservation accuracy of most codes. Filled symbols are pure N-body, while open include gas. Circles are ΛCDM, squares τCDM and triangles OCDM. *Right:* Likelihood estimates of the intercept $\sigma_{DM,15}$ derived by combining the log-normal PDFs of all the models using inverse variance (solid) or equal weighting (dashed), and for unbinned (light) or binned (bold) fits. The dotted line shows the summary result, Equation (1.16).

Coma-like cluster to a sample of 16 individual cluster simulations with nearly 1000 times improved mass resolution. Samples cover three cosmologies, epochs in the range $z \lesssim 1.5$, and three sets include a baryonic component gravitationally coupled to the dark matter.

The left panel of Figure 1.7 shows that there is remarkably good agreement in the dark matter virial relation fits among the samples. The slope is not particularly well determined, and tends to be somewhat steeper than the self-similar expectation of $1/3$. But the intercept $\sigma_{DM,15}$ is defined remarkably well. The statistical error in the very large Hubble Volume samples is extremely small (0.03%), much smaller than the typical energy accuracy ($\lesssim 1\%$) of the time integration schemes employed by N-body codes.

The right panel of Figure 1.7 shows estimates of $\sigma_{DM,15}$ based on summing log-normal PDFs from each sample fit, using either inverse variance weighting (solid) or linear weighting (dashed). To account for integration error, an uncertainty of 0.5% is added in quadrature to the statistical error in σ_{DM}. The resultant PDFs are reassuringly insensitive to the weighting scheme employed, but there is a slight shift between the binned (median 1077 km s^{-1}) and unbinned (median 1084 km s^{-1}) results. A Gaussian with mean and dispersion

$$\sigma_{DM,15} = 1080 \pm 10 \text{ km s}^{-1} \tag{1.16}$$

has a 90% confidence range that overlaps the estimates for both approaches. The uncertainty of 10 km s^{-1}, a one percent calibration of $\sigma_{DM,15}$, signals a high level of integrity in N-body solutions of strongly nonlinear clustering.

Table 1.4. *Dark Matter Virial Relation Calibrations*

Panel[a]	N_{cl}^b	N_{15}^c	$\sigma_{DM,15}^d$	Slope[d]
(a)	581866	450	1083.3 ± 0.3	0.3662 ± 0.0002
(b)	494803	450	1086.4 ± 0.3	0.3642 ± 0.0002
(c)	355	14,660	1080 ± 11	0.338 ± 0.006
(d)	1699	~ 100000	1085 ± 4	0.335 ± 0.002
(e)	8936	450	1084 ± 2	0.357 ± 0.001
(f)	1461	3400	1090 ± 5	0.343 ± 0.002
(g)	777	~ 15000	1075 ± 3	0.330 ± 0.003
(h)	158	~ 25000	1075 ± 14	0.337 ± 0.007
(i)	16	~ 300000	1121 ± 28	0.361 ± 0.032
(j)	601942	450	1085.6 ± 0.2	0.3642 ± 0.0002
(k)	263019	450	1083.7 ± 0.5	0.3651 ± 0.0003
(l)	703	4400	1085 ± 9	0.337 ± 0.004
(m)	221	~ 60000	1079 ± 11	0.339 ± 0.006

[a] Fig. 1.6.
[b] Number of clusters in sample.
[c] Number of particles in a $10^{15}\,h^{-1}M_\odot$ cluster.
[d] Quoted uncertainties are statistical.

1.5   Interpreting $n(T)$: An Example of Cosmological-Astrophysical Confusion

With well calibrated forms for the mass function and dark matter virial relations, the space density of dark matter potential wells can be accurately predicted from the linear fluctuation spectrum $\sigma(M)$. The β parameter then maps this distribution to an observable space density of temperature $n(T)$.

For a ΛCDM model with $\Omega_m = 0.3$ and $\sigma_8 = 0.9$, Evrard et al. (2002) show that $\beta = 0.92 \pm 0.06$ provides a good match to the combined observational samples of Markevitch (1998) and White (2000) that were employed in the σ_8 analysis of Pierpaoli, Scott, & White (2001). The quoted uncertainty in β is due to shot noise in the relatively small sample, which contains ~ 20 clusters hotter than 6 keV.

The mass normalization implied by combining $\beta = 0.92$ with the dark matter virial calibration is $M_{500}(6\text{ keV}) = 0.49 \times 10^{15}\,h^{-1}M_\odot$ (assuming a $c=5$ NFW profile for which $M_{200} = 1.38 M_{500}$). This value is appropriate for the assumed normalization $\sigma_8 = 0.9$. If a slightly different value of σ_8 had been chosen, then the mass scale at fixed space density would shift by an amount $d\ln M = 2.5\,d\ln \sigma_8$ (Fig. 1.4) across the entire observed range. Since the space density at a given temperature is fixed by the observations, the mass-temperature intercept required for the model to match the data will scale as

$$M_{500}(6\text{ keV}) = (0.64 \pm 0.06)\,\sigma_8^{5/2} \times 10^{15}\,h^{-1}M_\odot. \tag{1.17}$$

This can be alternately expressed as a constraint on the characteristic temperature (for $\Delta = 500$ in Eq. 1.12)

$$kT_{15} = (8.1 \pm 0.5)\,\sigma_8^{-5/3}\text{ keV} \tag{1.18}$$

or on the ratio of specific energies

$$\beta = (1.10 \pm 0.07)\, \sigma_8^{5/3}. \tag{1.19}$$

These relations all assume a fixed *shape* of the linear power spectrum, appropriate for a $\Omega_m = 0.3$ Universe. As Ω_m is varied, the mass function tilts, with lower values of Ω_m leading to flatter $n(M)$ and *vice versa*. The matter density is therefore mainly degenerate with the slope of the mass-temperature relation, but there is a normalization effect as well. The cumulative space density of clusters more massive than $\sim 10^{14.5}\, h^{-1} M_\odot$ is sensitive to the parameter combination $\sigma_8 \Omega_m^{0.6}$ (Pierpaoli et al. 2001). Roughly speaking, we can generalize the argument above by replacing σ_8 with $\sigma_8 (\Omega_m/0.3)^{0.6}$. A more careful approach would examine the full likelihood surface in the space of $\{\sigma_8, \Omega_m, kT_{15}, \alpha_T\}$, but this remains to be done.

1.5.1   A Cluster Energetics Problem?

Under the assumption that the principal mechanism of ICM heating is through grav-itationally induced shocks, existing gas dynamic simulations are able to provide an indepen-dent constraint on the specific energy ratio β. The shock-heating efficiency for this case is well calibrated by a set of 12 largely independent simulations of a single cluster (the "Santa Barbara" cluster of Frenk et al. 1999). The ensemble mean of the simulations is

$$\beta_{\rm sim} = 1.17 \pm 0.05. \tag{1.20}$$

A value greater than unity reflects incomplete thermalization of the ICM; the gas is partly supported by kinetic pressure of large-scale turbulent motions (Evrard 1990; Navarro et al. 1995; Rasia et al. 2004). When compared to the value derived from the space density argu-ment (Eq. 1.19), we infer that $\sigma_8 = 1.06 \pm 0.05$ is required for an ΛCDM model with $\Omega_m = 0.3$.

Such a high value of σ_8 has fallen out of favor recently. Analysis of microwave back-ground anisotropies from the *WMAP* mission (Spergel et al. 2003) yields $\sigma_8 = 0.90 \pm 0.1$ for an ΛCDM model with $\Omega_m = (0.29 \pm 0.04)(h/0.7)^2$ and a power-law primordial fluctuation spectrum with $n_s = 1$. When other CMB and large-scale structure data (2dF galaxy survey and Lyα forest observations) are added to the analysis of this model, the spectrum normal-ization drops to $\sigma_8 = 0.75^{+0.08}_{-0.07}$, while the matter density constraint is essentially unchanged.

There is a $\sim 3\sigma$ difference between this low value of σ_8 and the value derived from the cluster $n(T)$ argument above. What is wrong if σ_8 drops to 0.75? Well, lower σ_8 models make fewer massive clusters, so matching the observed space density of hot clusters requires lowering the β parameter to

$$\beta_{\rm WMAP} = 0.68 \pm 0.12. \tag{1.21}$$

Since the relation between $\sigma_{\rm DM}$ and mass is now fixed at the percent level, the only way to decrease β is to increase the ICM temperature at fixed mass. Relative to the case of purely gravitational heating, the boost in temperature required to match the lower *WMAP* normalization result would be

$$\frac{kT_{\rm WMAP}}{kT_{\rm grav}} = \frac{\beta_{\rm sim}}{\beta_{\rm WMAP}} = 1.72 \pm 0.30. \tag{1.22}$$

The extra energy, 0.7 ± 0.3 of that produced by gravity, is very large, $\sim 10^{63}\,$erg for the Coma cluster. To put this number in perspective, assume that the stellar mass in galaxies, $M_{\rm gal}$, is one-tenth the ICM mass (Lin, Mohr, & Stanford 2003). Then the overall stellar/black hole energy feedback efficiency (fraction of $M_{\rm gal}c^2$) required to provide the extra ICM heat is

$\sim 10^{-4}(kT/6\,\mathrm{keV})$. It is not at all clear whether the combined actions of known heating sources such as supermassive black hole-driven bubbles and supernova-driven winds, can produce an efficiency as large as this.

In addition, it is worth considering the other observable tracer of potential well depth, the galaxy velocity dispersion σ_{gal}. The ratio $\beta_{\mathrm{gal}} \equiv \sigma_{\mathrm{gal}}/(kT/\mu m_p)$ for clusters is measured to be 1.0 ± 0.1 (e.g., Xue & Wu 2000). The ratio of galaxy to dark matter specific energies is then

$$\left(\frac{\sigma_{\mathrm{gal}}}{\sigma_{\mathrm{DM}}}\right)^2 = \frac{\beta_{\mathrm{gal}}}{(1.10 \pm 0.07)\sigma_8^{5/3}} = 1.47 \pm 0.26 \qquad (1.23)$$

for the *WMAP* plus large-scale structure normalization. Apparently, galaxies, like the ICM, must end up hotter than the dark matter. Simulations of galaxy formation within clusters generally find the opposite; bright galaxies are slightly cooler than the mass (Springel et al. 2001).

Might this be an indication of new physics involving dark matter cooling? Perhaps, but this may be overreacting to what is now only a 3σ result. There may be more mundane explanations, especially if σ_8 relaxes upward. One possibility is that the temperatures determined from X-ray spectra are biased high relative to mass-weighted values, but deprojected cluster profiles do not support this idea (Allen et al. 2002), and simulations suggest that the effect works in the opposite direction (Mathiesen & Evrard 2001). Another possibility is that a number of smaller effects combine to produce the net offset. Three sources of $\sim 20\%$ bias could do the job, for example. Such a conspiratorial explanation has historical precedent in the original "beta-discrepancy" (Evrard 1990), where three reinforcing effects explain why values of β_{gal} are typically $\sim 50\%$ larger than values estimated from X-ray image profiles.

1.6   Discussion

The impressions I am trying to convey in this article are: (1) the statistical character of nonlinear structure is now very well established by simulations, and a population of structurally regular dark matter halos is expected at all redshifts; (2) a sound framework for extracting cosmological parameters from cluster surveys is thus in place; but (3) accurate constraints from such surveys demand a firm understanding of how observables relate to dark matter (along with, of course, a careful treatment of survey systematics).

The exercise of determining the power spectrum normalization σ_8 through the temperature space density $n(T)$ of clusters affords an example of astrophysical/cosmological confusion. Current observations place the constraint $\beta/\sigma_8^{5/3} = 1.10 \pm 0.07$ on an $\Omega_m = 0.3$ ΛCDM Universe. A normalization as low as $\sigma_8 = 0.75$ implies the existence of significant extra heat input to the ICM and/or new sources of bias in spectroscopic analysis of X-ray emission.

The era of large-N (10,000 or more) cluster samples is on the horizon. In the optical, the SDSS (Bahcall et al. 2003), 2dF (Couch, Colless, & De Propris 2004), and 2MASS (Kochanek et al. 2003) surveys will cover low redhifts, while the RCS2 (Gladders 2004) and other deep surveys will push beyond $z \approx 1$. In the X-rays, the XMM Cluster Survey (Romer et al. 2001) will identify large numbers of intermediate-redshift clusters with $T_X \gtrsim 4\,\mathrm{keV}$. The deepest reach will come from upcoming Sunyaev-Zel'dovich surveys (Holder et al. 2000; Kneissl et al. 2001; Carlstrom et al. 2002).

Theorists are gearing up to handle the data deluge. Survey design is guided by understanding the trade-offs in sample and shot noise errors (Hu & Kravtsov 2003). Projection and selection effects can be quantified using mock catalogs constructed from N-body simulations (Wechsler et al. 2004). The notion that surveys can be "self-calibrated" to non-standard astrophysical evolution (Majumdar & Mohr 2002; Hu 2003) is reinforcing optimism that the upcoming large cluster samples will constrain the equation of state parameter of the dark energy at the 10%-level or better (Haiman, Mohr, & Holder 2001; Molnar, Birkinshaw, & Mushotzky 2002; Weller, Battye, & Kneissl 2002).

Regardless of whether such parameter estimation forecasts are actually met, the next few decades promise to be a very exciting period for cosmology and galaxy cluster astrophysics.

Acknowledgements. Many thanks to John, Alan and (the other) Gus for putting together an outstanding program. I am grateful for their (and Luis Ho's) extreme patience in awaiting this contribution to the proceedings. Thanks also to my collaborators—John Bialek, Joe Mohr, Risa Wechsler, Martin White, Elena Rasia, Giuseppe Tormen, Lauro Moscardini, Mike Warren, Salmon Habib, Katrin Heitmann and members of the Virgo Consortium—for allowing me to present results of ongoing studies. This research is supported by the NSF and by NASA.

References

Adelberger, K. L., Steidel, C. C., Shapley, A. E., & Pettini, M. 2003, ApJ, 584, 45
Allen, S. W., Schmidt, R. W., & Fabian, A. C. 2002, MNRAS, 334, L11
Bahcall, N. A., & Fan, X. 1998, ApJ, 504, 1
Bahcall, N. A., et al. 2003, ApJS, 148, 243
Bennett, C. L., et al. 2003, ApJS, 148, 1
Bernardeau, F., Colombi, S., Gaztanaga, E., & Scoccimarro, R. 2002, Phys. Rep., 367, 1
Bertschinger, E. 1985, ApJS, 58, 39
——. 1998, ARA&A, 36, 599
Bialek, J., et al. 2004, in preparation
Bialek, J., Evrard, A. E., & Mohr, J. J. 2001, ApJ, 555, 597
Bode, P., Bahcall, N. A., Ford, E. B., & Ostriker, J. P. 2001, ApJ, 551, 15
Bond, J. R., Cole, S., Efstathiou, G., & Kaiser, N. 1991, ApJ, 379, 440
Bond, J. R., & Efstathiou, G. 1984, ApJ, 285, L45
Bond, J. R., Kofman, L., & Pogosyan, D. 1996, Nature, 380, 603
Borgani, S., et al. 2001, ApJ, 561, 13
——. 2004, MNRAS, in press (astro-ph/0310794)
Borgani, S., Girardi, M., Carlberg, R. G., Yee, H. K. C., & Ellingson, E. 1999, ApJ, 527, 561
Bower, R. G. 1991, MNRAS, 24, 332
Bower, R. G., Lucey, J. R., & Ellis, R. S. 1992, MNRAS, 254, 589
Briel, U. G., Henry, J. P., & Böhringer, H. 1992, A&A259, L31
Brüggen, M., & Kaiser, C. R. 2002, Nature, 418, 301
Bryan, G. L. 2000, ApJ, 544, L1
Bryan, G. L., & Norman, M. L. 1998, ApJ, 495, 80
Carlstrom, J. E., Holder, G. P., & Reese, E. D. 2002, ARA&A, 40, 643
Carroll, S. M., Press, W. H., & Turner, E. L. 1992, ARA&A, 30, 499
Cavaliere, A., & Fusco-Femiano, R. 1976, A&A, 49, 137
Cole, S., Lacey, C. G., Baugh, C. M., & Frenk, C. S. 2000, MNRAS, 319, 168
Couch, W. J., Colless, M. M., & De Propris, R. 2004, in Carnegie Observatories Astrophysics Series, Vol. 3:
    Clusters of Galaxies: Probes of Cosmological Structure and Galaxy Evolution, ed. J. S. Mulchaey, A. Dressler,
    & A. Oemler (Cambridge: Cambridge Univ. Press), in press
Davis, M., Efstathiou, G., Frenk, C. S., & White, S. D. M. 1985, ApJ, 292, 371
De Grandi, S., & Molendi, S. 2001, ApJ, 551, 153
Dressler, A. 1978, ApJ, 226, 55
Donahue, M., Voit, G. M., Gioa, I. M., Luppino, G., Hughes, J. P., & Stocke, J. T. 1998, ApJ, 502, 550
Efstathiou, G., Frenk, C. S., White, S. D. M., & Davis, M. 1988, MNRAS, 235, 715

Eke, V. R., Cole, S., & Frenk, C. S. 1996, MNRAS, 282, 263

Evrard, A. E. 1989, ApJ, 341, L71

——. 1990, ApJ, 363, 34

Evrard, A. E., et al. 2002, ApJ, 573, 7

Evrard, A. E., & Gioia, I. M. 2002, in Merging Processes in Galaxy Clusters, ed. L. Ferretti, I. M. Gioia, & G. Giovannini (Dodrecht: Kluwer), 253

Evrard, A. E., & Henry J. P. 1991, ApJ, 383, 95

Evrard, A. E., Metzler, C., & Navarro, J. F. 1996, ApJ, 469, 494

Finoguenov, A., Reiprich, T. H., & Böhringer, H. 2001, A&A, 368, 749

Frenk, C. S., et al. 1999, ApJ, 525, 554

Girardi, M., Borgani, S., Giuricin, G., Mardirossian, F., & Mezzetti, M. 2000, ApJ, 530, 62

Gladders, M. D. 2004, in Carnegie Observatories Astrophysics Series, Vol. 3: Clusters of Galaxies: Probes of Cosmological Structure and Galaxy Evolution, ed. J. S. Mulchaey, A. Dressler, & A. Oemler (Cambridge: Cambridge Univ. Press), in press

Governato, F., Babul, A., Quinn, T., Tozzi, P., Baugh, C. M., Katz, N., & Lake, G. 1999, MNRAS, 307, 949

Gunn, J. E., & Gott, J. R. 1972, ApJ, 176, 1

Haiman, Z., Mohr, J. J., & Holder, G. P. 2001, ApJ, 553, 545

Holder, G. P., Mohr, J. J., Carlstrom, J. E., Evrard, A. E., & Leitch, E. M. 2000, ApJ, 544, 629

Horner, D. J., Mushotzky, R. F., & Scharf, C. A. 1999, ApJ, 520, 78

Hu, W. 2003, Phys. Rev. D, 67, 081304

Hu, W., & Kravtsov, A. V. 2003, ApJ, 584, 702

Jenkins, A., et al. 1998, ApJ, 499, 20

Jenkins, A., Frenk, C. S., White, S. D. M., Colberg, J. M., Cole, S., Evrard, A. E., Couchman, H. M. P., & Yoshida, N. 2001, MNRAS, 321, 372

Kaiser N. 1986, MNRAS, 222, 323

——. 1991, ApJ, 383, 104

Kauffmann, G., Colberg, J. M., Diaferio, A., & White, S. D. M. 1999, MNRAS, 303, 188

Kneissl, R., Jones, M. E., Saunders, R., Eke, V. R., Lasenby, A. N., Grainge, K., & Cotter, G. 2001, MNRAS 328, 783

Kochanek, C. S., White, M., Huchra, J., Macri, L., Jarrett, T. H., Schneider, S. E., & Mader, J. 2003, ApJ, 585, 161

Kolb, E. W., & Turner, M. S. 1990, The Early Universe (Redwood City: Addison Wesley)

Lacey, C., & Cole, S. 1994, MNRAS, 271, 676

Lin, Y.-T., Mohr, J. J., & Stanford, S. A. 2003, ApJ, 591, 749

Lokas, E. L., & Hoffman, Y. 2000, ApJ, 542, L139

Majumdar, S., & Mohr, J. J. 2003, ApJ, 585, 603

Markevitch, M. 1998, ApJ, 504, 27

Mathiesen, B. F., & Evrard, A. E. 2001, ApJ, 546, 100

Mohr, J. J., Mathiesen, B., & Evrard, A. E. 1999, ApJ, 517, 627

Molnar, S. M., Birkinshaw, M., & Mushotzky, R. F. 2002, 570, 1

Narayan, R., & Medvedev, M. V. 2001, ApJ, 562, L129

Navarro, J. F., Frenk, C. S., & White, S. D. M. 1995, MNRAS, 275, 720

Nevalainen, J., Markevitch, M., & Forman, W. 1999, ApJ, 536, 73

Nichol, R. C. 2004, in Carnegie Observatories Astrophysics Series, Vol. 3: Clusters of Galaxies: Probes of Cosmological Structure and Galaxy Evolution, ed. J. S. Mulchaey, A. Dressler, & A. Oemler (Cambridge: Cambridge Univ. Press), in press

Pearce, F. R., Thomas, P. A., Couchman, H. M. P., & Edge, A. C. 2000, MNRAS, 317, 1029

Peebles, P. J. E. 1970, AJ, 75, 13

——. 1980, in The Large Scale Structure of the Universe (Princeton: Princeton Univ. Press)

——. 1993, in Principles of Physical Cosmology (Princeton: Princeton Univ. Press)

Pierpaoli, E., Scott, D., & White, M. 2001, MNRAS, 325, 77

Pratt, G. W., & Arnaud, M. 2002, A&A, 394, 375

Press, W. H., & Schechter, P. 1974, ApJ187, 425

Rasia, E., Tormen, G., & Moscardini, L. 2004, MNRAS, in press (astro-ph/0309405)

Reed, D., Gardner, J., Quinn, T., Stadel, J., Fardal, M., Lake, G., & Governato, F. 2003, MNRAS, 346, 565

Ricker, P. M., & Sarazin, C. L. 2002, AJ, 561, 621

Roettiger, K., Burns, J. O., & Loken, C. 1996, ApJ, 473, 651

Romer, A. K., Viana, P. T. P., Liddle, A. R., & Mann, R. G. 2001, ApJ, 547, 594

Rosati, P., Borgani, S., & Norman, C. 2002, ARA&A, 40, 539

Sanderson, A. J. R., Ponman, T. J., Finoguenov, A., Lloyd-Davies, E. J., & Markevitch, M. 2003, MNRAS, 340, 989

Scott, D., Silk, J., & White, M. 1995, Science, 268, 829

Seljak, U., & Zaldarriaga, M. 1996, ApJ, 469, 437

Sheth, R. K., Mo, H. J., & Tormen, G. 2001, MNRAS, 323, 1

Sheth, R. K., & Tormen, G. 1999, MNRAS, 308, 119

Shimizu, M., Kitayama, T., Sasaki, S., & Suto, Y. 2003, ApJ, 590, 197

Somerville, R. S., & Primack, J. R. 1999, MNRAS, 310, 1087

Spergel, D. N., et al. 2003, ApJS, 148, 175

Springel, V., White, S. D. M, Tormen, G., & Kauffmann, G. 2001, MNRAS, 328, 76

Stanford, S. A., Eisenhardt, P., & Dickinson, M. 1998, ApJ, 492, 461

Szapudi, I., Colombi, S., Jenkins, A., & Colberg, J. 2000, MNRAS, 313, 725

Thomas, P. A., Muanwong, O., Kay, S. T., & Liddle, A. R. 2002, MNRAS, 330, L48

Voit, G. M., Balogh, M. L., Bower, R. G., Lacey, C. G., & Bryan, G. L. 2003, ApJ, 593, 272

Wecshler, R. H., et al. 2004, in preparation

Weller, J., Battye, R. A., & Kneissl, R. 2002, Phys. Rev. Lett., 88, 1301

White, D. A. 2000, MNRAS, 312, 663

White, M. 2001, A&A, 367, 27

——. 2002, ApJS, 143, 241

White, S. D. M., Navarro, J. F., Evrard, A. E., & Frenk, C. S. 1993, Nature, 366, 429

Xue, Y.-J., & Wu, X.-P. 2000, ApJ, 538, 65

Zwicky, F. 1933, Helv. Phys. Acta, 6, 110

——. 1937, ApJ, 86, 217

2

Clusters of galaxies in the Sloan Digital Sky Survey

ROBERT C. NICHOL
Department of Physics, Carnegie Mellon University

Abstract
I review here past and present research on clusters and groups of galaxies within the Sloan Digital Sky Survey (SDSS). I begin with a short review of the SDSS and efforts to find clusters of galaxies using both the photometric and spectroscopic SDSS data. In particular, I discuss the C4 algorithm, which is designed to search for clusters and groups within a seven-dimensional (7-D) data space, i.e., simultaneous clustering in both color and space. The C4 catalog has a well-quantified selection function based on mock SDSS galaxy catalogs constructed from the Hubble Volume simulation. These simulations indicate that the C4 catalog is $> 90\%$ complete, with $< 10\%$ contamination, for halos of $M_{200} > 10^{14} M_\odot$ at $z < 0.14$. Furthermore, the observed summed r-band luminosity of C4 clusters is linearly related to M_{200}, with $< 30\%$ scatter at any given halo mass. I also briefly review the selection and observation of luminous red galaxies and demonstrate that these galaxies have a similar clustering strength as clusters and groups of galaxies. I outline a new collaboration planning to obtain redshifts for 10,000 luminous red galaxies at $0.4 < z < 0.7$ using the SDSS photometric data and the Anglo-Australian Telescope 2dF instrument. Finally, I review the role of clusters and groups of galaxies in the study of galaxy properties as a function of environment. In particular, I discuss the "star formation rate-density" and "morphology-radius" relations for the SDSS and note that both of these relationships have a *critical density* (or "break") at a projected local galaxy density of $\sim 1 h_{75}^{-2} \mathrm{Mpc}^{-2}$ (or between 1 to 2 virial radii). One possible physical mechanism to explain this observed critical density is the stripping of warm gas from the halos of infalling spiral galaxies, thus leading to a slow strangulation of star formation in these galaxies. This scenario is consistent with the recent discovery (within the SDSS) of an excess of "passive" or "anemic" spiral galaxies located within the infall regions of C4 clusters.

2.1 Introduction

As demonstrated by this conference—*Clusters of Galaxies: Probes of Cosmological Structure and Galaxy Evolution*—clusters and groups of galaxies have a long history as key tracers of the large-scale structure in the Universe and as laboratories within which to study the physics of galaxy evolution. At the conference, their important role as cosmological probes was reviewed by several speakers, including Gus Evrard and Alan Dressler. Therefore, I will not dwell on justifying the importance of clusters to cosmological research, but simply direct the reader to the reviews by these authors.

Instead, I provide below a brief overview of the Sloan Digital Sky Survey (SDSS), followed by a discussion of the cluster-finding algorithms (§2.2) used within the SDSS collaboration, and present new scientific results obtained from studies of galaxies as a function of environment (§2.5).

2.1.1   *The Sloan Digital Sky Survey*

The SDSS (York et al. 2000; Stoughton et al. 2002) is a joint multicolor (u, g, r, i, z) imaging and medium resolution ($R = 1800$; 3700 to 9100 Å) spectroscopic survey of the northern hemisphere using a dedicated 2.5 meter telescope located at the Apache Point Observatory near Sunspot, New Mexico. During well-defined (seeing $<1.''7$) photometric conditions, the SDSS employs a mosaic camera of 54 CCD chips to image the sky via the drift-scanning technique (see Gunn et al. 1998). These data are reduced at Fermilab using a dedicated photometric analysis pipeline (PHOTO; see Lupton et al. 2001), and object catalogs are obtained for spectroscopic target selection. During nonphotometric conditions, the SDSS performs multiobject spectroscopy using two bench spectrographs attached to the SDSS telescope and fed with 640 optical fibers. The other ends of these fibers are plugged into a pre-drilled aluminum plate that is bent to follow the 3-degree focal surface of the SDSS telescope. The reader is referred to Smith et al. (2002), Blanton et al. (2003), and Pier et al. (2003) for more details about the SDSS.

In this way, the SDSS plans to image the northern sky in five passbands, as well as obtain spectra for $\sim 10^6$ objects. In addition to the large amount of the data being collected, the quality of the SDSS data is high, which is important for many of the scientific goals of the survey. For example, the SDSS has a dedicated photometric telescope (Hogg et al. 2001), which is designed to provide the SDSS with a global photometric calibration of a few percent accuracy over the whole imaging survey. Also, SDSS spectra are spectrophotometrically calibrated.

The SDSS is now in production mode and has been collecting data for several years. As of January 2003, the SDSS had obtained 4470 deg^2 of imaging data (not unique area) and had measured a half million spectra. The SDSS has just announced its first official data release (see http://www.sdss.org/dr1/).

2.2   SDSS Cluster Catalogs

One of the fundamental science goals of the SDSS was to create new catalogs of clusters from both the imaging and spectroscopic data. In Table 2.1, I present a brief overview of past and present efforts within the SDSS collaboration as part of the SDSS Cluster Working Group. In this table, I provide an appropriate reference if available, a brief description of the cluster-finding algorithm, and the SDSS data being used to find clusters. Each of these algorithms has different strengthens and science goals, and a detailed knowledge of their selection functions is required before a fair comparison can be carried out between these different catalogs. The reader is referred to the review of Postman (2002) for more discussion of this point. However, Bahcall et al. (2003) have begun this process by performing a detailed comparison of the clusters found by both the AMF and maxBCG algorithms. The product of this work is a joint catalog of 799 clusters in the redshift range $0.05 < z < 0.3$ selected from ~ 400 deg^2 of early SDSS commissioning data.

As first outlined in Nichol et al. (2001), there are four main challenges to producing a robust, optically selected catalog of clusters. These are:

(1) To eliminate projection effects. This problem has plagued previous catalogs of clusters and

Table 2.1. *Overview of past and present efforts to find clusters and groups of galaxies within the SDSS collaboration. In the data column, "I" is for SDSS imaging data and "S" is for SDSS spectral data.*

Name	Reference	Data	Description
maxBCG	Annis et al. (2004)	I	Model colors of brightest cluster galaxies with z and look for E/S0 ridge-line
AMF	Kim et al. (2002)	I	Matched-filter algorithm looking for overdensities in luminosity and space
CE	Goto et al. (2002)	I	Color cuts, then uses SEXTRACTOR to find and deblend clusters
F-O-F	Berlind et al. (2004)	S	Friends-of-friends algorithm
Groups	Lee et al. (2004)	I	Mimics Hickson groups criteria
BH	Bahcall et al. (2003)	I	Merger of maxBCG and AMF
C4	Nichol et al. (2001)	I + S	Simultaneous clustering of galaxies in color and space

has been discussed by many authors (see, for example, Lucey et al. 1983; Sutherland 1988; Nichol et al. 1992; Postman, Huchra, & Geller 1992; Miller 2000).

(2) A full understanding of the selection function. This has been traditionally ignored for optical cluster catalogs (see Bramel, Nichol, & Pope 2000; Kochanek et al. 2003), but is critical for all statistical analyses using the catalog.

(3) To provide a robust mass estimator. Traditionally, this has been the Achilles' Heel of optical catalogs, as richness is a poor indicator of mass.

(4) To cover a large dynamic range in both redshift and mass.

I review below one SDSS cluster catalog with which I am involved in collaboration with Chris Miller at Carnegie Mellon University, which now meets these four challenges and is comparable in quality to the best X-ray catalogs of clusters (e.g., the REFLEX catalog of Böhringer et al. 2001).

2.3   The C4 Algorithm

The underlying hypothesis of the C4 algorithm is that a cluster or group of galaxies is a clustering of galaxies in both color and space. This is demonstrated in Figure 2.1, and the reader is referred to Gladders & Yee (2000) for a full discuss of all the evidence in support of this hypothesis. Therefore, by searching for clusters simultaneously in both color and space, the C4 algorithm reduces projection effects to almost zero, while still retaining much power for finding clusters (see Fig. 2.2).

2.3.1   *Overview of C4 Algorithm*

I present here a brief overview of the C4 algorithm. To date, the C4 algorithm has been applied to the SDSS main galaxy spectroscopic sample (Strauss et al. 2002) in the Early Data Release of the SDSS (see Stoughton et al. 2002; Gómez et al. 2003).

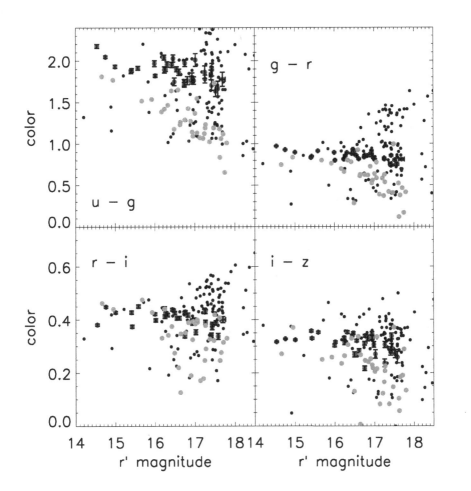

Fig. 2.1. The color-magnitude relations for a previously undiscovered $z = 0.06$ cluster in the Early Data Release of the SDSS detected by the C4 algorithm. The black dots are galaxies within an aperture of $1h^{-1}$ Mpc around the cluster center. Note the tight correlation in color of the points with error bars, which is the E/S0 ridge-line and is the signal the C4 algorithm uses to find clusters.

For each galaxy in the sample—called the "target" galaxy below—the C4 algorithm is performed in the following steps.

(1) A 7-D rectangular box is placed on the target galaxy. The center of this box is defined by the observed photometric colors (i.e., $u-g$, $g-r$, $r-i$, $i-z$), the Right Ascension, Declination, and redshift of the target galaxy. The width of the box is dependent on the redshift and photometric errors on the colors of the target galaxy. Once the box is defined, the number of neighboring galaxies is counted within the SDSS main galaxy sample (Strauss et al. 2002) inside this box, and this count is reported (see Fig. 2.2).

(2) The same 7-D box is placed on 100 randomly chosen galaxies, also taken from the SDSS main galaxy spectroscopic sample, that possess similar seeing and reddening values (see Fig.

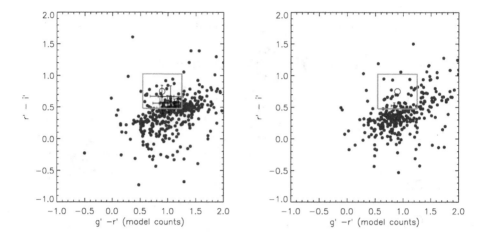

Fig. 2.2. Color-color plots for a sample cluster galaxy (*left*) and a randomly chosen field position (*right*). The black dots are all galaxies within the spatial part of the 7-D search box centered on the target galaxy (i.e., all galaxies satisfying the R.A., Dec., and z dimensions of the 7-D box). The points with error bars are now those galaxies which also lie within the color part of the 7-D search box; i.e., they are close in both space and color. The box is the size of the 7-D box in this color-color plane, and is much smaller than the scatter seen in the black points. For the random field position, there is only one point with an error bar inside the box compared to 10 around the cluster galaxy. Therefore, all projection effects have been eradicated as one does not expect false clustering in such a high-dimensional space.

2.2). For each of these 100 randomly chosen galaxies, the number of neighboring galaxies is counted (from the SDSS main galaxy sample) inside the box and a distribution of galaxy counts is constructed from these 100 randomly chosen galaxies.

(3) Using this observed distribution of galaxy counts, the probability of obtaining the observed galaxy count around the original target galaxy is computed.

(4) This exercise is repeated for all galaxies in the sample, and then all these probabilities are ranked.

(5) Using the "false discovery rate" (Miller et al. 2001) with an $\alpha = 0.2$ (i.e., only allowing a false discovery rate of 20%), a threshold in probability is determined that corresponds to galaxies that possess a high count of nearest neighbors in their 7-D box.

(6) All galaxies in our sample below this threshold in probability are removed, which results in the eradication of $\sim 80\%$ of all galaxies. By design, the galaxies that are removed are preferentially in low-density regions of this 7-D space (i.e., the field population). The galaxies that remain are called "C4 galaxies," which, by design, reside in high-density regions with neighboring galaxies that possess the same colors as the target galaxy. Figure 2.3 illustrates this step.

(7) Using only the C4 galaxies, the local density of all the C4 galaxies is determined using the distance to the 10th nearest neighbor.

(8) The galaxies are rank ordered based on these measured densities and then assigned to clusters based on this ranked list. This is the same methodology as used to create halo catalogs within N-body simulations (see Evrard et al. 2002).

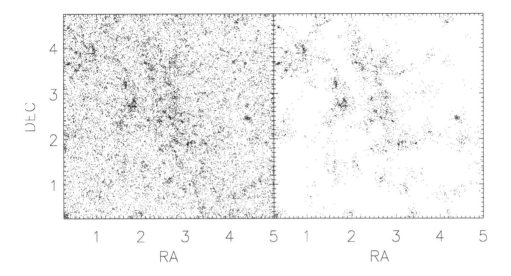

Fig. 2.3. The projected galaxy distribution in the simulations before (*left*) and after (*right*) the C4 algorithm has been run and a threshold applied to eliminate field-like galaxies. This illustrates the effect of step 6 in the algorithm.

(9) This results in a list of clusters, for which a summed total optical luminosity (z and r bands) and a velocity dispersion are computed (see below).

In summary, the C4 algorithm is a semi-parametric implementation of adaptive kernel density estimation. The key difference with this approach, compared to previous cluster-finding algorithms, is that it does not model either the colors of cluster ellipticals (e.g., Gladders & Yee 2000; Goto et al. 2003) or the properties of the clusters (e.g., Postman et al. 1996; Kepner et al. 1999; Kim et al. 2002). Instead, the C4 algorithm only demands that the colors of nearby galaxies are the same as the target galaxy. In this way, the C4 algorithm is sensitive to a diverse range of clusters and groups; for example, it would detect a cluster dominated by a extremely "blue" population of galaxies (compared to the colors of field galaxies).

2.3.2 Simulations of the SDSS Data

To address the four challenges given above, it is vital that we construct realistic simulations of the C4 cluster catalog. This has now been achieved through collaboration with Risa Wechsler, Gus Evrard, and Tim McKay at the University of Michigan. Briefly, mock SDSS galaxy catalogs have been created using the dark matter distribution from the Hubble Volume* simulations (Evrard et al 2002). The procedure for populating the dark matter distribution with galaxies is described in detail by Wechsler et al. (2004), but I present a brief overview here. The simulation we use provides the dark matter distribution of the full sky out to $z \approx 0.6$, where the dark matter clustering is computed within the light cone. Galaxies, with r-band luminosities following the luminosity function found by Blanton et al (2001) for the SDSS, are added to the simulation by choosing simulation particles (with

* http://www.mpa-garching.mpg.de/Virgo/hubble.html

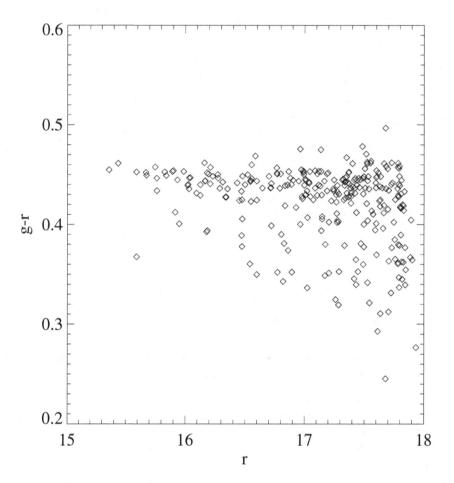

Fig. 2.4. The color-magnitude relations for a $z = 0.06$, $10^{15} M_\odot$ halo in the Hubble Volume simulation, where galaxies with SDSS colors and luminosities have been added to the simulation.

mass $\sim 2.2 \times 10^{12} M_\odot$ per particle) from a probability density function of local mass density that depends on galaxy luminosity. This luminosity-dependent probability density function is tuned so that the resulting galaxies match the luminosity-dependent two-point correlation function (Zehavi et al. 2002). Colors are then added to the simulation using the colors of actual SDSS galaxies with similar luminosities and local galaxy densities. The resulting catalog thus matches the luminosity function and color- and luminosity-dependent two-point correlation function of the SDSS data.

Therefore, these mock SDSS catalogs produce clusters of galaxies with a realistic color-magnitude diagram (compare Figs. 2.1 and 2.4). As we know the location of all the massive halos (i.e., the clusters) within the Hubble Volume simulation (see Jenkins et al. 2001 and Evrard et al. 2002), we can now use these mock catalogs to derive the selection function of

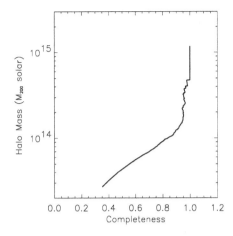
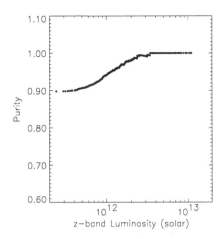

Fig. 2.5. *Left*: The completeness of the C4 catalog as a function of the dark matter halo mass (M_{200}) from the simulation. *Right*: The purity of the C4 catalog as a function of the summed total *r*-band luminosity of the cluster, as derived from our mock SDSS galaxy catalog.

the C4 algorithm. The benefit of this approach is that these mock catalogs contain realistic projection effects, as the galaxy clustering is constrained to match the real data, and the clusters (halos) possess realistic profiles, ellipticities, and morphologies. This is a significant advancement compared to past simulations used to quantify the selection function of cluster catalogs (see Postman et al. 1996; Bramel et al. 2000; Goto et al. 2002).

For comparison between the real and simulated data, two different cluster observables are used. The first is the bi-weighted velocity dispersion, and the second is the summed optical luminosity of the galaxies within the cluster. To calculate the velocity dispersion (σ_v), an iterative technique is performed using the robust bi-weighted statistics of Beers, Flynn, & Gebhardt (1990). The total optical luminosity of the clusters is determined by converting the apparent magnitudes of galaxies in the cluster to optical luminosities using the conversions in Fukugita et al. (1996). All magnitudes are also *k*-corrected according to Blanton et al. (2002) and extinction corrected according to Schlegel, Finkbeiner, & Davis (1998). Cluster membership is defined to be any galaxy within $4\sigma_v$ in redshift and within $1.5h^{-1}$ Mpc projected separation on the sky. For the halo catalog, the mass within 200 times the critical density (M_{200}), as determined from summing up all the dark matter particles within this radius around each halo (R_{200}; see Evrard et al. 2002), is used.

2.3.3 *Purity and Completeness: Challenges One and Two*

The C4 algorithm has been applied to these mock SDSS galaxy catalogs. In Figure 2.5, I show the purity of the matched C4 clusters found in the simulation as a function of the total *r*-band luminosity. Purity is defined to be the percentage of systems detected in the mock SDSS catalog, using the C4 algorithm, that match a known dark matter halo in the Hubble Volume simulation. Clearly, the purity of the C4 catalog remains high over nearly 2 orders of magnitude in luminosity and is a direct result of searching for clustering in a high-dimensional space where projection effects are rare.

In Figure 2.5, I also present the completeness of the C4 catalog, as a function of the dark halo mass (M_{200}), and demonstrate that the C4 catalog remains $> 90\%$ complete for systems with $M_{200} > 10^{14} M_\odot$. The accuracy of these completeness measurements greatly benefits from the large sample sizes available from the Hubble Volume simulation. However, for the lower-mass systems, there may be some incompletenesses in the original dark matter halo catalog because of the mass resolution of the Hubble Volume simulations. Furthermore, the purity and completeness of the C4 catalog are only a weak function of the input parameters (e.g., the false discovery rate threshold) and thus are robust against the exact parameter choices. In summary, using these mock SDSS catalogs, we have addressed the first two challenges given in §2.2: for $z < 0.14$ and halos of $M_{200} > 10^{14} M_\odot$, the C4 catalog is $> 90\%$ complete, with $< 10\%$ contamination.

2.3.4   *Mass Estimator and Dynamic Range: Challenges Three and Four*

In Figure 2.6, I present the correlation between the M_{200} dark matter halo mass (taken directly from Evrard et al. 2002) and the total summed r-band luminosities for 4734 clusters that match in the Hubble Volume simulation. As expected, these two quantities are linearly correlated over 2 orders of magnitude in both the halo mass and optical luminosity. The measured scatter for the whole data set is 25% (measured perpendicular to the best-fit line). This plot demonstrates that we have address the third and fourth challenges in §2.2.

In the future, the C4 algorithm will be extended to include: (1) higher-resolution dark matter simulations to test the sensitivity of the algorithm to lower-mass systems, (2) a range of cosmological simulations to probe our sensitivity to various cosmological parameters (e.g., σ_8), and (3) improved methods of populating the simulations with SDSS galaxies using higher-order statistics (e.g., the three-point correlation function).

2.4   Luminous Red Galaxies

I briefly review here forthcoming surveys of luminous red galaxies (LRGs), as these surveys will soon replace clusters as the most efficient tracers of the large-scale structure in the Universe. Such galaxies are selected to be dominated by an old stellar population (using the SDSS colors) and are luminous, so they can be seen to high redshift even in the SDSS photometric data (see Eisenstein et al. 2001 for details).

A preliminary analysis of the correlation functions for both a sample of LRGs, selected from the SDSS (Eisenstein et al. 2001), and normal SDSS galaxies (Zehavi et al. 2001) demonstrates that, as expected, the LRGs are more strongly clustered than normal galaxies. The LRG correlation function has an amplitude and scale length consistent with that measured for groups and clusters of galaxies (e.g., Peacock & Nicholson 1991; Nichol et al. 1992; Collins et al. 2000; Miller 2000; Nichol 2002). Therefore, by design, a large fraction of these LRGs must lie in dense environments, as their spatial distribution clearly traces the distribution of clusters and groups in the Universe. The main advantage of the LRG selection is that it does not depend upon the details of finding clusters of galaxies, and, therefore, their selection is more straightforward to model (see Eisenstein et al. 2001).

Due to the 45 minute spectroscopic exposure time of the SDSS, the SDSS only targets LRG candidates brighter than $r = 19.5$ mag (Eisenstein et al. 2001). This corresponds to a cut-off at $z \simeq 0.45$, but one can easily detect LRG candidates in the SDSS photometric

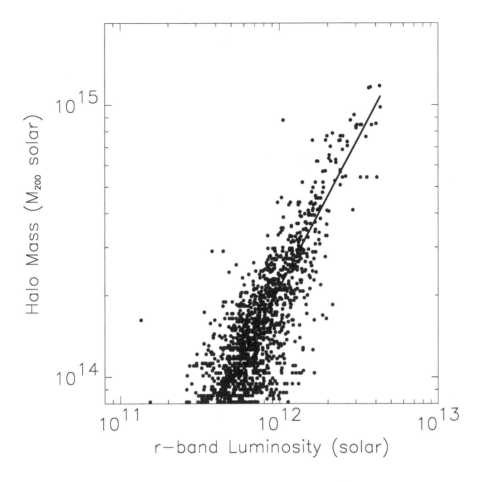

Fig. 2.6. The relation between the dark matter halo mass in our simulation and the total summed r luminosity. The line is the best fit, and the scatter, for a given mass, is $< 30\%$.

data to higher redshifts.* Therefore, using the unique 2dF multiobject spectrograph on the Anglo-Australian Telescope, we have begun a joint SDSS-2dF program to push the original SDSS LRG selection to higher redshift. At the time of writing, our initial observations have been very successful, with spectra for ~ 1000 LRGs in the redshift range of $0.4 < z < 0.7$. In Figure 2.7, I show the distribution in luminosity and redshift of these new SDSS-2dF LRGs and highlight that they cover a comparable range in their luminosities as the low-redshift LRGs. By the end of the SDSS-2dF LRG survey, we hope to have redshifts for 10,000 LRGs over this intermediate-redshift range. When combined with the low-redshift SDSS LRGs, we will be able to study the evolution in the properties and clustering of a single population of massive galaxies over half the age of the Universe.

---

* Beyond $z = 0.45$, the SDSS LRG selection becomes easier than at lower redshift (with less contamination) because of the fortuitous design of the SDSS filter system (see Eisenstein et al. 2001).

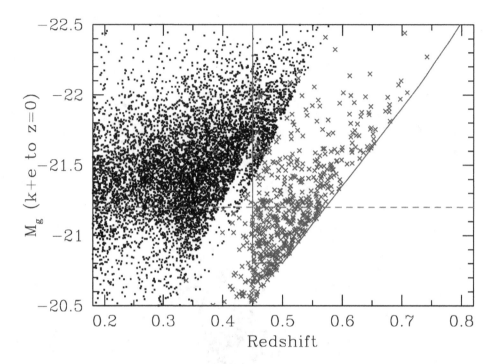

Fig. 2.7. The distribution in g-band luminosity and redshift for both the low-redshift LRGs (solid points) and the SDSS-2dF LRGs (crosses). The luminosities are k-corrected and corrected for passive evolution to $z = 0$. The solid lines show the expected selection boundaries for the SDSS-2dF survey, while the dashed line is the luminosity limit above which we expect the low-redshift SDSS LRG sample to be complete. Plot courtesy of Daniel Eisenstein and my SDSS-2dF colleagues.

2.5   Galaxy Properties as a Function of Environment

Clusters and groups of galaxies play an important role in studying the effects of environment on the properties of galaxies. With the SDSS data, it is now possible to extend such studies well beyond the cores of clusters into lower-density environments. Furthermore, the distance to the Nth nearest neighbor can be used to provide an adaptive measure of the local density of galaxies (see Dressler 1980; Lewis et al. 2002; Gómez et al. 2003). One can also use kernel density estimators (e.g., Eisenstein 2003).

There were many great talks and posters on the topic of galaxy evolution in clusters at this conference. For example, see the contributions by Bower, Davies, Dressler, Franx, Goto, Martini, N. Miller, Poggianti, Tran, and Treu. Also, I refer the reader to the work of Hogg et al. (2003), who is also using the SDSS data to study the effects of environment on the colors, surface brightnesses, morphologies, and luminosities of galaxies.

2.5.1   *Critical Density*

Is the star formation rate (SFR) of a galaxy affected by its environment? The answer appears to be yes, and was discussed by several authors at this conference. In particular,

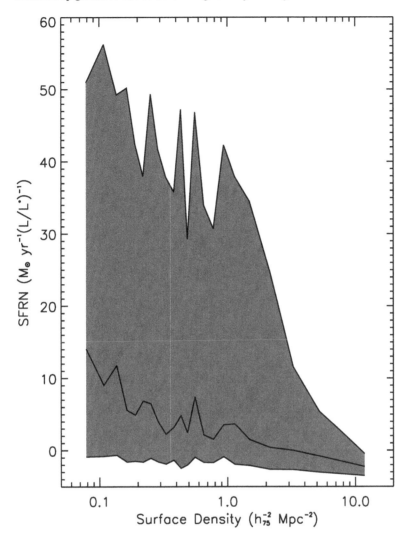

Fig. 2.8. The relation between SFR and density for a volume-limited sample of 8598 SDSS galaxies within $0.05 < z < 0.095$. The y-axis is the normalized SFR, i.e. the SFR divided by the z luminosity of the galaxy. The x-axis is the projected local surface density computed from the distance to the 10th nearest neighbor in a redshift shell of $\pm 1000\,km\,s^{-1}$. See Gómez et al. (2003) for details. The top of the shaded region is the 75th percentile of the distribution, while the bottom is the 25th percentile. The solid line through the shaded area is the median of the distribution. As one can see, beyond a density of $\sim 1\,h_{75}^{-2}\,Mpc^{-2}$, the tail of the distribution (75th percentile) is heavily curtailed in dense environments.

Gómez et al. (2003) find that the fraction of strongly star-forming galaxies in the SDSS decreases rapidly beyond a critical density of $\sim 1\,h_{75}^{-2}\,Mpc^{-2}$ (see also Lewis et al. 2002). This result is demonstrated in Figure 2.8 and appears to be the same for all morphological types (see Gómez et al. 2003 and below).

In Figure 2.9, we also show a preliminary SDSS *morphology-radius* relation based on

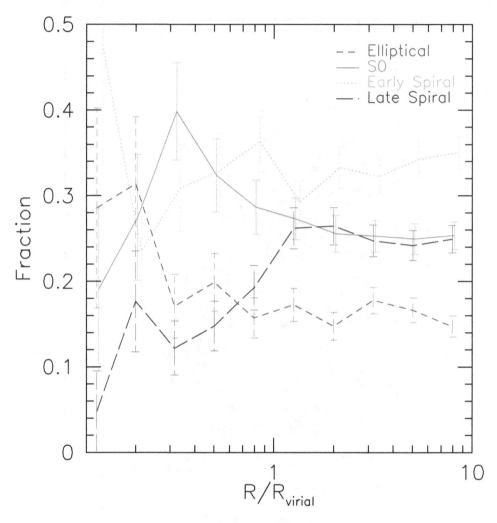

Fig. 2.9. The morphology-radius relation for C4 clusters in the SDSS. The morphologies were derived using the T_{auto} parameter discussed in the text, while the distance to the nearest C4 cluster (x-axis) has been scaled by the virial radius of that clusters. These plots were adapted from Gómez et al. (2003) and Goto et al. (2004).

the work of Goto et al. (2004). The morphological classifications used in this figure are based on the new and improved concentration index C_{in} (see Shimasaku et al. 2001) measurements of Yamauchi et al. (2004). Briefly, Yamauchi et al. compute their concentration index within two-dimensional, elliptical isotopes, which account for the observed ellipticity and orientation of the galaxy on the sky. This improvement helps prevent low-inclination galaxies (e.g., edge-on spirals) from being misclassified as early-type galaxies. Furthermore, Yamauchi et al. computes the "coarseness" of each galaxy, which is a measure of the residual variance after the best-fit two-dimensional galaxy model has been subtracted. This coarseness measurement can therefore detect the presence of spiral arms in a galaxy. These two measurements of the morphology are re-normalized (by their rms) and added to produce a final morphological parameter called T_{auto}. Yamauchi et al. have tested their algorithm extensively and have demonstrated that T_{auto} is more strongly correlated with visual morphological classifications than the normal C_{in} parameter, with a correlation coefficient with the visual morphologies of 0.89.

Using the T_{auto} classification, Goto et al. (2004) have separated SDSS galaxies into the four (traditional) morphological subsamples of ellipticals, lenticulars (S0s), early-type (Sa and Sb) spirals and late-type systems (Sc spirals and irregulars). This is presented in Figure 2.9. Clearly, the mapping between the T_{auto} parameter and these visually derived morphological classifications is not perfect, but such an analysis does allow for an easier comparison with previous measurements of the morphology-radius and morphology-density relations, and theoretical predictions of these relationships (see Benson et al. 2000).

The SDSS morphology-radius relation shown in Figure 2.9 remains constant at > 2 virial radii from C4 clusters. As expected, this corresponds to low-density regions ($< 1 h_{75}^{-2}$ Mpc^{-2}) in our volume-limited sample. This observation is consistent with previous determinations of the morphology-density and morphology-radius relations in that these functions are nearly constant at low densities (see Dressler 1980; Postman & Geller 1984; Dressler et al. 1997; Treu et al. 2003). At a radius of ~ 1 virial radius, we witness a change in the morphology-radius relation: we see a decrease in the fraction of late-type ($\sim$Sc) spiral galaxies with smaller cluster-centric radii. We also see some indication of a decrease in the early-type spirals. It is interesting to note that the critical density of $\sim 1 h_{75}^{-2}$ Mpc^{-2} seen in the SFR-density relation of Gómez et al. corresponds to a cluster-centric radius of between ~ 1 to 2 virial radii. The key question is: Are these two phenomena just different manifestations of the same physical process that is transforming both the morphology and the SFR of the galaxies at this critical density? I believe the jury is still out on this question.

I note here that Postman & Geller (1984) also reported a critical density (or "break") in their morphology-density relation at approximately the same density as seen in Figure 2.8, i.e. $\sim 3.5 h_{75}^{-3}$ Mpc^{-3}. Therefore, this "break" (or critical density) seen in the morphology-density relation appears to be universal, as it has been seen in two separate studies, which are based on different selection criteria and analysis techniques.

2.6   Strangulation of Star Formation

One possible physical model* for explaining the critical density seen at $\sim 1 h_{75}^{-2}$ Mpc^{-2} (or > 1 virial radius) in the SFR-density and morphology-radius relations is the stripping of the warm gas in the outer halo of infalling spiral galaxies via tidal interactions with the cluster potential. This process removes the reservoir of hydrogen that replenishes the gas in the cold disk of the galaxy, and thus slowly strangles (or starves) the star formation in the disk, leading to a slow death (see Larson, Tinsley, & Caldwell 1980; Balogh, Navarro, & Morris 2000; Diaferio et al. 2001). Recent N-body simulations of this model by Bekki, Shioya,

---

* See the reviews of Bower, Mihos, and Moore in this volume for a discussion of other physical mechanisms that can affect the properties of galaxies in clusters and groups of galaxies.

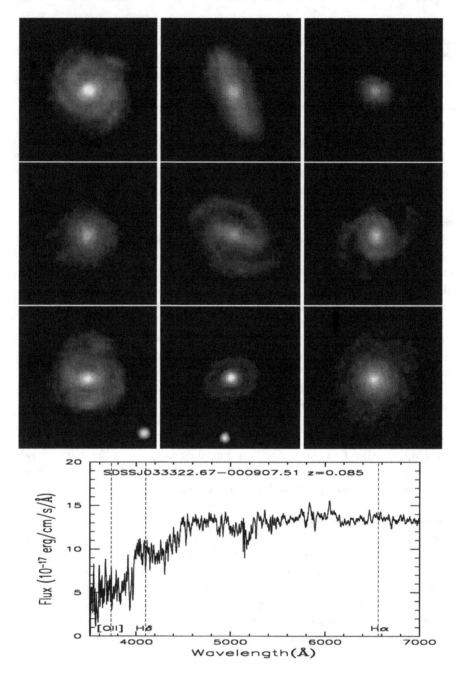

Fig. 2.10. *Top*: Nine images of "passive" spiral galaxies found by Goto et al. (2003) in their search for anemic galaxies in the SDSS database. Note the lack of "blue" H II, star-forming regions in the arms of these spirals. *Bottom*: The SDSS spectrum for one of these "passive" spirals (the galaxy in the top left-hand corner). Notice the lack of any emission lines indicative of ongoing star formation. These plots were taken from Goto et al. (2003).

& Couch (2002) demonstrate that it is viable and can happen at large cluster radii (or low densities).

One possible observational consequence of this strangulation of star formation is the existence of red or passive spiral galaxies, those that possess a spiral morphology but have no observed ongoing star formation. Such galaxies have been found already in studies of high-redshift clusters of galaxies (Couch et al. 1998) and have been known for some time at low redshift as "anemic" spirals (see van den Bergh 1991), although their true nature has been debated by many (see Bothun & Sullivan 1980; Guiderdoni 1987). Goto et al. (2003) has performed an automated searched for such "passive" or "anemic" spirals within the SDSS database by looking for galaxies with a high C_{in} value (indicating a face-on spiral galaxy; see Shimasaku et al. 2001) but with no detected emission lines in their SDSS spectra. In total, they found 73 such galaxies, which comprises only $0.28\% \pm 0.03\%$ of all spiral galaxies with the same C_{in} parameter values but with detected emission lines. I show some examples of these "passive" spirals in Figure 2.10.

The most interesting discovery of Goto et al. (2003) is the distribution of local densities for these "passive" spirals, which peaks at $\sim 1 h_{75}^{-2}\,\mathrm{Mpc}^{-2}$. Therefore, these galaxies appear to be preferential located close to the critical density discussed above for the SFR-density and morphology-radius relations, i.e. in the infall regions of C4 clusters. This is consistent with the observed decrease in the late-type (Sc) spiral galaxies witnessed in Figure 2.9.

In summary, the quality and quantity of the SDSS data allow us to study the environmental dependences of galaxy properties in greater detail than before. There does appear to be at least one critical density (at $\sim 1 h_{75}^{-2}\,\mathrm{Mpc}^{-2}$) affecting the properties of galaxies, and this could be due to the slow strangulation of star formation in these spiral galaxies as they fall into denser environments.

Acknowledgements. I would like to acknowledge the organizers of the conference for their invitation to participate and their hospitality in Pasadena. I also thank all my collaborators for allowing me to show their results and figures in this review. These include Chris Miller, Daniel Eisenstein, Risa Wechsler, Gus Evrard, Tim McKay, Tomo Goto, Michael Balogh, Ann Zabludoff, and my colleagues from both the SDSS-2dF LRG survey and SDSS. I thank Kathy Romer and Chris Miller for reading an earlier draft of this review.

Funding for the creation and distribution of the SDSS Archive has been provided by the Alfred P. Sloan Foundation, the Participating Institutions, the National Aeronautics and Space Administration, the National Science Foundation, the U.S. Department of Energy, the Japanese Monbukagakusho, and the Max Planck Society. The SDSS Web site is http://www.sdss.org/.

The SDSS is managed by the Astrophysical Research Consortium (ARC) for the Participating Institutions. The Participating Institutions are The University of Chicago, Fermilab, the Institute for Advanced Study, the Japan Participation Group, The Johns Hopkins University, Los Alamos National Laboratory, the Max-Planck-Institute for Astronomy (MPIA), the Max-Planck-Institute for Astrophysics (MPA), New Mexico State University, University of Pittsburgh, Princeton University, the United States Naval Observatory, and the University of Washington.

References

Annis, J., et al. 2004, in preparation
Bahcall, N. A., et al. 2003, ApJ, 585, 182
Balogh, M. L., Navarro, J. F., & Morris, S. L. 2000, ApJ, 540, 113
Beers, T. C., Flynn, K., & Gebhardt, K. 1990, AJ, 100, 32
Bekki, K., Shioya, Y., & Couch, W. J. 2002, ApJ, 577, 651
Benson, A. J., Baugh, C. M., Cole, S., Frenk, C. S., & Lacey, C. G. 2000, MNRAS, 316, 107
Berlind, A. A., et al. 2004, in preparation
Blanton, M. R., et al. 2001, AJ, 121, 2358
Blanton, M. R., Lin, H., Lupton, R. H., Maley, F. M., Young, N., Zehavi, I., & Loveday, J. 2003, AJ, 125, 2276
Böhringer, H., et al. 2001, A&A, 369, 826
Bothun, G. D., & Sullivan, W. T. 1980, ApJ, 242, 903
Bramel, D. A., Nichol, R. C., & Pope, A. C. 2000, ApJ, 533, 601
Collins, C. A., et al. 2000, MNRAS, 319, 939
Couch, W. J., Barger, A. J., Smail, I., Ellis, R. S., & Sharples, R. M. 1998, ApJ, 497, 188
Diaferio, A., Kauffmann, G., Balogh, M. L., White, S. D. M., Schade, D., & Ellingson, E. 2001, MNRAS, 323, 999
Dressler, A. 1980, ApJ, 236, 351
Dressler, A., et al. 1997, ApJ, 490, 577
Eisenstein, D. J. 2003, ApJ, 586, 718
Eisenstein, D. J., et al. 2001, AJ, 122, 2267
Evrard, A. E., et al. 2002, ApJ, 573, 7
Fukugita, M., Ichikawa, T., Gunn, J. E., Doi, M., Shimasaku, K., & Schneider, D. P. 1996, AJ, 111, 1748
Gladders, M. D., & Yee, H. K. C. 2000, AJ, 120, 2148
Gómez, P. L., et al. 2003, ApJ, 584, 210
Goto, T. et al. 2002, AJ, 123, 1807
——. 2003, PASJ, 55, 757
——. 2004, in preparation
Guiderdoni, B. 1987, A&A, 172, 27
Gunn, J. E., et al. 1998, AJ, 116, 3040
Hogg, D. W., et al. 2003, ApJ, 585, L5
Hogg, D. W., Finkbeiner, D. P., Schlegel, D. J., & Gunn, J. E. 2001, AJ, 122, 2129
Jenkins, A., Frenk, C. S., White, S. D. M., Colberg, J. M., Cole, S., Evrard, A. E., Couchman, H. M. P., & Yoshida, N. 2001, MNRAS, 321, 372
Kepner, J., Fan, X., Bahcall, N., Gunn, J., Lupton, R., & Xu, G. 1999, ApJ, 517, 78
Kim, R. S. J., et al. 2002, AJ, 123, 20
Kochanek, C. S., White, M., Huchra, J., Macri, L., Jarrett, T. H., Schneider, S. E., & Mader, J. 2003, ApJ, 585, 161
Larson, R. B., Tinsley, B. M., & Caldwell, C. N. 1980, ApJ, 237, 692
Lee, B., et al. 2004, ApJ, submitted
Lewis, I. J., et al. 2002, MNRAS, 334, 673
Lucey, J. R. 1983, MNRAS, 204, 33
Lupton, R. H., Gunn, J. E., Ivezić, Z., Knapp, G. R., Kent, S., & Yasuda, N. 2001, in Astronomical Data Analysis Software and Systems X, ed. F. R. Harnden, Jr., F. A. Primini, & H. E. Payne (San Francisco: ASP), 269
Miller, C. J. 2000, Ph.D. Thesis, Univ. of Maine
Miller, C. J., et al. 2001, AJ, 122, 3492
Nichol, R. C. 2002, in Tracing Cosmic Evolution with Galaxy Clusters, ed. S. Borgani, M. Marino, & G. Valdarini (San Francisco: ASP), 57
Nichol, R. C., et al. 2001, in Mining the Sky, ed. A. J. Banday, S. Zaroubi, & M. Bartelmann (Heidelberg: Springer-Verlag), 613
Nichol, R. C., Collins, C. A., Guzzo, L., & Lumsden, S. L. 1992, MNRAS, 255, 21P
Peacock, J. A., & Nicholson, D. 1991, MNRAS, 253, 307
Pier, J. R., Munn, J. A., Hindsley, R. B., Hennessy, G. S., Kent, S. M., Lupton, R. H., & Ivezić, Z. 2003, AJ, 125, 1559
Postman, M. 2002, in Tracing Cosmic Evolution with Galaxy Clusters, ed. S. Borgani, M. Marino, & G. Valdarini (San Francisco: ASP), 3
Postman, M., & Geller, M. J. 1984, ApJ, 281, 95
Postman, M., Huchra, J. P., & Geller, M. J. 1992, ApJ, 384, 404

Postman, M., Lubin, L. M., Gunn, J. E., Oke, J. B., Hoessel, J. G., Schneider, D. P., & Christensen, J. A. 1996, AJ, 111, 615

Schlegel, D. J., Finkbeiner, D. P., & Davis, M. 1998, ApJ, 500, 525

Shimasaku, K., et al. 2001, AJ, 122, 1238

Smith, J. A., et al. 2002, AJ, 123, 2121

Stoughton, C., et al. 2002, AJ, 123, 485 (erratum: 123, 3487)

Strauss, M. A., et al. 2002, AJ, 124, 1810

Sutherland, W. 1988, MNRAS, 234, 159

Treu, T., Ellis, R. S., Kneib, J.-P., Dressler, A., Smail, I., Czoske, O., Oemler, A., & Natarajan, P. 2003, ApJ, 591, 53

van den Bergh, S. 1991, PASP, 103, 390

Wechsler, R. H., et al. 2004, in preparation

Yamauchi, C., et al. 2004, in preparation

York, D. G., et al. 2000, AJ, 120, 1579

Zehavi, I., et al. 2002, ApJ, 571, 172

3

Clustering studies with the 2dF Galaxy Redshift Survey

WARRICK J. COUCH[1], MATTHEW M. COLLESS[2], and ROBERTO DE PROPRIS[2]
(1) School of Physics, University of New South Wales, Sydney, Australia
(2) Research School of Astronomy & Astrophysics, Australian National University, ACT, Australia

Abstract
The 2dF Galaxy Redshift Survey has now been completed and has mapped the three-dimensional distribution, and hence clustering, of galaxies in exquisite detail over an unprecedentedly large ($\sim 10^8\,h^{-3}\,\mathrm{Mpc}^3$) volume of the local Universe. Here we highlight some of the major results to come from studies of clustering within the survey: galaxy correlation function and power spectrum analyses and the constraints they have placed on cosmological parameters; the luminosity functions of rich galaxy clusters, their dependence on global cluster properties and galaxy type, and how they compare with the field; and the variation of galactic star formation activity with environment, both within clusters and in galaxy groups.

3.1 Introduction

Given the long and distinguished record the Carnegie Observatories have in the exploration of galaxies and determining their distances from us, it is perhaps fitting that this centennial celebration coincides with the emergence of the new generation of large galaxy redshift surveys that are now mapping the galaxy distribution over statistically representative ($\sim 10^8\,h^{-3}\,\mathrm{Mpc}^3$) volumes of the local Universe. The 2dF Galaxy Redshift Survey (2dFGRS; Colless et al. 2001), carried out on the 3.9 m Anglo-Australia Telescope at Siding Spring Observatory, Australia, is one such survey. Unlike the Sloan Digital Sky Survey (Nichol 2004), the 2dFGRS has already been completed, with its final observations taken on 11 April 2002. At that point it had obtained spectra for 270,000 objects, providing redshifts for 221,496 unique galaxies in the range $0.0 \leq z \leq 0.3$, with a median of $z = 0.11$.

The survey provides an almost complete sampling of the galaxy distribution down to an extinction-corrected limit of $b_J = 19.45$ mag over ~ 1800 square degrees of sky. This sky coverage is contained within two declination strips covering $75° \times 10°$ and $80° \times 15°$ in the NGP and SGP regions, respectively, plus 99 "random" 2-degree fields in the SGP. The contiguous coverage of the sky within the two strips, together with the almost complete ($\sim 95\%$) sampling of the galaxy distribution, has produced the highest fidelity three-dimensional (3-D) maps ever seen of the galaxy distribution and its large-scale structure. This is clearly revealed in the cone diagram shown in Figure 3.1, where the full range of structures (knots, filaments, voids) are resolved in fine and delicate detail.

One of the most conspicuous features of the galaxy distribution seen in Figure 3.1 is the *clustering* of galaxies, and the abundance of dense, "knotty" structures in which rich clusters are embedded. It is the quantification of this clustering and the rich wealth of data on clusters (and their environs) that the 2dFGRS has provided that is the focus of this paper.

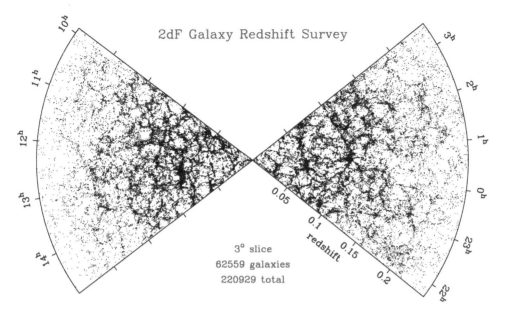

Fig. 3.1. A cone diagram based on a 3-degree slice taken through the NGP (*left*) and SGP (*right*) strips of the 2dFGRS.

Firstly, we briefly review the key findings to come from the precision measurements of galaxy clustering afforded by 2dFGRS, many of which have become the "flagship" results of the survey. We then describe the work that we have been doing on *known* rich clusters within the survey, using them as sites to study the galaxy luminosity function (LF) and galactic star formation in the densest environments, and hence via contrast with these same properties measured in lower-density regions within the survey, to draw conclusions as to their environmental dependence. Finally, we briefly mention work in progress on using the survey itself to generate a new 3-D-selected catalog of galaxy groups and clusters using automated and objective group-finding algorithms.

3.2   Galaxy Clustering: Key Results

In quantifying and characterizing the clustering seen in the 3-D galaxy distribution, 2dFGRS has realized significant advances, not just in the precision of the measurements, but also in extending them to much larger scales. The key to the latter is the much larger volumes that are (sparsely) probed by the random fields (cf. the strips on their own), allowing structure to be measured on scales up to $400\,h^{-1}$ Mpc.

The clustering over these scales has been measured statistically using both two-point correlation function and power spectrum analyses. Being based on redshift information, these functions are, of course, derived in terms of the redshift-space rather than the real-space positions of galaxies. Nonetheless, when used jointly and in combination with the matter power spectrum provided by cosmic microwave background measurements, they have yielded a number of new fundamental results on the origin of large-scale structure, the matter content and density of the Universe, and galaxy biasing; these can be summarized as follows (for further details, see the specific references quoted and also Colless 2004):

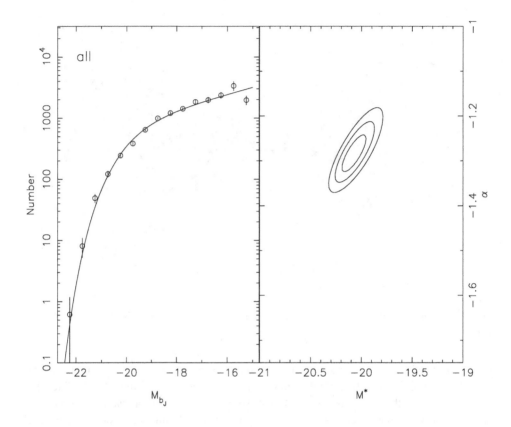

Fig. 3.2. *Left panel:* The composite LF measured for 60 known rich clusters in the 2dFGRS. *Right panel:* Error contours for the M^* and α parameter values from a Schechter function fit to this composite LF. (From De Propris et al. 2003.)

(1) Unambiguous detection of coherent collapse on large scales—manifested by a flattening of the 2-D correlation function in the line-of-sight direction at scales of $20–40\,h^{-1}$ Mpc—confirming structures grow via gravitational instability (Peacock et al. 2001).

(2) The detection in the power spectrum of "acoustic" oscillations due to baryon-photon coupling in the early Universe. Derivation from these oscillations of the baryon fraction: $\Omega_b/\Omega_m = 0.17 \pm 0.06$ (Percival et al. 2001).

(3) Measurement of Ω_m from both the power spectrum and the redshift-space distortions seen in the 2-D correlation function: $\Omega_m = 0.30 \pm 0.06$.

(4) Through comparison of the 2dFGRS power spectrum and the cosmic microwave background power spectrum, the first measurement of the galaxy bias parameter $b^* = 0.96 \pm 0.08$ (Lahav et al. 2002; see also Verde et al. 2002) and its variation with galaxy luminosity ($b/b^* = 0.85 + 0.15L/L^*$; Norberg et al. 2001) and type (Madgwick et al. 2004).

(5) Placement of a stronger limit on the neutrino fraction, $\Omega_\nu/\Omega_m < 0.13$, implying a limit on the mass of all neutrino species of $m_\nu < 1.8\,\mathrm{eV}$ (Elgarøy et al. 2002).

3.3 Cluster Luminosity Functions

The 2dFGRS includes an abundance of rich clusters, with 947 known clusters from the Abell (Abell 1958; Abell, Corwin, & Olowin 1989), APM (Dalton et al. 1997) and EDCC (Lumsden et al. 1992) catalogs identified and further characterized by De Propris et al. (2002) at a time when the survey was just half complete. Of these, 60 have since been used for a detailed study of the cluster LF (De Propris et al. 2003). Here the selection was restricted to clusters with $z < 0.11$ in order to sample well below the predicted M^*. A further criterion was that clusters must contain at least 40 confirmed members within the Abell radius ($1.5\,h^{-1}$ Mpc). Cluster membership had been determined previously in the De Propris et al. (2002) study using a "gapping" algorithm to isolate cluster galaxies in redshift space. The clusters were also chosen to ensure a range in velocity dispersion (and hence mass), richness, Bautz-Morgan (B-M) type, and structural morphology.

This sample was used to produce a series of "composite" LFs, with clusters and their member galaxies being combined to allow meaningful and statistically robust comparisons based on both global cluster properties and galaxy type. As a starting point, a composite LF was derived for the entire cluster sample containing 4186 members; this is shown in Figure 3.2. As can be seen, it provides a very high-quality "overall" cluster LF covering 7.5 magnitudes in luminosity ($-22 < M_{b_J} < -16$). It is well fit by a Schechter (1976) function with a characteristic magnitude of $M_{b_J}^* = -20.07 \pm 0.07$ and a faint-end slope of $\alpha = -1.28 \pm 0.03$ (right-hand panel of Fig. 3.2).

A series of composite LFs based on global cluster properties were also constructed, with the cluster sample divided into subsamples according to high ($\sigma_v \geq 800\,\mathrm{km\,s^{-1}}$)/low ($\sigma_v < 800\,\mathrm{km\,s^{-1}}$) velocity dispersion, rich/poor, "early" B-M (Types I, I–II, II)/"late" B-M (Types II–III, III), and with substructure/without substructure (see De Propris et al. 2003 for further details). Surprisingly, no statistically significant LF variation was seen between these different subsamples! The only conspicuous difference seen was between composite LFs formed for the inner (core) and outer regions of clusters, with the former having many more very bright galaxies than the latter.

Comparison of these composite cluster LFs, in particular the "overall" function shown in Figure 3.2, with their 2dFGRS field counterparts is readily available from the work of Madgwick et al. (2002). A Schechter function fit to their composite field LF over the same absolute magnitude range yields parameter values of $M_{b_J}^* = -19.79 \pm 0.07$ mag and $\alpha = -1.19 \pm 0.03$. Hence, taken at face value, it would appear that the cluster LF is 0.3 mag *brighter* in $M_{b_J}^*$ and has a steeper faint-end slope ($\alpha_{clus} - \alpha_{fld} = -0.1$).

However, clusters contain a very different morphological mix of galaxies compared to the low-density field (Oemler 1974; Dressler 1980), and before this difference in the LFs between the two can be interpreted as evidence for an environmental dependence, underlying LF-galaxy type effects (which have already been seen in the field; Folkes et al. 1999; Madgwick et al. 2002) need to be first investigated. This has been done on the basis of galaxy spectral type, paralleling the approach taken by Madgwick et al. (2002) for the field. Here, a galaxy's spectrum is typed on the basis of the relative strength of its first two principal components (from a principal component analysis analysis; Folkes et al. 1999), which represent the emission and absorption components within the spectrum. This is parameterized in terms of the quantity η, which is the linear combination of these two components: $\eta = a\,pc_1 - pc_2$. As might be expected, clusters and the field have quite a different mix of galaxies in terms of η, as can be seen in the upper panel of Figure 3.3. In line with the morphology-density

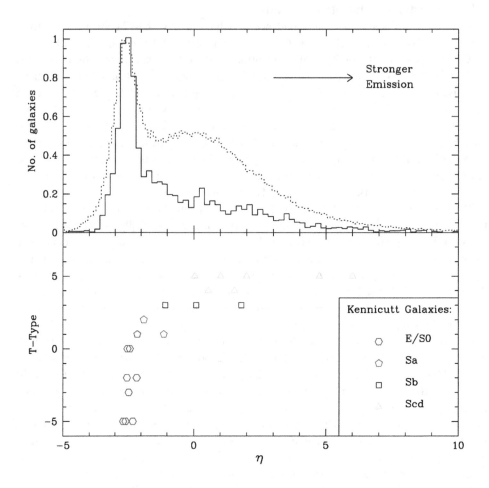

Fig. 3.3. *Upper panel:* Distribution of different spectral types (as measured by the η parameter; see text for details) within our clusters (*solid line*) and in the general field (*dotted line*). *Lower panel:* Relationship between galaxy morphology, represented by T type, and η. (From De Propris et al. 2003.)

relation (Dressler 1980) and the relationship between η and morphology (T type) that is seen in the bottom panel of Figure 3.3, clusters are dominated by galaxies with the lowest η values (absorption-line dominated, no emission), whereas the field contains a much larger proportion of galaxies with higher ($\eta > 0$) values, indicative of line emission in what are most likely late-type spirals. For the purposes of deriving LFs for different spectral types, Madgwick et al. (2002) divided the η scale into 4 intervals, with Type 1 galaxies being those in the range $-5 < \eta \leq -1.3$, Type 2 galaxies $-1.3 < \eta \leq 1.1$, Type 3 galaxies $1.1 < \eta \leq 3.4$, and Type 4 galaxies $\eta > 3.4$.

Using these same definitions, De Propris et al. (2003) derived composite LFs for the Type 1, Type 2, and Types 3+4 galaxies within their clusters; these are plotted in the left-

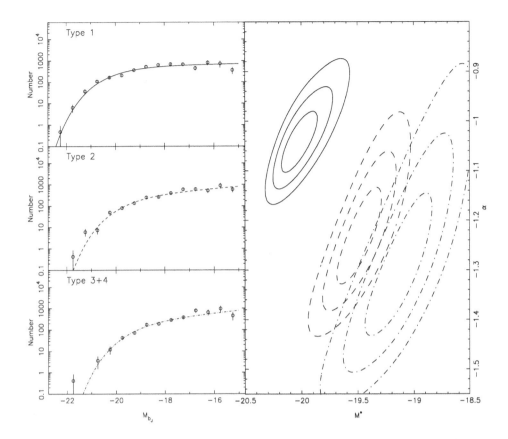

Fig. 3.4. *Left panel:* LFs for galaxies of different spectral types within clusters. *Right panel:* The 1, 2, and 3 σ error contours for the $M^*_{b_J}$ and α values from the Schechter function fits to these LFs. (From De Propris et al. 2003.)

hand panel of Figure 3.4. Here we see the same general trends that are seen in the field (Madgwick et al. 2002) in that earlier spectral types have LFs with brighter characteristic magnitudes and flatter faint-end slopes. This can be seen quantitatively in the right-hand panel of Figure 3.4. However, careful comparison of these LFs with their field counterparts (Fig. 3.5) reveals some subtle, but significant, differences, at a level of detail only a large survey like 2dFGRS can discern.

These differences are encapsulated in Figure 3.6, which shows the location in the $\alpha - M^*_{b_J}$ plane of the Type 1, Type 2, and Type 3+4 LFs both for clusters and the field, as well as the "overall" LFs based on all types. Clearly the biggest difference between cluster and field is for the early types, with the cluster Type 1 LF having a brighter $M^*_{b_J}$ (by ~ 0.5 mag) and a steeper faint-end slope ($\alpha_{cl} - \alpha_{fld} = -0.5$) compared to the field. In contrast, the cluster and field LFs for the latest types (Types 3 and 4) are, to within the errors, indistinguishable.

The type-dependent differences in the Schechter LF parameters that are seen between clusters and the field in Figure 3.6 also put the comparison of the "overall" LFs into perspective. Although there is a significant difference between the two, it is small in comparison to

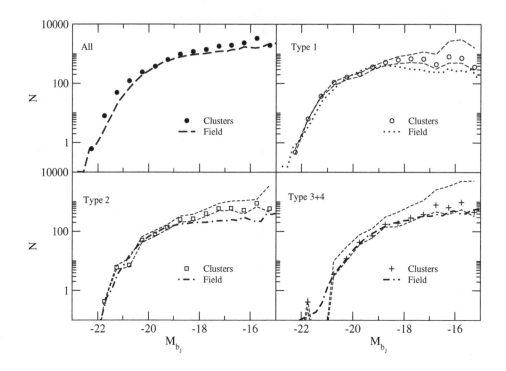

Fig. 3.5. Comparison of cluster and field LFs. The *points* represent the cluster data shown in the previous two figures. The *thick* dashed lines in each panel represent the 2dFGRS field data from Madgwick et al. (2002). The *thin* dashed lines provide representations of the cluster data with no completeness correction (*lower*) and maximum completeness correction (*upper*); see De Propris et al. (2003) for further details. (From De Propris et al. 2003.)

the differences seen between the early types (Types 1 and 2). This, together with the quite different mix of spectral types in clusters and in the field, would indicate that the similarity in the overall LFs is more by accident than design, with these mix and spectral type effects conspiring to produce LFs in the two different environments with characteristic magnitudes and faint-end slopes that are quite similar.

In order to understand these type-dependent LF differences between clusters and the field in the context of cluster galaxy and environment-dependent evolution, an initial attempt has been made to model them using a simple "closed box" approach (De Propris et al. 2003). This involves making the following basic assumptions: (1) cluster galaxies have evolved from field galaxies contained within the volume that collapses to become the cluster, (2) this evolution is characterized by a suppression of star formation activity and an accompanying change in spectral type, and (3) the number of galaxies is conserved (i.e., mergers are neglected). Under these assumptions, the type-specific field LFs are taken to be the initial LFs within the cluster volume, with the relative normalizations set by the mix of different types that are observed.

This very simple model has some success in that it reproduces the similar shape but

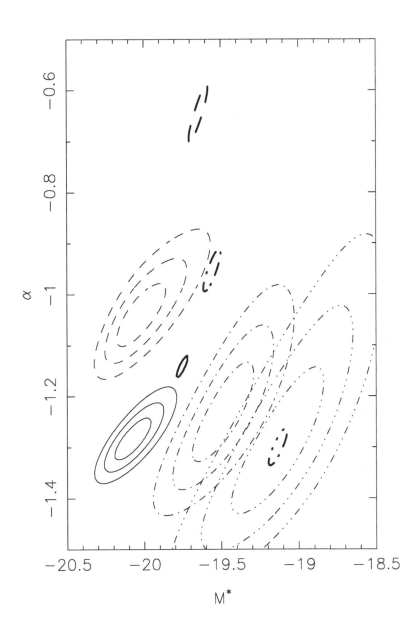

Fig. 3.6. Error ellipses for the Schechter function fits to the cluster (*thin* lines) and field (*thick* lines) LFs. *Dashed* lines: Type 1 galaxies; *dot-dashed* lines: Type 2 galaxies; *dotted-dashed* lines: Types 3+4. Only the 3 σ error ellipse is shown for the field data. (From De Propris et al. 2003.)

different normalization that is observed for the Type 3+4 LFs in clusters and in the field. Moreover, it does cause the initial field LFs of the Type 1 and 2 galaxies to evolve toward the forms they are observed to have in clusters. However, at a more detailed level, it fails in producing too many bright Type 2 cluster galaxies, too few very bright Type 1 cluster galaxies, and too few faint Type 1 cluster galaxies. Obvious refinements to this model (on which we are currently working) are to allow for luminosity-dependent fading to steepen the faint-end slopes of the earlier types, and to include mergers to explain the excess of very bright early types at the expense of their fainter counterparts.

3.4   Star Formation Versus Environment

The galaxy spectra obtained by the 2dFGRS are of sufficient quality not only to determine redshifts but also to measure spectral-line indices and derive astrophysical information about the galaxies themselves. Moreover, the spectra extend to sufficiently red wavelengths ($\lambda \approx 8500$ Å) to include the redshifted Hα emission (or absorption) line, thus providing a reliable means of measuring the overall star formation rate (SFR) within galaxies (Kennicutt 1992). In this section we describe how this has been used to track star formation activity as a function of environment, firstly from a cluster-centric point of view, and secondly in galaxy groups.

3.4.1   *Within Rich Clusters*

To date, the global SFR among galaxy populations has generally concentrated on the two extremes of galaxy environment: the low-density field and the dense cores of rich clusters. For the latter, attention has been very much drawn by the discovery of Butcher & Oemler (1978) that such systems harbored many more star-forming galaxies in the past. But galaxies in cluster cores comprise ony a small fraction of the stellar content of the Universe and may be subject to environmental effects that are peculiar to these very high-density environments (e.g., ram pressure stripping, galaxy "harassment," and tidal interactions; Dressler 2004). Much more pertinent to the evolution of the general galaxy population is the environment *between* cluster cores and the field, spanning 3 orders of magnitude in galaxy density and which remains largely unchartered in terms of tracking star formation.

As a first step toward redressing this situation, we have analyzed 17 $z \approx 0.05 - 0.1$ clusters from the De Propris et al. (2002) study, using them to trace the global galactic SFR continuously from their centers out to arbitrarily large radii. These clusters were chosen so that roughly half (10) had "high" velocity dispersions ($\sigma_v > 800\,\mathrm{km\,s^{-1}}$) and the other half had "low" velocity dispersions ($400 \leq \sigma_v \leq 800\,\mathrm{km\,s^{-1}}$). All the 2dFGRS galaxies within a projected distance of 20 Mpc from the centers of these clusters and within the range $0.06 < z < 0.10$ were selected for analysis. The restriction in redshift was to limit the effects of aperture bias and poor sky subtraction. A luminosity cutoff of $M_{b_J} = -19$ mag was then applied to the sample, this being the limit to which 2dFGRS is complete over the adopted redshift range. Finally, all galaxies for which there were continuum/line fitting problems at Hα or whose EW([N II] $\lambda6583$)/EW(Hα) ratios showed evidence of a nonstellar component (values >0.55) were excluded from the analysis. This resulted in a final sample of 11,006 galaxies.

The equivalent width of the Hα line (be it in absorption or emission) was measured using an automated Gaussian profile fitting algorithm, which simultaneously fitted and deblended the Hα line from the neighboring [N II] $\lambda6548$ Å and [N II] $\lambda6583$ Å lines. Since only equivalent width rather than line flux measurements are possible with the 2dFGRS spectra, we can only infer the SFR, μ, per unit luminosity: $\mu/L_{\mathrm{cont}} = 7.9 \times 10^{-42}$ EW(Hα), using Kennicutt's (1992) SFR-EW(Hα) relation. We then normalized this to a characteristic lu-

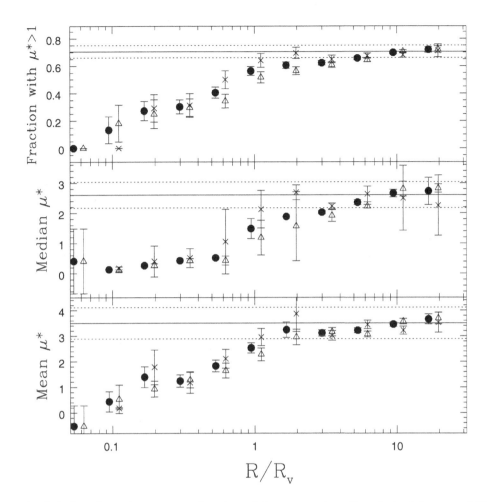

Fig. 3.7. Star formation rate as a function of cluster-centric radius, derived from our 2dF-GRS analysis. *Filled* points represent the full galaxy sample, while the *triangles* and *crosses* represent the "high"- and "low"-velocity dispersion clusters (see text for details). The *solid horizontal* lines represent the value of each statistic for our field galaxy sample. (From Lewis et al. 2002.)

minosity L^*: $\mu^* = \mu/(L_{\mathrm{cont}}/L^*) = 0.087\,\mathrm{EW(H\alpha)}$, where L^* is taken to be the knee in the luminosity function in the r' band (near rest-frame $H\alpha$), as determined by Blanton et al. (2001).

In Figure 3.7, the mean and median value of μ^* is plotted as a function of the projected cluster-centric radius. Here the latter has been normalized by the cluster virial radius, R_v, which has been calculated for each individual cluster and ranges from 1.4 to 2.4 Mpc. Also plotted in the top panel of Figure 3.7 is the fraction of galaxies with $\mu^* > 1\,M_\odot\,\mathrm{yr}^{-1}$, which represents the tail of the distribution, comprised of galaxies that are currently forming stars at a high rate relative to their luminosities. The solid horizontal line represents the values derived for the "field" galaxies within our sample, that is, galaxies that lie within the 20 Mpc selection radius, but which, from their redshifts, are identified as non-members. The bracketing dashed horizontal lines represent the $1\,\sigma$ standard deviation from field to field, giving some estimate of the cosmological variance in the field value.

Irrespective of which statistic is used, it is very clear that within the clusters μ^* falls significantly below the value of the field, the difference being at a maximum at the cluster center and then monotonically decreasing with increasing cluster-centric radius. *Importantly, convergence does not occur until $R > 3R_v$!* Hence, compared to the field, cluster galaxies differ in their mean star formation properties as far out as ~ 6 Mpc from their centers. Also of note is that these radial trends in μ^* appear to be insensitive to cluster mass, with there being no discernible difference between the "high"- and "low"-velocity dispersion clusters.

Such a radial analysis, however, is likely to be sub-optimal, since many of the clusters in our sample are clearly not spherically symmetric and show substructure. We have therefore analyzed μ^* as a function of the local projected galaxy density, Σ, based on the distance to the tenth nearest neighbor (Dressler 1980). The relationship between μ^* and Σ is shown in Figure 3.8, using the same three statistics that we used in the radial analysis. The vertical line shows the mean value of Σ within R_v, and once again the horizontal lines show μ^* and its $1\,\sigma$ variance for the field.

We see that cluster galaxy star formation is suppressed relative to the field, the difference being greatest at the highest local densities, and decreasing monotonically with decreasing density until the two converge at $\Sigma \approx 1.5$ galaxy Mpc^{-2}, a factor of ~ 2.5 times lower than the mean projected density of the cluster virialized region. Yet again the trend is no different in the "high"- and "low"-velocity dispersion clusters, indicating that SFRs depend only on the local density, regardless of the large-scale structure in which they are embedded. Further evidence that local density is the key variant is also provided by a plot based on just those galaxies with $R > 2R_v$, where the same trend observed for the full sample is seen. This would suggest that a more general view of star formation suppression be taken: that it will be low relative to the global average in *any* region where the local density exceeds a value of ~ 1 gal Mpc^{-2} ($M_b \leq -19$ mag).

It is well known that galaxy morphology is very strongly correlated with local galaxy density (Dressler 1980), and hence consideration needs to be given to what underlying contribution this has to the $\mu^* - \Sigma$ relationship seen in Figure 3.8. Using a simple model based on Dressler's morphology-density relation, we have calculated the expected variation in μ^*; this is represented by the *solid curves* in Figure 3.8. This appears to be shallower than the observed relation, suggesting that the morphology-density relation is distinct from the SFR-density relation, with additional processes operating at $\Sigma > 1$ galaxy Mpc^{-2} that drive μ^* significantly below that expected on the basis of the changing morphological mix. Indeed, it may well be the case that the suppression of star formation is the primary transformation that occurs with environment, with the change in morphological mix being a secondary effect (Poggianti et al. 1999; Shioya et al. 2002)

A complete description of this study and a full discussion of its implications can be found in Lewis et al. (2002).

3.4.2   *Within Galaxy Groups*

Extension of this analysis to the galaxy group environment has just recently become possible through the 2dFGRS being subjected to a 3-D "friends-of-friends" analysis to identify groups (and clusters) in both position and redshift space (Eke et al. 2004). This

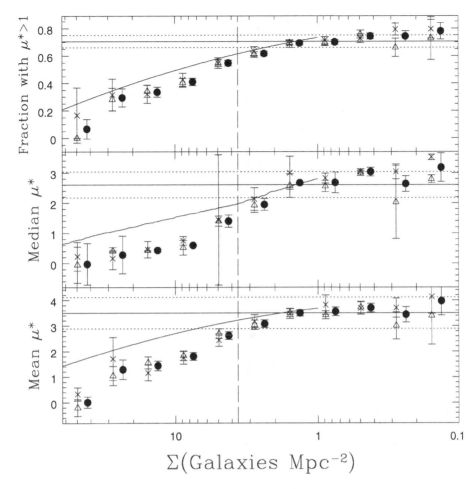

$\Sigma(\mathrm{Galaxies\ Mpc^{-2}})$

Fig. 3.8. Star formation rate as a function of local projected galaxy density, Σ. The *dashed* vertical line represents the mean value of Σ within R_v. The *curved* lines indicate the variation expected as a result of the varying morphological mix. (From Lewis et al. 2002.)

has produced a catalog containing $\sim 30,000$ groups with at least 2 members; their distribution within the two 2dFGRS strips is shown in Figure 3.9. The star formation properties of this sample are currently being analyzed, and we show here (courtesy of Dr. M. Balogh) a couple of very preliminary results that give the first indications of environmental trends in these much poorer, lower-velocity dispersion analogs to rich clusters.

Once again the Hα line has been used as a measure of SFR, with it being identified and its equivalent width measured in the spectra of all group members in identical fashion to that described above for cluster galaxies. An overall Hα equivalent width was then derived for each group taking the mean value of all its members. In Figure 3.10 this mean value is plotted as a function of group velocity dispersion, being represented by all the individual dots. To assess whether there is any overall trend, the data have been divided into equally populated

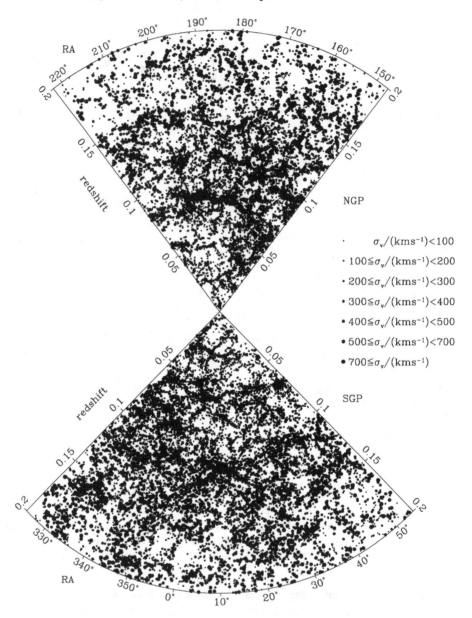

Fig. 3.9. A redshift slice showing the distribution of groups and clusters identified within the 2dFGRS using a 3-D "friends-of-friends" analysis (Eke et al. 2004). The estimated velocity dispersion is shown by the size of the dot.

velocity dispersion bins and averaged; the continuous line has been drawn through these average values. As a comparison, the mean Hα equivalent width for "field" galaxies—those galaxies identified as not belonging to any group within the friends-of-friends analysis—is shown as the horizontal dashed line.

It can be seen in Figure 3.10 that there is a noisy but significant trend of Hα emission (and

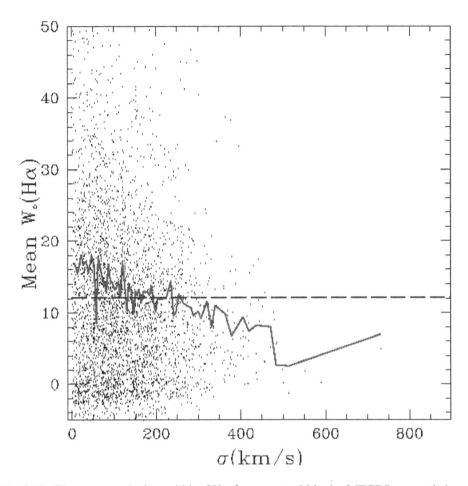

Fig. 3.10. The mean equivalent width of Hα for groups within the 2dFGRS versus their velocity dispersion. The *dots* represent individual groups whose values have been determined from their member galaxies; the *solid line* connects the mean values evaluated within equally populated velocity dispersion bins; the *dashed* horizontal line indicates the mean value for "field" galaxies.

hence SFR) becoming increasingly stronger with lower group velocity dispersion. Intriguingly, however, this trend intersects the horizontal "field" line at $\sigma \approx 250\,\mathrm{km\,s^{-1}}$, indicating that there are not only groups whose mean EW(Hα) is below that of the field (as might be expected from the SFR–Σ relation observed in the rich cluster study), but also groups whose mean EW(Hα) is *higher* than the field, hinting that star formation activity is for some reason enhanced (cf. the field) in these lower-velocity dispersion ($\sigma < 150\,\mathrm{km\,s^{-1}}$) groups.

The underlying reason for this may well be contained in Figure 3.11, where we show the mean EW(Hα) for just the *binary* groups within the $\sigma < 150\,\mathrm{km\,s^{-1}}$ regime, plotted as a function of their separation (in Mpc). Yet again, the equal-numbered bin averaging and

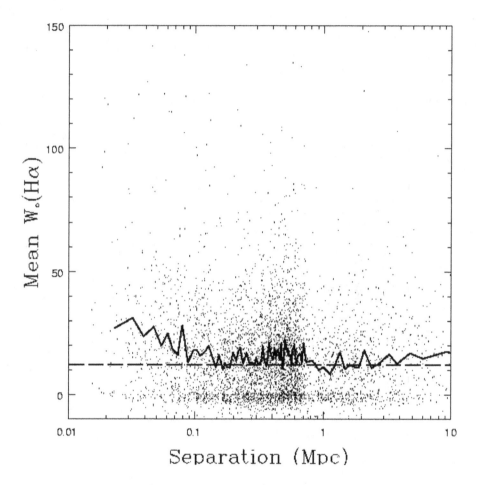

Fig. 3.11. The mean equivalent width of Hα for binary groups with velocity dispersions $\sigma < 150\,\mathrm{km\,s^{-1}}$, plotted as a function of the separation between the two galaxies. The *dots* represent individual groups; the *solid line* connects the mean values evaluated within equally populated galaxy separation bins; the *dashed* horizontal line indicates the mean Hα equivalent width for field galaxies.

the benchmark "field" value are represented in the same way as in Figure 3.10. We see that for separations in the range 0.1–$10\,h^{-1}$ Mpc, binary groups have, to within the fluctuations, the same mean EW(Hα) as the field. However, at separations below $0.1\,h^{-1}$ Mpc, the mean EW(Hα) for binary groups is seen to be significantly higher than the field. Although this is yet to be fully investigated, it is suggestive that the enhanced SFRs seen in the low-velocity dispersion groups (relative to the field) can be attributed to "close" binary systems where tidal interactions may well be responsible for this activity.

Acknowledgments. These results are presented on behalf of the 2dFGRS team: Ivan

Baldry, Carlton Baugh, Joss Bland-Hawthorn, Sarah Bridle, Terry Bridges, Russell Cannon, Shaun Cole, Matthew Colless, Chris Collins, Warrick Couch, Nicholas Cross, Gavin Dalton, Roberto De Propris, Simon Driver, George Efstathiou, Richard Ellis, Carlos Frenk, Karl Glazebrook, Edward Hawkins, Carole Jackson, Bryn Jones, Ofer Lahav, Ian Lewis, Stuart Lumsden, Steve Maddox, Darren Madgwick, Peder Norberg, John Peacock, Will Percival, Bruce Peterson, Wil Sutherland, and Keith Taylor. The 2dFGRS was made possible through the dedicated efforts of the staff of the Anglo-Australian Observatory, both in creating the 2dF instrument and in supporting it on the telescope. W.J.C. and R.D.P. acknowledge funding from the Australian Research Council throughout the course of this work.

References

Abell, G. O. 1958, ApJS, 3, 211

Abell, G. O., Corwin, H. C., & Olowin, R. 1989, ApJS, 70, 1

Blanton, M. R., et al. 2001, AJ, 121, 2358

Butcher, H., & Oemler, A., Jr. 1978, ApJ, 219, 18

Colless, M. 2004, in Carnegie Observatories Astrophysics Series, Vol. 2: Measuring and Modeling the Universe, ed. W. L. Freedman (Cambridge: Cambridge Univ. Press), in press

Colless, M., et al. 2001, MNRAS, 328, 1039

Dalton, G. B., Maddox, S. J., Sutherland, W. J., & Efstathiou, G. 1997, MNRAS, 289, 263

De Propris, R., et al. 2002, MNRAS, 329, 87

——. 2003, MNRAS, 342, 725

Dressler, A. 1980, ApJ, 236, 351

——. 2004, in Carnegie Observatories Astrophysics Series, Vol. 3: Clusters of Galaxies: Probes of Cosmological Structure and Galaxy Evolution, ed. J. S. Mulchaey, A. Dressler, & A. Oemler (Cambridge: Cambridge Univ. Press), in press

Eke, V. R., et al. 2004, MNRAS, submitted

Elgarøy, O., et al. 2002, Phys. Rev. Lett., 89, 061301

Folkes, S. R., et al. 1999, MNRAS, 308, 459

Kennicutt, R. C., Jr. 1992, ApJS, 79, 255

Lahav, O., et al. 2002, MNRAS, 333, 961

Lewis, I. J., et al. 2002, MNRAS, 334, 673

Lumsden, S. L., Nichol, R. C., Collins, C. A., & Guzzo, L. 1992, MNRAS, 258, 1

Madgwick, D. S., et al. 2002, MNRAS, 333, 133

——. 2004, MNRAS, submitted (astro-ph/0303668)

Nichol, R. C. 2004, in Carnegie Observatories Astrophysics Series, Vol. 3: Clusters of Galaxies: Probes of Cosmological Structure and Galaxy Evolution, ed. J. S. Mulchaey, A. Dressler, & A. Oemler (Cambridge: Cambridge Univ. Press), in press

Norberg, P., et al. 2001, MNRAS, 328, 64

Oemler, A., Jr. 1974, ApJ, 194, 1

Peacock, J. A., et al. 2001, Nature, 410, 169

Percival, W. J., et al. 2001, MNRAS, 327, 1297

Poggianti, B. M., Smail, I., Dressler, A., Couch, W. J., Barger, A. J., Butcher, H., Ellis, R. S., & Oemler, A. 1999, ApJ, 518, 576

Schechter, P. L. 1976, ApJ, 203, 279

Shioya, Y., Bekki, K., Couch, W. J., & De Propris, R. 2002, ApJ, 565, 223

Verde, L., et al. 2002, MNRAS, 335, 432

4

X-ray surveys of low-redshift clusters

ALASTAIR C. EDGE
Institute for Computational Cosmology, University of Durham

Abstract
The selection of clusters of galaxies through their X-ray emission has proved to be an extremely powerful technique over the past four decades. The growth of X-ray astronomy has provided the community with a steadily more detailed view of the intracluster medium in clusters. In this review I will assess how far X-ray surveys of clusters have progressed and how far they still have to travel.

4.1 Introduction

The principal baryonic component of clusters of galaxies is diffuse gas held in hydrostatic equilibrium in the gravitational potential of the cluster. This gas is hot ($10^7 - 10^8$ K), relatively dense ($10^{-4} - 10^{-2}$ atom cm^{-3}), and enriched with heavy elements (e.g., Fe of 0.3 Solar abundance). This combination results in significant X-ray emission through thermal bremsstrahlung radiation. Detailed X-ray observations of clusters can provide us with accurate total mass measurements, clues to the merger history of clusters, and a chemical record of the supernova ejecta that polluted the intracluster medium during the formation of the stars in the member galaxies.

The X-ray emission from clusters can also be exploited to select clusters irrespective of their member galaxies. While the optical selection of clusters is well established and understood, there are potential problems with projection and the imperfect scaling of the galaxy population to total cluster mass that make independent selection methods attractive.

There are four key considerations for any X-ray survey:

- **Spatial resolution** To capitalize on the extended nature of the X-ray emission in clusters, it is important to have sufficient spatial resolution to differentiate clusters from most other pointlike X-ray sources (i.e., stars and AGNs). On the other hand, the most nearby, diffuse clusters can be missed in the same way low-surface brightness galaxies may be missed in optical surveys.
- **Spectral resolution** Each class of X-ray source has a distinctive spectral signature (e.g., black body for white dwarfs), so information on the X-ray spectrum of each source can aid identification. The thermal nature of cluster spectra (temperatures mostly greater than 2 keV) give clusters relatively flat soft X-ray spectra (making them distinctive in the *ROSAT* survey), but the overall spectral shape is similar to most unabsorbed AGNs. Therefore, definitive cluster identification from spectral data alone requires many photons (>1,000), which is only feasible for the brightest detections (see Nevalainen et al. 2001 for an example using *XMM-Newton*).

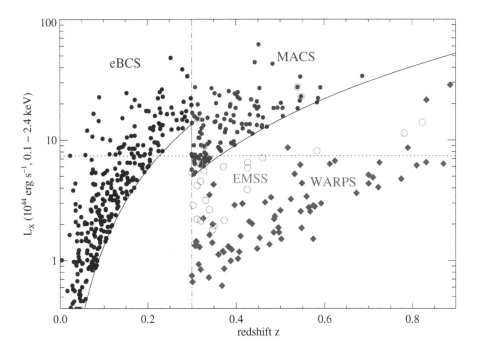

Fig. 4.1. The X-ray luminosity (0.1–2.4 keV) plotted against redshift for four X-ray samples: EMSS (Gioia et al. 1990), a serendipitous sample from *Einstein*; eBCS (Ebeling et al. 2000), a shallow, wide RASS survey; MACS (Ebeling et al. 2001), a deeper, wide RASS survey, and WARPS (Perlman et al. 2002) a much deeper serendipitous *ROSAT* survey.

- **Flux limit of the survey** This is a particularly important factor for cluster surveys as the flux-limited nature of X-ray samples translates to a selection over a wide range in redshifts, since the most luminous objects are being selected from a very much larger volume than the least luminous ones. This has its advantages but requires careful analysis. This is illustrated in Figure 4.1, where the X-ray luminosity is plotted against redshift for a variety of samples described later in the text.
- **Area of sky surveyed** In the ideal survey at any wavelength the aim is all-sky coverage. This has been achieved several times in X-ray astronomy with the earliest X-ray satellites scanning with collimated proportional counters ($\sim 1°$ resolution) and *ROSAT* with a soft X-ray imaging telescope ($\sim 1'$ resolution). This wide coverage comes at the expense of depth [the *ROSAT* All-Sky Survey reaches $\sim 10^{-12}$ erg s^{-1}cm^{-2} (0.5–2 keV)]. The alternative survey strategy is to select serendipitous sources in pointed imaging observations, as pioneered by the Extended *Einstein* Medium Sensitivity Survey (EMSS; Gioia et al. 1990). This allows much deeper surveys but at the expense of the area (and hence total volume) covered.

There are a number of cluster properties that can be used to constrain the nature and evolution of clusters.

- **Temperature** - For gas in hydrostatic equilibrium (which appears to hold for the majority of the volume of a cluster) the gas temperature and density can be used to directly determine the cluster mass.

- **Elemental abundances** The X-ray spectrum of clusters contains lines from a number of heavy elements. Most prominent of these is the iron 6.7 keV line. The inferred abundance ratios from X-ray spectra of O, S, Si and Fe can be used to determine the dominant supernova type (Loewenstein & Mushotzky 1996).

- **Surface brightness profiles** The distribution of gas in a cluster has a very significant effect on the total X-ray luminosity of the cluster, given that the intensity of emission is proportional to the density squared. Clusters with compact, dense cores (i.e., cooling flows; see Fabian 1994) are much more luminous that more extended clusters of the same measured X-ray temperature (Fabian et al. 1994) and can have an effect on the detection probability in X-ray surveys (Pesce et al. 1990). Also, the recent discovery of strong density discontinuities, termed "cold fronts" (Markevitch et al. 2000; Mazzotta et al. 2001) has highlighted the impact of past mergers on the intracluster medium. These factors make obtaining high-quality, high-resolution X-ray imaging a vital element of cluster studies.

Each of these requires either dedicated pointed observations or a survey drawn from the brightest serendipitous detections in pointed observations. The former is a relatively slow process requiring time allocation committees to put substantial resources into programs to observe "complete" samples. The latter is very slow given the area covered by sufficiently deep X-ray observations.

For the purposes of this review I will define "low redshift" as $z < 0.5$ and treat any paper presenting any new X-ray detection of cluster as a "survey."

In my talk I used the yardstick of exponential growth to judge progress in known numbers of X-ray emitting clusters which I modestly named "Edge's Law." This holds that for every decade of X-ray astronomy the number of clusters detected increases by an order of magnitude. I would like to stress that this was a narrative device and not a serious bid for future surveys in itself. That said, the rapid progress in cluster research in the past decades does require us to stand back and assess it as part of a larger picture.

4.2   An Historical Perspective

I would like to continue this review in a similarly light-hearted vein while giving the reader as comprehensive review as possible of X-ray cluster surveys.

4.2.1   *In the Beginning...*

X-ray astronomy began on the 18th of June 1962 with the detection of the X-ray background and Sco-X1 by a sounding rocket experiment (Giacconi et al. 1962). For shorthand in this review, this will be denoted as 0 Anno Giacconi (AG), and subsequent events will be quoted in these units.

During the first three years of sounding rocket experiments from 0 AG several cluster detections were in dispute [e.g., Coma claimed by Boldt et al. (1966) and discounted by Friedman & Byram (1967)], so the first unambiguous cluster detection came in 4 AG when Byram, Chubb, & Freidman (1966) detected M87/Virgo.

The numerous sounding rocket campaigns that occurred between 1962 and 1975 resulted in several more cluster detections (e.g., Perseus; Fritz et al. 1971) and the discovery that the X-ray emission in Coma was extended (Meekins et al. 1971). Unfortunately the collecting area and exposure time of these experiments ruled out the detection of all but the brightest few clusters.

On 12th December 1970 the first X-ray satellite, *UHURU*, was launched. The two-year lifetime of the mission allowed the whole sky to be scanned many times, producing the first true X-ray survey. The fourth and final UHURU catalog (4U; Forman et al. 1978) contains 52 clusters, several of which that were not known to be clusters at the time.

4.2.2 The End of the First Age of X-ray Astronomy?

UHURU was the first of a number of increasingly more complicated experiments that allowed further surveys and dedicated pointed observations. Most notable of these was *Ariel-V* (Cooke et al. 1978), which made the first iron line detection in a cluster (Mitchell et al. 1976).

The final mission in this series, *HEAO-1*, made the deepest X-ray survey (Piccinotti et al. 1982), which was the mainstay of X-ray astronomy for the two decades that followed. So at the end of this exciting period of X-ray astronomy, how well does "Edge's Law" stand up? At 17 AG a total of 95 clusters are known, which is well above the 50 required by this point.

4.3 X-ray Imaging Begins with *Einstein*

The launch of the first imaging X-ray satellite, *Einstein*, had a profound impact on our understanding of clusters.

4.3.1 Detailed Imaging and Spectra

The Imaging Proportional Counter (IPC) and High Resolution Imager (HRI) provided images of unprecedented quality (up to 5" FWHM). These two instruments provided a great deal of detailed information on individual clusters from targeted observations (Fabian et al. 1981; Jones & Forman 1984, 1999; Stewart et al. 1984; White, Jones, & Forman 1997). Most of the observations were of Abell clusters or other optically selected clusters, but radio galaxies in clusters were also targets. Given the nature of the targeted observations it is not possible to derive any stringent limits on the statistical properties of clusters, but a luminosity function was derived for Abell clusters (Burg et al. 1994).

Einstein also carried the first semi-conductor detector (the forerunner to today's CCDs), the Solid State Spectrometer (SSS) and a deployable Bragg Crystal, the Focal Plane Crystal Spectrometer (FPCS). Both of these instruments provided important results for clusters (Canizares et al. 1979; White et al. 1991), but were not used systematically.

4.3.2 The EMSS

The *Einstein* survey that has had the most impact on cluster research is undoubtedly the Extended *Einstein* Medium Sensitivity Survey (EMSS). By combining most of the serendipitous detections from IPC observations, it was possible to survey 980 □° and detect 99 clusters (Stocke et al. 1991). This sample has since been the basis for a great deal of work at all wavelengths (Donahue, Stocke, & Gioia 1992; Le Févre et al. 1994; Carlberg et al. 1997; Luppino et al. 1999)

4.4 The X-ray Dark Ages

The 1980's were a period of relative calm in X-ray astronomy. The only satellites launched between 1980 and 1990 were the ESA mission *EXOSAT* and two Japanese missions, *Tenma* and *Ginga*. All three of these satellites were designed for targeted follow-up of known objects, and only *EXOSAT* had any imaging capabilities. *EXOSAT* and *Ginga* contributed a significant number of accurate cluster temperature and iron abundance measurements (the data from *EXOSAT* making my thesis) but very few "new" detections.

This lull in proceedings did allow the previous scanning and *Einstein* surveys to be collated, and a complete sample of the brightest 55 clusters was compiled (Lahav et al. 1989). This sample has been used to determine the first cluster temperature function (Edge et al. 1990), the first correlation function from an X-ray sample (Lahav et al. 1989), and the fraction of cooling flows (Edge, Stewart, & Fabian 1992; Peres et al. 1998).

This "free-wheeling" in X-ray surveys leaves the number of clusters in 28 AG at 300, well short of the 630 required for my exponential growth. This gap did not last for long....

The German/UK/US satellite *ROSAT* was launched on the 1st of June 1990 (27.95 AG). The wide-field, soft X-ray imaging telescope of *ROSAT* was used to conduct a 6-month scanning survey of the whole sky, from August 1990 to February 1991, which detected in excess of 100,000 sources to a flux limit of $(0.3-1) \times 10^{-12}$ erg s^{-1}cm^{-2} (0.1–2.4 keV) (depth depending on position). From March 1991 to December 1998, *ROSAT* performed a series of pointed observations with the PSPC and HRI detectors. These observations targeted many known and recently detected clusters, as well as detecting a great many clusters serendipitously.

4.4.1   *ROSAT All-Sky Survey*

The *ROSAT* All-Sky Survey (RASS) is a resource that has still yet to fully exploited 12 years after it was completed. A number of coordinated cluster surveys were embarked upon as soon as the RASS ended. The understandably tight control over the RASS data release and the time-consuming nature of the optical follow-up of the clusters has meant a significant lag in the publication of these samples. Table 4.1 lists a representative set of RASS cluster surveys, both published and unpublished. There are other RASS studies containing clusters (e.g., the RBS, a complete sample of all RASS sources to a count rate limit of 0.2 PSPC count s^{-1}; Schwope et al. 2000), and studies of groups (e.g., RASSCALS; Mahdavi et al. 2000) and Hickson compact groups (Ebeling, Voges, & Böhringer 1994). Table 4.1 will be added to in the next few years by NORAS-2, REFLEX-2, and eMACS, which will extend each of the existing surveys to lower fluxes, but these will be reaching close to the intrinsic sensitivity limit of the majority of the RASS.

The wide variety of selection criteria, detection methods and areas covered are clear from Table 4.1. However, each of the larger samples (BCS, 1BS, Ledlow, REFLEX, and NEP) agree in their derived X-ray luminosity functions (Ebeling et al. 1997; de Grandi et al. 1999; Ledlow et al. 1999; Gioia et al. 2001; Böhringer et al. 2002), so these differences do not greatly affect the samples.

One important difference in the principal RASS samples is that one set (XBACS, BCS, and eBCS) is based on a selection using a Voronoi-Percolation-and-Tesselation (VTP) technique (Ebeling & Wiedenmann 1993) and the other (1BS, SGP, NORAS, and REFLEX) is based on a growth curve analysis (GCA) technique (Böhringer et al. 2001). The flux results from both methods agree within the errors, but only VTP acts as a detection algorithm, as the GCA method requires a set of input positions of potential clusters. This difference is relevant only for the most nearby, extended sources, which are not detected by detection algorithms tuned to search for point sources. VTP will reliably detect these, but, through lack

Table 4.1. *ROSAT Survey Samples*

Survey	Identification Paper	Flux Limit (erg s^{-1}cm^{-2})	Area ($\square°$)	Number Published?
XBACS	Abell clusters	5.0×10^{-12}	All-sky	276
	Ebeling et al. (1996)	(0.1–2.4 keV)		Y
BCS	Abell, Zwicky, extended	4.5×10^{-12}	13,578	199
	Ebeling et al. (1998)	(0.1–2.4 keV)		Y
RASS1BS	Abell, extended	3-4×10^{-12}	8,235	130
	de Grandi et al. (1999)	(0.5–2.0 keV)		Y
Ledlow	Abell $z < 0.09$	none	14,155	294
	Ledlow et al. (1999)			N
eBCS	Abell, Zwicky, extended	3.0×10^{-12}	13,578	299
	Ebeling et al. (2000)	(0.1–2.4 keV)		Y
HiFLUGS	All	20×10^{-12}	27,156	63
	Reiprich & Böhringer (2002)	(0.1–2.4 keV)		Y
NORAS	extended	3.0×10^{-12}	13,578	378
	Böhringer et al. (2000)	(0.1–2.4 keV)		Y
NEP	multiple	0.03×10^{-12}	80.7	64
	Gioia et al. (2001)	(0.5–2.0 keV)		Y
CIZA	CCD imaging, $|b| < 20°$	5×10^{-12}	14,058	73
	Ebeling, Mullis, & Tully (2002)	(0.1–2.4 keV)		Y
SGP	optical plates scans	3.0×10^{-12}	3,322	112
	Cruddace et al. (2002)	(0.1–2.4 keV)		Y
MACS	multiple, $z > 0.3$	1.0×10^{-12}	22,735	120
	Ebeling et al. (2001)	(0.1–2.4 keV)		N
REFLEX	multiple	3.0×10^{-12}	13,905	452
	Böhringer et al. (2001)	(0.1–2.4 keV)		N

of access to the full RASS data set, it was not run over the full sky during the compilation of the BCS. With all RASS data now in the public domain, this is now possible in principle.

The optical follow-up of clusters at redshifts above 0.3 requires additional optical imaging, as archival photographic plate material is too shallow to reliably detect cluster members. At the brighter flux limits (5×10^{-12} erg s^{-1}cm^{-2}) there are relatively few of these distant clusters [e.g., two in the BCS and RXJ1347–11 ($z = 0.45$) in RASS1BS], but this number increases with decreasing flux limit (e.g., there are seven in the eBCS). With these higher redshift, X-ray luminous clusters in mind, Harald Ebeling and I have searched the RASS-BSC sample (Voges et al. 1999) for $z > 0.3$ clusters using the UH 2.2 m telescope to a flux limit of 10^{-12} erg s^{-1}cm^{-2}, creating the MAssive Cluster Survey (MACS; Ebeling, Edge, & Henry 2001). To date, the sample contains 120 clusters with very few candidates left for imaging. The MACS sample has been extensively followed up at all wavelengths, including complete *VRI* imaging with the UH 2.2 m, multi-object spectroscopy with Keck, Gemini and CFHT, deep, wide-area imaging with Subaru/SUPRIMECAM, VLA imaging (Edge et al. 2003), Sunyaev-Zel'dovich observations (LaRoque et al. 2003), *Chandra* ob-

Table 4.2. *ROSAT Serendipitous Samples*

Survey	Selection Paper	Flux Limit (10^{-14} erg s^{-1}cm^{-2})	Area ($\square°$)	Number Published?
RIXOS	CCD imaging	3.0	15.8	25
	Mason et al. (2000)	(0.5–2.0 keV)		Y
WARPS-I	CCD imaging	6.5	14.1	25
	Perlman et al. (2002)	(0.5–2.0 keV)		Y
160sq.deg.	extent	3.0	158	203
	Vikhlinin et al. (1998)	(0.5–2.0 keV)		Y
SHARC-S	extent	3.9	17.7	16
	Collins et al. (1997)	(0.5–2.0 keV)		N
Bright SHARC	extent	16.3	179	37
	Romer et al. (2000)	(0.5–2.0 keV)		Y
RDCS	extent	3.0	50	103
	Borgani et al. (2001)	(0.5–2.0 keV)		N
ROXS	CCD imaging	2.0	4.8	57
	Donahue et al. (2002)	(0.5–2.0 keV)		Y
BMW	extent	~ 10	~ 300	~ 100
	Lazzati et al. (1999)	(0.1–2.4 keV)		N
XDCS	CCD imaging	3.0	11.0	15
	Gilbank et al. (2004)	(0.5–2.0 keV)		N
WARPS-II	CCD imaging	6.5	73	150
	Jones et al., in prep.	(0.5–2.0 keV)		N

servations, and Cycle 12 *HST*/ACS imaging. This sample represents more than an order of magnitude improvement in the number of distant, X-ray luminous clusters known (i.e., two EMSS clusters with $z > 0.4$, $L_x > 10^{45}$ erg s^{-1}, compared to 39 in MACS).

4.4.2   *ROSAT Pointed Observations*

The large field of view of the PSPC detector made it very efficient at detecting serendipitous X-ray sources within 15′ of the pointing position of the telescope. Given the substantial numbers of relatively deep observations during the *ROSAT* Pointed Phase an area of well over 500$\square°$ at high Galactic latitude has been covered by the central region of the PSPC with more than 10 ks exposure. This area is significantly reduced, as some targets are not suitable for serendipitous searches (e.g., nearby clusters, nearby galaxies, globular clusters, etc.), but the majority has been used in a series of surveys (listed in Table 4.2). As with the RASS samples, the selection strategies differ between surveys, but the results from each survey agree. For instance, the requirement of significant source extent used by SHARC certainly eases source selection, but at the potential loss of the most distant and/or compact, cooling flow clusters. These effects do not appear to have any great impact on the results.

The majority of the clusters selected in these surveys are relatively nearby ($z = 0.15–0.3$) and of low X-ray luminosity ($L_x = 10^{43-44}$ erg s^{-1}; 0.5–2 keV), but a few distant, luminous clusters are found (Ebeling et al. 2000, 2001), and in the deepest of the *ROSAT* serendipitous sample, RDCS (Rosati et al. 1998), there are several candidate clusters at $z > 1$ (Borgani et al. 2001).

The PSPC instrument eventually ran out of gas in mid-1994, but *ROSAT* continued to make observations with the HRI. While this instrument was less sensitive than the PSPC and covered a smaller area of sky, the excellent spatial resolution it provided has been used very effectively by the Brera Observatory group in the BMW survey (Lazzati et al. 1999; Panzera et al. 2003). While the combination of sensitivity and total area covered by the BMW survey will never match that of PSPC surveys (e.g., the 160 square degree survey; Vikhlinin et al. 1998), it does provide an important reliability test.

The full potential of the *ROSAT* pointed phase has yet to be tapped as all existing surveys have been restricted to the central $15'$–$20'$ radius where the point-spread function is best. While the flux sensitivity is poor in the outer parts of the detector, the brighter ($f_x > 10^{-13}$ erg s^{-1} cm^{-2}) sources can easily be detected. As noted above, the most time consuming part of the follow-up of X-ray selected cluster candidates at $z > 0.3$ is the deeper optical imaging required. This is exacerbated at lower X-ray fluxes by the fainter optical counterparts of all X-ray counterparts. The combination of the low-resolution X-ray imaging with deeper multicolor panoramic surveys such as SDSS, UKIDSS, and RCS2 will provide a "free" resource to identify cluster counterparts to these *ROSAT* sources.

So, how successful has *ROSAT* been overall in harvesting clusters? The RASS samples published or about to be published account for a total of around 1,200 new clusters, with a further 500–1,000 at lower fluxes. Add these to the serendipitous detections, $\sim$500 in the central region of the PSPC (of which $\sim$250 are published), $>$1,500 in the outer regions, and $\sim$300 clusters in the HRI. Therefore, the total after the *ROSAT* mission is $\sim$4,000 (when in 36.5 AG 4,500 would be required). It is worth noting that the majority of these clusters have yet to be identified and many may never be.

4.5 The Middle Age of X-ray Astronomy?

The arrival of the third decade of X-ray astronomy midway through the *ROSAT* mission coincided with advances in CCD detector technology that have allowed a vast increase in the power of X-ray spectroscopy. These advances, coupled with nested-mirror systems, mark a clear maturing in the field and a move away from large samples of objects with limited information to limited samples with very detailed information. This is an inevitable progression that emerging disciplines experience, radio astronomy being a prime example. With the advance of aperture synthesis, radio astronomers in $\sim$35 AJ (Anno Jansky) could obtain insights into the nature of individual sources and the focus moved away from surveys. In the past decade, radio astronomy has turned back to surveys (NVSS, FIRST, WENSS, 4MASS), and this will happen in X-ray astronomy (but hopefully in less than 25 years time!).

This trend for more detailed study has had a huge impact on cluster research, and the need for spatially resolved spectroscopy of clusters was apparent from the first X-ray detections of clusters. The nature of cluster surveys has also changed with a greater emphasis on understanding complete samples in many different wavelength regimes (e.g., Crawford et al. 1999; Giovannini, Tordi, & Ferretti 1999; Pimbblet et al. 2002). A sample of 200 clusters is a great resource but of little use without some information about the X-ray temperature, iron abundance, X-ray surface brightness profile, optical photometry and spectroscopy, or radio imaging. The availability of new optical, near-infrared, and radio surveys (e.g., SDSS, UKIDSS, NVSS, FIRST) will make the multiwavelength aspects of these studies much easier, but the need for further X-ray observations is hard to avoid.

4.5.1   ASCA *Observations*

The first step in this progression was the Japanese-US satellite *ASCA*. The nested, foil-replicated mirrors of *ASCA* resulted in a relatively asymmetric, broad point-spread function (2' FWHM), but the excellent performance of the SIS CCD detectors provided some very high-quality spectra for clusters (Mushotzky & Scharf 1997; Markevitch 1998; Fukazawa et al. 2000; Ikebe et al. 2002).

Over the course of the seven-year pointed phase, *ASCA* provided accurate temperatures and iron abundances for most of the 350 clusters observed. While few complete samples were observed, the *ASCA* data are an excellent complement to archival *ROSAT* observations. The notable exceptions to this are the flux-limited sample of 61 *ROSAT*-selected clusters (Ikebe et al. 2002) and the complete sample of $0.3 < z < 0.4$ EMSS clusters (Henry 1997) from which limits of the evolution of the cluster temperature function can be derived.

4.5.2   *The Unfulfilled Potential of* ABRIXAS

One of the most disappointing events in X-ray astronomy was the unfortunate failure of the German satellite *ABRIXAS* in June 1998. Its simple design and the track record of the team behind *ROSAT* meant the planned 3-year, all-sky survey *ABRIXAS* would have had a huge impact on X-ray astronomy. The survey depth envisioned of 1.5×10^{-13} erg s^{-1}cm^{-2} (0.5–2.0 keV) and 9×10^{-13} (2–12 keV) would have detected in excess of 20,000 clusters (i.e., more than the number required to keep pace with exponential growth).

4.5.3   Chandra *and* XMM-Newton

The launch of *Chandra* and *XMM-Newton* in 1999 has seen X-ray astronomy reach full maturity. The sub-arcsecond imaging of *Chandra* and unprecedented throughput of *XMM-Newton* have had a profound impact of our understanding of clusters (e.g., McNamara et al. 2000; Peterson et al. 2001; Allen, Schmidt, & Fabian 2002). The potential for surveys with both satellites is largely through serendipitous detections, but several important pointed surveys are being undertaken.

The only large *Chandra* serendipitous survey is CHamP (Wilkes et al. 2001), which will cover 14$\square°$ in 5 years and identify 8,000 X-ray sources of all types, of which 150–250 will be clusters (which will all be spatially resolved). The relatively small number of clusters makes this sample unlikely to set any strong cosmological constraints, but it will act as an excellent control sample for past and future samples to test how spatial resolution affects detection statistics.

XMM-Newton has a program similar to CHamP, the XID program that has three tiers: faint (10^{-15} erg s^{-1}cm^{-2}, 0.5$\square°$), medium (10^{-14} erg s^{-1}cm^{-2}, 3$\square°$), and bright (10^{-13} erg s^{-1}cm^{-2}, 100$\square°$). Again, like CHamP, the number of clusters detected in the XID program will be small (< 50), so from a purely cluster view-point is not particularly relevant. There are currently two dedicated serendipitous cluster surveys. One expands on the XID programme (Schwope et al. 2004) and the other (the X-ray Cluster Survey, XCS, Romer et al. 2001) aims to extract all potential cluster candidates from the *XMM-Newton* archive and compile a sample of $>5,000$ clusters from up to 1,000$\square°$ over the full lifetime of the satellite. The

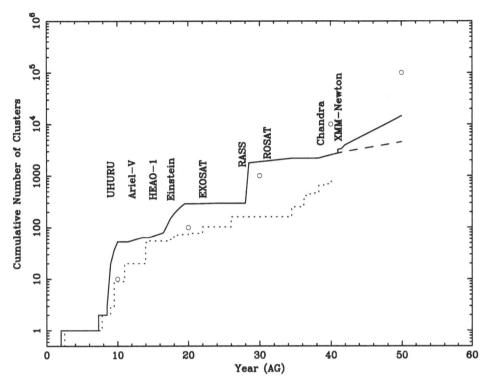

Fig. 4.2. The total number of clusters with X-ray detections known with time. The solid line marks the number detected or likely to be detected. The dotted line marks the detections published in the literature. The continuation above 40 AG is for an optimistic (solid) and pessimistic (dashed) assumption for the efficiency of *XMM-Newton* serendipitous surveys. The circles mark the number required for "Edge's Law" to hold.

contrast of XCS to CHamP and XID illustrates the huge increase in efficiency when one class of objects is chosen over the study of "complete" X-ray samples or contiguous area X-ray surveys, such as the XMM-LSS (Pierre et al. 2004), where the number of detected cluster is relatively small.

The principal pointed cluster surveys with *Chandra* and/or *XMM-Newton* target a sample of MACS clusters (Ebeling et al. 2001) with *Chandra* using GTO and GO time (PIs Van Speybroeck and Ebeling), a sample of REFLEX clusters with *XMM-Newton* in GO time (PI Böhringer) and a sample of SHARC clusters with *XMM-Newton* in GTO time (PI Lumb). Each of these projects is designed to determine the cluster temperature function, but will clearly have may other potential uses. These projects are all based on sub-samples of *ROSAT*-selected clusters to minimize the number of observations required. The reluctance of time allocation committees to devote time to complete samples in preference to the "exotica" (e.g., most distant, strongly lensing, etc., which predominate in successful proposals) is a hindrance to this "targeted" survey approach.

Table 4.3. *X-ray Missions*

Mission	Country	Status
ABRIXAS-II	German	renamed *ROSITA*
WFXT	Italian	rejected
PANORAM-X	ESA Flexi Mission	rejected
ROSITA	ESA ISS Mission	accepted phase A, could fly 2007
DUET	NASA Pathfinder	rejected

4.6   Can "Edge's Law" Hold?

The simple answer to this is question "No." Why should it? The observable Universe is finite so the number of clusters cannot grow exponentially indefinitely. Figure 4.2 shows the final version of the plot I showed during my talk with the cumulative number of clusters known with time. This illustrates the recent slowing of the number of clusters discovered and the increased lag between the detection and final publication of clusters.

As Simon White pointed out at the end of my talk, the case for ever-increasing sensitivity for the sake of it is a poor foundation for any field. The source counts for clusters crudely imply that every order of magnitude increase in number translates to an order of magnitude better sensitivity. So for the present, year 40 AG, the "Edge's Law" requirement would be equivalent to an all-sky survey to a flux of 10^{-13} erg s^{-1} cm^{-2} (0.5–2.0 keV). When I retire this will be 10^{-15} erg s^{-1} cm^{-2} (0.5–2.0 keV). This flux limit is reachable with current missions, but less than 1% of the sky could be covered.

There are, however, strong arguments for larger, deeper contiguous X-ray surveys than are available now or in the near future. The cosmological constraints that can be derived from the large-scale clustering of clusters, their mass function, and chemical evolution are complementary to those available elsewhere, most notably *WMAP* (Spergel et al. 2003). With the existence of large-area, multicolor optical and near-infrared surveys (e.g., SDSS, CHFTLS, RCS2, UKIDSS, Vista), the bottleneck of identification is eased and photometric redshifts will be sufficient for most purposes.

On a more practical level, the next generation of X-ray satellites, *XEUS* and *Constellation-X*, are optimized for the study of faint (10^{-14} erg s^{-1} cm^{-2}) sources. To get the best from the massive investment made in these missions, it would be sensible to have surveyed more than 1% of the sky to this depth.

Several proposals have been made to do this and are listed in Table 4.3. To date none of these missions is fully approved. The case for *DUET* was based on a survey of the SDSS area (Jahoda et al. 2003) and would have detected 20,000 clusters. It is likely that a proposal of this type will succeed (probably *ROSITA*) so some form of all/part-sky survey will have been performed by 50–55 AG. It is very unlikely that any mission (proposed or yet to be proposed) is likely to reach the limits required to keep above "Edge's Law," but the constant progress made in X-ray astronomy will see the number of clusters increase to well above 30,000 by 60 AG. This should be sufficient for cosmological work and more detailed studies with the next generation of X-ray satellites.

4.7   Conclusions

The success of X-ray surveys in the selection and understanding of clusters of galaxies over the past four decades has been remarkable. The statistical properties of X-ray selected samples of clusters have been used to determine cosmological parameters, and the detail found in individual clusters can be used to understand the evolution of that cluster. Current and planned surveys will build on these previous studies and will undoubtedly reveal further complexity in the intracluster medium, thereby refining our understanding of the astrophysics of these systems.

4.8   A Coda

As some of you will know, I was only able to attend the conference for one day due to the death of my father, David Edge. Fewer of you will know that one of my father's many legacies is one of the cornerstones of modern astronomy, the 3rd Cambridge Radio Catalogue (3C), which was his Ph.D. thesis. He always kept a keen interest in astronomy but moved on to become a leading figure in the sociology of science. I would like to thank Keith Taylor for his help on the evening before my departure and Richard and Barbara Ellis for providing the venue. Thanks too to all my friends and colleagues who have contacted me since then.

Acknowledgements. I owe Dave Gorman thanks for providing me with a presentation format that kept the audience awake. I am grateful to all those with whom I have worked on X-ray surveys, but particular thanks go to Harald Ebeling, whose contribution to the field is unrivaled.

References

Allen, S. W., Schmidt, R. W., & Fabian, A. C. 2002, MNRAS, 334, L11
Böhringer, H., et al. 2000, ApJS, 129, 435
——. 2001, A&A, 369, 826
——. 2002, ApJ, 566, 93
Boldt, E., McDonald, F. B., Riegler, G., & Serlemitsos, P. 1966, Phys. Rev. Lett., 17, 447
Borgani, S., et al. 2001, ApJ, 561, 13
Burg, R., Giacconi, R., Forman, W., & Jones, C. 1994, ApJ, 422, 37
Byram, E. T., Chubb, T. A., & Freidman, H. 1966, Science, 152, 66
Canizares, C. R., Clark, G. W., Markert, T. H., Berg, C., Smedira, M., Bardas, D., Schnopper, H., & Kalata, K. 1979, ApJ, 234, L33
Carlberg, R. G., et al. 1997, ApJ, 485, L13
Collins, C., Burke, D. J., Romer, A. K., Sharples, R. M., & Nichol, R. C. 1997, ApJ, 479, L117
Cooke, B. A., et al. 1978, MNRAS, 182, 489
Crawford, C. S., Allen, S. W., Ebeling, H., Edge, A. C., & Fabian, A. C. 1999, MNRAS, 306, 857
Cruddace, R., et al. 2002, ApJS, 140, 239
de Grandi, S., et al. 1999, ApJ, 514, 148
Donahue, M., et al. 2002, ApJ, 569, 689
Donahue, M., Stocke, J. T., & Gioia, I. M. 1992, ApJ, 385, 49
Ebeling, H., et al. 2000, ApJ, 534, 133
Ebeling, H., Edge, A. C., Allen, S. W., Crawford, C. S., Fabian, A. C., & Huchra, J. P. 2000, MNRAS, 318, 333
Ebeling, H., Edge, A. C., Böhringer, H., Allen, S. W., Crawford, C. S., Fabian, A. C., Voges, W., & Huchra, J. P. 1998, MNRAS, 301, 881
Ebeling, H., Edge, A. C., Fabian, A. C., Allen, S. W., Crawford, C. S., & Böhringer, H. 1997, ApJ, 479, L101
Ebeling, H., Edge, A. C., & Henry, J. P. 2001, ApJ, 553, 668
Ebeling, H., Jones, L. R., Fairley, B. W., Perlman, E., Scharf, C., & Horner, D. 2001, ApJ, 548, L23
Ebeling, H., Mullis, C. R., & Tully, B. R. 2002, ApJ, 580, 774

Ebeling, H., Voges, W., & Böhringer, H. 1994, ApJ, 436, 44

Ebeling, H., Voges, W., Böhringer, H., Edge, A. C., Huchra, J. P., & Briel, U. G. 1996, MNRAS, 281, 799

Ebeling, H., & Wiedenmann, G. 1993, Phys. Rev. E, 47, 704

Edge, A. C., Ebeling, H., Bremer, M., Röttgering, H., van Haarlem, M. P., Rengelink, R., & Courtney, N. J. D. 2003, MNRAS, 339, 913

Edge, A.C., Stewart, G. C., & Fabian, A. C. 1992, MNRAS, 258, 177

Edge, A. C., Stewart, G. C., Fabian, A. C., & Arnaud, K. A. 1990, MNRAS, 245, 559

Fabian, A. C. 1994, ARA&A, 32, 277

Fabian, A. C., Crawford, C. S., Edge, A. C., & Mushotzky, R. F. 1994, MNRAS, 267, 779

Fabian, A. C., Hu, E. M., Cowie, L. L., & Grindlay, J. 1981, ApJ, 248, 47

Forman, W., Jones, C., Cominsky, L., Julien, P., Murray, S., Peters, G., Tananbaum, H., & Giacconi, R., 1978, ApJS, 38, 357

Friedman, H., & Byram, E. T. 1967, ApJ, 147, 399

Fritz, G., Davidsen, A., Meekins, J. F., & Friedman, H. 1971, ApJ, 164, L81

Fukazawa, Y., Makishima, K., Tamura, T., Nakazawa, K., Ezawa, H., Ikebe, Y., Kikuchi, K., & Ohashi, T. 2000, MNRAS, 313, 21

Giacconi, R., et al. 1962, Phys. Rev. Lett., 9, 439

Gilbank, D. G., Bower, R. G., Castander, F. J., & Ziegler, B. L. 2004, MNRAS, submitted

Gioia, I. M., Henry, J. P., Mullis, C. R., Voges, W., Briel, U. G., Böhringer, H., & Huchra, J. P. 2001, ApJ, 553, L105

Gioia, I. M., Maccacaro, T., Schild, R. E., Wolter, A., Stocke, J. T., Morris, S. L., & Henry, J. P. 1990, ApJS, 72, 567

Giovannini, G., Tordi, M., & Ferretti, L. 1999, NewA, 4, 141

Henry, J. P. 1997, ApJ, 489, L1

Ikebe, Y., Reiprich, T., Böhringer, H., Tanaka, Y., & Kitayama, T. 2002, A&A, 383, 773

Jahoda, K., et al., 2003, AN, 324, 132

Jones, C., & Forman, W. 1984, ApJ, 276, 38

———., 1999, ApJ, 511, 65

Lahav, O., Edge, A. C., Fabian, A. C., & Putney, A. 1989, MNRAS, 238, 881

LaRoque, S. J., et al. 2003, ApJ, 583, 559

Lazzati, D., Campana, S., Rosati, P., Panzera, M. R., & Tagliaferri, G. 1999, ApJ, 524, 414

Ledlow, M., Loken, C., Burns, J. O., Owen, F. N., & Voges, W. 1999, ApJ, 516, L53

Le Févre, O., Hammer, F., Angonin, M. C., Gioia, I. M., & Luppino, G. A. 1994, ApJ, 422, L5

Loewenstein, M., & Mushotzky, R. F. 1996, ApJ, 466, 695

Luppino, G. A., Gioia, I. M., Hammer, F., Le Févre, O., & Annis, J. A. 1999, A&AS, 136, 117

Mahdavi, A., Böhringer, H., Geller, M. J., & Ramella, M. 2000, ApJ, 534, 114

Markevitch, M. 1998, ApJ, 504, 27

Markevitch, M., et al. 2000, ApJ, 541, 542

Mason, K. O., et al. 2000, MNRAS, 311, 456

Mazzotta, P., Markevitch, M., Vikhlinin, A., Forman, W. R., David, L. P., & Van Speybroeck, L. 2001, ApJ, 555, 205

McNamara, B. R., et al. 2000, ApJ, 534, L135

Meekins, J. F., Gilbert, F., Chubb, T. A., Friedman, H., & Henry, R. C. 1971, Nature, 231, 107

Mitchell, R. J., Culhane, J. L., Davison, P. J., & Ives, J. C. 1976, MNRAS, 176, 29

Mushotzky, R. F., & Scharf, C. A. 1997, ApJ, 482, 13

Nevalainen, J., Lumb, D., dos Santos, S., Siddiqui, H., Stewart, G. C., & Parmar, A. N. 2001, A&A, 374, 66

Panzera, M. R., Campana, S., Covino, S., Lazzati, D., Mignani, R. P., Moretti, A., & Tagliaferri, G. 2003, A&A, 399, 351

Peres, C., Fabian, A. C., Edge, A. C., Allen, S. W., Johnstone, R. M., & White, D. A. 1998, MNRAS, 298, 416

Perlman, E. S., Horner, D. J., Jones, L. R., Scharf, C. A., Ebeling, H., Wegner, G., & Malkan, M. 2002, ApJS, 140, 265

Pesce, J. E., Fabian, A. C., Edge, A. C., & Johnstone, R. M. 1990, MNRAS, 244, 58

Peterson, J. R., et al. 2001, A&A, 365, L104

Piccinotti, G., Mushotzky, R. F., Boldt, E. A., Holt, S. S., Marshall, F. E., Serlemitsos, P. J., & Shafer, R. A. 1982, ApJ, 253, 485

Pierre, M., et al. 2004, A&A, submitted (astro-ph/0305191)

Pimbblet, K. A., Smail, I., Kodama, T., Couch, W. J., Edge, A. C., Zabludoff, A. I., & O'Hely, E. 2002, MNRAS, 331, 333

Reiprich, T., & Böhringer, H. 2002, ApJ, 567, 716

Romer, A. K., et al. 2000, ApJS, 126, 209

Romer, A. K., Viana, P. T. P., Liddle, A. R., & Mann, R. G. 2001, ApJ, 547, 594

Rosati, P., Della Ceca, R., Norman, C., & Giacconi, R. 1998, ApJ, 492, L21

Schwope, A., et al. 2000, AN, 312, 1

Schwope, A., Lamer, G., Burke, D., Elvis, M., Watson, M.G., Schulze, M.P., Szokoly, G., & Urrutia, T., 2004, Proc. World Space Conf. Houston, October 2002, Adv. Space Res., in press (astro-ph/0306112)

Spergel, D. N., et al. 2003, ApJS, 148, 175

Stewart, G. C., Fabian, A. C., Jones, C., & Forman, W. 1984, ApJ, 285, 1

Stocke, J. T., Morris, S. L., Gioia, I. M., Maccacaro, T., Schild, R., Wolter, A., Fleming, T., & Henry, J. P. 1991, ApJS, 76, 813

Vikhlinin, A., McNamara, B. R., Forman, W., Jones, C., Quintana, H., & Hornstrup, A. 1998, ApJ, 498, 21

Voges, W., et al. 1999, A&A, 349, 389

White, D. A., Fabian A. C., Johnstone, R. M., Mushotzky, R. F., & Arnaud, K. 1991, MNRAS, 252, 72

White, D. A., Jones, C., & Forman, W. 1997, MNRAS, 292, 419

Wilkes, B. J., et al. 2001, in New Era of Wide Field Astronomy, ed. R. G. Clowes, A. J. Adamson, & G. E. Bromage (San Fransisco: ASP), 47

5

X-ray clusters at high redshift

PIERO ROSATI
European Southern Observatory

Abstract

Considerable observational progress has been made over the last decade in tracing the evolution of global physical properties of galaxy clusters in the Universe, as revealed by X-ray observations. Based on X-ray selected samples covering a wide redshift range, convincing evidence has emerged for modest evolution of both the bulk of the X-ray cluster population and their thermodynamical properties since redshift unity. With the advent of *Chandra* and *XMM-Newton*, and their unprecedented sensitivity and angular resolution, these studies have revealed the complexity of the thermodynamical structure of clusters and have been extended beyond redshift unity. By $z = 1$, clusters are found already in an advanced stage of formation, with processes responsible for metal enrichment and energy injection into the intracluster medium essentially already completed. Observations of their galaxy populations at $z \simeq 1.3$ have shown the first signature that we are approaching the epoch of formation of massive cluster galaxies. The overall observational scenario is consistent with hierarchical models of structure formation in a flat, low-density Universe with $\Omega_m \simeq 0.3$ and $\sigma_8 \simeq 0.7 - 0.8$ for the normalization of the power spectrum, although many details remain unknown regarding the formation of clusters in their cold and hot phase. The critical redshift range $z = 1.3 - 2$ needs to be explored in order to unveil these formation processes.

5.1   Introduction

Galaxy clusters form via the collapse of cosmic matter over a region of several megaparsecs. Cosmic baryons, which represent approximately 10%–15% of the mass content of the Universe, follow the dynamically dominant dark matter during the collapse through large scale filaments. As a result of adiabatic compression and of shocks generated by supersonic motions during shell crossing and virialization, a thin hot gas permeating the cluster gravitational potential well is formed. For a typical cluster mass of $10^{14} - 10^{15} M_{\odot}$ this gas, enriched with metals at some earlier epochs, reaches temperatures of several 10^7 K, becomes fully ionized, and therefore emits via thermal bremsstrahlung in the X-ray band. Since clusters arise from the gravitational collapse of rare, high peaks of primordial density perturbations, their number density is highly sensitive to specific cosmological models. The lower the density of the Universe (i.e., the matter density parameter $\Omega_m = \rho_m / \rho_{\mathrm{crit}}$), the higher the redshift at which the bulk of the cluster population forms.

Observations of clusters in the X-ray band provide an efficient and physically motivated method of identification, which allows (1) clusters to be unveiled out to $z > 1$, (2) the mass to be estimated from direct observables, and (3) a method to accurately compute the survey volume within which clusters are found. These are all essential ingredients to estimate how the cluster mass function varies with redshift, which is what cosmological models actually predict. For these reasons most of the cosmological studies based on clusters have used X-ray selected samples. X-ray studies of galaxy clusters thus provide an efficient way of mapping the overall structure and evolution of the Universe and an invaluable means of understanding the overall history of cosmic baryons.

X-ray cluster studies made substantial progress at the beginning of the 1990's with the advent of new X-ray missions. Firstly, the all-sky survey and the deep pointed observations conducted by the *ROSAT* satellite have been a goldmine for the discovery of hundreds of new clusters in the nearby and distant Universe. Follow-up studies with the *ASCA* and *Beppo-SAX* satellites revealed hints of the complex physics governing the intracluster gas. In addition to gas heating associated with gravitational processes, star formation processes and energy feedback from supernovae and galactic nuclear activity are now understood to play an important role in determining the thermal history of the intracluster medium (ICM), its X-ray properties, and its chemical composition. Studies utilizing the current new generation of X-ray satellites, *Chandra* and *XMM-Newton*, are radically changing our X-ray view of clusters. The large collecting area of *XMM-Newton*, combined with the superb angular resolution of *Chandra*, have started to unveil the interplay between the complex physics of the hot ICM and detailed processes of star formation associated with cool baryons. At very high redshifts (for clusters this means $z \gtrsim 1$), these new X-ray facilities have opened a window to cluster physical properties at lookback times that are critical for their formation history. These new data complement a wealth of information at optical/near-IR wavelengths, which we are now able to glean with the largest ground-based telescopes and the *Hubble Space Telescope (HST)*.

The scope of this article is to summarize the current observational status on the evolution of the X-ray cluster population out to $z \simeq 1$, and to provide some highlights of recent observations of the most distant X-ray clusters known to date. *

5.2   Evolution of the Cluster Abundance

An extensive review on X-ray cluster surveys over the last 20 years, as well as a discussion on basic methodologies adopted, can be found in Rosati, Borgani, & Norman (2002). A historical perspective and recent developments on cluster surveys at $z \lesssim 0.5$ can also be found in Edge (2004). Here, we give an up to date compilation of the latest results on the evolution of the cluster abundance out to $z \approx 1$.

The *ROSAT* All-Sky Survey was the first X-ray imaging mission to cover the entire sky, thus paving the way to large contiguous-area surveys of X-ray selected nearby clusters. To date, large cluster samples have been constructed in the northern (BCS, Ebeling et al. 2000; NORAS, Böhringer et al. 2000) and southern (the REFLEX cluster survey, Böhringer et al. 2001) hemisphere. In total, surveys covering more than 10^4 deg^2 have yielded over 1000 clusters, out to redshift $z \simeq 0.5$. A large fraction of these are new discoveries, whereas approximately one-third are identified as clusters in the Abell or Zwicky catalogs. A number of independent studies using different techniques and independent data sets have yielded local X-ray luminosity functions (XLF), in very close agreement with each other (see Rosati et al. 2002). This indicates that systematic effects associated with different selection functions are negligible for nearby samples, thus providing a robust local reference against which clus-

* Unless otherwise noted, we adopt the cosmological parameters $h = 0.65$, $\Omega_m = 0.3$, and $\Omega_\Lambda = 0.7$.

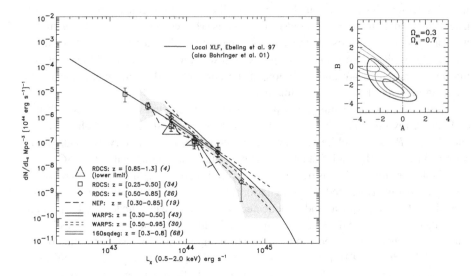

Fig. 5.1. *Left:* A compilation of X-ray luminosity functions of distant clusters from different surveys: RDCS, Rosati et al. 1998; NEP, Gioia et al. 2001; WARPS, Jones et al. 1998; 160 deg², Mullis et al. 2004 (an Einstein-de-Sitter Universe with $h = 0.5$ is adopted). *Right:* Maximum-likelihood contours (1, 2, and 3 σ) for the parameters A and B defining the XLF evolution as derived from the RDCS and EMSS samples: $\phi^* = \phi_0(1+z)^A$, $L^* = L_0^*(1+z)^B$ (see Eq. 5.1). (Updated from Rosati et al. 2002.)

ter evolution can be investigated. The cluster XLF is commonly modeled with a Schechter (1976) function:

$$\phi(L_X)dL_X = \phi^* \left(\frac{L_X}{L_X^*}\right)^{-\alpha} \exp(-L_X/L_X^*)\frac{dL_X}{L_X^*} \quad . \tag{5.1}$$

ROSAT/PSPC archival pointed observations were intensively used for serendipitous searches of distant clusters. A principal objective of all these surveys (RDCS, WARPS, SHARC, NEP, 160 deg², to mention just a few), which are now essentially completed, has been the study of the evolution of the XLF, stimulated by the early results of the Extended Medium Sensitivity Survey (EMSS, Gioia et al. 1990).

An updated compilation of the latest determinations of the XLF out to $z \approx 1$ is reported in Figure 5.1. The right panel shows best-fit evolutionary parameters obtained by a maximum-likelihood method, which compares the observed cluster distribution on the (L_X, z) plane with that expected from an XLF model: $\phi(L, z) = \phi_0(1+z)^A L^{-\alpha} \exp(-L/L^*)$, with $L^* = L_0^*(1+z)^B$, where A and B are two evolutionary parameters for density and luminosity, and ϕ_0 and L_0^* are the local XLF values (Eq. 5.1). Figure 5.1 shows an application of this method to the RDCS and EMSS sample and indicates that the no-evolution case ($A = B = 0$) is excluded at more than 3σ levels in both samples when the most luminous systems are included in the analysis, whereas the same analysis confined to clusters with $L_X < 3 \times 10^{44}$ erg s^{-1} yields an XLF consistent with no evolution.

In summary, by combining all the results from *ROSAT* surveys one obtains a consistent picture in which the comoving space density of the bulk of the cluster population is ap-

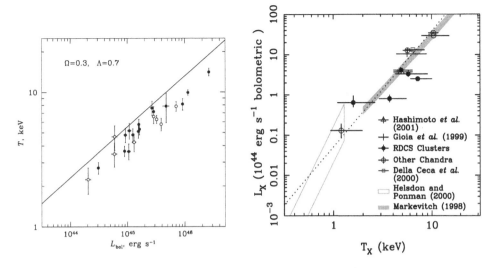

Fig. 5.2. *Left*: $L_X - T$ relation from Vikhlinin et al. (2002) (filled circles correspond to clusters at $0.4 < z < 0.7$, open circles to clusters at $z > 0.7$; the solid line is the correlation found at low redshifts by Markevitch 1998). *Right*: $L_X - T$ relation from the study of Holden et al. (2002).

proximately constant out to $z \simeq 1$, whereas the most luminous ($L_X \gtrsim L_X^*$), presumably most massive clusters were likely rarer at high redshifts ($z \gtrsim 0.5$). Significant progress in the study of the evolution of the bright end of the XLF would require a large solid angle and a relatively deep survey with an effective solid angle of $\gg 100$ deg^2 at a limiting flux of 10^{-14} erg cm^{-2} s^{-1}. This will be possible, to some extent, with similar serendipitous surveys based on *Chandra* and *XMM-Newton* pointings. The convergence of the results from several independent studies illustrates remarkable observational progress in determining the abundance of galaxy clusters out to $z \approx 1$. At the beginning of the *ROSAT* era, until the mid-nineties, controversy surrounded the usefulness of X-ray surveys of distant galaxy clusters, and many believed that clusters were absent at $z \approx 1$ (including many theorists!). The observational prejudice arose from an over-interpretation of the early results of the EMSS survey, which remain correct to date, but were thought to be in contrast with results from optical surveys without understanding that these studies were probing different sections of the cluster mass function at high redshift. Such a scenario of rapid evolution of the cluster mass function was also favored by a theoretical prejudice for an $\Omega = 1$ Universe (without a cosmological constant).

The study of scaling relations in clusters is an important diagnostic tool to understand their formation history and to test the validity of our simple assumptions on their thermodynamical properties. Specifically, with *Chandra* observations, we can now investigate for the first time the $L_X - T$ relation at high redshift. In Figure 5.2, we show results from two recent studies, which indicate a mild evolution of this relation, although they still disagree on its strength. These findings are in agreement with models that require a substantial injection of nongravitational energy (e.g., from supernovae or AGN activity) to explain the slope of the local $L_X - T$ relation (e.g., Evrard & Henry 1991; Cavaliere, Menci, & Tozzi 1998; Balogh, Babul, & Patton 1999; Ponman, Cannon, & Navarro 1999; Tozzi & Norman 2001; Voit et

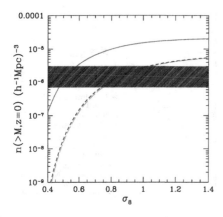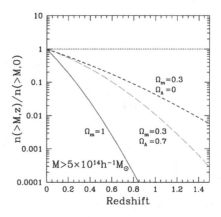

Fig. 5.3. The sensitivity of the cluster mass function to cosmological models. *Left*: The cumulative mass function at $z = 0$ for $M > 5 \times 10^{14} h^{-1} M_{\odot}$, for three cosmologies, as a function of σ_8, with shape parameter $\Gamma = 0.2$; solid line: $\Omega_m = 1$; short-dashed line: $\Omega_m = 0.3$, $\Omega_{\Lambda} = 0.7$; long-dashed line: $\Omega_m = 0.3$, $\Omega_{\Lambda} = 0$. The shaded area indicates the observational uncertainty in the determination of the local cluster space density. *Right*: Evolution of $n(>M, z)$ for the same cosmologies and the same mass limit, with $\sigma_8 = 0.5$ for the $\Omega_m = 1$ case and $\sigma_8 = 0.8$ for the low-density models. (Adopted from Rosati et al. 2002.)

al. 2002). The reader is referred to the articles by Mushotzky (2004) and Mulchaey (2004) for a discussion on scaling relations in clusters and groups.

5.3   Cosmology with X-ray Clusters

The space density of clusters in the local Universe has been used to measure the amplitude of density perturbations on ~ 10 Mpc scales, which is commonly parameterized by σ_8, the rms density fluctuation within a top-hat sphere of $8 h^{-1}$Mpc radius. Using only the number density of nearby clusters of a given mass M, one can constrain the amplitude of the density perturbation at the physical scale $R \propto (M/\Omega_m \rho_{\text{crit}})^{1/3}$ containing this mass. Since such a scale depends both on M and on Ω_m, the mass function of nearby ($z \lesssim 0.1$) clusters is only able to constrain a relation between σ_8 and Ω_m. In the left panel of Figure 5.3 we show that, for a fixed value of the observed cluster mass function, the implied value of σ_8 increases as the density parameter decreases (with a typical scaling $\sigma_8 \Omega_m^{\alpha} = 0.4 - 0.6$, with $\alpha \simeq 0.4 - 0.6$). Formal statistical uncertainties in the determination of σ_8 from different analyses are always far smaller, $\lesssim 5\%$, than the range of published values. This suggests that current discrepancies on σ_8 are likely to be ascribed to systematic effects, such as sample selection and different methods used to infer cluster masses. We refer the reader to the recent paper by Pierpaoli et al. (2003) for a review and a discussion on measurements of the cluster normalization.

The growth rate of the density perturbations depends primarily on Ω_m and, to a lesser extent, on Ω_{Λ}, at least out to $z \approx 1$, where the evolution of the cluster population is currently studied. Therefore, following the evolution of the cluster space density over a large redshift baseline, one can break the degeneracy between σ_8 and Ω_m. This is shown in the right panel of Figure 5.3: models with different values of Ω_m, which are normalized to yield the same number density of nearby clusters, predict cumulative mass functions that progressively differ by up to orders of magnitude at increasing redshifts. This plot also shows how future surveys, by amassing large samples of clusters at $z = 1 - 1.5$ (a daunting task indeed), will also be able to constrain the value of Ω_Λ by exploiting its dynamical effect on cluster formation (perturbations cease to grow at later epochs in the presence of a cosmological constant).

An estimate of the cluster mass function is reduced to the measurement of masses for a sample of clusters, stretching over a large redshift range, for which the survey volume is well known. Essentially four estimators of the cluster mass have been used to date: the cluster velocity dispersion, the gas temperature, the X-ray luminosity, and more recently the gravitational lensing shear strength (see Margoniner et al. 2004). The CNOC survey (e.g., Yee, Ellingson, & Carlberg 1996) still represents the state of the art of a study of the internal dynamics of a statistically complete sample of 16 clusters at $z < 0.55$. The extension of this method to large samples of distant clusters is extremely demanding from the observational point of view, which explains why it has not been pursued much further.

A conceptually robust and more popular method to estimate cluster masses is based on the X-ray measurement of the temperature of the intracluster gas. On the assumption that gas and dark matter particles share the same dynamics within the cluster potential well, the temperature T and the velocity dispersion σ_v are connected by the relation $k_B T = \beta \mu m_p \sigma_v^2$, where $\beta = 1$ would correspond to the case of a perfectly thermalized gas. If we assume spherical symmetry, hydrostatic equilibrium, and isothermality of the gas, the total cluster virial mass, M_{vir}, is related to the ICM temperature by

$$k_B T = \frac{1.38}{\beta} \left(\frac{M_{vir}}{10^{15} h^{-1} M_\odot} \right)^{2/3} [\Omega_m \Delta_{vir}(z)]^{1/3} (1+z) \, \text{keV} . \tag{5.2}$$

$\Delta_{vir}(z)$ is the ratio between the average density within the virial radius and the mean cosmic density at redshift z ($\Delta_{vir} = 18\pi^2 \simeq 178$ for $\Omega_m = 1$). Simulations have shown that cluster masses can be recovered from gas temperature with a $\sim 20\%$ precision, and the above formula holds with $0.9 \lesssim \beta \lesssim 1.3$ (e.g., Bryan & Norman 1998; Frenk et al. 1999; see also Evrard 2004). The main difficulty remains in connecting the theoretical *virial mass* with an observational equivalent, a topic that has become hotly debated by theorists and observers in recent times.

A number of measurements of cluster temperatures for flux-limited samples of clusters have been made over the last 15 years, using *ROSAT*, *Beppo-SAX*, and especially *ASCA*. With these data one can derive the X-ray temperature function (XTF), and hence the mass function using Equation 5.2. XTFs have been computed for both nearby (see Pierpaoli, Scott, & White 2001 for a recent review) and distant clusters (e.g., Henry 2000) and used to constrain cosmological models. The mild evolution of the XTF has been interpreted as a case for a low-density Universe, with $0.2 \lesssim \Omega_m \lesssim 0.6$. Besides questions related to the real validity Equation 5.2, a limitation of the XTF method so far has been the limited sample size (particularly at $z > 0.6$), as well as the lack of a homogeneous sample selection for local and distant clusters. In the next session, we illustrate how recent *Chandra* and *XMM-Newton* observations are contributing to change the observational scenario.

Another method to trace the evolution of the cluster number density is based on the XLF. The advantage of using X-ray luminosity as a tracer of the mass is that L_X is measured for

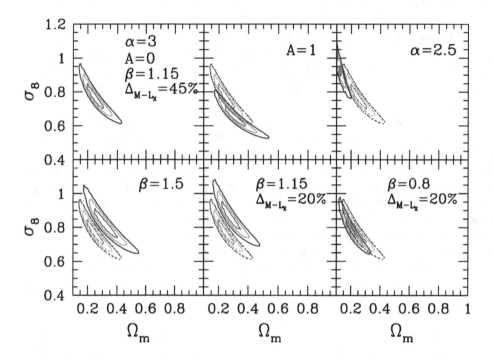

Fig. 5.4. Probability contours in the $\sigma_8 - \Omega_m$ plane from the evolution of the X-ray luminosity distribution of RDCS clusters. The shape of the power spectrum is fixed to $\Gamma = 0.2$ and a flat geometry is assumed (i.e., $\Omega_m + \Omega_\Lambda = 1$). The sequence of panels illustrates current systematics in converting cluster virial mass, M, into X-ray luminosity, L_X. The three parameters (A, α, β) describe the theoretical and observational uncertainties in converting cluster masses into temperatures ($T \sim M^{2/3}/\beta$), and temperatures into X-ray luminosities $[L_X \sim T^\alpha (1+z)^A]$. Δ_{M-L_X} is the observed scatter of the $M-L_X$ relation. The upper left panel shows the analysis corresponding to the choice of a reference parameter set. In each panel, these parameters are varied, with the dotted contours always showing the reference analysis. (From Rosati et al. 2002.)

a much larger number of clusters, which are homogeneously identified over a broad redshift baseline, out to $z \simeq 1.3$. This allows nearby and distant clusters to be compared within the same sample, i.e. with a single selection function. The potential disadvantage of this method is that it relies on the relation between L_X and $M_{\rm vir}$, which is based on additional physical assumptions and hence is more uncertain than the $M_{\rm vir} - \sigma_v$ or the $M_{\rm vir} - T$ relations. An empirical $L_X - M$ relation is well established (e.g., Reiprich & Böhringer 2002), although with some scatter. The recently improved determination of the $L_X - T$ relation, particularly its scatter and evolution out to $z \simeq 1$ (Holden et al. 2002, 2004; Vikhlinin et al. 2002; Jones et al. 2004), allows the XLF, $\phi(z)$, to be converted into an evolving mass function, $n(m, z)$, via $L_X \rightarrow T \rightarrow M_{\rm vir}$. A Press-Schechter-like analytical approach can then be used to constrain σ_8 and Ω_m (e.g., Borgani et al. 2001). Figure 5.4 shows the application of such a method using the RDCS cluster sample. The main uncertainty is due to "systematics" in estimating cluster masses; such an effect is illustrated by changing in turn the parameters defining the $M - L_X$ relation, such as the slope α and the evolution parameter A of the $L_X - T$ relation, the normalization β of the $M - T$ relation (see Eq. 5.2), and the overall scatter Δ_{M-L_X}. It is important to note that by varying these parameters within both the observational and the theoretical uncertainties, the matter density parameter remains confined in the range $0.1 \lesssim \Omega_m \lesssim 0.6$, with best parameters $\Omega_m \simeq 0.3$ and $\sigma_8 = 0.7 - 0.8$.

X-ray observations of clusters also offer a completely independent, and to some extent more robust, way of constraining Ω_m by providing a measure of the baryon mass fraction, which is essentially $f_{bar} = M_{gas}/M_{tot}$, since the intracluster gas dominates the baryonic mass content of rich clusters. If the baryon density parameter, Ω_{bar}, is known from independent considerations (e.g., by combining the observed deuterium abundance in high-redshift absorption systems with predictions from primordial nucleosynthesis), then the cosmic density parameter can be estimated as $\Omega_m = \Omega_{bar}/f_{bar}$ (e.g., White et al. 1993). For a value of the Hubble parameter $h \simeq 0.7$ and a typical measured value $f_{bar} \simeq 0.15$ (e.g., Evrard 1997; Ettori 2001), this method yields $\Omega_m \simeq 0.3 h_{70}^{-0.5}$, with a 50% scatter [using the currently most favored values of the baryon density parameter, $\Omega_{bar} \simeq 0.02\,h^{-2}$, as implied by primordial nucleosynthesis and by the spectrum of cosmic microwave background (CMB) anisotropies].

Under the assumption that the gas fraction, f_{gas}, does not evolve *intrinsically* with redshift, such a method has recently been extended by measuring the apparent redshift dependence of f_{gas}, which is mainly driven by a geometrical factor: $f_{gas}(z) \propto D_A^{3/2}(\Omega_m, \Omega_\Lambda, z)$, where D_A is the angular diameter distance to the clusters (Allen et al. 2002; Ettori et al. 2003). $f_{gas} =$ constant is a reasonable hypothesis, particularly for massive clusters, since these systems are expected to provide a fair sample of the matter content of the Universe, and mechanisms that would otherwise segregate baryons preferentially in cluster environments are not known. Results from the two main studies to date are shown in Figure 5.5 (see also Vikhlinin et al. 2003 for an alternative method). By using the most distant clusters, the leverage on Ω_Λ, or equivalently on the equation of state parameter w, increases (see right panel), although it becomes more difficult to estimate the total cluster mass due the inability to measure temperature profiles. Figure 5.5 shows how, by combining this method with independent cosmological constraints based on the CMB power spectrum and Type Ia supernovae, one can significantly narrow down the allowed region in the $(\Omega_m, \Omega_\Lambda)$ plane.

A more delicate issue is whether one can use the evolution of galaxy clusters for high-precision cosmology, say $\lesssim 10\%$ accuracy, particularly at the beginning of an era in which cosmological parameters can be derived rather accurately by combining methods that measure the global geometry of the Universe [the CMB spectrum (e.g., Spergel et al. 2003), Type Ia supernovae (e.g., Leibundgut 2001), and the large-scale distribution of galaxies (e.g., Peacock et al. 2001)]. Serendipitous searches of distant clusters from *XMM-Newton* and *Chandra* data will eventually lead to a significant increase of the number of high-z clusters with measured temperatures. Thus, the main limitation will lie in systematics involved in comparing the mass inferred from observations with that given by theoretical models. *Chandra* and *XMM-Newton* have revealed that physical processes in the ICM are rather complex. Our physical models and numerical simulations are challenged to explain the new level of spatial detail in the density and temperature distribution of the gas, and the interplay between heating and cooling mechanisms (see Donahue & Voit 2004). Such complexities need to be well understood physically before we can use clusters as high-precision cosmological tools.

The gas fraction $f_{gas}(z)$ method, on the other hand, is based on a limited number of assumptions, although it still requires long *Chandra* integrations. The promise of this tech-

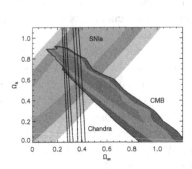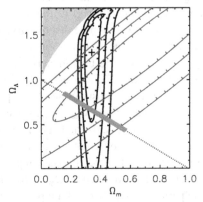

Fig. 5.5. Constraints on Ω_m and Ω_Λ from two independent analyses of the gas fraction and its redshift dependence, $f_{gas}(z)$. *Left*: From Allen, Schmidt, & Fabian (2002), who used a sample of *Chandra* clusters at $0.1 < z < 0.5$. *Right*: Adapted from Ettori, Tozzi, & Rosati (2003), who used the highest redshift sample of *Chandra* clusters available to date ($0.7 < z < 1.3, T > 4\,$keV). The shaded bar represents constraints from the recent *WMAP* results (Spergel et al. 2003).

nique is illustrated in Figure 5.5: the combination of these constraints with the recent exquisite *WMAP* results (Spergel et al. 2003) is sufficient to obtain a small allowed region around the current "concordance model," without the need of Type Ia supernovae, whose systematics are far from being under control.

5.4   Distant X-ray Clusters: the Latest View from *Chandra*

With its unprecedented angular resolution, the *Chandra* satellite has revolutionized X-ray astronomy, allowing studies with the same level of spatial detail as in optical astronomy. *Chandra* imaging of low-redshift clusters has revealed a complex thermodynamical structure of the ICM down to kiloparsec scales (e.g., Fabian et al. 2000; Markevitch et al. 2000). At high redshifts, deep *Chandra* images still have the ability to resolve cluster cores and to map ICM morphologies at scales below 100 kpc. Moreover, temperatures of major subclumps can be measured for the first time at $z > 0.6$, and a crude estimate of the ICM metalicity can be obtained even beyond $z = 1$.

Figure 5.6 is an illustrative example of the unprecedented view that *Chandra* can offer on distant clusters. We show nine archival images of clusters at $0.8 < z < 1.3$. The intensity (in grey scale) is proportional to the square root of the X-ray emission, so that they roughly map the gas density distribution in each cluster. The images are arranged in two redshift bins, with X-ray luminosities (and generally the gas temperature) increasing from left to right. One σ error bars for gas temperatures at these high redshifts range from 20% to 40%. A close inspection of these images reveal a deviation from spherical symmetry in all systems. Some of them are elongated or have cores clearly displaced with respect to the external diffuse envelope (e.g., Holden et al. 2002; Rosati et al. 2004).

Three of the most luminous clusters at $z \simeq 0.8$ (RXJ1716+6708: Gioia et al. 1999; RXJ0152.7–1357: Huo et al. 2004, Maughan et al. 2003; MS1054–0321: Jeltema et al. 2001, respectively the second, third, and fourth cutout in Figure 5.6) show a double core

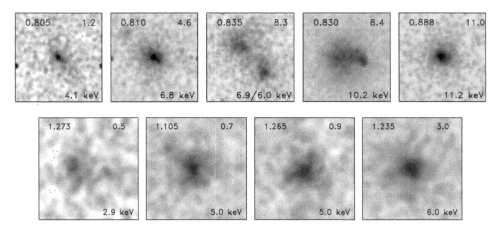

Fig. 5.6. *Chandra* archival images of nine distant clusters in the 0.5–2 keV band. Labels indicate redshifts (upper left), X-ray luminosities (upper right in the rest frame [0.5–2] keV band and in units of 10^{44} erg s^{-1}), and gas temperatures (bottom right). Images on the top row (2 Mpc across) are clusters at $0.8 < z < 0.9$ spanning a factor 10 in L_X; images on the bottom (1 Mpc across) are the four most distant clusters studied by *Chandra* to date at $1.1 < z < 1.3$. The X-ray emission has been smoothed at the same physical scale of 70 kpc (point sources in each field were removed).

structure both in the distribution of the gas and in their member galaxies. It is tempting to interpret these morphologies as the result of ongoing mergers, although no dynamical information has been gathered to date to support this scenario. In a hierarchical cold dark matter formation scenario, one does expect the most massive clusters at high redshift to be accreting subclumps of comparable masses, and the level of substructure to increase at high redshifts. With current statistical samples however, it is difficult to draw any robust conclusion on the evolution of ICM substructure.

The discovery and the study of systems beyond redshift unity have the strongest leverage on testing cosmological models. This, however, has been a challenging task with X-ray searches so far due to the limited survey areas covered at faint fluxes. The second row in Figure 5.6 shows the most distant clusters observed with *Chandra* to date, which were discovered in the *ROSAT* Deep Cluster Survey (Stanford et al. 2001, 2002; Rosati et al. 2004), at the very limit of the *ROSAT* sensitivity. RXJ0848.6+4453 and RXJ0848.9+4452 (the first and second image) are only 5′ apart on the sky (the Lynx field) and are possibly part of a superstructure at $z = 1.26$, consisting of two collapsed, likely virialized clusters (Rosati et al. 1999). These deep *Chandra* observations have yielded for the first time information on ICM morphologies in $z > 1$ clusters and allowed a measurement of their temperatures, which imply masses of $(0.5-3) \times 10^{14} h_{65}^{-1} M_\odot$ (Ettori et al. in prep.). For the hottest and most luminous of these systems (RDCS1252−2930), the Fe 6.7 keV iron line is clearly detected for the first time at $z > 1$. This implies a high metal abundance consistent with the local canonical value, $Z \simeq 0.3 Z_\odot$ (see Fig. 5.7). These findings can be used to set new constraints on the epoch of metal enrichment via star formation processes, as well as the time scale on which metals are spread across the ICM. Interestingly, we note that X-ray observations alone

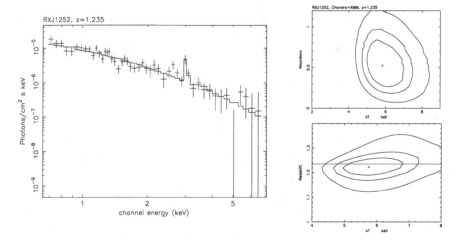

Fig. 5.7. *Left*: X-ray spectrum and best Raymond-Smith fit from *Chandra* observations (188 ks) of RDCS1252–2930 at z_{spec} = 1.235; a clear redshifted Fe 6.7 keV line is visible. *Right*: (top) best-fit temperature and metalicity of the gas obtained by combining *Chandra* and *XMM-Newton* (65 ks) data; (bottom) best-fit redshift of the ICM, with the horizontal shaded line marking the spectroscopic redshift based on 33 cluster members. (Adapted from Tozzi et al. 2003 and Rosati et al. 2004.)

are able to accurately yield the cluster redshift (bottom right panel in Fig. 5.7), a difficult task to achieve even with 8–10 m class telescopes.

In Figure 5.8, we show near-IR images of the four clusters at $z > 1$, overlaid with *Chandra* contours. Already at these large lookback times, surface brightness profiles of these systems are similar to those of low-redshift clusters. Moreover, the morphology of the gas, as traced by the X-ray emission, is well correlated with the spatial distribution of member galaxies, similar to studies at lower redshifts. This suggests that there are already at $z > 1$ galaxy clusters in an advanced dynamical stage of their formation, in which all the baryons (gas and galaxies) have had enough time to thermalize in the cluster potential well.

5.5   Galaxy Populations in the Most Distant Clusters

A discussion on galaxy populations in high-z clusters is beyond the scope of this article. We refer the reader to the reviews by Dressler (2004), Franx (2004), and Treu (2004) for both an historical and an up to date view on the evolution of cluster galaxies and their formation scenarios. In this section, we only give an example of how spectrophotometric studies of cluster galaxies at $z > 1$ can set important constraints on the mode and epoch of formation of early-type galaxies, owing to the strong leverage that these observations have on models at such large lookback times.

In Figure 5.9 (left panel), we show the evolution of the fundamental plane of early-type galaxies in clusters at increasing redshift (van Dokkum & Stanford 2003, and references therein). By measuring directly mass-to-light M/L ratios, one can disentangle the evolution of the galaxy luminosity function from the underlying mass function, thus providing stringent tests on hierarchical models of galaxy formation in which merging processes play a fundamental role. The data point at $z = 1.27$ is derived from the two most massive galaxies in RDCS0848+4453 (top left image in Fig. 5.8). We note how this data point, obtained

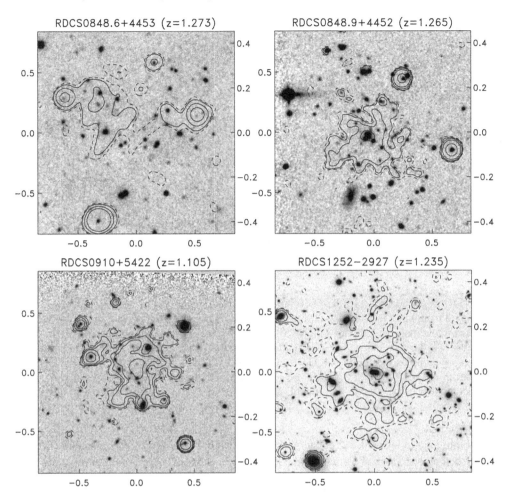

Fig. 5.8. *K*-band images of four *bona fide* clusters at $z > 1.1$ overlaid with contours mapping the X-ray emission detected by *Chandra* /ACIS-I (adapted from Stanford et al. 2001, 2002; Lidman et al. 2004; Rosati et al. 2004). Left and bottom axis give the scale in arcminutes; the right axis gives the scale in Mpc for our adopted cosmology.

by pushing spectroscopy with 10 m class telescopes to the ultimate limit, has the strongest leverage in discriminating among different models. The best-fit model, incorporating the "morphological evolution bias" (van Dokkum & Franx 2001), has a stellar formation red-shift $z_{form} \simeq 2.5$. In addition, spectrophotometric data of the brightest cluster galaxies indicate the presence young stellar populations (van Dokkum et al. 2001), with ages of 1–2 Gyr at the epoch of observations. These findings are corroborated by an independent spectroscopic study of the 10 brightest members in RDCS1252−2927 (out of the 33 spectroscopically confirmed to date; Demarco et al. 2004; Rosati et al. 2004). A stacking spectral analysis (Fig. 5.9, right) shows significant Hδ absorption, indicative of a poststarburst phase, i.e. the presence of stellar populations with ages of $\gtrsim 1$ Gyr. Moreover, color-magnitude diagrams (the so-called red sequence) in these two distant clusters appear to be flatter and

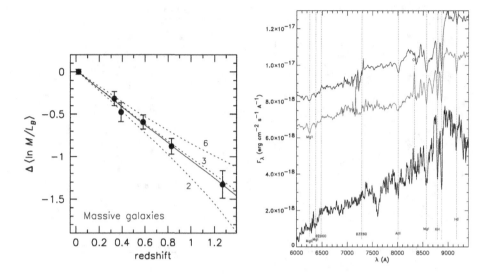

Fig. 5.9. *Left*: Evolution of the M/L ratio of early-type galaxies in clusters out to $z = 1.27$ (adapted from van Dokkum & Stanford 2003). The highest redshift data point is derived by the two most massive galaxies in RDCS0848+4453 (Fig. 5.8). Dashed lines are predictions from single-burst stellar population synthesis models with $z_{form} = 6$, 3, and 2; the solid line is a more complex model by van Dokkum & Franx (2001) with $z_{form} \simeq 2.5$. *Right*: Stacked spectrum of the 10 brightest early-type galaxies in the cluster RDCS1252–29 at $z = 1.235$. For comparison, the spectra of a local E and Sa galaxies are also shown at the top. (Adapted from Rosati et al. 2004.)

less tight than in low-redshift clusters. All these data indicate that at these redshifts we are approaching the epoch of formation massive cluster galaxies. These examples illustrate how spectrophotometric studies at these lookback times, when combined with *HST*/ACS imaging for morphological properties, hold promise to shed new light on their formation history.

5.6   Conclusions and Future Challenges

Considerable observational progress has been made in tracing the evolution of the global physical properties of galaxy clusters as revealed by X-ray observations. In the last five years, studies of distant clusters have not revealed any dramatic change in either their space density or their scaling relations (e.g., $L_X - T$). The *Chandra* satellite has given us the first view of the gas distribution in clusters at $z > 1$; their X-ray morphologies and temperatures show that they are already in an advanced stage of formation at these large lookback times. Moreover, X-ray spectroscopy has revealed that by this epoch the intracluster gas was already enriched with metals, with an iron abundance comparable with the local values.

These observations can be understood in the framework of hierarchical formation of cosmic structures, with a low density parameter, $\Omega_m \approx 1/3$, dominated by cold dark matter: structure formation started at early cosmic epochs and a sizable population of massive clusters was in place already at redshifts of unity. In addition, detailed X-ray observation of the intracluster gas show that the physics of the ICM needs to be regulated by additional nongravitational processes. The lack of strong evolution of cluster X-ray properties, as well

as the ICM metalicity, suggest that processes responsible for energy injection (star formation feedback, AGNs) and metal enrichment (supernova explosions) might be completed by
$z \simeq 1$.

In the opening era of high-precision cosmology, the role of clusters is still important
since they provide a completely independent method to constrain the universal geometry.
It remains remarkable that the evolution of the cluster abundance, the f_{gas} technique, the
CMB fluctuations, Type Ia supernovae and large-scale structure—all completely indepen-
dent methods—converge toward $\Omega_m \simeq 0.3$ in a spatially flat Universe ($\Omega_m + \Omega_\Lambda = 1$). Further
studies with the current new X-ray facilities will help considerably in addressing the issue
of systematics discussed above, although some details of the ICM in $z \gtrsim 1$ clusters, such as
temperature profiles, will remain out of reach until the next generation of X-ray telescopes.
Direct measurements of cluster masses at $z \gtrsim 1$ via gravitational lensing techniques will
soon be possible with the Advanced Camera for Surveys (Ford et al. 1998) onboard the
HST, which offers an unprecedented combination of sensitivity, angular resolution and field
of view.

The fundamental question remains as to the mode and epoch of formation of the ICM.
When and how was the gas preheated and polluted with metals? What is the epoch when
the first X-ray clusters formed, the epoch when the accreted gas thermalizes to the point at
which the clusters would lie on the $L_X - T$ relation? Are the prominent concentrations of
star-forming galaxies discovered at redshift $z \approx 3$ (Steidel et al. 1998, 2000) the progenitors
of the X-ray clusters we observed at $z \lesssim 1$? If so, cluster formation should have occurred in
the redshift range 1.5–2.5.

Although the redshift boundary for X-ray clusters has recently receded from $z = 0.8$ to
$z = 1.3$, a census of clusters at $z \simeq 1$ has just begun, and the search for clusters at $z > 1.3$
remains a serious observational challenge. Serendipitous searches for clusters based on
Chandra and *XMM-Newton* pointings (e.g., Romer et al. 2001; Boschin 2002), or large-area
surveys (e.g., Pierre et al. 2004) will allow good progress in this field. Using high-z radio
galaxies as signposts for protoclusters has been the only viable method so far to break this
redshift barrier and push it out to $z \simeq 4$. Recent work by G. Miley and collaborators has
been very successful in finding significant overdensities of Lyα-emitting galaxies around
selected radio galaxies, via narrow-band imaging techniques and subsequent spectroscopic
confirmation. These searches have also led to the discovery of extended Lyα nebulae around
distant radio galaxies (e.g., Venemans et al. 2002; Kurk et al. 2004), very similar to those
discovered by Steidel et al. (2000) in correspondence with large-scale structures at $z \simeq 3$.
The nature of such nebulae is still not completely understood; however, they could represent
the early phase of collapse of cool gas through mergers and cooling flows. If there is an
evolutionary link between "protoclusters" found around high-z radio galaxies and the X-ray
clusters at $z \simeq 1.2$, viable evolutionary tracks should be found linking the galaxy populations
in these systems, using their spectrophotometric and morphological properties.

Recent follow-up *Chandra* observations of high-redshift radio galaxies have revealed the
presence of diffuse X-ray emission, in addition to a central point source (3C 294 at $z = 1.786$:
Fabian et al. 2003; 4C 41.17 at $z = 3.8$: Scharf et al. 2004a, b). However, their spectral energy distribution and other energetic arguments indicate that the extended emission is likely
nonthermal, but rather due to inverse-Compton scattering of CMB photons by a population
of relativistic electrons associated with the radio source activity. The serendipitous detection of thermal ICM at $z > 1.5$ associated with $\sim L^*$ clusters remains extremely difficult, not

only for the lack of volume in current X-ray surveys, but also for the severe $(1+z)^4$ surface brightness dimming that affects X-ray observations. These limitations will be overcome by surveys exploiting the Sunyaev-Zel'dovich (1972) effect (e.g., Carlstrom, Holder, & Reese; Birkinshaw 2004), which will explore large volumes at $z > 1$. It is worth noting, however, that current sensitivity of Sunyaev-Zel'dovich observations is still not sufficient to detect any of the X-ray clusters at $z > 1$ shown in § 5.4, all having $L_X \lesssim L^* \approx 3 \times 10^{44}$ erg s^{-1}.

Near-IR large-area surveys remain a valid alternative to unveil a large number of clusters at $z \approx 1$. Surveys covering up to 10^3 deg^2 in the z band (Gladders & Yee 2000; Gladders 2004) are well underway and well suited to discover rare, massive clusters out to $z \simeq 1.2$. The next generation of large-area surveys in the *J, H,* and *K* bands (principally the UKIRT Infrared Deep Sky Survey; see http://www.ukidss.org) will push this boundary even further.

Ideally, the formation and evolution of galaxies in clusters should be linked to the evolution of the ICM, and the fact that we are still treating the two aspects as separate points to the difficulty in drawing a comprehensive, unified picture of the history of cosmic baryons in their cold and hot phase. Multiwavelength studies are undoubtedly essential to reach such a unified picture. A large-area (> 100 deg^2) joint survey combining Sunyaev-Zel'dovich, X-ray, and space-based near-IR observations (such experiments are being proposed or planned) could reveal the evolutionary trends in a number of independent physical parameters, including the cluster mass, the gas density and temperature, the underlying galactic mass, and star formation rates. Advances in instrumentation and observational technique will make this approach possible and will provide vital input for models of structure formation and independent constraints on the underlying cosmological parameters.

Acknowledgments. I am grateful to John Mulchaey, Alan Dressler, and Gus Oemler for organizing such a successful and stimulating meeting. I thank Stefano Borgani, Ricardo Demarco, Stefano Ettori, Vincenzo Mainieri, Colin Norman, Chris Mullis, and Paolo Tozzi for stimulating discussions and their help in the preparation of some of the figures presented here. I also thank Chris Lidman and Adam Stanford for their contributions and for permission to show results in advance of publication.

References

Allen, S. W., Schmidt, R. W., & Fabian, A. C. 2002, MNRAS, 334, L11
Balogh, M. L., Babul, A., & Patton, D. R. 1999, MNRAS, 307, 463
Birkinshaw, M. 2004, in Carnegie Observatories Astrophysics Series, Vol. 3: Clusters of Galaxies: Probes of Cosmological Structure and Galaxy Evolution, ed. J. S. Mulchaey, A. Dressler, & A. Oemler (Cambridge: Cambridge Univ. Press), in press
Böhringer, H., et al. 2000, ApJS, 129, 435
———. 2001, A&A, 369, 826
Borgani, S., et al. 2001, ApJ, 561, 13
Boschin, W. 2002, A&A, 396, 397
Bryan, G. K., & Norman, M. L. 1998, ApJ, 495, 80
Carlstrom, J. E., Holder, G. P., & Reese, E. D. 2002, ARA&A, 40, 463
Cavaliere, A., Menci, N., & Tozzi, P. 1998, ApJ, 501, 493
Demarco, R., Rosati, P., Lidman, C., Nonino, M., Mainieri, V., Stanford, A., Holden, B., & Eisenhardt, P. 2004, in Carnegie Observatories Astrophysics Series, Vol. 3: Clusters of Galaxies: Probes of Cosmological Structure and Galaxy Evolution, ed. J. S. Mulchaey, A. Dressler, & A. Oemler (Pasadena: Carnegie Observatories, http://www.ociw.edu/ociw/symposia/series/symposium3/proceedings.html)
Donahue, M., & Voit, G. M. 2004, in Carnegie Observatories Astrophysics Series, Vol. 3: Clusters of Galaxies: Probes of Cosmological Structure and Galaxy Evolution, ed. J. S. Mulchaey, A. Dressler, & A. Oemler (Cambridge: Cambridge Univ. Press), in press

Dressler, A. 2004, in Carnegie Observatories Astrophysics Series, Vol. 3: Clusters of Galaxies: Probes of Cosmological Structure and Galaxy Evolution, ed. J. S. Mulchaey, A. Dressler, & A. Oemler (Cambridge: Cambridge Univ. Press), in press

Ebeling, H., Edge, A. C., Allen, S. W., Crawford, C. S., Fabian, A. C, & Huchra, J.P. 2000, MNRAS, 318, 333

Edge, A. C. 2004, in Carnegie Observatories Astrophysics Series, Vol. 3: Clusters of Galaxies: Probes of Cosmological Structure and Galaxy Evolution, ed. J. S. Mulchaey, A. Dressler, & A. Oemler (Cambridge: Cambridge Univ. Press), in press

Ettori, S. 2001, MNRAS, 323, 1

Ettori, S., Tozzi, P., & Rosati, P. 2003, A&A, 398, 879

Evrard, A. E. 1997, MNRAS, 292, 289

——. 2004, in Carnegie Observatories Astrophysics Series, Vol. 3: Clusters of Galaxies: Probes of Cosmological Structure and Galaxy Evolution, ed. J. S. Mulchaey, A. Dressler, & A. Oemler (Cambridge: Cambridge Univ. Press), in press

Evrard, A. E., & Henry, J. P. 1991, ApJ, 383, 95

Fabian, A. C., et al. 2000, MNRAS, 318, L65

Fabian, A. C., Sanders, J. S., Crawford, C. S., & Ettori, S. 2003, MNRAS, 341, 729

Ford, H. C., et al. 1998, in Space Telescopes and Instruments V, ed. P. Y. Bely & J. B. Breckinridge, Proc. SPIE, 3356, 234

Franx, M. 2004, in Carnegie Observatories Astrophysics Series, Vol. 3: Clusters of Galaxies: Probes of Cosmological Structure and Galaxy Evolution, ed. J. S. Mulchaey, A. Dressler, & A. Oemler (Cambridge: Cambridge Univ. Press), in press

Frenk, C. S., et al. 1999, ApJ, 525, 554

Gioia, I. M., Henry J. P., Maccacaro, T., Morris, S. L., Stocke, J. T., & Wolter, A. 1990, ApJ, 356, L35

Gioia, I. M., Henry, J. P., Mullis, C. R,, Ebeling, H., & Wolter, A. 1999, AJ, 117, 2608

Gioia, I. M., Henry, J. P., Mullis, C. R,, Voges, W., & Briel, U. G. 2001, ApJ, 553, L109

Gladders, M. D. 2004, in Carnegie Observatories Astrophysics Series, Vol. 3: Clusters of Galaxies: Probes of Cosmological Structure and Galaxy Evolution, ed. J. S. Mulchaey, A. Dressler, & A. Oemler (Cambridge: Cambridge Univ. Press), in press

Gladders, M. D., & Yee, H. K. C. 2000, AJ, 120, 2148

Henry, J. P. 2000, ApJ, 534, 565

Holden, B., Stanford, S. A., Rosati, P., Eisenhardt, P., & Tozzi, P. 2004, in Carnegie Observatories Astrophysics Series, Vol. 3: Clusters of Galaxies: Probes of Cosmological Structure and Galaxy Evolution, ed. J. S. Mulchaey, A. Dressler, & A. Oemler (Pasadena: Carnegie Observatories, http://www.ociw.edu/ociw/symposia/series/symposium3/proceedings.html)

Holden, B., Stanford, S. A., Squires, G. K., Rosati, P., Tozzi, P., Eisenhardt, P., & Spinrad, H. 2002, AJ, 124, 33

Huo, Z.-Y., Xue, S.-J., Xu, H., Squires, G., & Rosati, P. 2004, AJ, submitted

Jeltema, T. E., Canizares, C. R., Bautz, M. W., Malm, M. R., Donahue, M., & Garmire, G. P. 2001, ApJ, 562, 124

Jones, L. R, et al. 2004, in Carnegie Observatories Astrophysics Series, Vol. 3: Clusters of Galaxies: Probes of Cosmological Structure and Galaxy Evolution, ed. J. S. Mulchaey, A. Dressler, & A. Oemler (Pasadena: Carnegie Observatories, http://www.ociw.edu/ociw/symposia/series/symposium3/proceedings.html)

Jones, L. R, Scharf, C., Ebeling, H., Perlman, E., Wegner, G., Malkan, M., & Horner, D. 1998, ApJ, 495, 100

Kurk, J. D., Pentericci, L., Röttgering, H. J. A., & Miley, G. K. 2004, MNRAS, in press

Leibundgut, B. 2001, ARA&A, 39, 67

Lidman, C., et al. 2004, in preparation

Margoniner, V., et al. 2004, in Carnegie Observatories Astrophysics Series, Vol. 3: Clusters of Galaxies: Probes of Cosmological Structure and Galaxy Evolution, ed. J. S. Mulchaey, A. Dressler, & A. Oemler (Pasadena: Carnegie Observatories, http://www.ociw.edu/ociw/symposia/series/symposium3/proceedings.html)

Markevitch, M. 1998, ApJ, 504, 27

Markevitch, M., et al. 2000, ApJ, 541, 542

Maughan B. J., Jones, L. R., Ebeling, H., Perlman, E., Rosati, P., Frye C., & Mullis, C. R. 2003, MNRAS, 587, 589

Mulchaey, J. S. 2004, in Carnegie Observatories Astrophysics Series, Vol. 3: Clusters of Galaxies: Probes of Cosmological Structure and Galaxy Evolution, ed. J. S. Mulchaey, A. Dressler, & A. Oemler (Cambridge: Cambridge Univ. Press), in press

Mullis, C., et al. 2004, in preparation

Mushotzky, R. F. 2004, in Carnegie Observatories Astrophysics Series, Vol. 3: Clusters of Galaxies: Probes of Cosmological Structure and Galaxy Evolution, ed. J. S. Mulchaey, A. Dressler, & A. Oemler (Cambridge: Cambridge Univ. Press), in press

Peacock, J. A., et al. 2001, Nature, 410, 169

Pierpaoli, E., Borgani, S., Scott, D., & White, M. 2003, MNRAS, 342, 163

Pierpaoli, E., Scott, D., & White, M. 2001, MNRAS, 325, 77

Pierre, M., et al. 2004, A&A, submitted

Ponman, T. J., Cannon, D. B., & Navarro, J. F. 1999, Nature, 397, 135

Reiprich, T. H., & Böhringer, H. 2002, ApJ, 567, 716

Romer, K., Viana, P., Liddle, A. R., & Mann, R. G. 2001, ApJ, 547, 594

Rosati, P., et al. 2004, in preparation

Rosati, P., Borgani, S., & Norman, C. 2002, ARA&A, 40, 539

Rosati, P., Della Ceca, R., Burg, R., Norman, C., & Giacconi, R. 1998, ApJ, 492, L21

Rosati, P., Stanford, S. A., Eisenhardt, P. R., Elston, R., Spinrad, H., Stern, D., & Dey, A. 1999, AJ, 118, 76

Scharf, C., et al. 2004a, in preparation

Scharf, C., Smail, I., Ivison, R. J., & Bower, R. G. 2004b, in Carnegie Observatories Astrophysics Series, Vol. 3: Clusters of Galaxies: Probes of Cosmological Structure and Galaxy Evolution, ed. J. S. Mulchaey, A. Dressler, & A. Oemler (Pasadena: Carnegie Observatories, http://www.ociw.edu/ociw/symposia/series/symposium3/proceedings.html)

Schechter, P. L. 1976, ApJ, 203, 297

Spergel, D. N., et al. 2003, ApJS, 148, 175

Stanford, S. A, Holden, B. P., Rosati, P., Eisenhardt, P. R., Stern, D., Squires, G., & Spinrad, H. 2002, AJ, 123, 619

Stanford, S. A., Holden, B., Rosati, P., Tozzi, P., Borgani, S., Eisenhardt, P. R., & Spinrad, H. 2001, ApJ, 552, 504

Steidel, C. C., Adelberger, K. L., Dickinson, M., Giavalisco, M., Pettini, M., & Kellogg, M. 1998, ApJ, 492, 428

Steidel, C. C., Adelberger, K. L., Shapley, A. E., Pettini, M., Dickinson, M., & Giavalisco, M. 2000, ApJ, 532, 170

Sunyaev, R. A., & Zel'dovich, Ya. B. 1972, Comments Astrophys. Space Phys., 4, 173

Tozzi, P., & Norman, C. 2001, ApJ, 546, 63

Tozzi, P., Rosati, P., Ettori, S., Borgani, S., Mainieri, V., & Norman, C. 2003, ApJ, 593, 705

Treu, T. 2004, in Carnegie Observatories Astrophysics Series, Vol. 3: Clusters of Galaxies: Probes of Cosmological Structure and Galaxy Evolution, ed. J. S. Mulchaey, A. Dressler, & A. Oemler (Cambridge: Cambridge Univ. Press), in press

van Dokkum, P. G., & Franx, M. 2001, ApJ, 553, 90

van Dokkum, P. G., & Stanford, S. A. 2003, ApJ, 585, 78

van Dokkum, P. G., Stanford, S. A., Holden, B. P., Eisenhardt, P. R., Dickinson, M., & Elston, R. 2001, ApJ, 552, L101

Venemans, B. P., et al. 2002, ApJ, 569, L11

Vikhlinin, A., et al. 2003, ApJ, 590, 15

Vikhlinin, A., VanSpeybroeck, L., Markevitch, M., Forman, W. R., & Grego, L. 2002, ApJ, 578, L107

Voit, G. M., Bryan, G. L., Balogh, M. L., & Bower, R. G. 2002, ApJ, 576, 601

White, S. D. M., Navarro, J. F., Evrard, A. E., & Frenk, C. S. 1993, Nature, 366, 429

Yee, H. K. C., Ellingson, E., & Carlberg, R. G. 1996, ApJS, 102, 269

6

The red sequence technique and high-redshift galaxy clusters

MICHAEL D. GLADDERS

The Observatories of the Carnegie Institution of Washington

Abstract

The advent of large-format CCD cameras on 4 m-class telescopes has led to a resurgence of interest in using optical and near-IR images for detecting galaxy clusters over a range of redshifts. Parallel to the development of new observing strategies has been the development of a number of sophisticated algorithms to analyze these imaging data in order to optimally select real clusters. One of the most successful algorithms, particularly at high redshifts, has been the so-called red sequence method. This method keys on the observational fact that all clusters possess a population of passively evolving early-type galaxies that form an identifiable color sequence in a color-magnitude diagram. This paper reviews the recent history of the development and application of red sequence methods, with particular focus on results from the Red-Sequence Cluster Survey (RCS). This survey, currently the largest moderately deep imaging survey designed to use this method, is particularly useful since its volume and redshift depth are comparable to those of the largest distant cluster surveys done at other wavelengths.

6.1 Cluster Surveys

The utility of optical imaging data for surveying large areas for galaxy clusters was recognized early on, and is epitomized by the work of Abell (1958). Abell's initial ground-breaking cluster catalog spurred an enormous amount of research on galaxy clusters, motivating both a wide array of other surveys (see Bahcall 1977 for a more complete list of early optical cluster catalogs) and a great deal of detailed research on cluster properties. More recently, X-ray surveys have come to be recognized as the standard method for galaxy cluster surveys, particularly for more distant clusters (e.g., the Extended Medium Sensitivity Survey cluster sample, Gioia et al. 1990; the *ROSAT* Distant Cluster Survey, Rosati et al. 1998, to name but a few among many examples). X-ray surveys are particularly attractive since the observational signature of a galaxy cluster is unique at X-ray wavelength. This results quite directly in remarkably clean, uncontaminated cluster catalogs (the dominant source of contamination being blends of unresolved X-ray point sources; see Vikhlinin et al. 1998 for a discussion). The selection functions, and hence the effective survey volume, can be readily quantified in X-ray surveys and are robust (e.g., Adami et al. 2000). These two properties, low contamination and well-determined selection functions, are of critical importance in large surveys, since the statistical analyses envisioned for future clusters samples (e.g., Haiman, Mohr, & Holder 2001; Levine, Schulz, & White 2002) are expected to be dominated by systematic uncertainties. The known projection-induced contamination of early, optically selected cluster catalogs (e.g., Lucey 1983; van Haarlem, Frenk, & White 1997), and difficulties in a ready quantification of selection functions for such surveys, led to these surveys falling out of favor coincident with the advent of X-ray methods. However, the development of large-format CCD cameras and sophisticated analysis methods has begun to reverse this trend.

The first significant effort to use CCDs and a sophisticated cluster-finding algorithm in order to locate distant galaxy clusters was the Palomar Distant Cluster Survey (PDCS: Postman et al. 1996). The PDCS used the matched-filter algorithm, in which information on the radial position and luminosity distribution of galaxies is used to discriminate projections from real clusters. A number of authors have since revised or extended the initial algorithm of Postman et al., and current variants include the adaptive matched filter (Kepner et al. 1999) and the hybrid matched filter (Kim et al. 2002). The matched-filter method has been demonstrated to be most successful at lower redshifts, via application to Sloan Digital Sky Survey (SDSS) data. There has been little application at $z > 0.5$, and little confirmation of initial PDCS cluster candidates at these distant redshifts. This result is not unexpected; the change in $m - M^*$ with redshift is most pronounced at lower redshifts, and hence one generally expects the luminosity filter to be most effective at lower redshifts.

An alternate algorithm was suggested by Gladders & Yee (2000). This method, dubbed the cluster red sequence (CRS) algorithm, and described in detail below, implicitly uses the concentration of cluster galaxies at a particular color as a tracer. Variants of this basic method have since been outlined by other authors, mostly in the context of the SDSS. These color-based methods include the "cut-and-enhance" algorithm of Goto et al. (2002) and the "MaxBCG" method of Annis et al. (1999). All use color information in some way to discriminate real structures from projections. Comparisons of some of these algorithms as applied to the SDSS show that they perform comparably well and similarly to matched-filter methods (Kim et al. 2002) at low redshifts.

For more distant clusters, efficiency of data acquisition becomes paramount (since cluster galaxies get very faint), and so methods that rely on multiple-filter data (such as the "C4" algorithm; see Nichol 2004) are likely to be less attractive. In this regard, the matched-filter methods may be the most attractive, since in principle imaging in only a single filter is required. However, in practice, most matched-filter surveys have used imaging in at least two filters, with results cross-checked between different wavelengths (e.g., Postman et al. 1996; Olsen et al. 2001), with the only exception being the Deeprange Survey of Postman et al. (2002). Moreover, considerations for the secondary use of wide-field imaging data often make at least one more filter preferable.

Arguably the most efficient optical method, at least in identification of initial candidates, is the background light fluctuation method of Dalcanton (1996), as used in the Las Campanas Distant Cluster Survey (González et al. 2001). In this method clusters are initially identified as unresolved enhancements in the extragalactic background light, which can then be followed up using modest aperture telescopes to acquire deeper direct imaging to gather photometric information and estimate redshifts. However, the advent of large-field mosaic cameras, large enough that multiple clusters are seen in any one image, mitigates this advantage significantly. With large cameras it is more efficient to gather the deeper data directly in survey mode.

6.2   The CRS Method in Detail

The CRS algorithm detailed in Gladders & Yee (2000) relies on two–filter imaging data to select clusters. Clusters are isolated in a 4-dimensional space of position (x, y), magnitude (m), and color (c). Using models of the expected evolution of the cluster red sequence, the algorithm specifically targets early-type galaxies. These objects offer a number of potential advantages as cluster tracers. Early-type galaxies evolve passively with redshift, and are about one magnitude brighter at $z = 1$ compared to present epoch (e.g., Schade, Barrientos, & López-Cruz 1997; van Dokkum et al. 1998). The morphology-density relation (Dressler 1980; Dressler et al. 1997) shows that the early-type galaxies are the most clustered cluster members. The presence of a red sequence is also a physical indicator of overdensity, much like the thermal bremsstrahlung emission from clusters seen in X-rays. Additionally, in good-quality imaging data, these galaxies can be morphologically selected, further enhancing their clustering signal. Early-type galaxies also dominate the bright end of the luminosity function in most clusters (e.g., Barger et al. 1998). Finally, the photometric evolution of passively evolving stellar populations is readily modeled, and hence photometric redshifts and other secondary measurements based on these galaxies are likely to be accurate.

Most importantly, the use of the CRS algorithm and appropriate filters should completely eliminate foreground projection, and mitigate background projection as well. This is illustrated in Figure 6.1, which shows the $B - R$ versus R color-magnitude diagram (CMD) for the inner 0.5 h^{-1} Mpc of $\sim$40 low-redshift, X-ray selected Abell clusters, taken from the data of López-Cruz (1997). Each cluster CMD has been k-corrected to the mean redshift of the sample, but has not been background corrected. The red sequence is obvious. As shown in Figure 6.1, because these filters span the 4000 Å break at this redshift, all foreground galaxies whose intrinsic spectra are no redder than the cluster early-type galaxies appear bluer than the red sequence in this CMD. A simple blue-side color cut thus eliminates all foreground galaxies. Background galaxies may appear either bluer or redder, but are generally fainter. The intrinsically redder (presumably early-type) galaxies at higher redshift appear redder than the cluster red sequence. Thus, a simple red-side color cut eliminates the most clustered background galaxies. The remaining galaxies, in a narrow color slice, are predominantly early-type galaxies at a particular redshift, with some contamination at fainter magnitudes by intrinsically bluer galaxies at higher redshift. This simple isolation of galaxies at a particular redshift is readily achieved at any redshift, so long as the filter pair spans the 4000 Å break.

The suggested variants of both the matched-filter and red sequence-based methods described above are often motivated by a desire to reduce the level of parameterization and hence sensitivity to model choices. In this context, it is worth noting that a mismatch between the assumed red sequence model and the actual colors of cluster galaxies does little to detection sensitivity in the CRS algorithm. Rather, such mismatches are likely to affect, in a systematic way, the estimation of cluster parameters such as the redshift. Fundamentally, the fact that any given cluster has a tighter color distribution than the field, regardless of its redshift, is what makes the CRS and similar algorithms work. Moreover, a tight red sequence is not generally required in order to detect a given cluster; even red sequences with scatters of half a magnitude (much larger than seen in any actual clusters) are detectable, because the field color scatter is even larger, and hence a color cut still enhances such a cluster against the more diverse background.

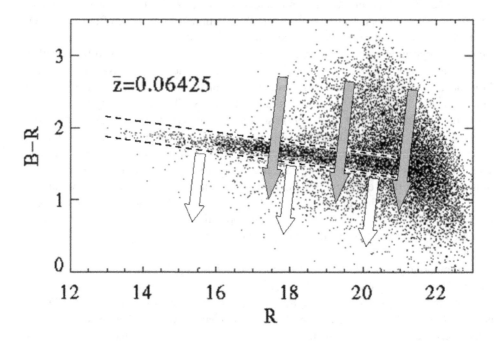

Fig. 6.1. Individual galaxies are shown as points. The red sequence of this aggregate cluster is obvious, and a narrow color slice bracketing it is shown by the dashed lines. The solid white arrows show the typical location of any normal galaxy foreground to the cluster, and the solid gray arrows similarly show the expected locations for background galaxies.

The CRS method has been tested in comparison to several other cluster-finding methods. A comparison to spectroscopically indicated structures in the CNOC2 survey (Yee et al. 2000) can be found in Gladders & Yee (2000), who showed that the method was able to delineate bound systems down to group sizes, had a contamination rate of 5% or less, even at these low mass thresholds, and provided quite precise redshift estimates. A preliminary study by Gladders & Donahue (2004) in several *ROSAT* Optical X-ray Survey (ROXS; Donahue et al. 2002) fields similarly shows that the CRS methods successfully finds all X-ray indicated structures. The most thorough analysis of the CRS method to date is by Gilbank (2003); his detailed comparison of the CRS method to both the matched-filter algorithm and X-ray selection shows that the CRS method is more complete and less contaminated by projection than the matched-filter method, and that it successfully recovers 100% of clusters that would be selected in an X-ray survey.

6.2.1   *Redshift Estimates*

Numerous authors, beginning with Visvanathan & Sandage (1977), have explored the possibility of using the details of the color-magnitude sequence as either a direct distance indicator, or more simply as a redshift indicator. This latter application is of interest in the context of cluster surveys since redshift and mass (the estimation of which usually depends

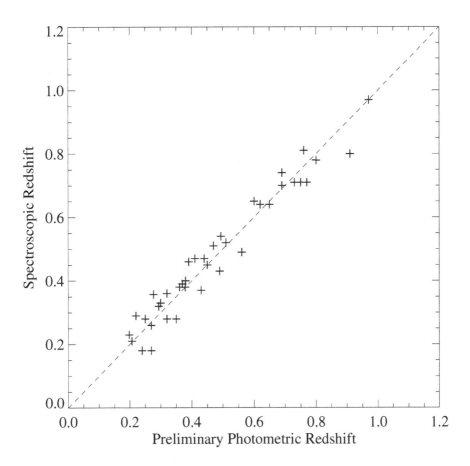

Fig. 6.2. A comparison of spectroscopic and photometric redshifts for an initial sample of 39 RCS cluster candidates. All clusters appear to be real (see the extensive discussion in Ellingson et al. 2004 for more details) and the photometric redshift uncertainty is about 0.05 across the entire sample. A restricted sample, consisting of clusters drawn only from patches with good photometric calibration, has a smaller redshift uncertainty of about 0.03.

on redshift) are the two most desirable basic properties to include in a cluster sample. The utility of the red sequence color as a redshift indicator has been noted at various redshifts. López-Cruz (1997), in a study of 45 Abell clusters mostly at $z < 0.1$, showed that the red sequence color provided a redshift estimate with an uncertainty of 0.008. Gladders & Yee (2000) similarly showed that the redshift uncertainty for poor systems at $0.12 < z < 0.55$ was 0.026. Smail et al. (1998), in a study of 10 X-ray selected clusters at $0.2 < z < 0.3$, similarly concluded that the red sequence provides an excellent and economical redshift indicator. Gilbank (2003) also examined the redshift estimates from both the matched-filter and CRS algorithms, and concluded that the CRS algorithm provides a much better redshift estimate. As shown in Figure 6.2, recent spectroscopy of clusters in the RCS (Ellingson et al. 2004) also illustrates the accuracy of the CRS redshift estimates. For clusters up to $z = 0.97$ (the highest-redshift RCS clusters with spectroscopic confirmation; see Barrientos et al. 2004) the CRS method provides redshifts with uncertainties less than 0.05, and as good as 0.03 in fields with final photometric calibration (Ellingson et al. 2004). With filters spanning the 4000 Å break at moderate redshift, the color change in magnitudes is roughly equal to the change in redshift, and hence a systematic cluster-to-cluster calibration uncertainty of only a few hundredths of a magnitude per filter (likely better than many extragalactic imaging projects) completely explains this scatter.

It is also worth noting that at higher redshifts the fundamental redshift accuracy achievable by this simple color-based method may be limited by uncertainties in the process of early-type galaxy formation. Figure 6.3 shows the likely effect of mismatches between the model used to compute redshifts and the actual scenario for formation of the cluster red sequence. Two lines are plotted, showing the systematic redshift error induced by using a simple instantaneous burst model with the burst formed at $z_f = 2.5$, when used to infer redshifts for clusters with actual formation redshifts of $z_f = 2$ and $z_f = 3$. In both cases there are systematic offsets in redshift that are less than about 0.01 at redshifts less than 0.8, increasing to offsets as large as 0.2 at the highest redshifts considered. At $z > 1$, the potential systematic redshift errors are thus likely to be at least as large as the random errors, and may well dominate. This dramatic increase in the redshift uncertainty is due to the small time difference between the observation epoch and formation epoch, and the bluing of the survey filters in the rest frame. This latter effect can be eliminated by reimaging the highest-redshift cluster candidates using redder filters. Moreover, the correct model can be chosen by calibrating to clusters with known redshifts. However, if the formation redshifts of clusters are stochastically distributed, then calibration of the photometric models to known clusters will not eliminate this extra source of uncertainty.

6.3   The Red Sequence and Cluster Confirmation

The detection of a red sequence in a cluster candidate selected by other means is arguably the observationally cheapest way to confirm the cluster identification. The data necessary to show a red sequence are modest in scope, and for even very distant systems can be readily acquired. Indeed, it is in the most distant clusters, to date typically selected by faint X-ray emission, where many of the arguments for the universality of the red sequence arise (Gladders & Yee 2000). Examples of clusters selected by a variety of means that show significant red sequences include X-ray selected clusters at $z > 1$ in the *ROSAT* Distant Cluster Survey (Rosati et al. 1999; Stanford et al. 2002; Holden et al. 2004), and $z \approx 1$ clusters selected using the matched-filter method applied to the ESO Imaging Survey (e.g., da Costa et al. 1999). The presence of an overdensity of red galaxies has been used extensively to argue for the presence of clusters around a number of distant AGNs (e.g., Dickinson 1995; Hall & Green 1998; Tanaka et al. 2000; Haines et al. 2001; Blanton et al. 2004) or to suggest the existence of clusters of galaxies associated with absorption-line complexes seen in the spectra of a background AGN (Liu et al. 2000). Lubin et al. (2000) used the presence of a spatially extended red sequence population to identify large-scale structure, on the super-cluster scale, associated with the $z \approx 0.91$ clusters Cl 1604+4304 and Cl 1604+4321.

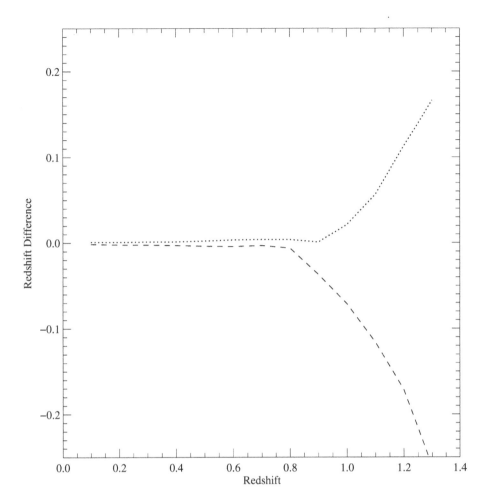

Fig. 6.3. The lines show the expected systematic redshift errors for mismatches between the actual red sequence models as compared to the fiducial model used in the CRS algorithm. The difference in estimated redshifts for $z_f = 2$ (dashed line) and $z_f = 3$ (dotted line) models when compared to the $z_f = 2.5$ model is illustrated.

6.4 The Red Sequence Cluster Survey

6.4.1 Survey Data and Design

The Red-Sequence Cluster Survey (RCS: Gladders 2002; Gladders & Yee 2004) is a $\sim$100 square degree R_C and z' imaging survey designed to detect and characterize galaxy clusters out to the redshift limits allowed by standard CCDs. In practice, the quickly falling response of CCDs at near-infrared wavelengths limits the bulk of the RCS cluster sample to below $z = 1.1$, with a tail of cluster candidates out to $z = 1.4$. The imaging for the RCS was carried out at the Canada-France-Hawaii Telescope (CFHT) from mid-1999 to early-2001, and at the Cerro Tololo Interamerican Observatory's (CTIO) 4 m telescope from early-2000 to late-2001. In both cases observations were made with large-format mosaic cameras—the

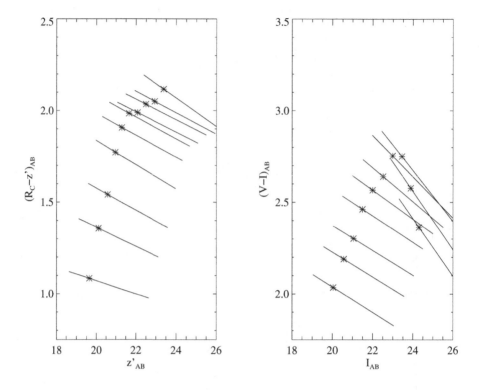

Fig. 6.4. Fiducial red sequences for cluster early-type galaxies at $0.5 < z < 1.4$, in steps of 0.1 in redshift, using the RCS R_C and z' filter pair (*left*) and the V and I filter pair (*right*) often used in previous optical cluster surveys. The star symbols indicate the location of M^* at each redshift.

CFH12K at CFHT (Cuillandre et al. 2000) and MOSAIC-II on the CTIO 4 m. Both data sets are of similar depth in both filters—with integration times anywhere from 12 to 25 minutes per filter—with 5 σ point source limits typically being two magnitudes past M^* at $z = 1$, for an early-type galaxy model with passive evolution. The CFHT data mostly have excellent seeing of $0\rlap{.}''6$–$0\rlap{.}''8$; the CTIO data are of more typical ground-based image quality, but similar in depth due to the larger aperture of the telescope and increases in exposure time. The RCS required, in total, 25 clear nights to complete, making it a remarkably efficient cluster survey given its total volume.

The RCS data were designed to optimally and efficiently sample the galaxy cluster population at moderate to high redshifts. As shown in Figure 6.4, the redshifts of early-type galaxies in these clusters are much better sampled by the R_C and z' filters, compared to the typical choice of V and I filters often used in other surveys (e.g., Postman et al. 1996; Donahue et al. 2002). On the other hand, the RCS redshift estimates at lower redshifts (less than $z = 0.5$) are less accurate than can be achieved using a bluer filter pair. Compare, for example, the redshift uncertainty of $\sim$0.05 seen in Figure 6.2 with the uncertainty of $\sim$0.026 found in an application of the CRS algorithm to CNOC2 $V - I$ data by Gladders &

Table 6.1. *Parameters for Simulated Clusters in RCS Tests*

Parameter	Symbol	Model Values	Notes
cluster redshift	z	0–1.4	
Abell Richness counts	N_{Abell}	35,44,56,72,93,120	Richness Classes 0–2
blue fraction	f_b	0.1,**0.45**,0.65,0.8,0.9	
NFW core scale radius	r_s	0.1,0.2,**0.3**,0.4,0.5	in h^{-1} Mpc
ellipticity	ϵ	0.0,0.2,**0.4**,0.6,0.8	ϵ at 1 h^{-1} Mpc
red seq. formation redshift	z_{RS}	2.0,**2.5**,3.0	lower limit in redshift
scatter in age prior to z_{RS}	Δt_{RS}	0.5,**1.0**,max	top-hat width in Gyr
LF R-band M^*	M_R^*	−22.75, **−22.25**, −21.75	$\alpha = -1.0$, $H_0 = 50$

The range or set of values used is indicated, with the "typical" value (if there is one) in boldface type.

Yee (2000). Recall, however, that in either filter pair two-filter photometric redshifts derived from colors are significantly more accurate than other methods.

6.4.2   Selection Functions

It has been argued for years that one of the attractions of X-ray selected samples is the relative simplicity of the selection functions for X-ray clusters. Landmark efforts to thoroughly model X-ray cluster selection functions range from the resampling treatment of Rosati et al. (1995), to the extensive Monte Carlo treatment of Vikhlinin et al. (1998), with the extensive and now complex simulations of the SHARC Survey by Adami et al. (2000) being the latest standard in this work. By no means are the Adami et al. simulations simple, and recent data on various X-ray selected samples (e.g., Lewis et al. 2002; Jones et al. 2004) clearly show that the complexity of the Adami et al. simulations is required to fully understand the selection functions for distant X-ray cluster populations.

In tandem with this realization of the actual complexity of X-ray cluster selection has been a growth in simulation methods to address the long-acknowledged complexity of optical selection. Complications in optical selection are obvious; depending on the method, optical surveys may be sensitive to details of the luminosity function, the radial galaxy distribution, the cluster ellipticity, details of the galaxy population, and evolution of these properties across both redshift and mass. This is in addition to selection effects induced by the characteristics of the observations, which are generally more complex (factoring in different filters, seeing, image depths, variations in depths, etc.) than the observational effects in X-ray data (primarily flux limits and spatial point-spread function variations). Nevertheless, several large surveys have now completed extensive modeling of optical selection functions. One notable example is the modeling of SDSS clusters sample, discussed in Nichol (2003). The particular advantage offered by the SDSS is that the "mock" catalogs used to measure the selection effects are very well constrained by the extensive SDSS data, and hence offer a thorough and robust platform in which to explore the effects of variations in a host of cluster properties, as well as variations due to different cluster-finding algorithms. At these redshifts the optical selection functions are at least as well constrained as X-ray selection functions.

A similar effort to model the RCS selection functions, using only the CRS algorithm (Gladders 2002; Gladders & Yee 2004), is illustrated in Figure 6.5. This shows a synopsis of the predicted detection rates of a few sets of typical clusters: both moderately poor and

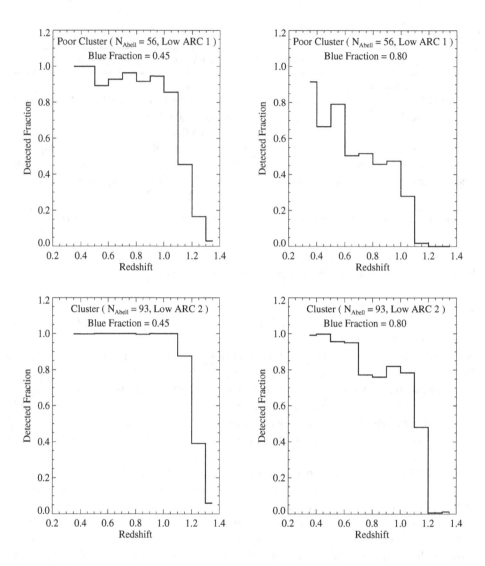

Fig. 6.5. The four panels show the recovered fraction of clusters in simulations for four representative clusters types, as indicated on each panel. Parameters not mentioned assume the standard value indicated in Table 6.1. Uncertainties due to the limited number of simulations are of order 5%–10%, as suggested by the variance in the plots.

rich clusters, with both moderate and high blue fractions, over the redshift range $0.3-1.4$. These simulations explored variations in a number of parameters possibly relevant for cluster detection; a listing of the parameters, as well as the range explored in each, can be found in Table 6.1. The variation in these parameters was chosen to more than bracket all existing observations of clusters, and hence generally include extremal models (very rich systems at low redshift with high blue fractions, for example), which do not appear to exist in the Universe. Clusters were embedded in realistic, evolving field galaxy models deduced from

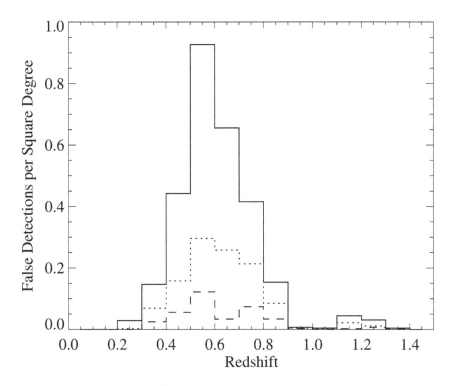

Fig. 6.6. The redshift histograms show the number of false-positive detections, per square degree, for selection thresholds of 3.0, 4.0, and 4.5 σ.

a combination of the CNOC2 survey's intermediate-redshift luminosity function (Lin et al. 1999) and a host of other magnitude, redshift, and color distributions from other surveys (see Gladders 2002 and Gladders & Yee 2004 for details). The field galaxies were spatially arranged to produce the observed CNOC2 correlation functions (Shepherd et al. 2001) using the simple algorithm of Soneira & Peebles (1978), but do not reflect in detail the higher-order clustering of clusters seen in the real Universe.

A simple synopsis of these extensive tests is that the RCS is expected to be 100 percent complete for galaxy clusters of richnesses equivalent to Abell Richness Class (ARC) 1 or greater (or, equivalently, of velocity dispersions ≥ 700 km s^{-1}; Yee & Ellingson 2003) to redshifts up to $z \approx 1.1$, with blue fractions as high as 0.65, regardless of the details of the spatial arrangement of the cluster. Yet higher blue fractions, generally higher than almost all observations, introduce some incompleteness above $z \approx 0.7$, for clusters with richnesses up to ARC 2; for the most massive clusters, with velocity dispersions in excess of 1000 km s^{-1}, the RCS should be complete to at least $z = 1.1$ regardless of all details of the galaxy population, provided that the optical mass-to-light ratios are within reason.

The false-positive rates for the RCS derived from these simulations are shown in Figure 6.6. Results are shown for three different significance cuts; for comparison, the significance cut used previously in the selection functions in Figure 6.5 is 3 σ. Based on initial

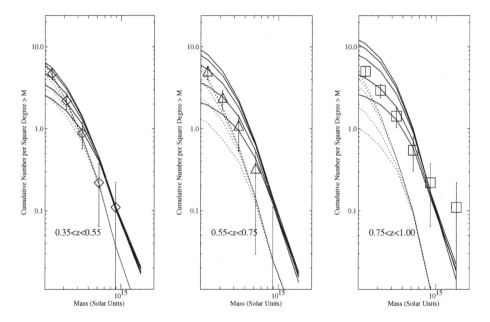

Fig. 6.7. The three panels show the deduced RCS cumulative cluster mass function in three redshift intervals, as indicated, from 2×10^{15} to 2×10^{14} M$_\odot$. The lower mass limit corresponds to the low end of Abell Richness Class 0. Solid lines show the predicted mass functions for the ΛCDM Virgo simulations, for various cluster blue fractions (0.1, 0.45, 0.65, 0.8 and 0.9). In general the data match the ΛCDM expectation quite well, particularly considering that the blue fraction is expected to evolve to higher values at higher redshifts. The dotted lines show the predicted mass functions for the τCDM Virgo simulations. τCDM is a much poorer fit to the data.

analyses of the RCS data, the false-positive rate at the 3 σ cut translates to about 5% contamination in real data. Note that this does not include contamination due to nearby projected structures, i.e. multiple groups along the same line of sight, with redshift separations less than the redshift uncertainty (see Fig. 6.2). However, even at twice the quoted contamination, the RCS has contamination rates competitive with X-ray surveys (e.g., Vikhlinin et al. 1998).

6.4.3   The RCS Mass Function

A preliminary measurement of the RCS mass function, using two of 20 survey patches and a galaxy richness-mass calibration derived from the CNOC1 survey (Yee & Ellingson 2003), is shown in Figure 6.7. RCS data are shown with random errors only, and are compared to an estimate of the mass distribution derived from the Virgo Hubble Volume simulations* (Evrard et al. 2002).

To make the comparison, we have taken the catalog of simulated cluster redshifts and velocity dispersions from the Virgo simulations and applied the RCS selection functions to

---

* The simulations used in this paper were carried out by the Virgo Supercomputing Consortium using computers based at the Computing Centre of the Max-Planck Society in Garching and at the Edinburgh Parallel Computing Centre. The data are publicly available at http://www.mpa-garching.mpg.de/NumCos.

predict which of the simulated clusters would be detected in a survey similar to the RCS. The richness-σ_1 relation deduced from the CNOC1 catalog (richness parameterized here by B_{gc}; see Yee & Ellingson 2003) is used to produce a B_{gc} value for each simulated cluster that is detected. The observed run of B_{gc} errors with redshift and B_{gc} in the real data is then used to randomly perturb the simulation B_{gc} values. Finally, we convert both the real and simulated B_{gc} measurements to virial masses, again using the relation from CNOC1. This last change is essentially cosmetic and serves simply to translate richness to more familiar masses.

The RCS data agree well with the predictions for a flat ΛCDM cosmology with $\Omega_m = 0.3$ and disagree with expectations for a $\Omega_m = 1.0$ τCDM cosmology. This is not unexpected given the wealth of other indicators favoring the former model (e.g., Spergel et al. 2003, and references therein). Note that in this comparison the range of redshifts considered ($0.35 < z < 1.0$) is well beyond the redshift range of the mass-calibrating sample ($0.15 < z < 0.55$), and so conclusions about cosmology based on the higher redshift evolution of the mass function must be made with care. Work is ongoing to extend the CNOC1 calibration to $z = 1$ using RCS clusters.

6.4.4   RCS High-redshift Clusters

Little is currently known about galaxy clusters at redshifts beyond 1. Detailed descriptions of several of the most notable clusters at $z \geq 1$ can be found in, for example, Holden et al. (2004) and Rosati (2004), but the current sample size of fewer than a dozen known clusters at these redshifts limits the global conclusions that can be drawn from these investigations. One of the primary goals of the RCS is to enable a more thorough study of high-redshift clusters by providing samples of hundreds, rather than handfuls, of target systems. An ongoing project using the Las Campanas telescopes is following up the 100 best RCS high-redshift cluster candidates. The basic data are deep I-band imaging with the Magellan telescopes, coupled to K-band imaging of matching depth from the du Pont Telescope. The observations are designed to be both deeper and redder than the initial RCS survey data.

An initial analysis of the first 12 candidates imaged in both filters yields the following basic results.

(1) The RCS false-positive rate at $z \geq 1$ is 10%–20%. This is somewhat higher than the nominal contamination rates discussed above, but not unexpected given recent observations of strong clustering in extremely red galaxies with colors corresponding to early-type galaxies at high redshift (Daddi et al. 2002). This strong evolution of the red galaxy correlation function was not included in the background models, and would produce more false-positive detections.

(2) The red sequence galaxies in the apparently real clusters display much greater photometric variation than is observed in the few massive clusters studied at similar redshifts (Stanford, Eisenhardt, & Dickinson 1998; van Dokkum et al. 2000) and display scatter more comparable to that seen in cluster candidates associated with AGNs (e.g., Tanaka et al. 2000). The richest and hence most massive clusters have photometric properties similar to that seen in clusters in the literature (see Barrientos et al. 2004) and similarly small color scatter, but poorer systems often show much larger scatter. A pair of CMDs for two clusters at $z \approx 1$ is shown in Figure 6.8; one cluster is massive (see the discussion on strong lensing below) and has a low scatter of ≤ 0.1 magnitudes in $I - K$, while the other cluster, of about half the richness, has a scatter of ~ 0.25 magnitudes. Under the standard interpretation (e.g., Bower,

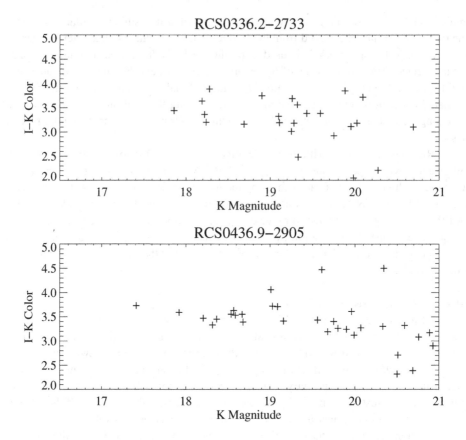

Fig. 6.8. The $I-K$ CMDs of two RCS clusters at $z \approx 1$ are shown. The cluster at the top is poorer and has a red sequence with a larger color scatter than the richer cluster at the bottom.

Lucey, & Ellis 1992), in which the color scatter is related to a combination of star formation synchronicity between galaxies and the age of galaxies, this indicates that the galaxies in the less massive clusters are either younger or have a more diverse star formation history. This has not been seen in prior studies, likely because the available targets at $z \geq 1$ from previous clusters surveys were limited to the few most massive clusters in a given survey volume. Notably, these results are broadly consistent with expectations from hierarchical clustering models, in which the seeds of galaxies in the densest portions of the Universe are thought to collapse first.

(3) The reddest red sequences seen to date are approximately one magnitude redder in $I-K$ than the spectroscopically confirmed $z = 0.97$ cluster RCS0439.6–2905 (Barrientos et al. 2004), and hence the RCS sample includes cluster candidates with photometric redshifts as distant as $z = 1.4$. Only a handful of such ultra-distant systems are expected in the 100 clusters follow-up, based on the selection functions discussed above.

Finally, note that initial analysis of *Chandra* data (discussed in detail in Hicks et al. 2004) on a variety of RCS clusters with redshifts estimated or spectroscopically confirmed to be in the range $0.65 < z < 1.3$ indicates some X-ray emission in all cases. Though these clusters are often underluminous in X-rays with respect to more local X-ray-optical correlations derived from X-ray selected cluster samples, the presence of X-rays in all cases verifies the efficacy of the CRS method, even at high redshift.

6.4.5   *RCS Strong Lensing Clusters*

One of the most striking initial results of the RCS is the large number of high-redshift lensing clusters seen in the survey (Gladders et al. 2003; Gladders, Yee, & Ellingson 2002). A total of five clusters with one or more arcs are seen in RCS survey data, with an additional three strong lensing clusters selected from the 100 cluster follow-up of $z \geq 1$ cluster candidates discussed above. Images of lensing clusters from both the primary and secondary samples are shown in Figures 6.9 and 6.10. Apart from the clear verification that the RCS does detect massive clusters, this sample of strong lensing clusters is particularly notable for two reasons. First, the redshift distribution of the primary sample is skewed toward higher redshifts, with all five clusters at $0.64 < z < 0.87$. Standard considerations for lensing by clusters leads to the expectation that most strong lens clusters should be at lower redshifts, since clusters at lower redshift are thought to be more massive and are geometrically favored for lensing. Thus, the absence of RCS lensing clusters at $z < 0.64$ indicates that something in addition to mass, likely associated with cluster formation, is boosting the lensing cross sections of these systems. Secondly, in both the primary and secondary sample there is a high proportion of multiple-arc clusters, again implying that the systems that do have lensing have large lensing probabilities. The physical cause of this lensing boost is at present unclear.

The redshift distribution seen in the RCS lenses is also interpretable as a very clear detection of evolution in the cluster population, albeit in a property other than simply mass. Evolution has been notoriously hard to detect in X-ray surveys (e.g., Holden et al. 2004, and references therein). This points to one of the potential advantages and complications of optical cluster surveys—that they tend to find a broad range of systems, not just those with strong X-ray emission. Or, put another way, optical cluster surveys probe much deeper into the mass function than X-ray surveys. When trying to understand questions of cluster evolution, this is a desirable feature.

One final implication of this result is that clusters with arcs likely represent a sample that is structurally biased. That is, the density profiles of such systems are likely not typical, and this makes it questionable whether detailed studies of individual lens systems (e.g., Treu 2004) can be used to probe clusters in general.

6.5   Summary and Future Directions

The future of optical surveys for galaxy clusters looks bright. At low redshifts the SDSS is providing an wealth of detailed information that will serve to anchor surveys toward higher redshifts. Multiple methods for filtering optical data to search for real clusters have been tested at lower redshifts, and many appear to work well. At higher redshifts the most successful method to date appears to be the CRS method, which keys on the excess of red galaxies seen in all clusters studied so far.

Photometric redshift estimates for clusters using only two filters are demonstrated to work well over the redshift range $0 < z < 1$. Testing of this in detail at higher redshifts awaits a larger sample of clusters with spectroscopy. At high redshifts, systematic effects due to variations in formation histories may be the limiting factor for this simple redshift estimator, and

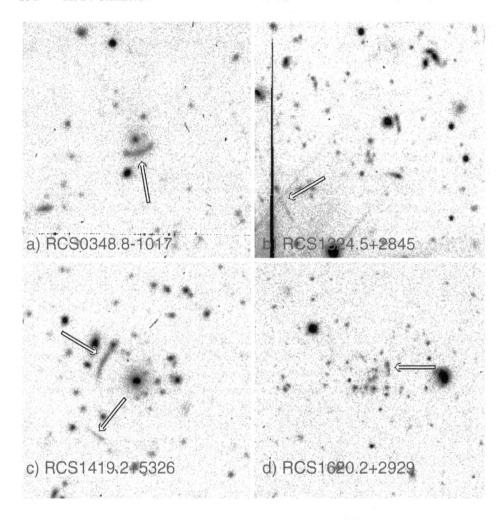

Fig. 6.9. The four panels show greyscale R_C-band images of the central $1' \times 1'$ of the four lensing clusters in the primary sample. A fifth cluster is described in detail in Gladders et al. (2002). The putative arc features are indicated by an arrow in each case.

at low redshifts the achievable accuracy is limited by systematic uncertainties in photometric calibration.

Calculations of selection functions for optically selected clusters are now as developed as those for X-ray selection. At higher redshifts, a lack of detailed data on many clusters sets some limits on the robustness of these calculations (a point that can be argued for X-ray selection as well), though new investigations will surmount this limitation shortly. Lingering concerns about the robustness of red sequence-based methods for finding high-blue fraction clusters have not yet been thoroughly addressed, though the computed selection functions include these effects and indicate that large blue fractions are only significant (in the RCS at least) for less massive systems with blue fractions in excess of 0.65—larger than indicated by most data.

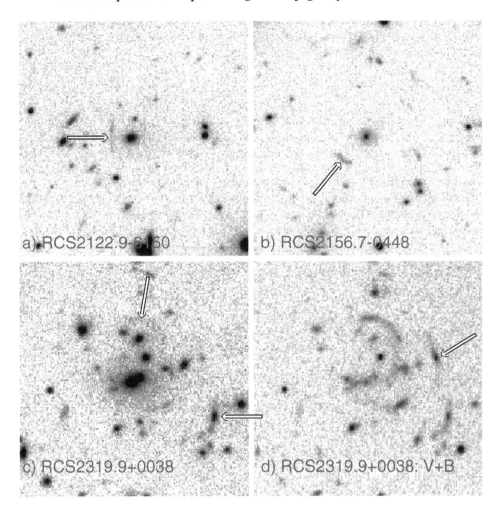

Fig. 6.10. Panels (*a*)–(*c*) show greyscale *I*-band images of the central $30'' \times 30''$ of the three lensing clusters in the secondary sample. The putative arc features are indicated by an arrow in each case. Panel (*d*) shows RCS2319.9+0038 in *V* and *B* light, from a summed image in which the arcs have similar signal-to-noise ratio in each filter.

The low contamination and high completeness of red sequence-based methods suggested by simulations is well borne out by observations. Cross comparison of X-ray selection and red sequence selection demonstrates that no X-ray selected clusters would be missed by this form of optical selection. Follow-up of the RCS, even to the highest redshifts in the RCS catalogs ($z \approx 1.4$) shows contamination rates comparable to X-ray selection.

Clusters have for some time been one of the prime targets for cosmological investigations, because the growth of clusters is thought to be extremely sensitive to parameters such as Ω_m and σ_8. Current discussions in the literature suggest that extremely large-area, high-redshift cluster surveys are needed in order to answer the most pressing cosmological questions. The demonstrated ability of red sequence-based cluster surveys to effectively find and characterize a clean cluster sample to very high redshifts makes such surveys an excellent tool for addressing these cosmological questions. To this end, a second-generation red sequence survey, envisioned as a 1000 square degree survey with selection functions similar to the RCS, is now underway at the CFHT. Coincident surveys at other wavelengths are highly desirable, and we plan to make the field locations for this next survey available soon in order to encourage other investigators to overlap with this survey. Unresolved questions regarding the efficacy of different methods for cluster selection, and, more importantly, questions regarding the correlations between mass-related observables, are best addressed using large, overlapping survey areas across a broad range of wavelengths.

Acknowledgements. It is a great pleasure to thank my RCS collaborators for their valuable insights and support during the RCS project. In particular, I am grateful to Howard Yee, Felipe Barrientos, Erica Ellingson, and Pat Hall for their strong involvement in this large project, without which it simply would not have happened. I am also grateful to my collaborators for allowing me to discuss several new RCS results in this review, prior to their publication elsewhere.

References

Abell, G. O. 1958, ApJS, 3, 211

Adami, C., Ulmer, M. P., Romer, A. K., Nichol, R. C., Holden, B. P., & Pildis, R. A. 2000, ApJS, 131, 391

Annis, J., et al. 1999, BAAS, 195, 1202

Bahcall, N. A. 1977, ARA&A, 15, 505

Barger, A. J., et al. 1998, ApJ, 501, 522

Barrientos, L. F., Manterola, M. C., Gladders, M. D., Yee, H. K. C., Infante, L., Hall, P., & Ellingson, E. 2004, in Carnegie Observatories Astrophysics Series, Vol. 3: Clusters of Galaxies: Probes of Cosmological Structure and Galaxy Evolution, ed. J. S. Mulchaey, A. Dressler, & A. Oemler (Pasadena: Carnegie Observatories, http://www.ociw.edu/ociw/symposia/series/symposium3/proceedings.html)

Blanton, E., Gregg, M. D., Helfand, D. J., Becker, R. H., & White, R. L. 2004, in Carnegie Observatories Astrophysics Series, Vol. 3: Clusters of Galaxies: Probes of Cosmological Structure and Galaxy Evolution, ed. J. S. Mulchaey, A. Dressler, & A. Oemler (Pasadena: Carnegie Observatories, http://www.ociw.edu/ociw/symposia/series/symposium3/proceedings.html)

Bower, R. G., Lucey, J. R., & Ellis, R. S. 1992, MNRAS, 254, 589

Cuillandre, J.-C., Luppino, G. A., Starr, B. M., & Isani, S. 2000, SPIE, 4008, 1010

da Costa, L. N., et al. 1999, A&A, 343, L29

Daddi, E., et al. 2002, A&A, 384, L1

Dalcanton, J. J. 1996, ApJ, 466, 92

Dickinson M. 1995, in A Fresh View of Elliptical Galaxies, ed. A. Buzzoni, A. Renzini, & A. Serrano (San Francisco: ASP), 283

Donahue, M., et al. 2002, ApJ, 569, 689

Dressler, A. 1980, ApJ, 236, 351

Dressler, A., et al. 1997, ApJ, 490, 577

Ellingson, E., et al. 2004, in preparation

Evrard, A. E., et al. 2002, ApJ, 573, 7

Gilbank, D. 2003, Ph.D. Thesis, Univ. Durham

Gioia, I. M., Henry, P. J., Maccacaro, T., Morris, S. L., Stocke, J. T., & Wolter, A. 1990, ApJ, 356, L35

Gladders, M. D. 2002, Ph.D. Thesis, Univ. of Toronto

Gladders, M. D., & Donahue, M. 2004, in preparation

Gladders, M. D., Hoekstra, H., Yee, H. K. C., Hall, P. B., & Barrientos, L. F. 2003, ApJ, 593, 48

Gladders, M. D., & Yee, H. K. C. 2000, AJ, 120, 2148

———. 2004, in preparation

Gladders, M. D., Yee, H. K. C., & Ellingson, E. 2002, AJ, 123, 1

González, A. H., Zaritsky, D., Dalcanton, J. J., & Nelson, A. 2001, ApJS, 137, 117

Goto, T., et al. 2002, AJ, 123, 1807

Haiman, Z., Mohr, J. J., & Holder, G. P. 2001, ApJ, 553, 545

Haines, C. P., Clowes, R. G., Campusano, L. E., & Adamson, A. J. 2001, MNRAS, 323, 688

Hall, P. B., & Green, R. F. 1998, ApJ, 507, 558

Hicks, A., et al. 2004, in preparation

Holden, B. P., Stanford, A., Rosati, P., Eisenhardt, P., & Tozzi, P. 2004, in Carnegie Observatories Astrophysics Series, Vol. 3: Clusters of Galaxies: Probes of Cosmological Structure and Galaxy Evolution, ed. J. S. Mulchaey, A. Dressler, & A. Oemler (Pasadena: Carnegie Observatories, http://www.ociw.edu/ociw/symposia/series/symposium3/proceedings.html)

Jones, L. R., Maughan, B. J., Ebeling, H., Scharf, C., Perlman, E., Limb, D., Gondoin, P., Mason, K. D., Cordova, F., & Priedhorsky, W. C. 2004, in Carnegie Observatories Astrophysics Series, Vol. 3: Clusters of Galaxies: Probes of Cosmological Structure and Galaxy Evolution, ed. J. S. Mulchaey, A. Dressler, & A. Oemler (Pasadena: Carnegie Observatories, http://www.ociw.edu/ociw/symposia/series/symposium3/proceedings.html)

Kepner, J., Fan, X., Bahcall, N., Gunn, J., Lupton, R., & Xu, G. 1999, ApJ, 517, 78

Kim, R. S. J., et al. 2002, AJ, 123, 20

Levine, E. S., Schulz, A. E., & White, M. 2002, ApJ, 577, 569

Lewis, A. D., Stocke, J. T., Ellingson, E., & Gaidos, E. J. 2002, ApJ, 566, 744

Lin, H., Yee, H. K. C., Carlberg, R. G., Morris, S. L., Sawicki, M., Patton, D. R., Wirth, G., & Shepherd, C. W. 1999, ApJ, 518, 533

Liu, M. C., Dey, A., Graham, J. R., Bundy, K. A., Steidel, C. C., Adelberger, K., & Dickinson, M. 2000, AJ, 119, 2556

López-Cruz, O. 1997, Ph.D. Thesis, Univ. of Toronto

Lubin, L. M., Brunner, R., Metzger, M. R., Postman, M., & Oke, J. B. 2000, ApJ, 531, L5

Lucey, J. R. 1983, MNRAS, 204, 33

Nichol, R. C. 2004, in Carnegie Observatories Astrophysics Series, Vol. 3: Clusters of Galaxies: Probes of Cosmological Structure and Galaxy Evolution, ed. J. S. Mulchaey, A. Dressler, & A. Oemler (Cambridge: Cambridge Univ. Press), in press

Olsen, L. F., et al. . 2001, A&A, 380, 460

Postman, M., Lauer, T. R., Oegerle, W., & Donahue, M. 2002, ApJ, 579, 93

Postman, M., Lubin, L. M., Gunn, J. E., Oke, J. B., Hoessel, J. G., Schneider, D. P., & Christensen, J. A. 1996, AJ, 111, 615

Rosati, P. 2004, in Carnegie Observatories Astrophysics Series, Vol. 3: Clusters of Galaxies: Probes of Cosmological Structure and Galaxy Evolution, ed. J. S. Mulchaey, A. Dressler, & A. Oemler (Cambridge: Cambridge Univ. Press), in press

Rosati, P., Della Ceca, R., Burg, R., Norman, C., & Giacconi, R. 1995, ApJ, 445, L11

Rosati, P., Della Ceca, R., Norman, C., & Giacconi, R. 1998, ApJ, 492, L21

Rosati, P., Stanford, S. A., Eisenhardt, P. R., Elston, R., Spinrad, H., Stern, D., & Dey, A. 1999, AJ, 118, 76

Schade, D., Barrientos, L. F., & López-Cruz, O. 1997, ApJ, 477, L17

Shepherd, C. W., Carlberg, R. G., Yee, H. K. C., Morris, S. L., Lin, H., Sawicki, M., Hall, P. B., & Patton, D. R. 2001, ApJ, 560, 72

Smail, I., Edge, A. C., Ellis, R. S., & Blandford, R. D. 1998, MNRAS, 293, 124

Soneira, R. M., & Peebles, P. J. E. 1978, AJ, 83, 845

Spergel, D. N., et al. 2003, ApJS, 148, 175

Stanford, S. A., Eisenhardt, P. R., & Dickinson, M. 1998, ApJ, 492, 461

Stanford, S. A., Holden, B., Rosati, P., Eisenhardt, P. R., Stern, D., Squires, G., & Spinrad, H. 2002, AJ, 123, 619

Tanaka, I., Yamada, T., Aragón-Salamanca, A., Kodama, T., Miyaji, T., Ohta, K., & Arimoto, N. 2000, ApJ, 528, 123

Treu, T. 2004, in Carnegie Observatories Astrophysics Series, Vol. 3: Clusters of Galaxies: Probes of Cosmological Structure and Galaxy Evolution, ed. J. S. Mulchaey, A. Dressler, & A. Oemler (Cambridge: Cambridge Univ. Press), in press

van Dokkum, P. G., Franx, M., Kelson, D., Illingworth, G. D., Fisher, D., & Fabricant, D. 1998, ApJ, 504, L17

van Dokkum, P. G., Franx, M., Fabricant, D., Illingworth, G. D., & Kelson, D. 2000, ApJ, 541, 95

van Haarlem, M. P., Frenk, C. S., & White, S. D. M. 1997, MNRAS, 287, 817

Vikhlinin, A., McNamara, B. R., Forman, W., Jones, C., Quintana, H., & Hornstrup, A. 1998, ApJ, 502, 558

Visvanathan, N., & Sandage, A. 1977, ApJ, 216, 214

Yee, H. K. C., et al. 2000, ApJS, 129, 475

Yee, H. K. C., & Ellingson, E. 2003, ApJ, 585, 215

7

Probing dark matter in clusters

IAN SMAIL

Institute for Computational Cosmology, University of Durham, UK

Abstract

Gravitational lensing provides a unique probe of the distribution of mass in clusters on scales from 10 kpc out to the turn-around radius at ~ 5 Mpc. I review how lensing has been used to investigate the total mass in the central regions of clusters, before discussing more recent work tracing the form of the mass distribution in cluster cores, and finally touching on how lensing is being used to map the environment of galaxies and determine how this affects their dark matter halos and star formation histories.

7.1 Introduction

To produce an Einstein ring image of a suitably positioned distant source, a foreground gravitational lens has to focus multiple light paths onto the observer. To achieve this, the mass density in front of the source must exceed a critical mass surface density of around $0.5 \, \mathrm{g \, cm^{-2}}$ (or $2.4 \times 10^9 \, M_\odot \, \mathrm{kpc^{-2}}$). The projected mass in the central regions of the most massive, intermediate-redshift clusters is of the order of $\sim 10^{14} \, M_\odot$ within a radius of 50–100 kpc, and these regions therefore exceed the density threshold required for producing multiple images of cosmological sources.

The study of gravitational lensing by clusters is a relatively young field; the first papers presenting observational evidence for gravitational lensing by clusters were published in the late 1980s (Soucail et al. 1988; Hammer et al. 1989; Lynds & Petrosian 1989; Wambsganss et al. 1989). The field developed rapidly in the 1990s (e.g., Kochanek 1990; Tyson, Wenk, & Valdes 1990; Blandford & Narayan 1992), producing a near-exponential growth in the number of refereed publications (Fig. 7.1), which now approaches ~ 100 papers per year.

The growth of research on cluster lenses reflects their usefulness as a tool to tackle a wide variety of questions in observational cosmology, including constructing mass-selected samples of distant clusters, surveys of the faintest galaxies, active galactic nuclei, or supernovae in the distant Universe, constraining the properties of dark matter, confirming the presence of mass over-densities associated with high-z structures, searches for dark halos, and tests of cosmological parameters. This review discusses how lensing can be used to address three problems that are relevant for this Symposium:

- Comparing X-ray and lensing masses for clusters
- Measuring the mass profile of clusters
- Relating galaxy properties to their local mass density

For simplicity, the content of this written review has a stronger focus on published results

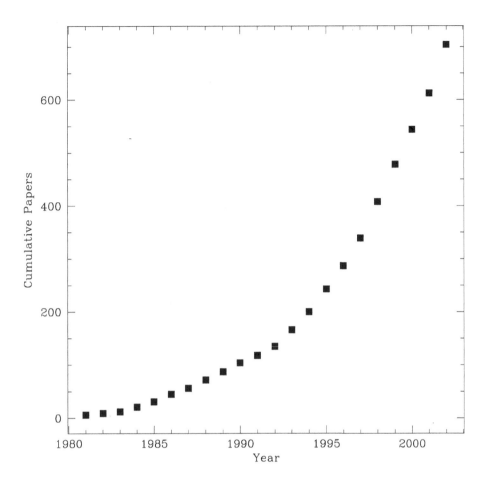

Fig. 7.1. Cumulative refereed publications dealing with gravitational lensing by clusters as listed on ADS (January 2003). The accelerated growth after 1993 reflects the refurbishment of *HST* and the development of new techniques for extracting the 2-dimensional mass distributions from weak shear observations.

than did the oral presentation at the Symposium. For consistency with the usage in the reviewed work, we assume $H_0 = 50 \, \mathrm{km \, s^{-1} \, Mpc^{-1}}$ and $q_0 = 0.5$ throughout.

7.2 Lensing Methods

The first goal of any cluster lensing study is to derive an accurate understanding of the amount and distribution of mass within the cluster. For some applications this is also the only goal; however, it is frequently just a step on the way to exploiting the cluster lens for other purposes. Techniques to determine the mass distribution within a cluster lens can be crudely divided into two different approaches: parametric and non-parametric methods.

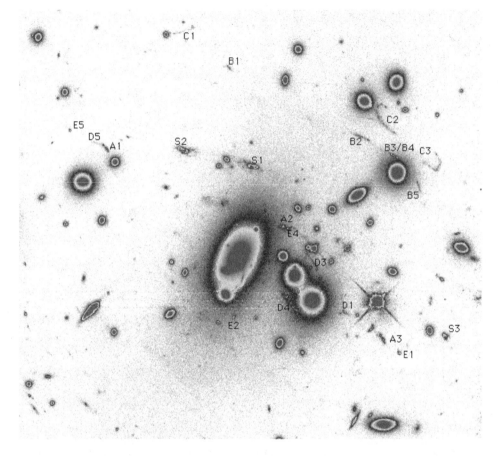

Fig. 7.2. The core region of the $z = 0.31$ cluster AC 114 as seen in a deep *HST* WFPC2 F702W image. This graphically demonstrates the power of *HST* for morphologically identifying the multiple images of background galaxies, such as C1-C2-C3 at $z = 2.86$ or the $z = 1.87$ galaxy S1-S2-S3 (Smail et al. 1995; Campusano et al. 2001). I mark the various images of the six multiply imaged background galaxies (S, A–E), which are visible through the cluster core. The field shown is $50'' \times 50''$ in extent, and the lensed features precisely constrain the distribution of mass within the cluster core on scales from $\sim 20 - 150$ kpc (Natarajan et al. 1998).

7.2.1   *Parametric Lens Modeling*

One cause of the rapid rise of publications dealing with lensing by clusters was the refurbishment of the *Hubble Space Telescope* (*HST*) and the installation of the Wide Field and Planetary Camera 2 (WFPC2) in 1993. This has produced a uniquely powerful instrument for lensing studies, combining excellent image quality, depth, and image stability. The fine spatial resolution of *HST* means we can compare the morphologies of candidate multiply imaged galaxies (Fig. 7.2) to confirm their identifications (e.g., Smail et al. 1995; Tyson et al. 1998; Bézecourt et al. 1999). In addition, it facilitates the identification of faint, lensed features against the crowded background of diffuse halos of the elliptical galaxies in the cores of massive clusters. The sensitivity, resolution, and stability of *HST* are also well suited to searching for the weaker signals of gravitational lensing on much larger scales around clusters, by detecting and reliably measuring the shapes of large numbers of weakly distorted arclets.

The first parametric studies of the mass distribution within clusters, however, predated *HST* and had to rely on deep ground-based imaging (Hammer et al. 1989). These studies reconstructed the dark matter distribution within clusters using simple parametric representations of the mass, guided by the distribution and extent of the light in these systems (e.g., Kneib et al. 1993). With higher-quality observations from *HST*, such studies have now become much more refined, yielding constraints from both strong and weak lensing regimes and allowing for a large number of mass components within the clusters (Kneib et al. 1996; Abdelsalam, Saha, & Williams 1998; Tyson et al. 1998).

These models rely on the identification of multiply imaged background sources, where we can see N images of the same galaxy. Each multiply imaged source provides $3N$ observables: position + flux of each of the images. There are 3 unknowns for the source (its position on the source plane and flux), and hence $3(N-1)$ constraints are available to model the lens. The best-studied strong lenses, those with deep *HST* imaging, can have several multiply imaged background sources. For example, Figure 7.2 shows the central region of the cluster lens AC 114. This area contains 21 images of six different background galaxies, three quintets and three triplets, at redshifts between $z = 1.50$ and $z = 3.35$ (Campusano et al. 2001).

Using these multiply imaged sources, as well as the weak shear field on somewhat larger scales, a model can then be constructed using a series of large-scale potentials to represent the cluster (and any large sub-clumps) and galaxy-scale mass components tied to individual cluster members. The positions, masses, orientations, ellipticities, and radial profiles of these components can all be fit, although to restrict the number of free parameters the masses of the galaxies are usually required to follow a simple scaling relation based upon their luminosities, and their other parameters are similarly fixed to those of their light distributions (Natarajan et al. 1998). Such models have now been constructed for over a dozen clusters (Kneib et al. 1996; Smith et al. 2001, 2004; Figs. 7.3 and 7.4), and the bulk of this review deals with results derived from this "strong" lensing technique.

7.2.2 Non-parametric Techniques

The development of non-parametric techniques to recover the mass distribution in clusters from observations of the weakly sheared arclets provided a second major impetus to work on gravitational lensing by clusters (Kaiser & Squires 1993; see also Tyson et al. 1990). These techniques use the coherence of the weak distortion imprinted on the shapes of faint, randomly oriented background galaxies to recover the lensing signal and hence trace the mass distribution out to the turn-around radius, ~ 5 Mpc or $\sim 0.25°$ for a massive cluster at $z \approx 0.2$. These techniques have been applied to clusters from $z \approx 0.1$ to $z \approx 1$ and provide our only "true" images of the distribution of mass in clusters. They are also well adapted to searches for distant clusters by identifying the lensing signature they produce in the shapes of faint field galaxies, as demonstrated by several speakers at this Symposium (see Margoniner 2004 and Margoniner et al. 2004).

Recent refinements have included more complex approaches to measuring galaxy shapes and correcting for the aberrations induced by the telescope optics (Kaiser, Squires, & Broadhurst 1995; Refregier & Bacon 2003), the incorporation of weak and strong lensing constraints within the reconstruction, and improvements to the stability of the techniques in

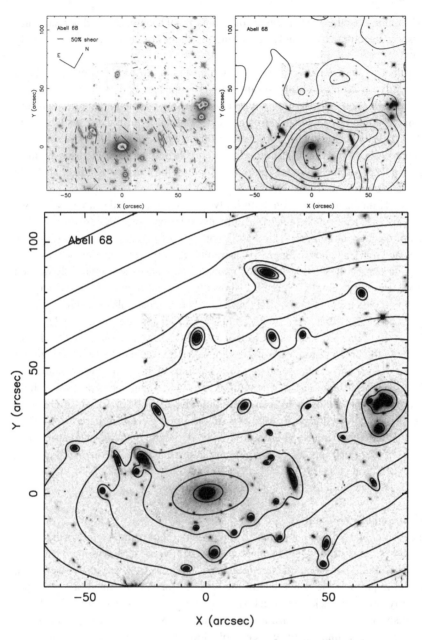

Fig. 7.3. Three representations of the $z = 0.21$ cluster Abell 68, one of the X-ray luminous clusters included in the *HST* lensing survey of Smith et al. (2004). These panels each show the F702W exposure of the cluster core. In the top-left panel, the *HST* image is overlaid with a vector field representing the lensing-induced shear signal. This is constructed by simply averaging the shapes of faint galaxies seen through the cluster core. The shear field identifies a saddle region in the potential between two main mass concentrations. This bimodality is less obvious, but still visible, in the X-ray image of this cluster, which is shown as the contour map in the top-right panel. The bottom panel illustrates, as a logarithmically spaced contour plot, the best-fitting mass model constructed for the cluster by Smith et al. (2004). The secondary clump to the north-west of the central cD contains roughly 30% of the total mass of the core.

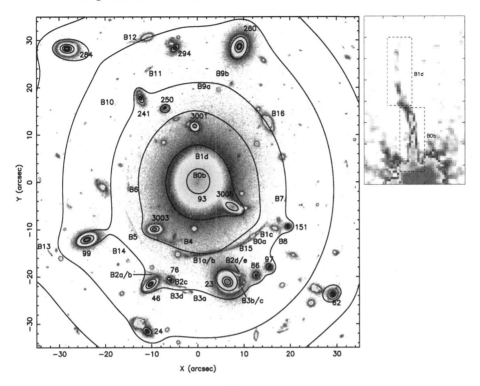

Fig. 7.4. An *HST* WFPC2 F702W image of a $70'' \times 70''$ region in the core of Abell 383, overlaid with isodensity contours illustrating the lens model of Smith et al. (2001). The alphanumeric labels identify multiply imaged background galaxies and arclets used in the lens modeling, while the numerical labels show the galaxy-scale masses incorporated in the model. The central galaxy shows what appears to be a dust lane across its center, as well as two radial features, B0b and B1d, which are highlighted in the panel to the right. These features are interpreted as radial arcs associated with the same $z \approx 1.0$ background galaxies that produce the main giant arcs (B0 and B1). These radial arcs provide strong constraints on the form of the mass profile in the very center of the cluster. The mass contours in the main panel correspond to projected surface densities of $(0.3, 0.4, 0.6, 0.8, 1.1, 1.5) \times 10^{10} M_{\odot} \, \text{kpc}^{-2}$. See Smith et al. (2001), from which this figure is adapted, for more details.

the presence of noise (e.g., Seitz, Schneider, & Bartelmann 1998; Marshall et al. 2002). Non-parametric mass reconstructions are also possible based upon the magnification of the background field population, although to date these have not been so widely developed (Dye et al. 2002).

7.2.3 Cosmic Magnifying Glasses

The magnification produced by cluster lenses can be used as an aid in studying any population of extragalactic sources that is faint, distant, and relatively numerous. Indeed, lensing-boosting is now becoming a standard technique in studies of very faint populations in the distant Universe. For example, the quality of the rest-frame UV spectroscopy of the galaxy cB58 at $z = 2.72$ rivals that of the best-studied galaxies in the local Universe, as a result of its substantial magnification by a foreground cluster lens (Pettini et al. 2002).

Similarly, narrow-band searches for emission-line galaxies at the highest redshifts have also benefited from the added boost of a foreground lens (Hu et al. 2002), while studies of intrinsically faint sources selected in the near-infrared or submillimeter wavebands have provided unique insights into the properties of these populations (Cowie, Barger, & Kneib 2002; Smail et al. 2002; Smith et al. 2002; Wehner, Barger, & Kneib 2002).

Abell 2218 provides a good example of how versatile a single well-constrained lens can be. This rich cluster lies at $z = 0.18$, and its core contains at least nine recognizable multiply imaged background sources, which produce in total 29 images. As a result the mass model for this cluster has been very well constrained, allowing it to be used for detailed studies of the properties of very faint and very distant galaxies. The nine multiply imaged sources lie in a region of the sky directly behind the cluster core, which is roughly $10'' \times 10''$ in size. This area contains an $I = 23.5$ mag $z = 2.52$ star-forming galaxy (Ebbels et al. 1996), an $I = 26$ mag submm-bright "extremely red object" also at $z \approx 2.6$ (van der Werf et al. 2004), and finally an $I = 30$ mag proto-galactic clump at $z = 5.6$ (Ellis et al. 2001), all of which are magnified by factors of 10–30. Just outside this area lies an $I = 21.5$ mag spiral galaxy at $z = 1.03$ with a resolved rotation curve (Swinbank et al. 2004). The discovery and detailed study of these galaxies have benefited substantially from their position behind the core of one of the most massive and best-constrained cluster lenses on the sky. Such regions provide an unique opportunity to study ultra-faint galaxies in a level of detail that will be otherwise unavailable until the advent of 30-m telescopes.

7.3   Comparing X-ray and Lensing Masses

Gravitational lensing provides a completely independent method to estimate the mass of clusters, free of many of the uncertainties associated with using indirect tracers. For example, when using the dynamics of cluster members, orbital characteristics and the virial state of the sample have to be assumed. Or, when modeling the X-ray emission from the intracluster medium, hydrostatic conditions have to be adopted and also a high degree of symmetry in the system has to be assumed to reliably deproject the mass distribution (Brainerd et al. 1998). Lensing, of course, has its own drawbacks: the shear signal is weak, and corrections need to be applied to remove instrumental effects. The strength of the lensing signal also depends not only on the mass of the cluster, but also on the redshifts of the background population that are used in the analysis. These are typically very faint galaxies for which redshifts have to be statistically estimated based on photometric redshift surveys of regions such as the Hubble Deep Field. Nevertheless, lensing provides an important, independent technique to derive cluster masses, and so there is considerable interest in comparing the various mass estimators, in particular lensing and X-ray estimates, to see how well they agree.

A number of early comparisons of the X-ray and lensing mass estimates of clusters on < 1 Mpc scales, mostly from strong lensing constraints, found significant differences, with the lensing-derived masses typically being a factor of 2–3 larger than the X-ray masses (Miralda-Escudé & Babul 1995; Wu & Fang 1997). Explanations for this offset included the projection onto the cluster cores of unrelated structures that artificially increase the lensing-derived masses, or proposed mechanisms to provide nonthermal pressure support to reduce the X-ray estimates.

- AGREE, $M_{lens}/M_X \approx 1$: MS 1008–12, $z = 0.31$; RX J1347–11, $z = 0.45$; MS 0016+16, $z = 0.55$ (Fischer & Tyson 1997; Schindler et al. 1997; Sahu et al. 1998; Allen, Schmidt, & Fabian 2002; Ettori & Lombardi 2003; Worrall & Birkinshaw 2003).
- DISAGREE, $M_{lens}/M_X \approx 2-3$: Abell 1689, $z = 0.18$; MS 0440–02, $z = 0.19$; MS 1008–12, $z = 0.31$; Cl 0500–24, $z = 0.32$; Abell 370, $z = 0.37$; Cl 0024+16, $z = 0.39$ (Gioia et al. 1998; Ota, Mitsuda, & Fukazawa 1998; Soucail et al. 2000; Athreya et al. 2002; Xue & Wu 2002; Ota et al. 2004).

The lists above give a random selection of X-ray/lensing mass comparisons from the recent literature. This shows that the situation is still confused, with groups claiming either agreement or disagreement between the two estimates, sometimes for the same clusters. Many of these studies can be criticized for adopting overly simplistic models for their lensing or X-ray analyses, or for concentrating on the same small set of well-known cluster lenses. However, the criticisms are not unique to one or other set of studies; rather, the situation seems to reflect a real diversity in the cluster properties.

What is the origin of disagreements in the mass determinations for some clusters? Does it reflect an oversimplification or misapplication of assumptions in one or other of the mass estimates, or a real physical effect? Allen (1998) answered these questions by reassessing the X-ray data for a mixed sample of 13 clusters (all known strong lenses) containing both examples where the lensing and X-ray masses agreed, and those where they disagreed.

Allen used *ROSAT* HRI X-ray images and *ASCA* temperatures for the clusters. To determine the mass profile of these clusters, he applied a standard X-ray deprojection analysis, which required an isothermal X-ray temperature profile, and accounted for the contribution from any cold gas in the cluster core to the measured X-ray temperatures. Combining these mass estimates with those from simple lens models for the clusters, he found a wide range in M_{lens}/M_X (Fig. 7.5). However, the clusters with centrally peaked X-ray emission, which have high central gas densities and short inferred cooling times for the gas (I will use the possible misnomer "cooling flow" to identify this class), do have $M_{lens}/M_X \approx 1$, while those clusters with lower central gas densities give discrepant values, $M_{lens}/M_X \approx 2-4$ (Fig. 7.5). Short gas cooling times have previously been interpreted as an indication of a relaxed system, suggesting that the non-cooling flow clusters may be dynamically more active. The cooling flow and non-cooling flow clusters are also distinguished by the size of the core radii in the cluster potentials, which are inferred from the deprojection analysis (Allen 1998). Cooling flow clusters typically have inferred core sizes of ~ 50 kpc, where as non-cooling flow clusters have best-fitting cores of > 250 kpc. These apparently large cores could be spurious and instead reflect the breakdown of hydrostatic equilibrium in the centers of these clusters, probably due to dynamical activity associated with mergers. Indeed, deprojecting the non-cooling flow clusters with a fixed ~ 50 kpc core radius produces mass estimates in much better agreement with those from lensing, $M_{lens}/M_X \approx 1.2 \pm 0.3$.

A comparison of the X-ray and lensing mass estimates (from both strong and weak lensing measurements) as a function of scale on which the masses are estimated is shown in Figure 7.5; to assure the reliability of both estimates, only cooling flow clusters are included at small scales. This shows good agreement between the two estimates over a very wide range in scale.

Thus, the origin of the disagreements between lensing and X-ray mass estimates appears to be the inappropriate application of X-ray analysis techniques, which assume hydrostatic equilibrium, to the core regions of clusters that may be far from equilibrium (most probably

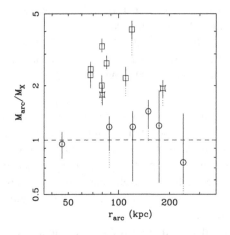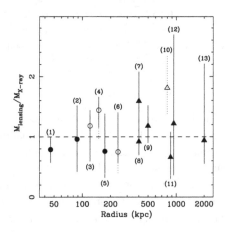

Fig. 7.5. The *left* panel shows the distribution on the $M_{lens}/M_X - r_{arc}$ plane for those clusters with strongly peaked central X-ray emission ("cooling flow"; circles) and those with less-peaked X-ray emission (squares). The former systems show good agreement between the two mass estimates, whereas those with less centrally concentrated X-ray emission typically show low X-ray-estimated masses compared to their lensing values. The *right* panel shows a compilation of M_{lens}/M_X estimates from the literature as a function of the scale of the measurements (see Allen 1998 for the meaning of the symbols and the references to the original work). At small radii, only clusters that exhibit strongly peaked X-ray emission are included in the figure. This figure demonstrates reasonable agreement between X-ray and lensing mass estimates over nearly 3 orders of magnitude in scale. Both figures are taken from Allen (1998).

due to mergers). This problem is exacerbated by comparing the two estimates on small scales, where the X-ray analysis is particularly sensitive to these model assumptions.

Allen's (1998) conclusions prompt two important questions: (1) How frequently do the X-ray and lensing mass estimates for cluster cores disagree?" and (2) Can these discrepancies be directly related to merging in the clusters? To answer these requires a joint lensing and X-ray analysis of an unbiased sample of clusters. Smith et al. (2004) have undertaken such a survey using *HST* imaging and *Chandra* X-ray observations of 10 X-ray luminous clusters ($L_X \geq 8 \times 10^{44}$ erg s^{-1}) in a narrow redshift range around $z = 0.2$, selected from the XBACS sample (Ebeling et al. 1996). The *HST* imaging provides the basis for detailed lensing models of the clusters, which can be compared to the results from the spectral analysis of the *Chandra* X-ray data on ~ 1 Mpc scales. The lensing and X-ray results for one of the clusters from the sample are shown in Figure 7.3, illustrating the level of detail available for the comparisons of the different baryonic and non-baryonic components within the clusters.

The Smith et al. (2004) survey answers both of the questions raised above. They find mass substructures (consisting of $\geq 10\%$ of the cluster mass) in half of the clusters in their sample (see also Dahle et al. 2002). The lensing and X-ray mass estimates are in good agreement for the regular clusters, but those clusters that exhibit substructure have X-ray temperatures measured across the central 1 Mpc that are $\sim 30\%$ hotter at a fixed mass than the more regular clusters. The effectiveness of even relatively minor mass mergers in heating the intracluster gas and the prevalence of these minor mergers in the most X-ray luminous clusters clearly raises concerns over attempts to use X-ray observations to infer the detailed mass distribution in any clusters that show even hints of structural complexity. This is particularly true of clusters at higher redshifts, where the effects of infall are likely to have been more extreme. To test this expectation, a similar joint *HST*/*Chandra* analysis is being performed by the same group on an identically selected sample of X-ray luminous clusters at $z = 0.4$.

7.4 Cluster Mass Profiles

The next step beyond measuring the amount of mass in clusters with lensing is to derive their radial profiles. This endeavor has been given fresh impetus by the claims of a "universal mass profile" derived from N-body simulations of halos in a cold dark matter (CDM) cosmogony (Navarro, Frenk, & White 1997, hereafter NFW). On large scales the NFW profile falls off as $\rho \propto r^{-3}$, steeper than isothermal ($\rho \propto r^{-2}$), while on small scales the NFW profile flattens to $\rho \propto r^{-1}$, although it retains a central cusp. The exact details of the form of the mass profile are the subject of debate within the theoretical community (e.g., Ghigna et al. 2000), but the central claim that CDM halos have a relatively well-determined profile has not been contradicted and provides an observationally testable prediction for which lensing is particularly well suited.

One class of gravitationally lensed features are particularly useful for studying the mass distribution in the central regions of lensing clusters. These are the radial arcs found in the very centers of a handful of cluster lenses, mostly from *HST* imaging: MS 2137.3–23 (Fort et al. 1992; Mellier, Fort, & Kneib 1993), Abell 370 (Smail et al. 1996), AC 114 (Natarajan et al. 1998). While the giant, tangential arcs (e.g., B0/B1 in Fig. 7.4) provide a precise estimate of the total enclosed mass in the cluster centers, the positions of the radial arcs are sensitive to the local gradient of the mass profile, allowing us to model the mass on scales of $< 100\,\text{kpc}$ (Williams, Navarro, & Bartelmann 1999).

The mass profile derived from the lensing model of Abell 383 (Fig. 7.4) by Smith et al. (2001) on scales $\geq 10\,\text{kpc}$ is dominated by the cluster-scale potential, the core radius of the cluster-scale mass distribution ($\sim 46\,\text{kpc}$) being consistent with previous estimates of core radii from lensing studies. The shape of Abell 383's mass profile is most precisely measured within $r \approx 100\,\text{kpc}$, where constraints from the multiple images are available, and Smith et al. (2001) estimate the total projected mass within the radius of the giant arc (65 kpc) to be $(3.5 \pm 0.1) \times 10^{13}\,M_\odot$. The density profile of Abell 383 at the positions of the radial arcs ($\sim 20\,\text{kpc}$) is $d(\log\rho)/d(\log r) = -1.29 \pm 0.03$, intermediate between the NFW and Ghigna et al. (2000) predictions. However, the observed gradient includes a contribution from the baryonic component associated with the central galaxy of the cluster (Fig. 7.4), which is not included in the predictions of the dark-matter-only simulations. The unknown contribution of the central galaxy makes it nontrivial to compare the observations and predictions (see also Gavazzi et al. 2003).

To circumvent this problem, Sand, Treu, & Ellis (2002) and Sand et al. (2004) have been gathering velocity dispersion information for the central galaxies in clusters with radial arcs. This allows them to model and remove the contribution from the stars in the central galaxy, leading to much flatter mass profiles for the underlying dark matter, $d(\log\rho)/d(\log r) < -0.9$, in apparent conflict with the theoretical predictions for the dark matter profiles. Again, however, the comparison with theory is nontrivial, as it is not clear how the dark matter distribution in the cluster core would react to the growth of the baryonic component of

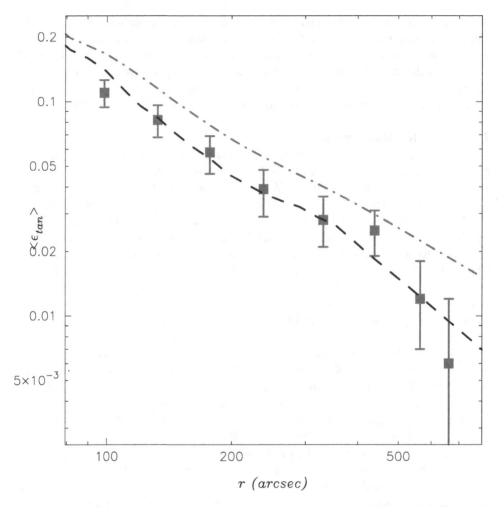

Fig. 7.6. The tangential shear profile ($\langle \epsilon_{\text{tan}} \rangle$) around Cl 0024+16 from an analysis of the *HST* WFPC2 and STIS imaging of the cluster (Kneib et al. 2004). Isothermal (dot-dashed) and NFW (dashed) profiles are plotted; both are normalized to the strong-lens model of the cluster core. It is clear that the isothermal model grossly overpredicts the strength of the shear field out to the limit of the data.

the central galaxy, if this was included in the theoretical simulations. Nevertheless, the necessary observations now exist, and the hope is that the modelers will rise to the challenge and produce more sophisticated "predictions" of the mass profile in cluster cores.

On larger scales, where the contribution from dissipational baryons should be less of a concern, the mass profile of clusters is being studied using weak shear techniques, primarily through comparisons of the observed and predicted shear profiles. Ground-based, panoramic imaging of three intermediate-redshift clusters is discussed by Clowe & Schneider (2001, 2002), who conclude that their data are insufficient to strongly differentiate between a singular isothermal sphere and a NFW profile on scales of 0.13–2 Mpc. To improve on this situation, Kneib et al. (2004) obtained a sparse mosaic of *HST* WFPC2 and STIS pointings covering a 5 Mpc region around the $z = 0.39$ cluster Cl 0024+16. Using the higher-fidelity imaging available from space, they were able to trace the mass profile from the core out to ~ 4 Mpc and strongly reject an isothermal mass distribution, finding instead that the NFW model provides an acceptable description of the mass profile on Mpc scales in at least this cluster (Fig. 7.6). Clearly, it is important to extend this approach to a larger sample of clusters to observationally determine the true *universality* of the mass profile.

7.5   Relating Galaxy Properties to their Dark Matter Environment

The final section of this review focuses on the application of gravitational lensing to obtain an even more detailed view of the mass distribution within clusters: on the scales of galaxy halos.

The weak shear induced in the shapes of background field galaxies by a foreground cluster lens reflects not only the large-scale distribution of mass within the cluster, but also the mass associated with individual cluster members. The latter can be studied by *stacking* the weak shear signal around each galaxy, where the contribution from the cluster-scale potential acts as an incoherent source of noise in the galaxy's reference frame, allowing the lensing contribution of the halos of the ensemble of galaxies to be recovered.

Natarajan et al. (1998) constructed a detailed lens model for the cluster AC 114 based on a wide-field *HST* mosaic. Their model incorporated small-scale perturbations associated with the halos of individual galaxies, and they then employed this model to constrain the properties of different populations of cluster galaxies. By adopting simple scaling relations between the radial extent and central velocity dispersion of the galaxies, both scaling with their optical luminosities, it is possible to constrain these parameters for a fiducial L^* cluster galaxy (Brainerd, Blandford, & Smail 1996).

They were able to show that the dark matter halo of a typical L^* spheroidal cluster galaxy is roughly 15 kpc in extent, much smaller than the ≥ 150 kpc inferred for similar luminosity galaxies in the field (Brainerd et al. 1996). This strongly suggested that the high-density cluster environment is truncating the dark matter halos of the galaxies. Unfortunately, the sample of early-type galaxies was too small to convincingly detect differences in the halo properties of elliptical and lenticular galaxies, although there was a hint that the halos of S0s could be less extended than those of ellipticals at a fixed luminosity.

Natarajan et al. (2002) extended this analysis to a larger sample of intermediate-redshift clusters for which *HST* imaging was available and which exhibit strongly lensed features suitable for constructing a detailed mass model. The best-fitting galaxy parameters from these mass models confirmed that typical L^* cluster galaxies are more compact than their field counterparts. However, when the truncation radii of the galaxies are plotted against the central density of their host clusters, a more intriguing correlation is found (Fig. 7.7). We see that the clusters with higher central densities have galaxy populations with smaller dark matter halos. The slope of this correlation agrees with that expected from tidal stripping of the halos by the cluster potential. This is arguably one of the few direct pieces of evidence for the interaction of galaxies with their environment. If correct, it underlines the importance of halo stripping on the dynamical evolution of galaxies lying within the infall regions of clusters. Relating the stripping of dark matter halos to the star formation history of the galaxies residing within them will provide important insights into how galaxies react to changes in their halos and also how they interact with the wider dark matter environment (Gray et al. 2002, 2004).

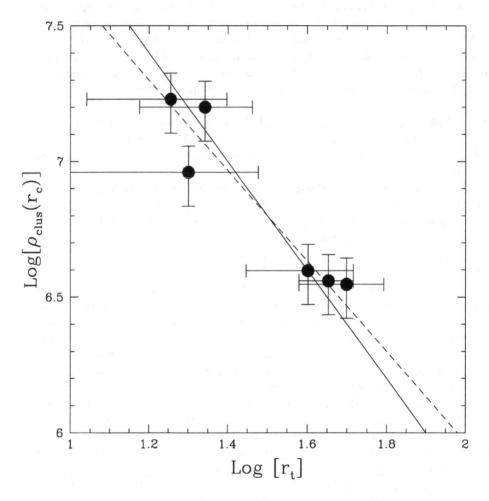

Fig. 7.7. The tidal radii of galaxies as a function of the central mass density in the clusters in which they reside. The dashed line shows the least-squares fit to the data, which gives a best-fit power-law index ($r_t^* \propto \rho^\beta$) of $\beta = -0.6 \pm 0.2$, in good agreement with the theoretical prediction of $\beta = -0.5$ from a tidal-stripping model (Natarajan, Kneib, & Smail 2002).

7.6   Summary

Gravitational lensing provides an increasingly important tool for investigating the distribution of mass in clusters. The information from lensing studies can be compared and contrasted to other tracers of cluster mass to address a range of questions about the formation and evolution of clusters and their component galaxies.

Comparison of X-ray and lensing mass estimates within the virial radii of rich clusters shows reasonable agreement for systems where both approaches can be reliably employed. However, there is evidence that X-ray measurements of the masses of merging systems are systematically in error, even for relatively minor mergers (1:10 mass ratios).

On large scales the projected mass profile of clusters appears to roughly follow the form predicted from hierarchical models using a CDM power spectrum. The form of the mass profile in the cores of the halos is also similar to the theoretical predictions, but this may be a coincidence as the observations indicate substantially flatter dark matter profiles than theory predicts when the dominant baryonic component (associated with the central galaxy) is removed. However, predictions are based on dark-matter-only simulations, and hence the behavior of the combined baryonic and dark matter profile has not yet been accurately modeled.

Finally, the behavior of the massive halos associated with individual galaxies within clusters suggests that these are affected by their environment. There are indications that this behavior follows that expected from tidal stripping in the cluster potential well. Future work on this issue, focusing on the effects on the star formation history of the galaxies residing in the halos, promises to produce important constraints on galactic ecology.

Acknowledgements. I thank John, Alan and Gus for organizing a very memorable meeting and a fitting tribute to OCIW's centenary. I also wish to thank Steve Allen, Meghan Gray, and David Sand for generously allowing me to present their latest work in the oral version of this review, and Harald Ebeling, Richard Ellis, Jean-Paul Kneib, Priya Nataraajan, and Graham Smith for allowing me to present results from our research programs. I acknowledge support through a Royal Society University Research Fellowship and a Philip Leverhulme Prize Fellowship.

References

Abdelsalam, H. M., Saha, P., & Williams, L. L. R. 1998, AJ, 116, 1541

Allen, S. W. 1998, MNRAS, 296, 392

Allen, S. W., Schmidt, R. W., & Fabian, A. C. 2002, MNRAS, 335, 256

Athreya, R. M., Mellier, Y., van Waebeke, L., Pelló, R., Fort, B., & Dantel-Fort, M. 2002, A&A, 384, 743

Bézecourt, J., Kneib, J.-P., Soucail, G., Ebbels, T. M. D. 1999, A&A, 347, 21

Blandford, R. D., & Narayan, R. 1992, ARA&A, 30, 311

Brainerd, T. G., Blandford, R. D., & Smail, I. 1996, ApJ, 466, 623

Brainerd, T. G., Wright, C. O., Goldberg, D. M., & Villumsen, J. V. 1998, ApJ, 502, 505

Campusano, L. E., Pelló, R., Kneib, J.-P., Le Borgne, J.-F., Fort, B., Ellis, R. S., Mellier, Y., & Smail, I. 2001, A&A, 378, 394

Clowe, D., & Schneider, P. 2001, A&A, 379, 384

——. 2002, A&A, 395, 385

Cowie, L. L., Barger, A. J., & Kneib, J.-P. 2002, ApJ, 123, 2197

Dahle, H., Kaiser, N., Irgens, R. J., Lilje, P. B., & Maddox, S. J. 2002, ApJS, 139, 313

Dye, S., et al. 2002, A&A, 386, 12

Ebbels, T. M. D., Le Borgne, J.-F., Pelló, R., Ellis, R. S., Kneib, J.-P., Smail, I., & Sanahuja, B. 1996, MNRAS, 281, L75

Ebeling, H., Voges, W., Böhringer, H., Edge, A. C., Huchra, J. P., & Briel, U. G. 1996, MNRAS, 281, 799

Ellis, R. S., Santos, M. R., Kneib, J.-P., & Kuijken, K. 2001, ApJ, 560, L119

Ettori. S., & Lombardi, M. 2003, A&A, 398, L5

Fischer, P., & Tyson, A. J. 1997, AJ, 114, 14

Fort, B., Le Fèvre, O., Hammer F., & Cailloux, M. 1992, ApJ, 399, L125

Gavazzi, R., Fort, B., Mellier, Y., Pelló, R., & Dantel-Fort, M. 2003, A&A, 403, 11

Ghigna, S., Moore, B., Governato, F., Lake, G., Quinn, T., & Stadel, J. 2000, ApJ, 544, 616

Gioia, I. M., Shaya, E. J., Le Fèvre, O., Falco, E. E., Luppino, G. A., & Hammer, F. 1998, ApJ, 497, 573

Gray, M. E., et al. 2004, in preparation

Gray, M. E., Taylor, A. N., Meisenheimer, K., Dye, S., Wolf, C., & Thommes, E. 2002, ApJ, 568, 141

Hammer, F., Rigaut, F., Le Fèvre, O., Jones, J., & Soucail, G. 1989, A&A, 208, L7

Hu, E. M., Cowie, L. L., McMahon, R. G., Capak, P., Iwamuro, F., Kneib, J.-P., Maihara, T., & Motohara, K. 2002, ApJ, 568, L75

Kaiser, N., & Squires, G. 1993, ApJ, 404, 441

Kaiser, N., Squires, G., & Broadhurst, T. 1995, ApJ, 484, 460

Kneib, J.-P., Ellis, R. S., Smail, I., Couch, W. J., & Sharples, R. M. 1996, ApJ, 471, 643

Kneib, J.-P., Hudelot, P., Ellis, R. S., Treu, T., Smith, G. P., Marshall, P., Czoske, O., Smail, I., & Natarajan, P. 2004, ApJ, submitted

Kneib, J.-P., Mellier, Y., Fort, B., & Mathez, G. 1993, A&A, 273, 367

Kochanek, C. S. 1990, MNRAS, 247, 135

Lynds, R., & Petrosian, V. 1989, ApJ, 336, 1

Margoniner, V. E. 2004, in Carnegie Observatories Astrophysics Series, Vol. 3: Clusters of Galaxies: Probes of Cosmological Structure and Galaxy Evolution, ed. J. S. Mulchaey, A. Dressler, & A. Oemler (Pasadena: Carnegie Observatories, http://www.ociw.edu/ociw/symposia/series/symposium3/proceedings.html)

Margoniner, V. E., et al. 2004, in Carnegie Observatories Astrophysics Series, Vol. 3: Clusters of Galaxies: Probes of Cosmological Structure and Galaxy Evolution, ed. J. S. Mulchaey, A. Dressler, & A. Oemler (Pasadena: Carnegie Observatories, http://www.ociw.edu/ociw/symposia/series/symposium3/proceedings.html)

Marshall, P. J., Hobson, M. P., Gull, S. F., & Bridle, S. L. 2002, MNRAS, 335, 1037

Mellier, Y., Fort, B., & Kneib J.-P. 1993, ApJ, 407, 33

Miralda-Escudé, J., & Babul, A., 1995, 449, 18

Natarajan, P., Kneib, J.-P., & Smail, I. 2002, ApJ, 580, L11

Natarajan, P., Kneib, J.-P., Smail, I., & Ellis, R. S. 1998, ApJ, 499, 600

Navarro, J. F., Frenk, C. S., & White, S. D. M. 1997, ApJ, 490, 493

Ota, N., Hattori, M., Pointecouteau, E., & Mitsuda, K., 2004, A&A, submitted

Ota, N., Mitsuda, K., & Fukazawa, Y. 1998, ApJ, 495, 1700

Pettini, M., Rix, S. A., Steidel, C. C., Adelberger, K. L., Hunt, M. P., & Shapley, A. E. 2002, ApJ, 569, 742

Refregier, A., & Bacon, D. 2003, MNRAS, 338, 48

Sahu, K. C., et al. 1998, ApJ, 492, L125

Sand, D. J., Treu, T., Smith, G. P., & Ellis, R. S. 2004, ApJ, submitted

Sand, D. J., Treu, T., & Ellis, R. S. 2002, ApJ, 574, L129

Schindler, S., Hattori, M., Neumann, D. M., & Böhringer, H. 1997, A&A, 317, 646

Seitz, S., Schneider, P., & Bartelmann, M. 1998, A&A, 337, 325

Smail, I., Couch, W. J., Ellis, R. S., & Sharples, R. M. 1995, ApJ, 440, 501

Smail, I., Dressler, A. Kneib, J.-P., Ellis, R. S., Couch, W. J., Sharples, R. M., & Oemler, A., Jr. 1996, ApJ, 469, 508

Smail, I., Ivison, R. J., Blain, A. W., & Kneib, J.-P. 2002, MNRAS, 331, 495

Smith, G. P., et al. 2002, MNRAS, 330, 1

Smith, G. P., Kneib, J.-P., Ebeling, H., Czoske, O., & Smail, I. 2001, ApJ, 552, 493

Smith, G. P., Kneib, J.-P., Smail, I., Ebeling, H., Mazzotta, P., & Czoske, O. 2004, ApJ, in preparation

Soucail, G., Mellier, Y., Fort, B., Mathez, G., & Cailloux, M. 1988, A&A, 191, L19

Soucail, G., Ota, N., Böhringer, G., Czoske, O., Hattori, M., & Mellier, Y. 2000, A&A, 355, 433

Swinbank, A. M., et al. 2004, ApJ, submitted

Tyson, J. A., Kochanski, G. P., & dell'Antonio, I. P. 1998, ApJ, 498, L107

Tyson, J. A., Wenk, R. A., & Valdes, F. 1990, ApJ, 349, L1

van der Werf, P., Knudsen, K. K., Kneib, J.-P., & Smail, I. 2004, in preparation

Wambsganss, J., Giraud, E., Schneider, P., & Weiss, A. 1989, ApJ, 337, L73

Wehner, E. H., Barger, A. J., & Kneib, J.-P. 2002, ApJ, 577, L83

Williams, L. L. R., Navarro, J. F., & Bartelmann, M. 1999, ApJ, 527, 535

Worrall, D. M., & Birkinshaw, M. 2003, MNRAS, 340, 1261

Wu, Z.-P., & Fang, L.-Z. 1997, ApJ, 483, 62

Xue, S.-J., & Wu, X.-P. 2002, ApJ, 576, 152

8

Clusters of galaxies: an X-ray perspective

RICHARD F. MUSHOTZKY

NASA/Goddard Space Flight Center

Abstract

There has been extensive recent progress in X-ray observations of clusters of galaxies with the analysis of the entire *ASCA* database and recent new results from *Beppo-SAX*, *Chandra*, and *XMM-Newton*. The temperature profiles of most clusters are isothermal from 0.05–0.6 R_{viral}, contrary to theoretical expectations and early results from *ASCA*. Similarly, the abundance profiles of Fe are roughly constant outside the central regions. The luminosity-temperature relation for a very large sample of clusters show that $L_X \propto T^3$ over the whole observable luminosity range at low redshift, but the variance increases at low luminosity, explaining the previously claimed steepening at low luminosity. Recent accurate cluster photometry in red and infrared passbands have resulted in much better correlations of optical and X-ray properties, but there is still larger scatter than one might expect between total light and X-ray temperature and luminosity. The velocity dispersion and the X-ray temperature are strongly correlated, but the slope of the relation is somewhat steeper than expected. The surface brightness profiles of clusters are very well fit by the isothermal β model out to large radii and show scaling relations, outside the central regions, consistent with a Λ-dominated Universe.

At high masses the gas mass fraction of clusters is quite uniform and is consistent with the low WMAP value of Ω_m. The recent analysis of cluster mass-to-light ratio and the mass-to-light ratio of stars indicates that the ratio of gas to stellar mass is $\sim$10:1 in massive clusters. There is an apparent decrease in gas mass fraction and increase in stellar mass fraction at lower mass scales, but the very flat surface brightness of the X-ray emission makes extension of this result to large scale lengths uncertain. The normalization of the scaling of mass with temperature, derived from measurements of density and temperature profiles and assuming hydrostatic equilibrium, is lower than predicted from simulations that do not include gas cooling or heating and has a slightly steeper slope. Detailed *Chandra* and *XMM-Newton* imaging spectroscopy of several clusters show that the form of the potential is consistent with the parameterization of Navarro, Frenk, & White (1997) over a factor of 100 in length scale and that there is no evidence for a dark matter core. *Chandra* X-ray images have revealed rather complex internal structures in the central regions of some clusters, which are probably due to the effects of mergers; however, their nature is still not completely clear.

There are now more than 100 clusters with well-determined Fe abundance, several with accurate values at redshifts $z \approx 0.8$, with little or no evidence for evolution in the Fe abundance with redshift. There is real variance in the Fe abundance from cluster to cluster, with a trend for clusters with higher gas densities to have higher Fe abundances. The Si, S, and

Ni abundances do not follow patterns consistent with simple sums of standard Type Ia and Type II supernova, indicating that the origin of the elements in clusters is different from that in the Milky Way. The Si/Fe abundance rises with cluster mass, but the S/Fe ratio does not. The high Ni/Fe ratio indicates the importance of Type Ia supernovae. *XMM-Newton* grating spectra of the central regions of clusters have derived precise O, Ne, Mg, and Fe abundances. *XMM-Newton* CCD data are allowing O abundances to be measured for a large number of clusters.

8.1   Introduction

Clusters of galaxies are the largest and most massive collapsed objects in the Universe, and as such they are sensitive probes of the history of structure formation. While first discovered in the optical band in the 1930's (for a review, see Bahcall 1977a), in same ways the name is a misnomer since most of the baryons and metals are in the hot X-ray emitting intracluster medium and thus they are basically "X-ray objects." Studies of their evolution can place strong constraints on all theories of large-scale structure and determine precise values for many of the cosmological parameters. As opposed to galaxies, clusters probably retain all the enriched material created in them and being essentially closed boxes they provide an unbiased record of nucleosynthesis in the Universe. Thus, measurement of the elemental abundances and their evolution provide fundamental data for the origin of the elements. The distribution of the elements in clusters reveals how the metals were removed from stellar systems into the intergalactic medium (IGM). Clusters should be fair samples of the Universe, and studies of their mass and their baryon fraction reveal the gross properties of the Universe as a whole. Since most of the baryons are in the gaseous phase and clusters are dark matter dominated, the detailed physics of cooling and star formation are much less important than in galaxies. This makes clusters much more amenable to detailed simulations than galaxies or other systems in which star formation has been an overriding process. Detailed measurements of their density and temperature profiles allow an accurate determination of the dark matter profile and total mass. While gravity is clearly dominant in massive systems, much of the entropy of the gas in low-mass systems maybe produced by nongravitational processes.

Clusters are luminous, extended X-ray sources and are visible out to high redshifts with present-day technology. The virial temperature of most groups and clusters corresponds to $kT \approx (2-100) \times 10^6$ K (velocity dispersions of 180–1200 km s^{-1}), and while lower mass systems certainly exist, we usually call them galaxies. Most of the baryons in groups and clusters of galaxies lie in the hot X-ray emitting gas, which is in virial equilibrium with the dark matter potential well [the ratio of gas to stellar mass is $\sim(2-10)$:1; Ettori & Fabian 1999]. This gas is enriched in heavy elements (Mushotzky et al. 1978) and is thus the reservoir of stellar evolution in these systems. The presence of heavy elements is revealed by line emission from H and He-like transitions, as well as L-shell transitions of the abundant elements. Most clusters and groups are too hot to have significant line emission from C or N, but all abundant elements with $Z > 8$ (O) have strong lines from H and He-like states in the X-ray band, and their abundances can be well determined.

Clusters of galaxies were discovered as X-ray sources in the late 1960's (see Mushotzky 2002 for a historical review), and large samples were first obtained with the *Uhuru* satellite in the early 1970's (Jones & Forman 1978). Large samples of X-ray spectra and images were first obtained in the late 1970's with the *HEAO* satellites (see Forman & Jones 1982 for

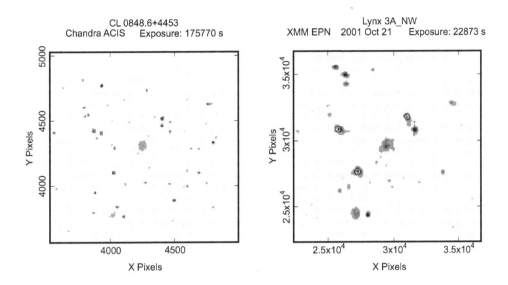

Fig. 8.1. *Chandra* (left panel) and *XMM-Newton* (right panel) images of the Lynx region (Stanford et al. 2001), which contains three high-redshift clusters. The *Chandra* image has ∼170 ks and the *XMM-Newton* image ∼20 ks exposure.

an early review). The early 1990's brought large samples of high-quality images with the *ROSAT* satellite and good quality spectra with *ASCA* and *Beppo-SAX*. In the last three years there has been an enormous increase in the capabilities of X-ray instrumentation due to the launch and operation of *Chandra* and *XMM-Newton*. Both *Chandra* and *XMM-Newton* can find and identify clusters out to $z > 1.2$, and their morphologies can be clearly discerned to $z > 0.8$ (Fig. 8.1). The cluster temperatures can be measured to $z \approx 1.2$, and *XMM-Newton* can determine their overall chemical abundances to $z \approx 1$ with sufficiently long exposures (very recently the temperature and abundance of a cluster at $z = 1.15$ was measured accurately in a 1 Ms *XMM-Newton* exposure; Hasinger et al. 2004). Temperature and abundance profiles to $z \approx 0.8$ can be well measured and large samples of X-ray selected clusters can be derived. *Chandra* can observe correlated radio/X-ray structure out to $z > 0.1$ and has discovered internal structure in clusters. The *XMM-Newton* grating spectra can determine accurate abundances for the central regions of clusters, in a model independent fashion, for O, Ne, Mg, Fe, and Si.

It is virtually impossible to give a balanced review of the present observational state of X-ray cluster research, with more than 100 papers published each year. I will not say much about those issues for which we have had detailed talks at this meeting: cooling flows, high-redshift clusters and evolution, X-ray data and the Sunyaev-Zel'dovich effect, radio source interaction, X-ray selected active galaxies in clusters, X-ray emission from groups, and detailed comparison of masses derived from lensing and X-ray observations. Other areas, such as the presence of nonthermal emission and the existence of very soft components, were not discussed. Even limiting the talk this much results in an abundance of material. However, for the purposes of continuity, I have included some material that overlaps with the reviews on chemical abundance given by Renzini (2004) and on groups by Mulchaey (2004). This review does not consider work published after February 2003.

8.2   Temperature Structure of Clusters

As discussed in detail by Evrard (2004), we now have a detailed understanding of the formation of the dark matter structure for clusters of galaxies. If gravity has completely controlled the formation of structure, one predicts that the gas should be in hydrostatic equilibrium with the vast majority of the pressure being due to gas pressure. If this is true, its density and temperature structure provide a detailed measurement of the dark matter distribution in the cluster. Recent theoretical work has also taken into account other process such as cooling and turbulence, which can be important. The fundamental form of the Navarro, Frenk, & White (1997; hereafter NFW) dark matter potential and the assumption that the fraction of the total mass that is in gas is constant with radius result in a prediction, both from analytic (Komatsu & Seljak 2001) and numerical modeling (Loken et al. 2002), that the cluster gas should have a declining temperature profile at a sufficiently large distance from the center (in units of $R/R_{\rm virial}$). The size of the temperature drop in the outer regions is predicted to be roughly a factor of 2 by $R/R_{\rm virial} \approx 0.5$, which is consistent with the *ASCA* results of Markevitch (1998). However, there is considerable controversy about the analysis and interpretation of temperature profiles before *XMM-Newton* and *Chandra*. Results from both *ASCA* (Kikiuchi et al. 1999; White & Buote 2000) and *Beppo-SAX* (Irwin & Bregman 2000; De Grandi & Molendi 2002), indicate either isothermal gas or a temperature gradient in the outer regions of some "cooling flow" clusters. *XMM-Newton* is perfect for resolving this controversy, having a much better point spread function than *ASCA* and much more collecting area than *Beppo-SAX* and *Chandra*, and having a larger field of view than *Chandra*. However, there is a selection effect due to the smaller *XMM-Newton* field of view than *ASCA*, and in order to go out to the virial radius in one pointing one must observe clusters at $z > 0.1$.

There are several published temperature profiles from *XMM-Newton* (Tamura et al. 2001; Majerowicz, Neumann, & Reiprich 2002; Pratt & Arnaud 2002) and I have analyzed several other moderate redshift clusters and others were presented at this conference (Jones et al. 2004). With the exception of one object (A1101S; Kaastra et al. 2001) all the published *XMM-Newton* profiles are consistent with isothermal profiles out to $R/R_{\rm virial} \approx 0.5$ (Fig. 8.2), which is in strong disagreement with the numerical and analytic modeling. This sample of ~ 12 objects is highly biased to smooth, centrally condensed clusters (with the exception of Coma, which has been known to be isothermal from the early work of Hughes et al. 1988). The data for A2163 are consistent with a temperature drop at even larger radii (Pratt, Arnaud, & Aghanim 2002), but the relatively high *XMM-Newton* background makes the results somewhat uncertain. The origin of the difference between some of the *Beppo-SAX*, *ASCA*, and *XMM-Newton* results is not clear. It is possible that there is a difference between the low-z systems studied by *Beppo-SAX* and *ASCA* and the higher-redshift systems studied by *XMM-Newton* and/or a selection effect in the objects so far analyzed with *XMM-Newton*. While the *Chandra* data do not go out to very large length scales (Allen, Schmidt, & Fabian 2002), analysis of 2 $z \approx 0.7$ clusters with *Chandra* (Ettori & Lombardi 2003) also show isothermal profiles.

We must now take seriously the disagreement between theory and observation in the tem-

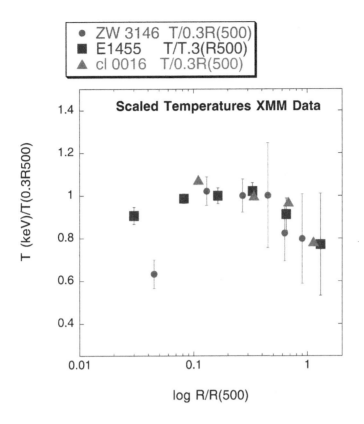

Fig. 8.2. Normalized temperature profiles of three moderate redshift clusters derived from *XMM-Newton* EPIC observations. The ratio of the temperature in a radial bin vs. the radius in units of R_{500} is plotted.

perature profiles in comparing cluster properties with simulations. Another serious issue is the inability of theoretical models to match the observed temperature drops in the centers of the "cooling flow" clusters. The question is then, what is the origin of the discrepancy? Several possibilities are that the form of the theoretical potential is incorrect, that the gas distribution is not calculated correctly, or that physics other than gravity needs to be included. As shown below (§8.8) the form of the potential from X-ray imaging spectroscopy agrees quite well with the NFW potential, which is consistent with the analytic work. *ROSAT* and *XMM-Newton* analysis of X-ray surface brightness distributions (§8.5) shows that the β model is a good description of the X-ray surface brightness at large radii. This leaves us with the possibility that additional physics is needed. Recent analysis of *Chandra* data (cf. Markevitch et al. 2003) strongly constrains the effects of conduction, which will tend to make isothermal spectra, while the inclusion of cooling and heating in the theoretical models (Loken et al. 2002) does not seem to affect the temperature profile significantly. Thus, the origin of this severe discrepancy is not currently known.

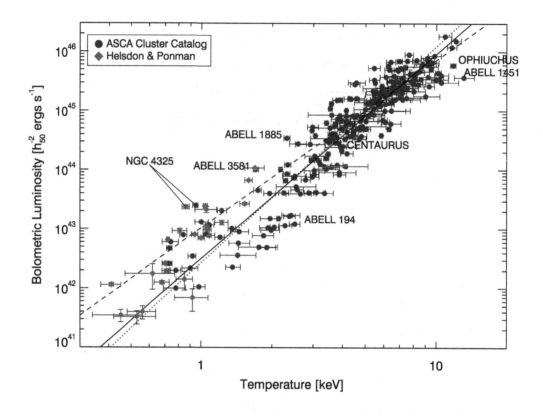

Fig. 8.3. X-ray luminosity vs. X-ray temperature, derived from *ASCA* observations of ~270 clusters (Horner 2001). The best-fit $L_X \propto T^3.4$ for the overall sample (solid line), while the best fit for clusters of luminosity less than $2/times 10^{44}$ ergs s^{-1} is drawn as a dotted line

8.3   Luminosity-Temperature Relation for Clusters

As pointed out by Kaiser (1986), simple scaling relations predict that the cluster luminosity should scale as the temperature squared (T^2). To see this, note that the X-ray luminosity should scale as the density squared times the volume times the gas emissivity, $L_X \propto \rho^2 V \Lambda$. The mass of gas scales like $\sim \rho V$, and it is assumed that the total mass M_T scales as M_{gas}. Since the emissivity for bremsstrahlung, the prime cooling mechanism in gas hotter than 2 keV, scales as $T^{0.5}$ (Sutherland & Dopita 1993), one has $L_X \propto M\rho T^{0.5}$. Finally, since, theoretically, the total mass scales as $T^{1.5}$, one has $L_X \propto \rho T^2$. The other free parameter, the average density, is related to the mass and collapse epoch of the cluster.

It has been known for 20 years (Mushotzky 1984) that the actual relationship between temperature and luminosity is steeper than the simplest theoretical prediction. Recently, Horner et al. (2004) have examined the $L_X - T$ relation using the largest sample of clusters to date (270 clusters taken from the *ASCA* database). In this sample one finds that, over a factor of 10^4 in luminosity, the luminosity scales as T^3. As one goes to lower luminosities there is a wider range of luminosity at a fixed temperature (Fig. 8.3), but there is no need to change the scaling law. This increase in variance probably explains the steeper fit at low luminosity found by Helsdon & Ponman (2000). This continuity is rather strange, since at $T < 2$ keV the cooling function changes sign and scales more like T^{-1}, and thus the theoretical relation between L_X and T changes slope.

There have been many papers written about the origin of the discrepancy, but the main conclusion is that it is due to the breaking of scaling laws via the inclusion of physics other than gravity. The same physics that helps to explain the deviation of entropy in groups, such as heating and cooling, can also explain the slope and normalization of the $L_X - T$ relation (see Mulchaey 2004 and Borgani et al. 2002). Another indication of this scale breaking is the relative low level of evolution in the $L_X - T$ relation out to $z \approx 1$ (Borgani et al. 2002) which is not what is predicted in simple theories of cluster evolution, since objects at $z \approx 1$ are predicted to be denser and have a higher temperature for a fixed mass. Simple scaling predicts that $T \propto M^{1.5}(1+z)$, and thus one predicts $L \propto T^2(1+z)^{0.5}$ at a fixed mass, which is not seen (but see Vikhlinin et al. 2002 for a different opinion).

It was pointed out by Fabian et al. (1994) that high central density, short cooling time clusters (alias "cooling flow" clusters) have a considerably higher luminosity for their temperature than non-cooling flow systems. This result is confirmed in the larger Horner et al. (2004) sample. Markevitch (1998) removed the high-central surface brightness central regions from these clusters and found that the scatter in the $L_X - T$ relationship was much reduced and the fit was flatter than T^3. If the scatter in the $L_X - T$ relationship was due to cool gas in the center of the cooling flow clusters, one should expect that the *ROSAT* luminosities, which are very sensitive to low-temperature gas, would be systematically larger than the luminosities calculated from isothermal fits to the *ASCA* data. However, Horner et al. (2004) find that the bolometric luminosities obtained by *ASCA* are in very good agreement with the *ROSAT* results. This indicates that the central luminosity "excess" is not due to cool gas, as was originally shown in the *ASCA* data for the Centaurus cluster (Ikebe et al. 1999) and recently shown in detail by *XMM-Newton* spectroscopy of many cooling flow clusters (Peterson et al. 2003). Horner et al. (2004) find that the most reasonable explanation for the higher luminosity of the cooling flow clusters is due to their higher central density in the core. This result is consistent with the detailed analysis of cluster surface brightness profiles by Neumann & Arnaud (2001) (see §8.5). It thus seems that the scatter in the $L_X - T$ relation at high temperatures is due to differing cluster central gas densities, while the scatter at low temperatures is due to different "amounts" of additional, nongravitational physics.

8.4   Optical Light, Velocity Dispersion, and X-ray Properties

It has been known since the early *Uhuru* results (Jones & Forman 1978) that there is a great degree of scatter in the correlation between cataloged optical properties, such as Abell richness, and X-ray properties, such as luminosity and temperature. The best correlations between optical and x-ray properties seen in the early data were between central galaxy density and X-ray luminosity (Bahcall 1977b), and between X-ray temperature and optical velocity dispersion (Edge & Stewart 1991). The wide scatter is nicely illustrated in Figure 5 of Borgani & Guzzo (2001)which shows that the Abell counts are only weakly related to total mass, while the x-ray luminosity is strong correlated.

Bird, Mushotzky, & Metzler (1995) showed that much of the scatter in the temperature - velocity dispersion correlation was due to undersampled optical data and velocity substructure in the clusters. More recent optical and X-ray work (Girardi et al. 1998; Horner et al. 2004) shows that when the velocities of a sufficient number of galaxies in a cluster

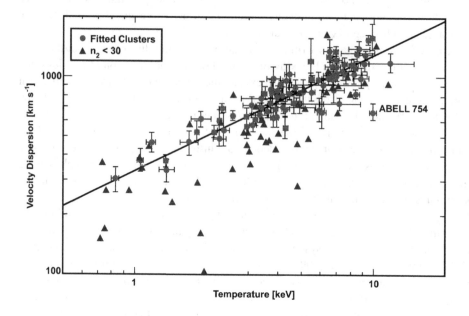

Fig. 8.4. X-ray temperature vs. velocity dispersion, taken from Horner (2001). The triangles represent points with fewer than 30 galaxies per cluster. Note that these points contribute much of the variance in the fit.

are measured (one needs more than 30 galaxies) (Fig. 8.4) there is a tight relation between velocity dispersion and temperature of the form $\sigma \propto T^{0.59\pm0.03}$, consistent with the work of Bird et al. (1995) and close to the theoretical slope of 0.5. This has been confirmed in an infrared-selected sample by Kochanek et al. (2003). The normalization of this relation at high temperatures agrees with theoretical work (Evrard 2004), and thus one has to conclude that low-velocity dispersion clusters are too hot for their dispersion, or that low-temperature clusters have too low a dispersion for their temperature. The fact that clusters have very small radial velocity dispersion gradients (Biviano & Girardi 2003) or temperature gradients (§8.2) makes comparison of the average temperature and velocity dispersion meaningful. This variation with temperature of the velocity dispersion to temperature ratio will also change the effective X-ray vs. optically determined mass by a factor of 50% over the full mass range of clusters.

Recent 2MASS work by Kochanek et al. (2003) shows that, if the "optical" data are handled carefully (e.g., accurate photometry, well-defined selection criteria, observing in a red passband, etc.), there is a strong relation between the total light in a cluster and the X-ray temperature and luminosity (also see Yee & Ellingson 2003). However, while the correlations are much better than in previous work, the scatter in the relation is large, almost a factor of 10 in light at a fixed X-ray temperature or luminosity, or, alternatively, a factor of ~2 in temperature at a fixed optical luminosity. Thus, one expects that optical and X-ray catalogs of clusters might differ considerably depending on where the cuts are made. There is no evidence for either optically or X-ray quiet clusters, but there is evidence for relatively optically or X-ray bright objects. The nature and origin of this variance is not understood at present, but, given the quality of modern data, this variance seems to be real, rather than due to measurement uncertainties. Assuming that the X-ray properties accurately trace mass, the K-band light is a mass indicator accurate to 50% (Lin, Mohr, & Stanford 2003). The converse test, estimating the mass from the optical data and comparing it to the X-ray data, shows large scatter (Yee & Ellingson 2003), where the temperature data are taken from the literature. If it is indeed the case that there is a large variation in the ratio of optical light to X-ray temperature, this indicates that there is a considerable variance in cluster mass-to-light ratio at a fixed mass. This would be a major challenge to structure formation theories.

8.5 Surface Brightness Profiles

It has been known since the pioneering work of Jones & Forman (1984) that the surface brightness profiles of most clusters can be well fit, at large radii, by the "isothermal" β model, $S(r) = S_0(1 + (r/a)^2)^{(-3\beta+0.5)}$, with a central excess above the β model in cooling flow clusters. As seen in *ROSAT* data for high-redshift systems (Vikhlinin, Forman, & Jones 1999), the β model fits amazingly well out to the largest radii measurable for massive clusters. The fitted values of β are smaller for low-mass systems (Helsdon & Ponman 2000; Mulchaey et al. 2003), but there are two selection effects that make the interpretation of this result difficult. First of all, because of their low surface brightness, the group profile hits the background at relatively small distances from the center, and thus one does not detect low-mass systems out to large fractions of the virial radius. This can introduce a bias to the fitted values of β. Secondly, the effects of the central galaxy on the surface brightness is often not well determined from *ROSAT* PSPC data (Helsdon & Ponman 2003)and thus, frequently, the structural parameters are not well constrained. This latter effect is not present in *XMM* or *Chandra* data.

Neumann & Arnaud (2001) have pointed out that the surface brightness profiles of high-temperature clusters remain self-similar as a function of mass and redshift, as expected from cold dark matter models (see also Vikhlinin et al. 1999). Since the conversion from angle to distance depends on the cosmology, they have been able to show that the change of profile with redshift is most consistent with a Λ dominated cosmology. The homology of the profiles is only applicable outside of the central 100 kpc, as inside this radius there are often large deviations from the scaling laws. However, in order to achieve the scaling they require that the relationship of gas mass to temperature be $M_{gas} \propto T^2$, steeper than the theoretical scaling between total mass and temperature (i.e., $M_{total} \propto T^{1.5}$). Since the surface brightness profiles scale according to the predicted evolution from the cold dark matter models, the lack of evolution in the luminosity-temperature law must be a cosmic conspiracy between the cosmological model and the change of density with redshift. The prediction is that the emission measure of the gas scales as $EM \propto \beta f_{gas}^2 \Delta^{1.5}(1+z)^{9/2}(kT)^{0.5}h^3$, where Δ is the overdensity of the cluster and f_{gas} is the fraction of mass that is in gas (Arnaud, Aghanim, & Neumann 2002).

There are "single" clusters that are not well fit by the β model. The most obvious example is MS 1054–0321 at $z = 0.82$ (Jeltema et al. 2001), which is much more concentrated than a β model. This is not a function of redshift, since many clusters at $z > 0.6$ are well fit by the β model.

8.6   Mass of Baryons and Metals and How They Are Partitioned

The two main baryonic components of clusters are the X-ray emitting gas and the stars, since the total contribution from cold gas and dust is very small. The major uncertainty in the relative baryonic contribution is due to the uncertainty in the transformation from light to mass for the stars. Recent work from large optical surveys (Bell et al. 2003) shows that the mass-to-light ratio of stars changes as a function of galaxy but is $\sim$3.5 in the Sloan g band for a bulge-dominated population. Using this value and the mean mass-to-light ratio of clusters $\sim$240 (Girardi et al. 2002), the stars have $\sim$0.015 of the total mass. The gas masses have been well determined from *ROSAT* data (Ettori & Fabian 1999; Allen et al. 2002) and scatter around $f_{\text{gas}} \approx 0.16 h_{70}^{-0.5}$. Thus, the gas-to-stellar mass ratio is $\sim$10:1, and the total baryon fraction is almost exactly consistent with the recent *WMAP* results for the Universe as a whole. Since it is thought that clusters are representative of the Universe as a whole, this suggests that the vast majority of baryons in the Universe do not lie in stars. Turning this around, one can use the baryonic fraction in clusters as a bound on Ω_{m} (White et al. 1993). The most recent analysis using this technique finds $\Omega_{\text{m}} < 0.38 h_{70}^{-0.5}$ (Allen et al. 2002), in excellent agreement with the *WMAP* data. It is interesting to note that the high baryonic fraction in clusters has been known for over 10 years and was one of the first strong indications of a low Ω_{m} Universe. Since it is thought that the baryonic fraction in clusters should not evolve with redshift, derivation of the baryonic abundance in high-z clusters, which depends on the luminosity distance, provides a strong constraint on cosmological parameters (Ettori & Tozzi and Rosati 2003).

The mean metallicity of the gas in clusters is $\sim$1/3 solar (see §8.10), while that of the stars may be somewhat larger. If we assume 1/2 solar abundance for the stars, than $\sim$85% of the metals are in the gas phase. Since all the metals are made in stars, which lie primarily in galaxies, this implies that most of metals have either been ejected or removed from the galaxies. Since the stellar mass is dominated by galaxies near $L^{\star}$, which have a mean escape velocity, today, of >300 km s^{-1}, this implies very strong galactic winds at high redshift. This scenario is consistent with the results of Adelberger et al. (2003) on the high-redshift, rapidly star-forming U and B-band drop-out galaxies, which all have large-velocity winds.

Analysis of the gas mass fraction in groups and clusters (Sanderson et al. 2003) indicates that the fraction apparently drops at lower masses by a factor of 2–3, with the reduction setting in at a mass scale corresponding to 1–3 keV at 0.3 R_{200}. In addition, the stellar mass-to-light ratio decreases by 60% over the same mass range (Marinoni & Hudson 2002), and thus in groups the gas-to-stellar mass ratio is only (1–2):1 at 0.3 R_{200}, considerably smaller than in clusters. However, there is a serious problem for groups in evaluating both the gas and stellar masses at large radii (see Fig. 10 in Mulchaey et al. 2003), and this result should be taken with some caution. In particular, the X-ray surface brightness distribution of groups is often very flat, and extrapolating from 0.3 R_{200} to R_{200} is rather risky. However, if these trends are real, this would indicate that groups are truly baryon poor, that the baryons have been pushed out of the group, or that the gas has been puffed up. If the gas has been puffed up, this is consistent with the somewhat high temperatures of groups compared to their optical galaxy velocity dispersions, indicative of extra heat deposited in the gas, which both heats it and "puffs" it up (see discussion in the review by Mulchaey 2004).

8.7   Mass Scaling Laws

Detailed theoretical work has verified that clusters should satisfy the virial theorem, and thus their mass should scale as $M \propto TR$, with $R \propto T^{1/2}$, and thus $(1+z)M^{2/3} \propto T$ (e.g., Eke, Navarro, & Frenk 1998), with the normalization being set by theory and the value of the cosmological parameters (Evrard 2003). The first test of this relation (Horner, Mushotzky, & Scharf 1999) found a scaling that was somewhat steeper, with $M \propto T^{1.7}$, and a normalization that was 40% lower than predicted. Finoguenov, Reiprich, & Böhringer (2001) and Reiprich & Böhringer (2002) have confirmed these results with more uniform samples, and higher quality, spatially resolved spectra. Recent *Chandra* results (Allen, Schmidt, & Fabian 2001) are also consistent with the Horner et al. (1999) finding. *XMM-Newton* data for A1413 (Pratt & Arnaud 2002) show that the normalization scaling is not only violated by the sample, but by individual objects. The normalization in the Reiprich & Böhringer (2002) sample agrees with theoretical expectations at the high-mass end. This indicates that lower-temperature clusters are less massive than expected on the basis of their temperature, consistent with the trend seen in the velocity dispersion-temperature relation. Recently, it has been pointed out (Shimizu et al. 2003) that the combination of the scaling of mass by $M \propto T^{1.7}$ and the gas mass fraction scaling as $T^{1/3}$ (a reasonable fit to the Sanderson et al. 2003 data) can reproduce the observed $L_X \propto T^3$ relationship. Theoretical calculations that include the effects of cooling (Thomas et al. 2002) seem to be consistent with the lower normalization, but so far the slope difference has not been explained.

8.8   Form of the Potential

As discussed extensively in this conference, the form of the potential in clusters should be determined by the distribution of dark matter. Recent numerical work seems to validate the NFW potential, and much has been made of the fact that low-mass and low-surface brightness galaxies do not seem to follow this form in their central regions. Recent *Chandra* and *XMM-Newton* observations (Allen et al. 2002; Arabadjis, Bautz, & Garmire 2002; Pratt & Arnaud 2002) have been able to determine extremely accurate mass profiles via spatially resolved X-ray spectroscopy and the assumption of hydrostatic equilibrium. Perhaps the best documented of these examples are the *Chandra* data for Abell 2029 (Lewis, Buote, & Stocke 2003), in which the profile is determined over a factor of 100 in length scale, from 0.001–0.1 characteristic lengths of the NFW profile, with essentially no deviation from the NFW prediction. This striking result is also seem in other *Chandra* results in the cores of clusters. The data show that the central regions of clusters tend to have rather steep density profiles in the innermost radii, indicating that whatever causes the deviation of the form of the potential in dwarf galaxies does not occur in clusters. This results strongly constrains interacting dark matter models (Bautz & Arabadjis 2004). A survey of *Chandra* central mass profiles is made somewhat difficult because of the possibly complex nature of the IGM in the central regions of many clusters, and the exact slope and normalization of the mass depends on the details of the thermal model used. However, if the data are of sufficiently high signal-to-noise ratio, the form of the mass profile can be determined precisely. I anticipate quite a few exciting new results in this area; preliminary results, presented in several conferences, indicate a predominance of steep mass profiles with slopes close to the NFW level, but with some scatter.

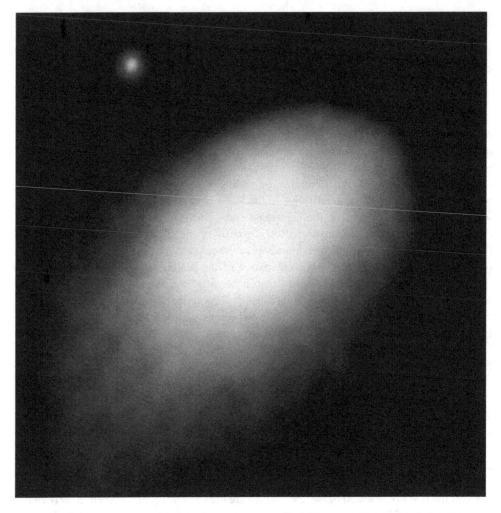

Fig. 8.5. The *Chandra* image of a cold front in Abell 2142 (courtesy of the *Chandra* Science Center). Note the very sharp boundary to the north, which is not an artifact of the image processing.

8.9   Merges, Structures, etc.

The early *Einstein* Observatory images of clusters (Henry et al. 1979) showed that a substantial fraction of the X-ray images were not simple, round systems, but often complex in form and sometimes even double. This observation is consistent with the idea that clusters form in a hierarchical fashion via mergers, and that the complex systems are in the process of merging. The fact that mergers are actually occurring, rather than the complex structures in the images being simply projection effects, was indicated by complexity in the temperature structure of many of these systems shown by the *ROSAT* (Briel & Henry 1994) and *ASCA* (Markevitch 1996) data. The details of the nature of this process have had to wait until the precise *Chandra* spectral images showed the full range of complexity. While "textbook" examples of merger shocks have been seen (e.g., 1E 0657−56; Markevitch et al.

2002), many of the objects show only subtle temperature variations (e.g., Sun et al. 2002). These variations have only shown up in the most recent, very high-resolution numerical simulations, indicating the non-intuitive nature of these data.

The recent spectacular *XMM-Newton* temperature image of the Perseus cluster (Churazov et al. 2003) illustrates the wealth of detail that is now possible to obtain. It is interesting that these spectral images do not show the numerous "cold spots" that are predicted in cluster simulations that include cooling (Motl et al. 2004). The ability to obtain spectral images has also revealed "hidden mergers." Both the Coma and Ophiuchus clusters, the hottest nearby systems, show smooth, regular X-ray images; however, X-ray temperature maps show strong spatial variations (Arnaud et al. 2001; Watanabe et al. 2001). So far the data on abundance variations in the mergers is sparse, but the abundances seem uniform, within errors, in Coma and may vary by less than a factor of $\sim$2 in Ophiuchus. It seems as if many of the large-scale length, non-cooling flow clusters are recent mergers.

One of the surprises of the *Chandra* data was the discovery of surface brightness discontinuities in the surface brightness — the so-called cold fronts (Fig. 8.5; Vikhlinin & Markevitch 2002). These cold fronts are apparently contact discontinuities, across which the pressure is smoothly varying but the density and temperature change discontinuously by factors of $\sim$2. They can occur in "pure hydro" numerical simulations (Bialek, Evrard, & Mohr 2002). Their relative frequency is a indication of the merger rate. However, the details (e.g., temperature drop, size of region, etc.) and their relation to merger dynamics are not certain (Fujita et al. 2002). The stability of the cold fronts, their sizes, and shapes are indications of the strength of the magnetic field, velocity vector of the merger, and the amount of turbulence (Mazzota, Fusco-Femiano, & Vikhlinin 2002) in the cluster gas. It is clear that there is much to learn from further studies of these unexpected structures, but they already confirm that the gas is usually not strongly shocked, nor highly turbulent.

8.10   Abundances

As indicated above, most of the metals in the cluster lie in the hot, X-ray emitting gas. Thus, in order to understand the formation and evolution of the elements one must determine accurate abundances, the abundance distribution in the gas, and its evolution with cosmic time. Before giving the results it is important to remind the reader that the measurement of abundances in the cluster gas via X-ray imaging spectroscopy is a robust process. Most of the baryons and metals are in the hot gas, and the spectral signature of the heavy elements are relatively strong H and He-like lines (Fig. 8.6). This is a well-understood emission mechanism, with little or no radiative transfer difficulties. Because of the high temperature and short spallation times, dust is destroyed rapidly and thus is not a problem. The deep potential well captures an integrated record of all the metals produced, and thus the derived abundances are true averages of the metal production process. All the abundant elements from oxygen to nickel can have their abundances determined. Direct measurement of the electron temperature from the form of the continuum and from ratios of H to He-like lines ensures small systematic errors in the abundances. The strongest lines in the spectrum of hot clusters are due to Fe and Si, followed by O, S and Ni. The emission from Ne and Mg is blended with Fe L-shell lines from Fe XVII–XXIV at the resolution of X-ray CCDs, and the lines from Ca and Ar are weak. With present-day technology, one can measure Fe to $z \approx 1$ and Si to $z \approx 0.4$, and can thus obtain a true measure of the metal formation mech-

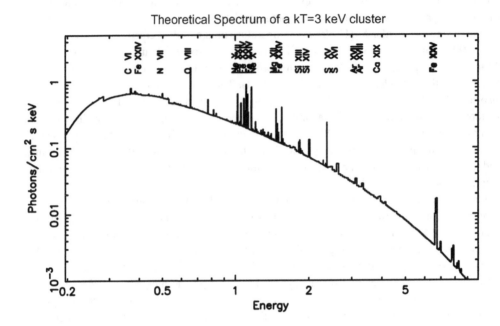

Fig. 8.6. Theoretical spectrum of an isothermal plasma with $kT \approx 3$ keV, with many of the strong transitions labeled; however, many of the transitions from the same ion (e.g., multiple lines from Fe XXIV) have been suppressed. Note the strong lines from all the abundant elements. The data are plotted in the usual way for X-ray astronomers, photons cm^{-2} s^{-1} keV^{-1} vs. keV, which emphasizes the dynamic range of X-ray spectroscopy.

anism and its evolution. For much of the rationale and background for cluster abundance measurements, see Renzini's (2004) review in this volume.

Recently (Baumgartner et al. 2004; Horner et al. 2004), a uniform analysis of the *ASCA* cluster database of 270 clusters has been performed, which updates previous work (e.g., Mushotzky et al. 1996; Fukazawa et al. 2000) on cluster abundances. Horner et al. (2004) and Baumgartner et al. (2004) measure the average cluster Fe, Si, S, and Ni abundances with no spatial information. They find (Fig. 8.7) that the Fe abundance is not the same for all clusters, but shows a small spread of a factor of ~2. In agreement with Fabian et al. (1994), the cooling flow clusters show, on average, a higher Fe abundance. There is no evidence for any evolution in the Fe abundance out to $z \approx 0.5$ on the basis of *ASCA* data. Recent *XMM-Newton* and *Chandra* results (Jones et al. 2004; Mushotzky, private communication) show no evolution in the Fe abundance to $z \approx 0.8$. This lack of evolution indicates that the metals are created at $z > 1.3$ for a Λ cosmology (I have added in the lifetime of the A stars that would be visible for the massive amount of star formation necessary to produce the observed metals). Since the vast majority of the metals are in the gas, the rate of specific star formation (e.g., the rate per unit visible stars) would have to be enormous to produce

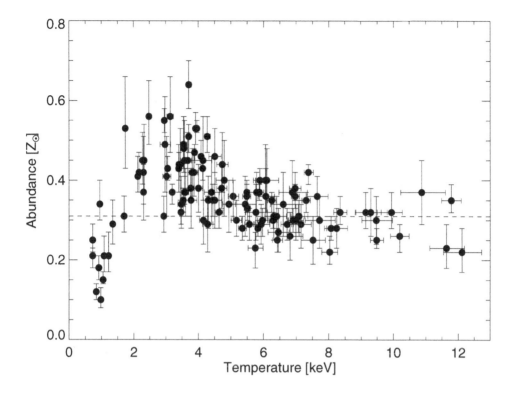

Fig. 8.7. Fe abundance vs. temperature for the *ASCA* cluster database (Horner 2001). The dashed line shows the average for clusters with temperatures greater than 5 keV. Notice that there is little variation at high temperatures, but there is a systematic rise in abundance and then fall at low temperatures.

the elements if it were to occur at $z < 2$. The Fe abundance is weakly correlated with the temperature, reaching a maximum at $kT \approx 2-3$ keV, but is more or less constant for $kT > 4.5$ keV clusters. Given the accuracy of recent plasma codes, the "peak" in abundance, which occurs at a temperature range where both Fe L and K lines contribute to the abundance determination, is almost certainly a real effect. The physical origin of the variance in Fe abundance and the trends are unknown. However, since there are trends in the apparent ratio of starlight to gas (§8.6), this may be the cause. Further progress in this area requires an enhancement of the original work of Arnaud et al. (1992), which found a correlation between the light in elliptical galaxies and the total mass of Fe in the cluster.

The distribution of the elements in a cluster determines the total amount of material and gives clues as to how the material was deposited in the IGM. The previous generation of X-ray satellites (*ASCA*, *ROSAT*, and *Beppo-SAX*) derived abundance profiles of Fe in >20 clusters (Finoguenov, David, & Ponman 2000; Irwin & Bregman 2000; White & Buote 2000; De Grandi et al. 2003). However, these results did not always agree (different analysis of the same data and comparison of data from the same object from different satellites produced different results). There was a tendency for cooling flow clusters to have high central Fe abundances and larger total abundances, suggesting a different origin of IGM enrichment in the central regions, the effects of mixing by mergers on the Fe abundance profile, or a physical difference in the origin of the metals in cooling flow clusters.

XMM-Newton and *Chandra* data have much smaller systematic errors and much better signal-to-noise ratios than the data from the earlier observatories. Early results are available for ∼15 systems — most are isochemical at large radii, with several having gradients in the central 100 kpc. The *Chandra* and *XMM-Newton* data are well resolved and show that the abundance gradients are quite concentrated toward the center (cf. David et al. 2001; Tamura et al. 2001). For a few objects the profiles reach to near the virial radius (e.g., Zw 3146, Cl 0016 (Mushotzky priv. comm.) , and A 1835; Majerowicz et al. 2002), two of which (Zw 3156 and A 1835) are massive cooling flows do not show abundance gradients outside 100 kpc. Numerical evaluation of the observed Fe abundance gradients (De Grandi et al. 2003) shows that most of the variation in the average Fe abundance between the cooling flow and non-cooling flow clusters is not due to differences in the Fe gradients. The "excess" amount of Fe in the central regions seen in the cooling flow systems is correlated with the presence of a cD galaxy, and the mass of "excess" Fe is roughly consistent with its being produced in the stars in the central cD galaxy. This is rather unexpected, since isolated elliptical galaxies have only ∼1/5 of the Fe that should have been produced by the stars (Awaki et al. 1994).

The fact that gradients do not dominate the average abundance allows a direct interpretation of the *ASCA* average abundances. The *ASCA* database of ∼270 X-ray spectra allow determination of average Fe, Si , S, and Ni in clusters of galaxies (Baumgartner et al. 2004). However, the signal-to-noise ratio for most of the individual clusters is not adequate to derive robust S or Ni abundances, and 20–40 objects in each temperature bin must be added together to derive average values and their variation with temperature. Since cluster mass is directly related to the temperature and line strength is also directly connected to temperature, this is the natural space for averaging. As originally pointed out by Fukazawa et al. (2000), as T increases, Si/Fe increases. However, the new data show that S remains roughly constant versus temperature. Baumgartner et al. (2004) also find that the Ni/Fe ratio is approximately 3 times solar. While these are very surprising results, they are similar to previous analysis of smaller *ASCA* and *XMM-Newton* data sets. The S/Fe, Si/Fe, and Ni/Fe ratios depend on the relative abundance of the types of supernovae (SNe). Type Ia SNe produce mostly Fe and Ni, while Type II SNe produce a wide range of elements but large ratios of the α elements (O, Ne, and Si) to Fe. Si and S are produced via very similar mechanisms, and at first sight it is hard to understand how they could have different abundance patterns. In addition, in the Milky Way, S almost always directly tracks Si. The fact that both Si/Fe and S/Fe drop as Fe increases shows that there is indeed a difference in the mechanisms producing the metals as a function of mass scale. It seems rather unexpected that the ratio of Type II to Type Ia SNe in the stars that live in cluster galaxies should change with the mass scale of the cluster. However, the high Ni/Fe ratio indicates that Type Ia SNe are important in the production of Fe, at least in the central regions of clusters (Dupke & White 2000), and this high ratio does not allow a simple variation in SN type with cluster mass to readily explain the abundances patterns seen in the ASCA data.

XMM-Newton data allow the measurement of O abundances for a reasonable sample of objects for the first time. The best sample published to date is based on the high-resolution RGS data (Peterson et al. 2003). They find that the O/Fe ratio varies by a factor of ∼2 from cluster to cluster, with no apparent correlation with temperature. Analysis of *XMM-Newton* CCD data taken over a larger scale (the RGS data sample only the central $1'-2'$ of the cluster) confirm this variance. As noted in Gibson, Loewenstein, & Mushotzky (1997), the elemental abundance ratios averaged over the cluster do not agree with any simple ratio of Type Ia to Type II SNe. However, it is clear that over 90% of the O, Ne, and Mg must originate in Type II SNe.

The new *XMM-Newton* O abundances further strengthen this conclusion. However, some of the difficulties may be caused by differential abundance gradients of different elements. There are strong indications from *ASCA* data (Fukazawa et al. 2000; Finoguenov et al. 2001) that the Fe/Si ratio rises in the cluster centers, consistent with the cD galaxy being a source of Fe-rich material, probably due to Type Is SNe. However, the new *XMM-Newton* data show that O does not follow this pattern. It is clear that more work is necessary with larger samples and abundance profiles before we can obtain a clear picture of the metal enrichment process in clusters.

8.11 Conclusion

The progress in this field in the last 10 years has been amazing. The X-ray properties of objects at redshift $z \approx 0.8$ are routinely measured, and clusters are now X-ray detected at $z > 1.15$. The use of clusters for cosmology, an area covered in the volume by Freedman (2004), is exploding. The physics of clusters and groups holds the key to understanding the origin and evolution of structure and the origin of the elements. It was the cluster data that first showed that most of the baryons and metals in the Universe are in the hot phase, and that the baryonic Universe, as seen by our eyes, is only a shadow of the real Universe. In the next few years we will continue to obtain vast amounts of new data from *Chandra* and *XMM-Newton*, and much of the present observations will be analyzed, interpreted, and new patterns found. There are over 400 *Chandra* and *XMM-Newton* observations of clusters and groups in the database so far, with many more to be observed over the lifetimes of these telescopes. I anticipate many major new discoveries based on these instruments. Furthermore, the launch of *Astro-E2* in 2005 will allow detailed measurements of cluster turbulence, accurate abundances of many elements outside the cluster cores, and direct measures of the thermodynamic properties of the gas.

The field has benefited enormously from the synergistic interaction of theory and observation. Most theorists and observers are now aware of the major issues and the current observational capabilities. Looking beyond the next few years, I anticipate that a major new X-ray survey, perhaps 30 times better than *ROSAT*, will fly, producing an extremely large and uniform cluster catalog complete out to $z \approx 0.7$. In the more distant future, the *Constellation-X* mission will provide precision temperatures and abundances out to the highest redshifts that clusters exist.

Acknowledgements. I would like to thank my long-time collaborators and students at Goddard for their major contribution to this work: Keith Arnaud, Wayne Baumgartner, Don Horner, Mike Loewenstein, and John Mulchaey. I would like to thank the *Chandra* and *XMM-Newton* projects for their major efforts in developing, launching, and operating these amazing instruments. I would also like to thank the *ASCA* team for their pioneering efforts in the first X-ray imaging spectroscopy mission. I thank M. Arnaud and D. Neumann for communicating results ahead of publication. I also thank the organizers, especially John Mulchaey, for an exciting and stimulating meeting.

References

Adelberger, Kurt L., Steidel, C. C., Shapley, A. E., & Pettini, M. 2003, ApJ, 584, 45

Allen, S. W., Schmidt R. W., & Fabian A. C. 2001, MNRAS, 328, L37

——. 2002, MNRAS, 334, L11

Arabadjis, J. S., Bautz, M. W., & Garmire, G. P. 2002, ApJ, 572, 66

Arnaud, M., et al. 2001, A&A, 365, 67

Arnaud, M., Aghanim, N., & Neumann, D. M. 2002, A&A, 389, 1

Arnaud, M., Rothenflug, R., Boulade, O., Vigroux, L., & Vangioni-Flam, E. 1992, A&A, 254, 49

Awaki, H., et al. 1994, PASJ, 46, L65

Bahcall, N. A. 1977a, ARA&A, 15, 505

——. 1977b, ApJ, 218, 9

Baumgartner, W., et al. 2004, in preparation

Bautz, M. W., & Arabadjis, J. S. 2004, in Carnegie Observatories Astrophysics Series, Vol. 3: Clusters of Galaxies: Probes of Cosmological Structure and Galaxy Evolution, ed. J. S. Mulchaey, A. Dressler, & A. Oemler (Pasadena: Carnegie Observatories, http://www.ociw.edu/ociw/symposia/series/symposium3/proceedings.html)

Bell, E. F., McIntosh, D. H., Katz, N., & Weinberg, M. D. 2003, ApJ, 585, L117

Bialek, J. J., Evrard, A. E., & Mohr, J. J. 2002, ApJ, 578, L9

Bird, C. M., Mushotzky, R. F., & Metzler, C. A. 1995, ApJ, 453, 40

Biviano, A., & Girardi, M. 2003, ApJ, 585, 205

Borgani S., Governato F., Wadsley, J., Menci, N., Tozzi, P., Quinn, T., Stadel, J., & Lake, G. 2002, MNRAS, 336, 409

Borgani, S., & Guzzo, L. 2001, Nature, 409, 39

Briel, U. G., & Henry, J. P. 1994, Nature, 372, 439

Churazov, E., Forman, W., Jones, C., & Böhringer, H. 2003, ApJ, 590, 225

David, L. P., Nulsen, P. E. J., McNamara, B. R., Forman, W., Jones, C., Ponman, T., Robertson, B., & Wise, M. 2001, ApJ, 557, 546

De Grandi, S., et al. 2003, Ringberg Conference Proceedings (XXX)

De Grandi, S., & Molendi, S. 2002, ApJ, 567, 163

Dupke, R. A., & White, R. E., III 2000, ApJ, 528, 139

Edge, A. C., & Stewart, G. C. 1991, MNRAS, 252, 428

Eke, V. R., Navarro, J. F., & Frenk, C. S. 1998, ApJ, 503, 569

Ettori, S., & Fabian, A. C 1999, MNRAS, 305, 834

Ettori, S., & Lombardi, M. 2003, A&A, 398, L5

Ettori, S., Tozzi, P., & Rosati, P. 2003, A&A, 398, 879

Evrard, A. E. 2004, in Carnegie Observatories Astrophysics Series, Vol. 3: Clusters of Galaxies: Probes of Cosmological Structure and Galaxy Evolution, ed. J. S. Mulchaey, A. Dressler, & A. Oemler (Cambridge: Cambridge Univ. Press), in press

Fabian, A. C., Crawford, C. S., Edge, A. C., & Mushotzky, R. F. 1994, MNRAS, 267, 779

Finoguenov, A., David, L. P., & Ponman, T. J. 2000, ApJ, 544, 188

Finoguenov, A., Reiprich, T. H., & Böhringer, H., 2001, A&A, 330, 749

Forman, W., & Jones, C. 1982, ARA&A, 20, 547

Freedman, W. L., ed. 2004, Carnegie Observatories Astrophysics Series, Vol. 2: Measuring and Modeling the Universe (Cambridge: Cambridge Univ. Press), in press

Fujita, Y., Sarazin, C. L., Nagashima, M., & Yano, T. 2002, ApJ, 577, 11

Fukazawa, Y., Makishima, K., Tamura, T., Nakazawa, K., Ezawa, H., Ikebe, Y., Kikuchi, K., & Ohashi, T. 2000, MNRAS, 313, 21

Gibson, B. K., Loewenstein, M., & Mushotzky, R. F. 1997, MNRAS, 290, 623

Girardi, M., Giuricin, G., Mardirossian, F., Mezzetti, M., & Boschin, W. 1998, ApJ, 505, 74

Girardi, M., Manzato, P., Mezzetti, M., Giuricin, G., & Limboz, F. 2002, ApJ, 569, 720

Hasinger, G., et al. 2004, in preparation

Helsdon, S. F., & Ponman, T. J. 2000, MNRAS, 315, 356

——. 2003, MNRAS, 340, 485

Henry, J. P., Branduardi, G., Briel, U., Fabricant, D., Feigelson, E., Murray, S., Soltan, A., & Tananbaum, H. 1979, ApJ, 234, L15

Horner, D. J. 2001, Ph.D. Thesis, Univ. Maryland

Horner, D. J., et al. 2004, in preparation

Horner, D. J., Mushotzky, R. F., & Scharf, C. A. 1999, ApJ, 520, 78

Hughes, J. P., Yamashita, K., Okumura, Y., Tsunemi, H., & Matsuoka, M. 1988, ApJ, 327, 615

Ikebe, Y., Makishima, K., Fukazawa, Y., Tamura, T., Xu, H., Ohashi, T., & Matsushita, K. 1999, ApJ, 525, 58

Irwin, J. A., & Bregman, J. N. 2000, ApJ, 538, 543

Jeltema, T. E., Canizares, C. R., Bautz, M. W., Malm, M. R., Donahue, M., & Garmire, G. P. 2001, ApJ, 562, 124

Jones, C., & Forman, W. 1978, ApJ, 224, 1

——. 1984, ApJ, 276, 38

Jones, L. R., et al. 2004, in Carnegie Observatories Astrophysics Series, Vol. 3: Clusters of Galaxies: Probes of Cosmological Structure and Galaxy Evolution, ed. J. S. Mulchaey, A. Dressler, & A. Oemler (Pasadena: Carnegie Observatories, http://www.ociw.edu/ociw/symposia/series/symposium3/proceedings.html)

Kaastra, J. S., Ferrigno, C., Tamura, T., Paerels, F. B. S., Peterson, J. R., & Mittaz, J. P. D. 2001, A&A, 365, L99

Kaiser, N. 1986, MNRAS, 222, 323

Kikuchi, K., Furusho, T., Ezawa, H., Yamasaki, N. Y., Ohashi, T., Fukazawa, Y., & Ikebe, Y. 1999, PASJ, 51, 301

Kochanek, C. S., White, M., Huchra, J., Macri, L., Jarrett, T. H., Schneider, S. E., & Mader, J. 2003, ApJ, 585, 161

Komatsu, E., & Seljak, U. 2001, MNRAS, 327, 1353

Lewis, A. D., Buote, D. A., & Stocke, J. T. 2003, ApJ, 586, 135

Lin, Y.-T., Mohr, J. J., & Stanford, S. A. 2003, ApJ, in press (astro-ph/0304033)

Loken, C., Norman, M. L., Nelson, E., Burns, J., Bryan, G. L., & Motl, P. 2002, ApJ, 579, 571

Majerowicz, S., Neumann, D. M., & Reiprich, T. 2002, A&A, 394, 77

Marinoni, C., & Hudson, M. J. 2002, ApJ, 569, 101

Markevitch, M. 1996, ApJ, 465, L1

——. 1998, ApJ, 504, 27

Markevitch, M., et al. 2003, ApJ, 586, L19

Markevitch, M., Gonzalez, A. H., David, L., Vikhlinin, A., Murray, S., Forman, W., Jones, C., & Tucker, W. 2002, ApJ, 567, L27

Mazzotta, P., Fusco-Femiano, R., & Vikhlinin, A. 2002, ApJ, 569, L31

Motl, P. M., Burns, J. O., Loken, C., Norman, M. L., & Bryan, G. 2004, ApJ, in press (astro-ph/0302427)

Mulchaey, J. S. 2004, in Carnegie Observatories Astrophysics Series, Vol. 3: Clusters of Galaxies: Probes of Cosmological Structure and Galaxy Evolution, ed. J. S. Mulchaey, A. Dressler, & A. Oemler (Cambridge: Cambridge Univ. Press), in press

Mulchaey, J. S., Davis, D. S., Mushotzky, R. F., & Burstein, D. 2003, ApJS, 145, 39

Mushotzky, R. F. 1984, Physica Scripta, 7, 157

——. 2002, in A Century of Space Science, ed. J. A. M. Bleecker, J. Geis, & M. Huber (Dordrecht: Kluwer), 473

Mushotzky, R. F., Loewenstein, M., Arnaud, K. A., Tamura, T., Fukazawa, Y., Matsushita, K., Kikuchi, K., & Hatsukade, I. 1996, ApJ, 466, 686

Mushotzky, R. F., Serlemitsos, P. J., Boldt, E. A., Holt, S. S., & Smith, B. W. 1978, ApJ, 225, 21

Navarro, J. F., Frenk, C. S., & White, S. D. M. 1997, ApJ, 490, 493 (NFW)

Neumann, D. M., & Arnaud, M. 2001, A&A, 373, L33

Peterson, J. R., Kahn, S. M., Paerels, F. B. S., Kaastra, J. S., Tamura, T., Bleeker, J. A. M., Ferrigno, C., & Jernigan, J. G. 2003, ApJ, 590, 207

Pratt, G. W., & Arnaud, M. 2002, A&A, 394, 375

Pratt, G. W., Arnaud, M., & Aghanim, N. 2002, in Tracing Cosmic Evolution with Galaxy Clusters, ed. S. Borgani, M. Mezzetti, & R. Valdarnini (San Francisco: ASP), 433

Reiprich, T. H., & Böhringer H. 2002, ApJ, 567, 716

Renzini, A. 2004, in Carnegie Observatories Astrophysics Series, Vol. 3: Clusters of Galaxies: Probes of Cosmological Structure and Galaxy Evolution, ed. J. S. Mulchaey, A. Dressler, & A. Oemler (Cambridge: Cambridge Univ. Press), in press

Sanderson, A. J. R., Ponman, T. J., Finoguenov, A., Lloyd-Davies, E. J., & Markevitch, M. 2003, MNRAS, 340, 989

Shimizu, M., Kitayama, T., Sasaki, S., & Suto, Y. 2003, ApJ, 590, 197

Stanford, S. A., Holden, B., Rosati, P., Tozzi, P., Borgani, S., Eisenhardt, P. R., & Spinrad, H. 2001, ApJ, 552, 504

Sun, M., Murray, S. S., Markevitch, M., & Vikhlinin, A. 2002, ApJ, 565, 867

Sutherland, R. S., & Dopita, M. A. 1993, ApJS, 88, 253

Tamura, T., Bleeker, J. A. M., Kaastra, J. S., Ferrigno, C., & Molendi, S. 2001, A&A, 379, 107

Thomas, P. A., Muanwong, O., Kay, S. T., & Liddle, A. R. 2002, MNRAS, 330, L48

Vikhlinin, A., Forman, W., & Jones, C. 1999, ApJ, 525, 47

Vikhlinin, A. A., & Markevitch, M. L. 2002, Astron. Lett., 28, 495

Vikhlinin, A., VanSpeybroeck, L., Markevitch, M., Forman, W. R., & Grego, L. 2002, ApJ, 578, L107

Watanabe, M., Yamashita, K., Furuzawa, A., Kunieda, H., & Tawara, Y. 2001, PASJ, 53, 605

White, D. A., & Buote, D. A. 2000, MNRAS, 312, 649

White, S. D. M., Navarro, J. F., Evrard, A. E., & Frenk, C. S. 1993, Nature, 366, 429

Yee, H. K. C., & Ellingson, E. 2003, ApJ, 585, 215

9

Cool gas in clusters of galaxies

MEGAN DONAHUE and G. MARK VOIT
Space Telescope Science Institute and Michigan State University

Abstract

Early X-ray observations suggested that the intracluster medium cools and condenses at the centers of clusters, leading to a cooling flow of plasma in the cluster core. The increased incidence of emission-line nebulosity, excess blue light, AGN activity, and molecular gas in the cores of clusters with short central cooling times seemed to support this idea. However, high-resolution spectroscopic observations from *XMM-Newton* and *Chandra* have conclusively ruled out simple, steady cooling flow models. We review the history of this subject, the current status of X-ray observations, and some recent models that have been proposed to explain why the core gas does not simply cool and condense.

9.1   A Census of Cool Gas

Clusters of galaxies have very deep potential wells with virial velocities equivalent to temperatures of $10^7 - 10^8$ K. Gravitationally driven processes like accretion shocks and adiabatic compression should therefore heat gas accumulating within a cluster to X-ray emitting temperatures. Spectroscopic X-ray observations show that most of a cluster's gas is indeed near the virial temperature $T_{\rm vir} = \mu m_p \sigma_{\rm 1D}^2/k$, equivalent to $7.1 \times 10^7 \, \sigma_{1000}^2$ K or $6.2 \, \sigma_{1000}^2$ keV, where σ_{1000} is the line-of-sight velocity dispersion in units of $1000 \, {\rm km \, s^{-1}}$ (Sarazin 1986).

Roughly 10%–20% of the baryons associated with clusters have a temperature significantly less than the virial temperature, qualifying as "cool gas" for the purposes of this review. Much of this gas would be considered quite hot in other astrophysical contexts, but in order to be cooler than the virial temperature today, it must either have avoided the gravitational heating experienced by the rest of the cluster or it must have significantly cooled after entering the cluster.

A large proportion of this cool gas is only moderately cooler than the virial temperature. In the central $\sim10\%$ of many clusters, corresponding to gas masses of $10^{11} - 10^{13} \, M_\odot$, temperatures dip to $\sim T_{\rm vir}/2$. Because this gas is dense enough to radiate an energy equivalent to its thermal energy in less than a Hubble time, astronomers have long speculated that it cools and contracts, forming a "cooling flow" of condensing gas in the cluster core (Cowie & Binney 1977; Fabian & Nulsen 1977; Mathews & Bregman 1978).

Gas much cooler than the virial temperature is also seen in clusters. For example, all the stars in a cluster's galaxies are made of such gas, implying that at least some cooling and condensation must have occurred during the assembly of the cluster. Applying a standard mass-to-light ratio, one finds that $\sim 0.2 h^{3/2}$ of a cluster's baryons are "cool gas" of this kind

Fig. 9.1. Hubble Heritage image of NGC 1275.

(Arnaud et al. 1992; White et al. 1993; with $h = H_0/100$ km s^{-1} Mpc^{-1}). While it may seem strange to include stars in a census of cool intracluster gas, the total mass of stars does serve as a lower limit to the amount of gas that passed through a cold phase at some point in the cluster's past.

Many clusters also host optical emission-line nebulae within their cores that appear to be associated with the cooler ($\sim T_{vir}/2$) X-ray emitting gas (Fabian & Nulsen 1977; Ford & Butcher 1979; Cowie et al. 1983; Hu, Cowie, & Wang 1985; Heckman et al. 1989; Crawford & Fabian 1992; Donahue, Stocke, & Gioia 1992; Crawford 2004). One could even say that Carnegie Observatories initiated the study of cool gas in clusters. Hubble & Humason (1931) noted that NGC 1275, the central galaxy in the Perseus cluster, had a discrepant color index because of its strong emission spectrum, saying that "it could be classified as an elliptical nebula that has broken up without the formation of spiral arms." Later, Baade & Minkowski (1954) noted that NGC 1275 was unusual among Seyfert galaxies because its emission lines were not restricted to the nuclear regions. Lynds (1970) eventually imaged this amazing Hα emission-line nebula using an interference filter. Figure 9.1 shows a recent Hubble Heritage close-up of NGC 1275, featuring a hint of spiral structure, complex dust lanes, and evidence for recent star formation.

The total amount of $\sim 10^4$ K gas in such nebulae is a mere $\sim 10^4 - 10^7 M_\odot$ (Heckman et al. 1989), but this nebulosity may be only the glowing skin surrounding considerably larger masses of much cooler gas. Clusters with Hα emission also have closely associated H$_2$ emission (Elston & Maloney 1994; Jaffe & Bremer 1997; Falcke et al. 1998; Donahue et al. 2000; Jaffe, Bremer, & van der Werf 2001; Edge et al. 2002). Furthermore, recent CO observations of a few cluster indicate that they may contain up to $10^{9-11.5} M_\odot$ in the form of cool molecular gas (Edge 2001).

The primary question concerning cool gas in clusters is whether these pieces—cool X-ray gas, stars, nebulae, molecular clouds—all fit together into a single coherent picture of condensation and star formation. If so, then studies of cluster cores may have much to teach us about the processes that govern galaxy formation. In this review, we will first recap the cooling flow hypothesis, now over 25 years old, suggesting that X-ray gas should cool and flow into cluster cores (see also Fabian, Nulsen, & Canizares 1984, 1991; Fabian 1994). Then we will present evidence showing that simple cooling flows, in which cooling proceeds unopposed by heating or feedback, do not occur (Molendi & Pizzolato 2001; Peterson et al. 2001, 2003). Supernovae and AGN activity must provide at least some feedback during the history of the cluster. In fact, *the global properties of clusters cannot be understood without accounting for radiative cooling and subsequent feedback* (Lewis et al. 2000; Pearce et al. 2000; Voit & Bryan 2001; Voit et al. 2002). Conduction may also suppress cooling in cluster cores (Bertschinger & Meiksin 1986; Bregman & David 1988; Sparks, Macchetto, & Golombek 1989), and this possibility has received renewed attention in recent years (Malyshkin 2001; Narayan & Medvedev 2001; Fabian, Voigt, & Morris 2002; Voigt et al. 2002). However, we do not yet know which is the dominant mechanism opposing cooling—feedback, conduction, or perhaps a combination of the two (Ruskowkski & Begelman 2002; Brighenti & Mathews 2003). We close the review by summarizing a few clues that might help answer this question.

9.2 The Cooling Flow Hypothesis

The road from the discovery of hot gas in clusters to the cooling flow hypothesis was rather short. Clusters of galaxies were first confirmed to be sources of X-ray emission in 1971 by the *UHURU* satellite (Gursky et al. 1971). Thermal emission from hot intracluster gas seemed like a natural interpretation (Lea et al. 1973; Lea 1975) given the extent of the emission (e.g., Forman et al. 1972; Kellogg et al. 1972) and the spectrum (e.g., Gorenstein et al. 1973; Davidsen et al. 1975; Kellogg, Baldwin, & Koch 1975), but it was not confirmed until the 6.7 keV iron-line complex from helium-like and hydrogen-like ions was discovered in the Perseus cluster by Mitchell et al. (1976) using *Ariel V*, and in Virgo, Perseus, and Coma by Serlemitsos et al. (1977) using *OSO-8*.

Simple calculations of radiative cooling at the centers of clusters like Perseus revealed that the cooling time, t_c, was probably less than a Hubble time (Cowie & Binney 1977; Fabian & Nulsen 1977). These authors suggested that, in the absence of a compensating heat source, the core gas ought to cool and condense at the cluster's center. Thus, the centers of all clusters with $t_c < H_0^{-1}$ soon became known as "cooling flows," even though there was not yet any firm evidence for either cooling or flowing. The main piece of circumstantial evidence was the close association between a short central cooling time and the presence of an optical emission-line nebula at the cluster's center, presumed to be generated by gas cooling through $\sim 10^4$ K. Hu et al. (1985) showed that these nebulae are frequently found in clusters with $t_c \lesssim H_0^{-1}$, but never in clusters with $t_c > H_0^{-1}$.

A simple estimate of the implied cooling rate can be drawn from the X-ray luminosity of the cooling region by assuming the gas cools from the virial temperature at constant pressure:

$$\dot{M}_X \approx \frac{2}{5} \frac{\mu m_p}{k T_X} L_X(< r_c) \ . \tag{9.1}$$

Here, $L_X(< r_c)$ is the X-ray luminosity coming from inside the cooling radius r_c, at which $t_c \approx H_0^{-1}$. Estimates for $\dot{M}$ derived from X-ray imaging often exceed $100 M_\odot \, \mathrm{yr}^{-1}$ (Fabian et al. 1984), even approaching $1000 M_\odot \, \mathrm{yr}^{-1}$ in some extreme cases (e.g., White et al. 1994).

The X-ray surface brightness distributions of cooling flow clusters are inconsistent with steady flows in which $d\dot{M}/dr = 0$ because such flows produce exceedingly strong central peaks in brightness. To obtain better-fitting surface brightness profiles, cooling flow modelers allowed for spatially distributed mass deposition that led to a decline in $\dot{M}$ as the flow approached $r = 0$ (Fabian et al. 1981; Stewart et al. 1984). Models of this kind fit the data best if $\dot{M}(r) \propto r$ (Fabian et al. 1984), implying that the flow must be inhomogeneous, with a range of cooling times at any given radius, because only a subset of the inflowing gas manages to condense within each radial interval (e.g., Thomas, Fabian, & Nulsen 1987). However, the overall $\dot{M}$ values derived from such models are similar to the simple estimates based on $L_X(< r_c)$.

Individual X-ray emission lines could, in principle, be used to estimate the rate at which matter is cooling (Cowie 1981). For cooling at constant pressure, the luminosity of emission line i is

$$L_i = \dot{M} \frac{5k}{2\mu m} \int \frac{\epsilon_i(T)}{\Lambda(T)} dT, \tag{9.2}$$

where T is the plasma temperature, $\epsilon_i(T)/\Lambda(T)$ is the fraction of the cooling emissivity function owing to emission line i as a function of T, μm is the mean mass per particle, and k is the Boltzmann constant. In the steady cooling flow model, this expression is integrated from $T = 0$ to $T = T_{hi}$. There were two high-resolution spectrometers on board the *Einstein* Observatory, and results (with rather low signal-to-noise ratio) from both of those spectrometers seemed to confirm the rates inferred from X-ray surface brightness distributions (Canizares et al. 1982; Canizares, Markert, & Donahue 1988; Mushotzky & Szymkowiak 1988).

9.3   The Trouble with Cooling Flows

X-ray astronomers have historically been quite fond of the cooling flow hypothesis but have had trouble convincing colleagues who work in other wavebands because no one has ever found a central mass sink containing the $\dot{M}_X H_0^{-1} \approx 10^{11-13} M_\odot$ implied by the simplest interpretation of the X-ray observations. Now that *Chandra* and *XMM-Newton* are providing high-resolution spectra of cluster cores, X-ray astronomers themselves have become convinced that cooling flows are not that simple, if indeed they occur at all, because the cooling rates derived from spectroscopy do not agree with simple cooling flow predictions.

9.3.1   The Mass-Sink Problem

The trouble with cooling flows began when optical observers could not locate all the stars that ought to be formed in the prodigious cooling flows ($> 100 M_\odot \, \mathrm{yr}^{-1}$) of some clusters (Fabian et al. 1991). Star formation rates derived from observations of excess blue light and Hα nebulosity, assuming a standard initial mass function, amounted to only $\lesssim 0.1 \dot{M}_X$ (Johnstone, Fabian, & Nulsen 1987; McNamara & O'Connell 1992; Allen 1995; Cardiel, Gorgas, & Aragón-Salamanca et al. 1995, 1998). While it remains possible in principle that star formation in cooling flows is heavily skewed toward unobservable low-mass stars

(Fabian, Nulsen, & Canizares 1982), there is still no compelling theoretical justification for this idea.

Initial enthusiasm about the Hα emission representing $\sim 10^4$ K cooling flow gas (e.g., Cowie, Fabian, & Nulsen 1980) abated when it was realized that the $\dot{M}$ implied by the Hα luminosity in some clusters was $\sim 10^2 \dot{M}_X$ (Cowie et al. 1983; Heckman et al. 1989). Models have been proposed in which the Hα is boosted by absorption of EUV and soft X-ray emission from cooling gas (Voit & Donahue 1990; Donahue & Voit 1991) or by cooling through turbulent mixing layers (Begelman & Fabian 1990). However, it now seems likely that most of the Hα emission comes from photoionization by OB stars (Johnstone et al. 1987; Voit & Donahue 1997; Cardiel et al. 1998; Crawford et al. 1999).

Hope for a solution to the mass-sink problem rose with the apparent discovery of excess soft X-ray absorption in cooling flow clusters, which would require $\sim 10^{12} M_\odot$ of cold gas distributed over the central ~ 100 kpc (White et al. 1991; Allen et al. 1993). Yet, dogged pursuit of this cold gas by radio astronomers failed to find either 21 cm emission (Dwarakanath, van Gorkom, & Owen 1994; O'Dea, Gallimore, & Baum 1995; O'Dea, Payne, & Kocevski 1998) or CO emission (O'Dea et al. 1994; Braine et al. 1995) with the necessary covering factor and beam temperature. Some clusters do have significant amounts of molecular gas, but detections so far generally find it only within the central ~ 20 kpc (Donahue et al. 2000; Edge 2001; Edge et al. 2002).

One explanation for the undetectability of the cooling flow sink is that this gas may become so cold that it produces no detectable emission (Ferland, Fabian, & Johnstone 1994, 2002). However, cold clouds bathed in the X-rays found in cluster cores must reradiate the X-ray energy they absorb in some other wave band. At minimum, these clouds should have an observable warm skin of detectable H I if they do indeed cover the central regions of clusters (Voit & Donahue 1995). Cold clouds with a low covering factor may still evade current radio observations but would not produce appreciable soft X-ray absorption.

Soft X-ray absorption itself is probably now a phenomenon that no longer needs explaining. Recent cluster observations with *Chandra* and *XMM-Newton* are failing to confirm the levels of absorption suggested by lower-resolution X-ray observations (McNamara et al. 2000; Blanton, Sarazin, & McNamara 2003; Peterson et al. 2003). If these observations are correct, then there is no evidence at all, in any waveband, for a large mass sink in cooling flow clusters.

9.3.2 The Spectroscopic $\dot{M}$ Problem

A recent breakthrough in X-ray astronomy is reframing the whole debate about cooling flows. In a simple, steady-state cooling flow one expects to see emission from gas over the entire range of temperature from T_{vir} to the sink temperature, whatever that may be. Because the thermal energy lost as gas cools from T to $T - \Delta T$ is proportional to ΔT, the luminosity coming from gas within that temperature interval is expected to be $\Delta L \propto \dot{M}\Delta T$. Thus, X-ray spectroscopy of the emission lines characteristic of gas at each temperature can be used to test whether $\Delta L/\Delta T$ is constant with temperature (Cowie et al. 1980). For example, we can use Fe XVII to track gas at $\lesssim 10^7$ K, O VIII to track gas at $\lesssim 2 \times 10^7$, and UV observations of O VI to track gas at $\sim 10^6$ K. Figure 9.2 shows the predicted spectrum if the cooling gas is assumed to be an inhomogeneous (multiphase) medium, as inferred from $\dot{M} \propto r$, that cools at constant pressure.

High-resolution spectroscopic observations with *XMM-Newton* and *Chandra* are now re-

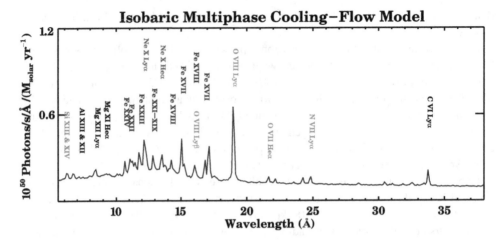

Fig. 9.2. Spectrum emitted by gas cooling from 6 keV at constant pressure. Because the gas recombines as it cools, the relative strengths of the emission lines reveal how much gas cools through each temperature. (Figure from Peterson et al. 2003.)

vealing a deficit of emission from gas below $\sim T_{\rm vir}/3$, relative to this predicted spectrum. Peterson et al. (2003) compiled Reflection Grating Spectra (RGS) spectra of 12 cooling flow clusters, the single largest collection to date. We plot an example from the Perseus cluster in Figure 9.3. None of the clusters hotter than 4 keV show evidence for Fe XVII emission from gas below 1 keV, and Fe XVII is weaker than expected in cluster with global temperatures of 2–4 keV. This line does appear in the spectra of supernova remnants, so its absence in cluster spectra is not a shortcoming of the plasma codes or the detectors. Furthermore, the early *XMM-Newton* RGS results (Peterson et al. 2001) have been confirmed by *Chandra* grating spectroscopy (e.g., Hicks et al. 2002). Gas at $\lesssim 1$ keV apparently does not exist in the amounts predicted by simple cooling flow models. Even the data from instruments with lower spectral resolution, such as the ACIS-S detector on board *Chandra*, suggest significantly lower mass cooling rates than obtained from previous analyses of *ROSAT* and *ASCA* data (e.g., McNamara et al. 2000; Wise & McNamara 2001; Lewis, Stocke, & Buote 2002). Faint detections and strong limits on O VI emission from the *FUSE* satellite (Oegerle et al. 2001) also imply lower mass cooling rates (Fig. 9.4).

Two *ad hoc* models for cool gas do fit the high-resolution observations obtained with the *XMM-Newton* RGS instrument reasonably well (Kaastra et al. 2001; Peterson et al. 2001, 2003). One is a two-temperature model, in which some gas is at $T_{\rm vir}$ and some is at $\sim T_{\rm vir}/2$. The other is a modified cooling flow model, in which the amount of cooling gas tapers off from $T_{\rm vir}$ to a minimum temperature $\sim T_{\rm vir}/3$ (Peterson et al. 2003). Because the temperature floor in these models seems to scale with $T_{\rm vir}$, it would appear that whatever prevents the gas from cooling further is sensitive to the depth of the cluster potential.

The assumption that cooling flows contain inhomogeneous, multiphase gas, as implied by their surface brightness profiles, has also been called into question. *XMM-Newton* observations of M87, at the center of the nearest cooling flow cluster, indicate that the surrounding intracluster medium consists of a single temperature plasma, except for those regions of the cluster associated with the M87 radio source (Matsushita et al. 2002).

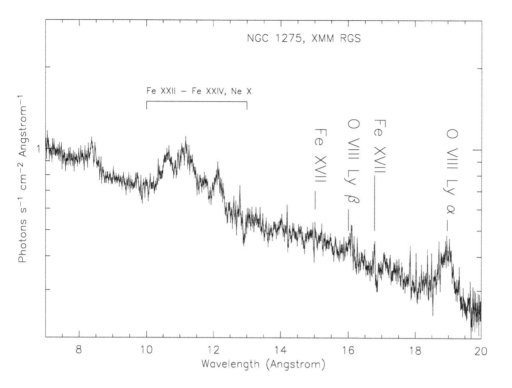

Fig. 9.3. Figure based on *XMM-Newton* RGS data for NGC 1275 in the Perseus cluster. The O VIII Lyα and Lyβ lines were detected, but no Fe XVII is apparent at the expected wavelengths of 15.014 Å or 16.78 Å. (Data courtesy J. Peterson; Peterson et al. 2003.)

9.3.3 *Time for a New Name*

What should we call these clusters in which gas no longer appears to be cooling and flowing? The close association between short central cooling times, Hα nebulosity, and H_2 emission strongly suggests that something unusual is happening in their cores. Star formation in some cases is rapid enough to qualify as a starburst (e.g., McNamara & O'Connell 1992; Cardiel et al. 1995), even though it cannot solve the mass-sink problem. The goings-on in the cores of these clusters certainly qualify as an important astrophysical puzzle that may have far reaching implications for galaxy formation. However, as we search for a new name for "cooling flow" clusters, we should perhaps settle for an observable, such as "cool core" clusters, as has been also suggested by others (Molendi & Pizzolato 2001).

Adopting a name less freighted with theoretical assumptions might promote more balanced consideration of alternatives to the cooling flow hypothesis. Any successful model must explain the following features of cool core clusters:

- The apparent lack of a mass sink comparable to $\dot{M}_X H_0^{-1}$.
- The positive core temperature gradients extending to $\sim 10^2$ kpc in clusters with $t_c < H_0^{-1}$.
- The frequent incidence of emission-line nebulae, dust lanes, and molecular gas in clusters with $t_c < H_0^{-1}$ and their absence in clusters with $t_c > H_0^{-1}$.

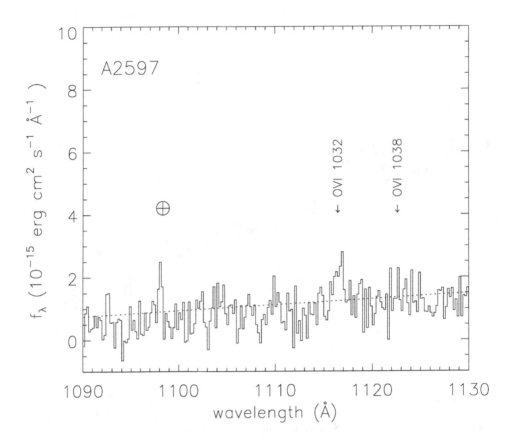

Fig. 9.4. *FUSE* detection of O VI in the central 36 kpc of the cooling flow cluster Abell 2597. The line flux is consistent with the luminosity expected from $\sim 40 M_\odot$ of gas cooling through $\sim 10^6$ K. (Figure from Oegerle et al. 2001.)

- The tendency for radio sources to be present in clusters with $t_c < H_0^{-1}$.

In light of the new X-ray observations, many of the competing ideas that have previously received less attention and testing than the cooling flow hypothesis are now being revisited. The next section discusses how feedback from supernovae and AGNs might limit the amount of gas that condenses in clusters, and the following section outlines the potentially important role of electron thermal conduction.

9.4   The Galaxy-Cluster Connection

Simple cooling flows may be disproven, but cooling in general plays a major role in determining the global X-ray properties of clusters. Cosmological models of cluster formation that do not include radiative cooling and the ensuing feedback processes fail to produce realistic clusters (Lewis et al. 2000; Pearce et al. 2000; Muanwong et al. 2001; Voit & Bryan 2001). The most glaring failure is in predictions of the L_X-T_X relation. Models without galaxy formation predict $L_X \propto T_X^2$ (Kaiser 1986; Borgani et al. 2001; Muanwong et al. 2001), while observations indicate $L_X \propto T_X^3$ (Mushotzky 1984; Edge & Stewart 1991; David et al. 1993; Markevitch 1998; Arnaud & Evrard 1999; Novicki, Sornig, & Henry 2002). Ignoring cooling and feedback also causes problems with the slope and normalization of the $M_{\rm vir}$-T_X relation between virial mass and temperature (Horner, Mushotzky, & Scharf 1999; Nevalainen, Markevitch, & Forman 2000; Finoguenov, Reiprich, & Böhringer 2001), which is a fundamental ingredient in efforts to constrain cosmological parameters with cluster observations.

Recent work has shown that tracing the development of intracluster entropy is a powerful way to understand how cooling, supernova feedback, and perhaps energy injection by AGNs conspire to determine both the L_X-T_X and $M_{\rm vir}$-T_X relations of present-day clusters (Ponman, Cannon, & Navarro 1999; Bryan 2000; Voit & Bryan 2001; Voit et al. 2002, 2003; Wu & Xue 2002a,b). Here we briefly outline some connections between a cluster's galaxies and its intracluster medium and show how these connections manifest themselves in the intracluster entropy distribution. Then we focus on some particular models for how AGNs might quench cooling in clusters.

9.4.1 The Theoretical Cooling Flow Problem

Cosmological models for cluster formation that do not include cooling are clearly too simplistic because they do not spawn galaxies. Radiative cooling initiates galaxy birth but is responsible for the now-classic overcooling problem (White & Rees 1978; Cole 1991). If no form of feedback opposes cooling, then at least 20% of the baryons in the Universe, and maybe more, should have condensed into stars. Yet, the observed fraction of baryons in stars is $\lesssim 10\%$ (see Fig. 9.5; Balogh et al. 2001). This overcooling problem is even more acute in clusters, where primordial densities are higher, enabling even more of the baryons to condense.

One could also call this problem the "theoretical cooling flow problem" because far too many baryons cool and condense if there is no heat source to compensate for radiative cooling. Supernova feedback is generally assumed to provide the requisite heat to halt overcooling in galaxies, although the precise mechanism remains murky (e.g., Kay et al. 2002). However, supernovae might not provide enough heat to halt overcooling in clusters, where the binding energy per particle exceeds the mean supernova energy per particle (~ 1 keV), as measured from the intracluster metallicity (e.g., Finoguenov, Arnaud, & David 2001). Thus, feedback from AGNs may be necessary to suppress cluster cooling flows.

9.4.2 Cooling, Feedback, and Intracluster Entropy

The slope of the observed L_X-T_X relation has long been assumed to reflect an early episode of feedback that imposed a universal entropy floor throughout the intergalactic medium (Evrard & Henry 1991; Kaiser 1991). An entropy floor steepens the L_X-T_X relation from $L_x \propto T_X^2$ to $L_X \propto T_X^3$ because the extra entropy stiffens the intracluster medium against compression. Lower temperature clusters with shallower potential wells therefore have a harder time compressing their core gas, leading to lower core densities and smaller X-ray luminosities than expected in models without cooling and feedback.

Measurements of intracluster entropy in the vicinity of the X-ray core radius support this notion because they indicate elevated entropy levels in groups and poor clusters, corresponding to $T n_e^{-2/3} \approx 100 - 150 \, {\rm keV \, cm}^2$ (Ponman et al. 1999; Lloyd-Davies, Ponman, & Cannon

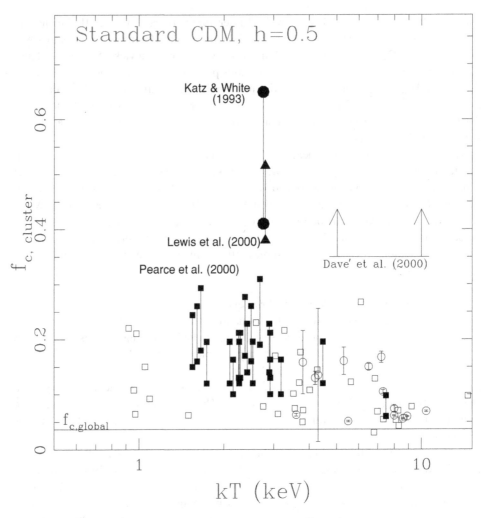

Fig. 9.5. The global overcooling problem. High-resolution cosmological simulations including cooling, represented by the labeled solid points, predict that at least 20% of the Universe's baryons should have condensed into stars or cold clouds, if feedback is ineffective. However, the global condensed baryon fraction $f_{c,global}$ inferred from large-scale surveys is ∼5%–10%, depending on the initial mass function, and the condensed baryon fractions inferred from cluster observations (empty squares) are ∼10%–20%. (Figure from Balogh et al. 2001.)

2000). In order to produce such an entropy floor through supernova heating alone, a large proportion of the available supernova energy is needed (Kravtsov & Yepes 2000). Even then, the required supernova heating efficiency may be unrealistic, in which case additional heat input from AGNs would be required (Valageas & Silk 1999; Wu, Fabian, & Nulsen 2000). However, there is another way to interpret these entropy measurements that does not involve global heating of the intergalactic medium.

Instead, the L_X-T_X relation may reflect a conspiracy between cooling and feedback that

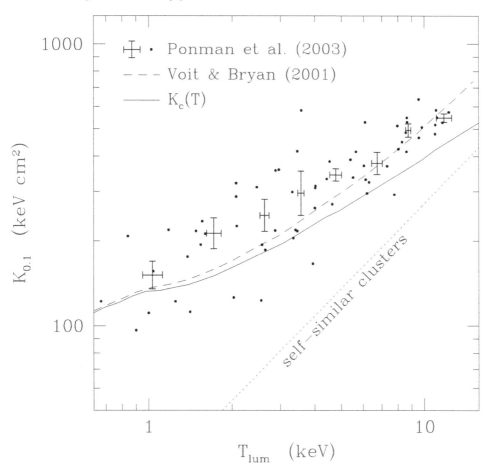

Fig. 9.6. Relationship between core entropy and the cooling threshold. Each point with error bars shows the mean core entropy $K_{0.1}$, measured at $0.1 r_{200}$, for eight clusters within a given bin of luminosity-weighted temperature T_{lum}, and small circles show measurements for individual clusters (Ponman et al. 2003). The dotted line shows a self-similar relation calibrated using the median value of $K_{0.1}$ measured in simulation L50+ of Bryan & Voit (2001), which does not include cooling or feedback. The solid line shows the cooling threshold $K_c(T)$, defined to be the entropy at which the cooling time equals 14 Gyr, assuming the cooling function of Sutherland & Dopita (1993) for 0.3 solar metallicity. The dashed line shows the entropy at $0.1 r_{200}$ in the model of Voit & Bryan (2001) when this cooling function is used.

regulates the core entropy of clusters and groups (Voit & Bryan 2001). Figure 9.6 shows measurements of core entropy from Ponman, Sanderson, & Finoguenov (2003) along with the locus in T-$T n_e^{-2/3}$ space at which the cooling time equals a Hubble time. The way in which core entropy tracks this locus suggests that gas with a short cooling time is eliminated from clusters by a combination of cooling and feedback.

A parcel of gas with entropy $(T n_e^{-2/3})$ below this threshold must condense unless feedback intervenes. If feedback is effective, then it will raise the entropy of the gas parcel until it exceeds the threshold, where it is no longer subject to cooling. If feedback is ineffective, then most of the parcel's gas will cool and condense. Either way, both cooling and feedback remove gas from the region below the threshold, establishing a core entropy at the level of the threshold.

This mechanism explains why simulations that include cooling produce clusters with reasonably realistic L_X-T_X and M_{vir}-T_X relations, regardless of the efficiency of feedback (Muanwong et al. 2001; Borgani et al. 2002; Davé, Katz, & Weinberg 2002). However, the amount of baryons that end up in galaxies is very sensitive to how feedback is implemented (Kay, Thomas, & Theuns 2003). Thus, it would appear that cooling is essential to a proper understanding of cluster properties and that the details of how cooling flows are suppressed are crucial to understanding hierarchical galaxy formation in the context of clusters.

9.4.3   AGNs and Cooling Flows

Many cooling flow clusters also contain radio sources indicative of recent nuclear activity (e.g., Burns 1990). This close association between AGNs and clusters with short central cooling times supports the idea that feedback from AGNs helps to suppress cooling. Some authors have proposed that radiation from the active nucleus heats the cluster core (e.g., Ciotti & Ostriker 1997, 2001), but far more attention has been paid to the possibility that radio jets somehow heat the intracluster medium (e.g., Binney & Tabor 1995; Churazov et al. 2001; Soker et al. 2001; Brüggen & Kaiser 2002; Reynolds, Heinz, & Begelman 2002). Such heating was originally not considered to be a viable solution to the mass-sink problem because the total amount of energy needed to stabilize a strong cooling flow is quite large ($\sim 10^{62}$ erg), and the spatial deposition of that heat would need to be precisely matched to local cooling rates in order to maintain thermal stability (Fabian 1994). However, *Chandra* and *XMM-Newton* observations showing widespread interactions between radio plasma and the intracluster medium (e.g., Fabian et al. 2000; McNamara et al. 2000) have stimulated new interest in connections between radio jets and cooling flows.

The high spatial resolution of the *Chandra* observations reveals that jets do not simply shock-heat the surrounding intracluster medium, because the gas surrounding the lobes appears somewhat cooler and denser than the undisturbed gas farther from the lobes (McNamara et al. 2000). Thus, because cluster cores do not appear to be shock heated, most of the recent theoretical models have focused on mixing and turbulent heating stirred up as the buoyant radio plasma rises through the intracluster medium (e.g., Quilis, Bower, & Balogh 2001; Brüggen & Kaiser 2002; Reynolds et al. 2002). Both mixing and heating raise the entropy of the core gas, consequently raising its cooling time as well. These models circumvent the local fine-tuning problem by distributing heat over a large region through convection, and they add additional thermal energy to the core beyond that supplied by the AGN itself by mixing the core gas with overlying gas of higher entropy.

However, not all clusters with short central cooling times have obvious nuclear activity. Thus, if AGN heating is the solution to the cooling flow puzzle, then it must be episodic. A recent model by Kaiser & Binney (2003) shows how the central entropy profile would evolve under episodic heating. Because cooling rates rise dramatically as isobaric gas cools to lower temperatures, an episodically heated medium usually contains very little gas below $\sim T_{vir}/3$. When the central gas reaches this temperature it is assumed to cool very quickly to even cooler temperatures and accrete onto the AGN, triggering another episode of heating. This feature of episodic heating may explain the absence of line emission from colder gas in cool core clusters.

9.5 The Revival of Conduction

During the first two decades of the cooling flow hypothesis, the idea that electron thermal conduction might somehow suppress cooling was a minority viewpoint, despite the fact that it has many attractive features. Because conduction carries heat from warmer regions to cooler regions, it naturally directs thermal energy into regions that would otherwise condense. Also, it taps the vast reservoir of thermal energy in the intracluster medium surrounding the cluster core, which is more than sufficient to resupply the radiated energy.

Many models invoking conduction have been developed (e.g., Tucker & Rosner 1983; Bertschinger & Meiksin 1986; Bregman & David 1988; Rosner & Tucker 1989; Sparks 1992), but conduction has often been dismissed as a global solution on the grounds that it is not stable enough to preserve the observed temperature and density gradients for periods of order $\gtrsim 1$ Gyr (Cowie & Binney 1977; Fabian 1994). The heat flux from unsaturated conduction proceeding uninhibited by magnetic fields is $\kappa_s \nabla T$, with $\kappa_s \approx 6 \times 10^{-7} T^{5/2} \, \mathrm{erg \, cm^{-1} \, s^{-1} \, K^{-7/2}}$, the so-called Spitzer rate (Spitzer 1962). Because of this extreme sensitivity to temperature, it is difficult for radiative cooling and conduction to achieve precise thermal balance with a globally stable temperature gradient (Bregman & David 1988; Soker 2003). However, any mechanism that places cool gas at the center of a cluster, such as a merger of a gas-rich galaxy with the central cluster galaxy, sets up a temperature gradient that would cause uninhibited conduction to proceed until either the cool gas has evaporated or the hot gas has condensed (Sparks et al. 1989). As long as a temperature gradient exists, a certain amount of conduction has to occur.

In order for a standard, steady cooling flow alone to produce the temperature gradients observed in cool core clusters, conduction must be highly suppressed by at least 2 orders of magnitude below the Spitzer rate, presumably by tangled magnetic fields (Binney & Cowie 1981; Fabian et al. 1991). Yet, recent theoretical analyses of conduction have concluded that this level of suppression is unrealistically high (Malyshkin 2001; Malyshkin & Kulsrud 2001; Narayan & Medvedev 2001). These studies suggest that magnetic field tangling may only suppress conduction by a factor ~ 3–10, implying that it may be important in the cores of clusters.

This finding, coupled with the X-ray spectroscopic observations showing little evidence for cooling gas, has helped to spur a remarkable revival of the idea of conduction, with notable assistance from some of its harshest earlier critics (Fabian et al. 2002; Voigt et al. 2002; Zakamska & Narayan 2003; but see Loeb 2002). One can analyze the observed temperature gradients of clusters by defining an effective conduction coefficient $\kappa_{\mathrm{eff}}(r) \equiv L(< r)/4\pi r^2 (dT/dr)$ that would lead to balance between radiative cooling and conductive heating. The values of κ_{eff} measured at ~ 100 kpc in cool core clusters are typically $\sim (0.1 - 0.3)\kappa_S$, suggesting that electron thermal conduction is a plausible mechanism for counteracting radiative cooling over much of the region where $t_c < H_0^{-1}$ (Fig. 9.7).

Even though conduction may be important at ~ 100 kpc, the required effective conductivity exceeds the Spitzer rate at radii ~ 10 kpc (Ruszkowski & Begelman 2002; Voigt et al. 2002), a result presaged by the analysis of Bertschinger & Meiksin (1986). Thus, a modest amount of feedback may be necessary to offset cooling in the centers of cool core clusters. Hybrid models involving conduction in the outer parts of the core and AGN feedback in the inner parts have been developed by Ruszkowski & Begelman (2002) and Brighenti & Mathews (2003).

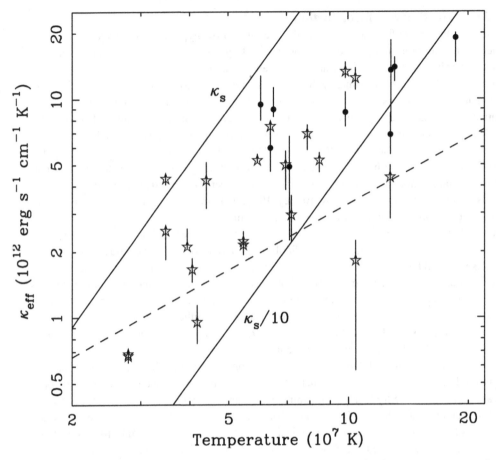

Fig. 9.7. Effective conduction coefficients κ_{eff} required for conduction to compensate for radiative cooling within the central regions of clusters, plotted as a function of cluster temperature. The required conductivity generally does not exceed the Spitzer rate κ_S at radii ~ 100 kpc, implying that conduction is potentially important in cluster cores. (Figure from Fabian et al. 2002.)

9.6   Paths to a Resolution

Observations from the present generation of X-ray telescopes have dethroned the cooling flow hypothesis, but what will take its place? Star formation, radio jets, and conduction may all have important roles to play in the development of cluster cores. Conduction is notoriously hard to test because the rate at which it proceeds depends on the unknown geometry of intracluster magnetic fields and uncertain factor by which these fields suppress heat flow. Looking for hallmarks of episodic feedback, from both AGNs and supernovae, may be more fruitful, at least in the short term.

If feedback is episodic, then the state of the central intracluster medium should be closely related to other goings-on in the cluster core. Thus, it would be interesting to test whether the $\sim T_{\text{vir}}/3$ scaling of the minimum plasma temperature apparent in the early sample of *XMM-Newton* clusters from Peterson et al. (2003) holds for a large sample of cool core clusters with various levels of core activity. How do the X-ray emission-line spectra of clusters with radio-loud nuclei differ from those of clusters with radio-quiet nuclei? Are there any correlations between X-ray line emission and the presence of obvious star formation or emission-line nebulae? Episodic heating also leads to a predictable pattern in the evolution of the core entropy distribution (Kaiser & Binney 2003). Thus, studying the core entropy distributions of a large sample of clusters may reveal a telltale pattern of entropy evolution with time.

In order to look for evidence of a feedback duty cycle in cluster cores and to study how their properties depend on AGN and star formation activity, we are now in the midst of an archival *Chandra* study of cluster cores. The result of this program will be a publicly available library of entropy distributions showing how the entropy of intracluster gas depends on radius and enclosed gas mass within that radius (Horner et al., in preparation.) We are focusing on entropy because it is the thermodynamic quantity most closely related to heat input and radiative cooling. We invite all who are interested in the vexing problem of cooling flows to take advantage of this database.

References

Allen, S. W. 1995, MNRAS, 276, 947
Allen, S. W., Fabian, A. C., Johnstone, R. M., White, D. A., Daines, S. J., Edge, A. C., & Stewart, G. C. 1993, MNRAS, 262, 901
Arnaud, M., & Evrard, A. E. 1999, MNRAS, 328, L37
Arnaud, M., Rothenflug, R., Boulade, O., Vigroux, L., & Vangioni-Flam, E. 1992, A&A, 254, 49
Baade, W. & Minkowski, R. 1954, ApJ, 119, 215
Balogh, M. L., Pearce, F. R., Bower, R. G., & Kay, S. T. 2001, MNRAS, 326, 1228
Begelman, M. C., & Fabian, A. C. 1990, MNRAS, 244, 26P
Bertschinger, E., & Meiksin, A. 1986, ApJ, 306, L1
Binney, J., & Cowie, L. L. 1981, ApJ, 247, 464
Binney, J., & Tabor, G. 1995, MNRAS, 276, 663
Blanton, E. L., Sarazin, C. L., & McNamara, B. R. 2003, ApJ, 585, 227
Borgani, S., Governato, F., Wadsley, J., Menci, N., Tozzi, P., Lake, G., Quinn, T., & Stadel, J. 2001, ApJ, 559, L71
Borgani, S., Governato, F., Wadsley, J., Menci, N., Tozzi, P., Quinn, T., Stadel, J., & Lake, G. 2002, MNRAS, 336, 409
Braine, J., Wyrowski, F., Radford, S. J. E., Henkel, C. & Lesch, H. 1995, A&A, 293, 315
Bregman, J. N., & David, L. P. 1988, ApJ, 326, 639
Brighenti, F., & Mathews, W. G. 2003, ApJ, 587, 580
Brüggen, M, & Kaiser, C. R. 2002, Nature, 418, 301
Bryan, G. 2000, ApJ, 544, L1
Bryan, G. L., & Voit, G. M. 2001, ApJ, 556, 590
Burns, J. O. 1990, AJ, 99, 14
Canizares, C. R., Clark, G. W., Jernigan, J. G., & Markert, T. H. 1982, ApJ, 262, 33
Canizares, C. R., Markert, T. H., & Donahue, M. E. 1988, in Cooling Flows in Clusters and Galaxies, ed. A. C. Fabian (Dordrecht: Kluwer), 63
Cardiel, N., Gorgas, J., & Aragón-Salamanca, A. 1995, MNRAS, 277, 502
——. 1998, MNRAS, 298, 977
Churazov, E., Brüggen, M., Kaiser, C. R., Böhringer, H., & Forman, W. 2001, 554, 261
Ciotti, L., & Ostriker, J. P. 1997, ApJ, 487, 105
——. 2001, ApJ, 551 131
Cole, S. 1991, ApJ, 367, 45
Cowie, L. L. 1981, in X-ray Astronomy with the Einstein Satellite, ed. R. Giacconi (Dordrecht: Reidel), 227
Cowie, L. L., & Binney, J. 1977, ApJ, 215, 723
Cowie, L. L., Fabian, A. C., & Nulsen, P. E. J. 1980, ApJ, 191, 399
Cowie, L. L., Hu, E. M., Jenkins, E. B., & York, D. G. 1983, ApJ, 272, 29
Crawford, C. S. 2004, in Carnegie Observatories Astrophysics Series, Vol. 3: Clusters of Galaxies: Probes of

Cosmological Structure and Galaxy Evolution, ed. J. S. Mulchaey, A. Dressler, & A. Oemler (Pasadena: Carnegie Observatories, http://www.ociw.edu/ociw/symposia/series/symposium3/proceedings.html)

Crawford, C. S., Allen, S. W., Ebeling, H., Edge, A. C., & Fabian, A. C. 1999, MNRAS, 306, 857

Crawford, C. S., & Fabian, A. C. 1992, MNRAS, 259, 265

Davé, R., Katz, N., & Weinberg, D. H. 2002, ApJ, 579, 23

David, L., Slyz, A., Jones, C., Forman, W., Vrtilek, S. D., & Arnaud, K. A. 1993, ApJ, 412, 479

Davidsen, A., Bowyer, S., Lampton, M., & Cruddace, R. 1975, ApJ, 198, 1

Donahue, M., Mack, J., Voit, G. M., Sparks, W., & Elston, R., Maloney, P. R. 2000, ApJ, 545, 670

Donahue, M., Stocke, J. T., & Gioia, I. M. 1992, 385, 49

Donahue, M. & Voit, G. M. 1991, ApJ, 381, 361

Dwarakanath, K. S., van Gorkom, J. H., & Owen, F. N. 1994, ApJ, 432, 469

Edge, A. C. 2001, MNRAS, 328, 762

Edge, A. C. & Stewart, G. C. 1991, MNRAS, 252, 414

Edge, A. C., Wilman, R. J., Johnstone, R. M., Crawford, C. S., Fabian, A. C., & Allen, S. W. 2002, MNRAS, 337, 49

Elston, R. & Maloney, P. 1994, in Infrared Astronomy with Arrays, the Next Generation, ed. I. S. McLean (Astrophysics and Space Science Library, 190), 169

Evrard, A. E., & Henry, J. P. 1991, ApJ, 383, 95

Fabian, A. C. 1994, ARA&A, 32, 277

Fabian, A. C. et al. 2000, MNRAS, 318, L65

Fabian, A. C., Hu, E. M., Cowie, L. L., & Grindlay, J. 1981, ApJ, 248, 47

Fabian, A. C. & Nulsen, P. E. J. 1977, MNRAS, 180, 479

Fabian, A. C., Nulsen, P. E. J., & Canizares, C. R. 1982, MNRAS, 201, 933

——. 1984, Nature, 310, 733

——. 1991, A&ARv, 2, 191

Fabian, A. C., Voigt, L. M., & Morris, R. G. 2002, MNRAS, 335, L71

Falcke, H., Rieke, M. J., Rieke, G. H., Simpson, C., & Wilson, A. S. 1998, ApJ, 494, L155

Ferland, G. J., Fabian, A. C., & Johnstone, R. M. 1994, MNRAS, 266, 399

——. 2002, MNRAS, 333, 876

Finoguenov, A., Arnaud, M. & David, L. P. 2001, ApJ, 555, 191

Finoguenov, A., Reiprich, T. H., & Böhringer, H. 2001, A&A, 368, 749

Ford, H. C., & Butcher, H. 1979, ApJS, 41, 147

Forman, W., Kellogg, E., Gursky, H., Tananbaum, H., & Giacconi, R. 1972, ApJ, 178, 309

Gorenstein, P., Bjorkholm, P., Harris, B., & Harnden, F. R., Jr. 1973, ApJ, 183, L57

Gursky, H., Kellogg, E., Murray, S., Leong, C., Tananbaum, H., & Giacconi, R. 1971, ApJ, 167, L81

Heckman, T. M., Baum, S. A., van Breugel, W. J. M., & McCarthy, P. 1989, ApJ, 338, 48

Hicks, A. K., Wise, M. W., Houck, J. C., & Canizares, C. R. 2002, ApJ, 580, 763

Horner, D. J., Mushotsky, R. F., & Scharf, C. A. 1999, ApJ, 520, 78

Hu, E. M., Cowie, L. L, & Wang, Z. 1985, ApJS, 59, 447

Hubble, E., & Humason, M. L. 1931, ApJ, 74, 43

Jaffe, W., & Bremer, M. N. 1997, MNRAS, 284, L1

Jaffe, W., Bremer, M. N., & van der Werf, P. P. 2001, MNRAS, 324, 443

Johnstone, R. M., Fabian, A. C., & Nulsen, P. E. J. 1987, MNRAS, 224, 75

Kaastra, J. S., Ferrigno, C., Tamura, T., Paerels, F. B. S., Peterson, J. R., & Mittaz, J. P. D. 2001, A&A, 365, L99

Kaiser, C. R., & Binney, J. 2003, MNRAS, 338, 837

Kaiser, N. 1986, MNRAS, 222, 323

——. 1991, ApJ, 383, 104

Katz, N., & White, S. D. M. 1993, ApJ, 412, 455

Kay, S. T., Pearce, F. R., Frenk, C. S., & Jenkins, A. 2002, MNRAS, 330, 113

Kay, S. T., Thomas, P. A., & Theuns, T. 2003, MNRAS, 343, 608

Kellogg, E., Baldwin, J. R., & Koch, D. 1975, ApJ, 199, 299

Kellogg, E., Gursky, H., Tananbaum, H., Giacconi, R., & Pounds, K. 1972, ApJ, 174, L65

Kravtsov, A., & Yepes, G. 2000, MNRAS, 318, 227

Lea, S. M. 1975, ApL, 16, 141

Lea, S. M., Silk, J. I., Kellogg, E., & Murray, S. 1973, ApJ, 184, L111

Lewis, A. D., Stocke, J. T., & Buote, D. A. 2002, ApJ, 573, 13

Lewis, G. F., Babul, A., Katz, N., Quinn, T., Hernquist, L., & Weinberg, D. H. 2000, ApJ, 536, 623

Lloyd-Davies, E. J., Ponman, T. J., & Cannon, D. B. 2000, MNRAS, 315, 689

Loeb, A. 2002, NewA, 7, 279

Lynds, R. 1970, ApJ, 159, L151

Malyshkin, L. 2001, ApJ, 554, 561

Malyshkin, L., & Kulsrud, R. 2001, ApJ, 549, 402

Markevitch, M. 1998, ApJ, 504, 27

Mathews, W. G., & Bregman, J. N. 1978, ApJ, 224, 308

Matsushita, K., Belsole, E., Finoguenov, A., & Böhringer, H. 2002, A&A, 386, 77

McNamara, B. R., et al. 2000, ApJ, 534, L135

McNamara, B. R., & O'Connell, R. W. 1992, ApJ, 393, 579

Mitchell, R. J., Culhane, J. L., Davison, P. J. N., & Ives, J. C. 1976, MNRAS, 175, 29P

Molendi, S., & Pizzolato, F. 2001, ApJ, 560, 194

Muanwong, O., Thomas, P. A., Kay, S. T., Pearce, F. R., & Couchman, H. M. P. 2001,ApJ, 552, L27

Mushotzky, R. F. 1984, Physica Scripta, T7, 157

Mushotzky, R. F. & Szymkowiak, A. E. 1988, in Cooling Flows in Clusters and Galaxies, ed. A. C. Fabian (Dordrecht: Kluwer), 53

Narayan, R., & Medvedev, M. V. 2001, ApJ, 562, L129

Nevalainen, J., Markevitch, M., & Forman, W. 2000, ApJ, 532, 694

Novicki, M. C., Sornig, M., & Henry, J. P. 2002, AJ, 124, 2413

O'Dea, C. P., Baum, S. A., Maloney, P. R., Tacconi, L., & Sparks, W. B. 1994, ApJ, 422, 467

O'Dea, C. P., Gallimore, J. F., & Baum, S. A. 1995, AJ, 109, 26

O'Dea, C. P., Payne, H. E., & Kocevski, D. 1998, AJ, 116, 623

Oegerle, W. R., Cowie, L., Davidsen, A., Hu, E., Hutchings, J., Murphy, E., Sembach, K., & Woodgate, B. 2001, ApJ, 560, 187

Pearce, F. R., Thomas, P. A., Couchman, H. M. P., & Edge, A. C. 2000, MNRAS, 317, 1029

Peterson, J. R. et al. 2001, A&A, 365, L104

Peterson, J. R., Kahn, S. M., Paerels, F. B. S., Kaastra, J. S., Tamura, T., Bleeker, M., Ferrigno, C., & Jernigan, J. 2003, ApJ, 590, 207

Ponman, T. J., Cannon, D. B., & Navarro, J. F. 1999, Nature, 397, 135

Ponman, T. J., Sanderson, A. J. R., & Finoguenov, A. 2003, MNRAS, 343, 331

Quilis, V., Bower, R. G., & Balogh, M. L. 2001, MNRAS, 328, 1091

Reynolds, C. S., Heinz, S., & Begelman, M. C. 2002, ApJ, 332, 271

Rosner, R., & Tucker, W. H. 1989, 338, 761

Ruszkowski, M., & Begelman, M. C. 2002, ApJ, 581, 223

Sarazin, C. L. 1986, Rev. Mod. Phys., 58, 1

Serlemitsos, P. J., Smith, B. W., Boldt, E. A., Holt, S. S., & Swank, J. H. 1977, ApJ, 211, L63

Soker, N. 2003, MNRAS, 342, 463

Soker, N., White, R. E., David, L. P., & McNamara, B. R. 2001, ApJ, 549, 832

Sparks, W. B. 1992, ApJ, 399, 66

Sparks, W. B., Macchetto, F., & Golombek, D. 1989, ApJ, 345, 153

Spitzer, L., Jr. 1962, Physics of Fully Ionized Gases (New York: Interscience, 2nd Edition)

Stewart, G. C., Canizares, C. R., Fabian, A. C., & Nulsen, P. E. J. 1984, ApJ, 278, 536

Sutherland, R. S., & Dopita, M. A. 1993, ApJS, 88, 253

Thomas, P. A., Fabian, A. C., & Nulsen, P. E. J. 1987, MNRAS, 228, 973

Tucker, W. H., & Rosner, R. 1983, ApJ, 267, 547

Valageas, P., & Silk, J. 1999, A&A, 350, 725

Voigt, L. M., Schmidt, R. W., Fabian, A. C., Allen, S. W., & Johnstone, R. M. 2002, MNRAS, 335, L7

Voit, G. M., Balogh, M. L., Bower, R. G., Lacey, C. G., & Bryan, G. L. 2003, ApJ, 593, 272

Voit, G. M., & Bryan, G. 2001, Nature, 414, 425

Voit, G. M., Bryan, G. L., Balogh, M. L., & Bower, R. G. 2002, ApJ, 576, 601

Voit, G. M. & Donahue, M. 1995, ApJ, 360, L15

——. 1990, ApJ, 452, 16

——. 1997, ApJ, 486, 242

White, D. A., Fabian, A. C., Allen, S. W., Edge, A. C., Crawford, C. S., Johnstone, R. M., Stewart, G. C., & Voges, W. 1994, MNRAS, 269, 589

White, D. A., Fabian, A. C., Johnstone, R. M., Mushotzky, R. F., & Arnaud, K. A. 1991, MNRAS, 252, 72

White, S. D. M., Navarro, J. F., Evrard, A. E., & Frenk, C. S. 1993, Nature, 366, 429

White, S. D. M., & Rees, M. J. 1978, MNRAS, 183, 541

Wise, M. W., & McNamara, B. R 2001, in Two Years of Science with Chandra, Abstracts from the Symposium held in Washington, DC 5-7 Sept 2001

Wu, K. K. S., Fabian, A. C., & Nulsen, P. E. J. 2000, MNRAS, 318, 889

Wu, X.-P., & Xue, Y.-J. 2002a, ApJ, 569, 112

——. 2002b, ApJ, 572, 19

Zakamska, N. L., & Narayan, R. 2003, ApJ, 582, 162

10

Using the Sunyaev-Zel'dovich effect to probe the gas in clusters

MARK BIRKINSHAW
Department of Physics, University of Bristol

Abstract

The thermal Sunyaev-Zel'dovich effect is an important probe of clusters of galaxies, and has the attractive property of being proportional to the thermal energy content of the intracluster medium. With the assistance of X-ray data, the effect can be used to measure the number of hot electrons in clusters, and thus measure cluster baryon contents. Cluster absolute distances and other structural parameters can also be measured by combining thermal Sunyaev-Zel'dovich, X-ray, and other data. This review presents an introduction to the effect, shows some representative results, and sketches imminent developments.

10.1 Introduction

Ever since the cosmic microwave background radiation was discovered and interpreted as a thermal signal coming from the epoch of decoupling (Dicke et al. 1965; Penzias & Wilson 1965), it has been seen as a major cosmological tool. High-quality information about the background radiation is now available. The *COBE* mission demonstrated that the spectrum of the microwave background radiation is precisely thermal, with a temperature $T_{rad} = 2.728 \pm 0.002$ K (where necessary, all errors have been converted to $\pm 1\sigma$), and a maximum distortion characterized by a Comptonization parameter (§ 10.2.1) $\bar{y} < 1.5 \times 10^{-5}$ or a chemical potential $|\mu| < 9 \times 10^{-5}$ (Fixsen et al. 1996). *COBE* also demonstrated that the background contains small brightness fluctuations induced by density perturbations at decoupling with the character expected from simple inflation models (Gorski et al. 1996; Hinshaw et al. 1996; Wright et al. 1996).

More recently, *WMAP* has measured the power spectrum of these fluctuations over a wide range of multipole orders, l, and confirmed the existence of the peak in the power spectrum at $l \approx 200$, which the BOOMERanG (de Bernardis et al. 2000) and MAXIMA (Hanany et al. 2000) experiments used to show that the Universe is close to flat (Jaffe et al. 2001). Fits to the *WMAP* power spectrum have determined several cosmological parameters with good accuracy, notably the density parameter in baryons, $\Omega_b = 0.047 \pm 0.006$, the density parameter in matter, $\Omega_m = 0.29 \pm 0.06$, and the total density parameter $\Omega_{total} = 1.02 \pm 0.02$ (Bennett et al. 2003; Spergel et al. 2003). This review will adopt a value 72 ± 2 km s^{-1} for the Hubble constant, derived from the *WMAP* results and the Hubble Key Project (Freedman et al. 2001), and will ignore the contribution of neutrinos to the Universe's dynamics since the *WMAP* results suggest that Ω_ν is negligible.

Additional brightness structures are induced in the microwave background radiation by massive objects between the epoch of decoupling and the present. The most important of these brightness structures is the thermal Sunyaev-Zel'dovich effect (Sunyaev & Zel'dovich 1972), which arises from the inverse-Compton scattering of the microwave background radiation by hot electrons in cluster atmospheres. Several other, but lower-amplitude, structures are also induced by clusters. Recent attention has focussed on those generated by the kinematic Sunyaev-Zel'dovich effect and the Rees-Sciama effect (Rees & Sciama 1968), particularly in the version that arises from the motions of clusters of galaxies across the line of sight (Pyne & Birkinshaw 1993; Molnar & Birkinshaw 2000).

The thermal Sunyaev-Zel'dovich effect was first detected at a high level of significance in 1978 (Birkinshaw, Gull, & Northover 1978). As instrumentation developed through the 1980's and 1990's it became a routine matter to measure the Sunyaev-Zel'dovich effects of the richest clusters of galaxies (e.g., Myers et al. 1997). Nevertheless, the use of the effect to explore the physics of the intracluster medium is still in its infancy because measurements remain slow. This has often led to reports about the effect, from individual clusters, rather than papers giving measurements for substantial cluster samples. With the development of dedicated telescopes the situation is now changing, and the effect is becoming an important feature of cluster studies.

In this review I describe the physics underlying the Sunyaev-Zel'dovich effect (§ 10.2), and then show how measurements of the effect can be used to extract quantities of physical interest about the clusters and the gas that they contain in a relatively model-independent way (§ 10.3) before describing the methods used to measure the effect and their limitations and future prospects (§ 10.4). There have been several reviews of the Sunyaev-Zel'dovich effect in recent years (Rephaeli 1995a; Birkinshaw 1999; Carlstrom, Holder, & Reese 2002), and these should be read to gain a more complete picture of Sunyaev-Zel'dovich effect research.

10.2   The Physics of the Sunyaev-Zel'dovich Effect

10.2.1   The Thermal Sunyaev-Zel'dovich Effect

The Sunyaev-Zel'dovich effects arise because hot electrons in the atmospheres of clusters of galaxies provide a significant optical depth to microwave background photons. A cluster of galaxies containing an X-ray emitting atmosphere with a electron temperature $T_e = 7 \times 10^7$ K ($k_B T_e = 6$ keV) and average electron density $\overline{n}_e = 10^3$ m^{-3} will have an optical depth for inverse-Compton scattering of

$$\tau_e = \overline{n}_e \sigma_T L \approx 10^{-2} \quad , \tag{10.1}$$

where $\sigma_T = 6.65 \times 10^{-29}$ m^2 is the Thomson scattering cross-section, for a path length $L \approx$ 1 Mpc through the cluster. Since electrons in the intracluster medium have higher mean energies than the photons, the 1% of the photons that are scattered gain energy on average. The mean fractional frequency change in a scattering is

$$\frac{\Delta \nu}{\nu} = \frac{k_B T_e}{m_e c^2} \approx 10^{-2} \quad , \tag{10.2}$$

so that the fractional change in the specific intensity of the microwave background radiation, as seen through the cluster, relative to directions far from the cluster, is

$$\frac{\Delta I_\nu}{I_\nu} \propto \tau_e \frac{\Delta \nu}{\nu} \sim 10^{-4} \quad . \tag{10.3}$$

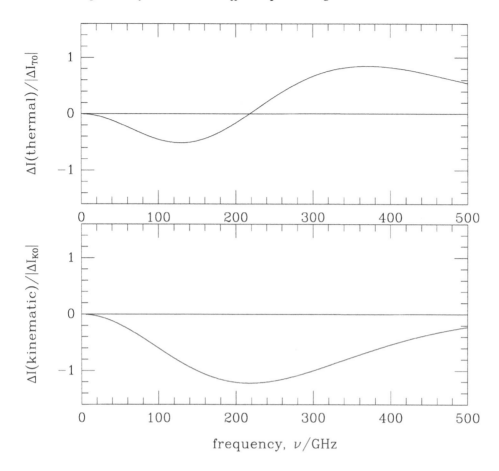

Fig. 10.1. The thermal (upper plot) and kinematic (lower plot) Sunyaev-Zel'dovich effects as functions of frequency, expressed in terms of the specific intensity change, ΔI_ν, that they produce. The largest negative and positive intensity changes in the thermal effect occur at 129 and 370 GHz, while the zero of the thermal effect and the largest negative kinematic effect occur at 218 GHz.

In the Rayleigh-Jeans part of the microwave background spectrum, where $I_\nu \propto \nu^2$, the constant of proportionality is -2. The full spectrum of the thermal effect is shown in the upper panel of Figure 10.1 in terms of the specific intensity change, ΔI_ν, normalized by

$$\Delta I_{T0} = \frac{2h}{c^2} \left(\frac{k_B T_{rad}}{h} \right)^3 y \quad , \tag{10.4}$$

where the Comptonization parameter,

$$y = \int n_e \sigma_T dz \left(\frac{k_B T_e}{m_e c^2} \right) \quad , \tag{10.5}$$

measures the strength of the scattering

The spectrum in Figure 10.1 has an unusual shape for an astronomical source. Below about 218 GHz (the frequency of peak intensity of the microwave background) clusters of galaxies appear as decrements, while at higher frequencies they are brightness enhancements. Even weak thermal Sunyaev-Zel'dovich effects could be identified against the background of primary fluctuations by observing over a wide range of frequencies and making use of this strong spectral signature. This will be possible with, for example, the data from the *Planck* satellite (Trauber 2001).

At low electron temperatures the spectrum of the thermal Sunyaev-Zel'dovich effect is independent of T_e, but as $k_B T_e$ rises above about 5 keV, the increasing fraction of electrons with speeds approaching the speed of light causes significant spectral modifications from relativistic effects (Rephaeli 1995b). In the limit of extreme relativistic electrons the spectrum becomes that of the (inverted) microwave background itself, since scattered photons are moved to energies far outside the microwave band. This nonthermal Sunyaev-Zel'dovich effect has been discussed as a probe of radio galaxy electron populations (McKinnon, Owen, & Eilek 1990), but is better observed from the scattered photons in the X-ray band than in the unscattered photons of the radio band (e.g., Hardcastle et al. 2002).

The amplitude of the thermal Sunyaev-Zel'dovich effect can be expressed in terms of the change in total intensity, ΔI_ν, at frequency ν, the change in the apparent thermodynamic temperature of a Planckian spectrum, ΔT_ν, or the brightness temperature change

$$\Delta T_{\text{RJ},\nu} = \frac{c^2}{2k_B \nu^2} \Delta I_\nu = f_T(\nu) T_{\text{rad}} y \quad, \tag{10.6}$$

where the spectral function $f_T(\nu)$, in the limit of low T_e, has the Kompaneets form

$$f_T(\nu) = \frac{x^2 e^x}{(e^x - 1)^2} \left(x \coth \left(\frac{1}{2} x \right) - 4 \right) \quad, \tag{10.7}$$

and

$$x = \frac{h\nu}{k_B T_{\text{rad}}} \tag{10.8}$$

is a dimensionless measure of frequency.

It is clear from Equation 10.6 that the brightness temperature effect depends only on quantities intrinsic to the cluster, and is therefore redshift independent. This property of redshift independence means that clusters of galaxies can be studied in the thermal Sunyaev-Zel'dovich effect at any redshift where they have substantial atmospheres. The wide range of redshifts over which the thermal Sunyaev-Zel'dovich effect can be seen makes the effect an important probe of cluster evolution.

Practical telescope systems observe a quantity that is a fraction of the total flux density of a cluster, with the fraction depending principally on the method of observation and the cluster angular size, and often being almost constant over a wide range of redshifts (as in Fig. 10.4). The total cluster flux density at frequency ν is

$$\Delta S_\nu = \int \Delta I_\nu \, d\Omega \propto \frac{\int n_e T_e dV}{D_A^2} \quad, \tag{10.9}$$

where the integrals are over the solid angle of the cluster and the cluster volume. ΔS_ν decreases as the inverse square of the angular diameter distance, D_A, rather than the inverse square of the luminosity distance. This can be thought of as indicating that Sunyaev-Zel'dovich effect luminosities increase as $(1+z)^4$, because the luminosity depends on the energy density of the cosmic microwave background that is available to be scattered. If a given cluster of galaxies were to be moved from low to high redshift, its flux density would first decrease, and then increase beyond the redshift of minimum angular size, or $z = 1.62$ in the ΛCDM cosmology adopted here.

Thus at mm wavelengths, the brightest sources in the sky would be clusters of galaxies if clusters had the same atmospheres in the past as they have today. That the brightest mm-wave sources are not clusters does not place a strong constraint on cluster evolution because $D_A(z)$ is a weak function of redshift at $z > 1.6$: from $z = 1.6$ to 9.8 it decreases by only a factor of 2.

10.2.2   The Kinematic Sunyaev-Zel'dovich Effect

If a cluster of galaxies is not at rest in the Hubble flow then the pattern of radiation illuminating the atmosphere in its rest frame is anisotropic, and scattering adds a kinematic Sunyaev-Zel'dovich effect to the thermal effect described in § 10.2.1 (Sunyaev & Zel'dovich 1972; Rephaeli & Lahav 1991). The kinematic effect has an amplitude that is proportional to the cluster's peculiar (redshift-direction) velocity, and has a spectrum that is that of the primary perturbations in the microwave background radiation (Fig. 10.1, lower panel), where

$$\Delta I_{K0} = \tau_e \frac{v_z}{c} \frac{2h}{c^2} \left(\frac{k_B T_{rad}}{h} \right)^3 . \tag{10.10}$$

The similarity of spectrum makes it difficult to distinguish the kinematic effect from primary fluctuations in the microwave background radiation. This is especially a problem since the kinematic effect is smaller than the thermal effect by a factor

$$\frac{\Delta T_{RJ,kinematic}}{\Delta T_{RJ,thermal}} = 0.085 \left(v_z/1000 \text{ km s}^{-1} \right) \left(k_B T_e/10 \text{ keV} \right)^{-1} \tag{10.11}$$

at low frequencies. Observation near the null of the thermal Sunyaev-Zel'dovich effect, at $v_0 = 218$ GHz, gives the greatest contrast for the kinematic effect, but it is still a small quantity compared to the confusion from primary structure in the microwave background radiation on the angular scales of interest (a few arcmin). This limits the velocity accuracy attainable for any cluster of galaxies at moderate redshift to about ± 150 km s^{-1}, even with perfect data.

The relative contributions of the thermal and kinematic Sunyaev-Zel'dovich effects to the power spectrum of the microwave background radiation have been calculated by a number of authors (e.g., da Silva et al. 2000; Molnar & Birkinshaw 2000; Springel, White, & Hernquist 2000). The thermal Sunyaev-Zel'dovich effect is expected to dominate on sufficiently small angular scales (Fig. 10.2), but on average the kinematic effect makes a contribution only about 1% that of the thermal effect. Both effects are larger than the lensing effect of moving clusters of galaxies.

10.3   Uses of the Sunyaev-Zel'dovich Effect in Cluster Studies

10.3.1   Thermal Energy Content of Clusters

The total flux density of the thermal Sunyaev-Zel'dovich effect from a cluster is given by Equation 10.9, which can be rewritten

$$\Delta S_\nu \propto \frac{U_e}{D_A^2} , \tag{10.12}$$

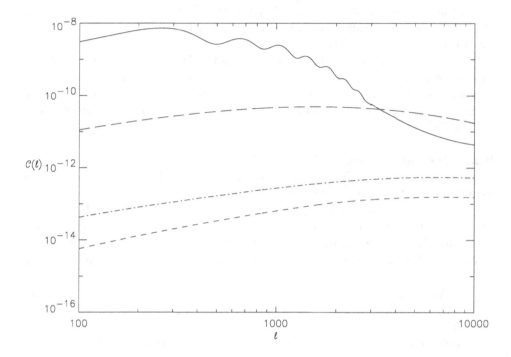

Fig. 10.2. The lensed primary power spectrum for a ΛCDM cosmology with $\Omega_\Lambda = 0.8$, $\Omega_{CDM} = 0.2$, and $n = -1.4$ (solid line), with the power spectrum expected from the thermal Sunyaev-Zel'dovich effect (long dashed line), kinematic Sunyaev-Zel'dovich effect (dash-dot line), and the Rees-Sciama effect from moving clusters (short dashed line) effects. (From Molnar & Birkinshaw 2000.)

where U_e is the total thermal energy content of electrons in the intracluster medium. That is, measurements of the redshift and thermal Sunyaev-Zel'dovich effect of a cluster allow us to infer the thermal energy content of the intracluster medium without needing to know the structure of the density or temperature of this gas.

The thermal Sunyaev-Zel'dovich effect therefore can be used as a calorimeter, measuring the integrated heating to which the cluster gas has been subjected, although a correction has to be made for energy lost by (principally X-ray) radiation (Lapi, Cavaliere, & De Zotti 2004). If the gas is approximately in hydrostatic equilibrium in the cluster, this thermal energy content should be a good measure of the gravitational energy of the cluster, and so a survey for thermal Sunyaev-Zel'dovich effects should naturally pick out the deepest gravitational potential wells in the Universe, provided that the ratio of intracluster gas mass to total mass does not vary too much from cluster to cluster.

10.3.2   *The Baryonic Mass Content of Clusters*

If the integrated flux density of the thermal Sunyaev-Zel'dovich effect and the X-ray spectrum of a cluster are both available, then we can rewrite Equation 10.12 as

$$\Delta S_\nu \propto \int d\Omega \int dz\, n_e\, T_e \propto N_e\, \bar{T_e} \quad .$$

(10.13)

If the mass-weighted mean electron temperature, $\overline{T_e}$, can be approximated by the emission-measure weighted temperature measured by X-ray spectra, then ΔS_ν measures the total electron count in the intracluster medium, N_e. A metalicity measurement for the gas, again from the X-ray spectrum, allows N_e to be converted into the baryonic mass of the intracluster medium. This is usually far larger than the stellar mass content, and so is a good measure of the total baryonic content of the cluster. X-ray imaging data can be used to estimate cluster total masses, using the assumption of hydrostatic equilibrium (Fabricant, Lecar, & Gorenstein 1980), and so the baryonic mass fraction of clusters can be measured.

Since X-ray images and spectra, and Sunyaev-Zel'dovich effects, can be measured from massive clusters to $z \approx 1$, a history of the baryonic mass fraction of clusters can be constructed, although this is currently biased to the high-mass end of the cluster population. The result, for the 10–20 clusters for which this has been done to date (Grego et al. 2001), is that this ratio does not change significantly with redshift, remaining close to the result $\Omega_b/\Omega_m = 0.16 \pm 0.04$ deduced from the *WMAP* results or primordial nucleosynthesis. This suggests that clusters are fair samples of the mass of the Universe out to the largest redshifts at which this test has been applied: there is no strong segregation of dark and baryonic matter during cluster collapse.

10.3.3   *Cluster Lensing and the Sunyaev-Zel'dovich Effects*

It is interesting to consider the possibility of combining Sunyaev-Zel'dovich effect data with gravitational lensing, rather than X-ray, data, to study baryonic mass fractions. If the ellipticity field $\epsilon_i(\theta)$ has been measured (and corrected to estimate the shear distortion field), then the surface mass density of a lensing cluster is given by an integral

$$\Sigma = -\frac{1}{\pi}\Sigma_{\rm crit} \int d^2\theta' \, K_i(\theta',\theta)\,\epsilon_i(\theta') \quad , \tag{10.14}$$

where $\Sigma_{\rm crit}$ is the critical surface density, and the kernels K_i are angular weighting functions. Clearly, the surface mass density is a linear (though nonlocal) function of the observable quantity, $\epsilon_i(\theta)$. Since the Sunyaev-Zel'dovich effect is a linear measure of the projected baryonic mass density (if the cluster is isothermal), the ratio of a Sunyaev-Zel'dovich effect map to a mass map derived from lensing should give a good measure of the (projected) radial dependence of baryonic mass fraction. This should be consistent with the result obtained by applying the standard assumption of hydrostatic equilibrium to the X-ray data, but is less susceptible to errors arising from the unknown structure of the cluster along the line of sight.

The earliest suggestion of the use of lensing and Sunyaev-Zel'dovich effect data together to study clusters appears to have been made by Ostriker & Vishniac (1986) in the context of the quasar pair $1146+111B,C$, but little work on the comparison has been done to date, although Doré et al. (2001) and Zaroubi et al. (2001) have shown that the addition of lensing data allows the three-dimensional structure of clusters to be investigated.

A comparison of lensing and X-ray derived masses has been made for the inner 250 kpc of cluster CL 0016+16 (Fig. 10.3) by Worrall & Birkinshaw (2004). Within this radius, set by the limited lensing data, the masses are

$$M_{\rm tot} = (2.7 \pm 0.9) \times 10^{14}\ {\rm M}_\odot \tag{10.15}$$

and

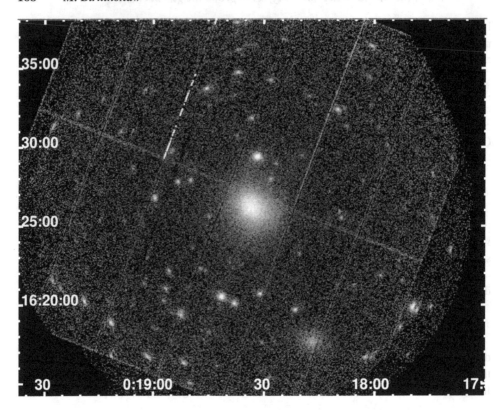

Fig. 10.3. A vignetting-corrected $0.3 - 5.0$ keV image of CL 0016+16, from an *XMM-Newton* observation (Worrall & Birkinshaw 2004). A quasar $3'$ north of the cluster, and a second cluster, $9'$ to the south-west, are at similar redshifts to CL 0016+16.

$$M_{tot} = (2.0 \pm 0.1) \times 10^{14} \, M_\odot \quad , \tag{10.16}$$

found from lensing (Smail et al. 1997, converted to our cosmology) and X-ray data, respectively. The results agree to within the limited accuracy of the lensing mass. A combination of these data with the central Sunyaev-Zel'dovich effect for the cluster, of -1.26 ± 0.07 mK, then leads to a distance-independent measure of the baryonic mass fraction within the central part of the cluster of 0.13 ± 0.02, close to the value implied by cosmological parameters.

10.3.4   *Cluster Structures*

The structures of cluster atmospheres are generally studied from their X-ray images. The X-ray surface brightness

$$b_X \propto \int n_e^2 \Lambda(T_e) \, dz \quad , \tag{10.17}$$

where $\Lambda(T_e)$ is the emissivity function. The X-ray image can be inverted to derive an electron density profile, $n_e(r)$, for the cluster atmosphere under assumptions about the shape of the atmosphere and the absence of clumping, which would cause $\overline{n_e^2} > (\overline{n_e})^2$. Since the Sunyaev-Zel'dovich effects are a projection of n_e, rather than the more complicated quantity $n_e^2\Lambda(T_e)$, they should give a cleaner measure of the structure of the atmosphere. However, the restricted angular dynamic range of measurements of Sunyaev-Zel'dovich effects (§ 10.4), and lower signal-to-noise ratio and angular resolution of Sunyaev-Zel'dovich effect maps relative to what is possible with long X-ray exposures, means that X-ray based structures are superior to those derived entirely from the Sunyaev-Zel'dovich effects. Furthermore, since the Sunyaev-Zel'dovich effects are line-of-sight integrals of electron pressure, they are insensitive to subsonic structures within a cluster, such as cold fronts, which are easily seen on *Chandra* X-ray images (e.g., Markevitch et al. 2000).

The different n_e dependencies of the Sunyaev-Zel'dovich effect and X-ray surface brightness also implies that clusters of galaxies have larger angular sizes in Sunyaev-Zel'dovich effects than in the X-ray, and that the X-ray emission is more sensitive to cluster cores, while the Sunyaev-Zel'dovich effects are more weighted toward the low-density envelopes. However, the low surface brightness of cluster envelopes, and low signal-to-noise ratio even in the centers of most Sunyaev-Zel'dovich effect maps, means that the X-ray data are superior out to the largest radii to which gas has been detected.

10.3.5   Cluster Distances

Perhaps the most widely discussed use of the thermal Sunyaev-Zel'dovich effect has been to measure the distances of clusters of galaxies. In its simplest form, the method is to compare the X-ray surface brightness of a cluster on some fiducial line of sight,

$$b_{X0} \propto n_{e0}^2 \Lambda(T_{e0}) L \quad , \tag{10.18}$$

with the thermal Sunyaev-Zel'dovich effect on the same line of sight,

$$\Delta T_{RJ,0} \propto n_{e0} T_{e0} L \quad , \tag{10.19}$$

and eliminate the scale (perhaps central) electron density in the cluster and find an absolute measurement of the path length along some fiducial line of sight

$$L \propto \frac{\Delta T_{RJ,0}^2}{b_{X0}} \cdot \frac{\Lambda(T_{e0})}{T_{e0}^2} \quad . \tag{10.20}$$

This path length can then be compared with the angular size of the cluster to infer the cluster's angular diameter distance, under the assumption that the cluster is spherical. This technique has been used for a number of clusters (e.g., Hughes & Birkinshaw 1998; Mason, Myers, & Readhead 2001; Reese et al. 2002). A recent recalculation of the distance to CL 0016+16 (Fig. 10.3) using this technique gave $D_A = 1.16 \pm 0.15$ Gpc (Worrall & Birkinshaw 2004), and implies a Hubble constant of $68 \pm 8 \pm 18$ km s^{-1} Mpc^{-1}.

While single cluster distances are likely to be error-prone, because of their unknown three-dimensional shapes (and hence the large systematic error on the result for CL 0016+16), a suitably chosen sample of clusters, without an orientation bias, can be used to map the Hubble flow to $z \approx 1$, and hence to measure a number of cosmological parameters. Molnar, Birkinshaw, & Mushotzky (2002) have shown that a set of about 70 clusters could provide useful measure of the equation-of-state parameter, w, as well as the Hubble constant. A recent review of the state of measurement of $D_A(z)$ using this technique is given by Carlstrom et al. (2002).

It is critical that the absolute calibrations of the X-ray and Sunyaev-Zel'dovich effect data are excellent, and that cluster substructure is well modeled, if distances are to be estimated in this way. Since clusters are relatively young structures, and likely to be changing significantly with redshift, variations in the amount of substructure with redshift might be a significant source of systematic error. High-quality X-ray imaging and spectroscopy are therefore essential if the distances obtained are to be reliable.

Finally, a crucial element of this technique is that the set of clusters used should be free from any biasing selection effect. Perhaps the most important such effect is that of orientation: if clusters are selected by any surface brightness sensitive criterion, then they will tend to be preferentially aligned relative to the line of sight. Since the distance measurement technique relies on comparing the line-of-sight depth of the cluster, L, with the cross line-of-sight angular size, θ, such a selection will inevitably induce a bias into the measured $D_A(z)$ function.

Clumping of the intracluster medium and a number of other problems can also cause biases in the results. Comprehensive reviews of such biases can be found in most papers on Hubble constant measurements using this technique, and in Birkinshaw (1999).

10.3.6   *Cluster Peculiar Velocities*

The kinematic Sunyaev-Zel'dovich effect (§ 10.2.2) is heavily confused by primary anisotropies in the microwave background radiation since cluster peculiar velocities are expected to be less than 100 km s^{-1}. It seems likely, therefore, that measurement of the kinematic effect will only be possible on a statistical basis, by comparing the brightness of the microwave background radiation toward the clusters of galaxies with positions in the field. Since the kinematic effect, like the thermal effect, is redshift independent, such a statistical measurement would be an important check on the changing velocities of clusters with redshift as structure develops. Efforts to measure the kinematic effect continue, although only relatively poor limits on cluster velocities, of order 1000 km s^{-1}, have been obtained to date (Holzapfel et al. 1997a; LaRoque et al. 2004).

10.3.7   *Cluster Samples*

Since the Sunyaev-Zel'dovich effects are redshift-independent in ΔT_{RJ} terms, and the observable ΔS_ν are almost redshift-independent, it follows that Sunyaev-Zel'dovich effect surveys should be more effective than X-ray or optical surveys at detecting clusters of galaxies at large redshift. Such samples of clusters, selected by their Sunyaev-Zel'dovich effects, will be unusually powerful in being almost mass limited, and counts of the number of such clusters as a function of redshift provide a good method of measuring σ_8 (Fan & Chiueh 2001).

Such samples are also ideal for work on the distance scale, and hence the measurement of the equation-of-state parameter, w.

10.4   Instruments and Techniques

Many observations of the thermal Sunyaev-Zel'dovich effect have followed its first reliable detection, in Abell 2218 (Birkinshaw et al. 1978). While the earliest observations were made with single-dish radiometers, more recently work has been done with interferometers and bolometer arrays. The quality of the observational data is advancing rapidly, so that the review of the observational situation that I wrote in 1999 (Birkinshaw 1999) grossly underrepresents the number of good detections and images of the effect.

The most effective observations of the Sunyaev-Zel'dovich effects over the past few years have used interferometers, notably the Ryle telescope (Jones et al. 1993) and the BIMA and OVRO arrays (Carlstrom, Joy, & Grego 1996). Since the thermal Sunyaev-Zel'dovich effects of rich clusters typically have angular sizes exceeding 1′, most of the correlated signal appears on the shortest ($< 2000\lambda$) baselines. These baselines are usually underrepresented in the interferometers, so that all these instruments had to be retrofitted by increasing the observing wavelength, λ, or altering the array geometry. The effectiveness of the changes is demonstrated by the large number of high signal-to-noise ratio detections of clusters of galaxies that interferometers have produced (e.g., Joy et al. 2001; Cotter et al. 2002; Grainge et al. 2002).

The major advantage of interferometers is their ability to reject contaminating signals from their surroundings and the atmosphere, and to spatially filter the data to exclude radio point sources in the fields of the clusters. Extremely long interferometric integrations are possible before parasitic signals degrade the data. The limitation, at present, is that the retrofitted arrays are not ideally matched to the purpose of mapping the Sunyaev-Zel'dovich effect. This limitation is being addressed by the construction of a new generation of interferometers, including AMI (Kneissl et al. 2001), SZA (Mohr et al. 2002), and AMiBA (Lo et al. 2001), which will have enough sensitivity to undertake deep blank-field surveys for clusters at $z > 1$.

Single-dish radiometer systems, particularly when equipped with radiometer arrays, are efficient for finding strong Sunyaev-Zel'dovich effects. Their large filled apertures can integrate the signal over much of the solid angle of a cluster. However, spillover and residual atmospheric noise limit the length of useful integrations and hence the sensitivity that can be achieved (Birkinshaw & Gull 1984). Since practical systems always involve differencing between on-source and off-source sky regions, this technique, just as interferometry, is sensitive only to clusters smaller than some maximum angular size, and hence at sufficiently large redshift. Finally, there is no simple way of removing the signals of contaminating (and often variable) radio sources that appear superimposed on the Sunyaev-Zel'dovich effect. This further restricts the set of clusters that can be observed effectively.

Modern antennas provide new opportunities for radiometer systems since they have superior spillover characteristics and can be equipped with radiometer arrays. The 100-meter Green Bank Telescope is an obvious example, and the OCRA project (§ 10.4.2) is intending to provide a fast survey capability for this and other large telescopes.

Bolometers are capable of measuring the spectrum of the Sunyaev-Zel'dovich effect above ~ 90 GHz, and could separate the thermal Sunyaev-Zel'dovich effect from the associated kinematic effect or from primary anisotropy confusion. The intrinsic sensitivity of bolometers should allow fast measurements of targeted clusters, or high-speed surveys of large regions of sky. However, bolometers are exposed to a high level of atmospheric and other environmental signals, and so the differencing scheme used to extract sky signals from the noise must be of high quality.

There have been two attempts to use the spectral capability of bolometers to determine cluster peculiar velocities (Holzapfel et al. 1997a; LaRoque et al. 2004), and a similar experiment can use the spectral distortion of the thermal effect to measure the temperature of the microwave background radiation in remote parts of the Universe and test whether $T_{rad} \propto (1+z)$ (Battistelli et al. 2002). Newer bolometer arrays, such as BOLOCAM (Glenn et al. 1998) and ACBAR (Romer et al. 2001), should allow fast surveys for clusters and could make confusion-limited measurements of the cluster velocity if the flux scale near 1 mm is well calibrated.

10.4.1 Recent Advances

A major advance over the past few years has been the change from making and reporting Sunyaev-Zel'dovich effects for individual clusters to reporting for samples of clusters. The samples most favored are based on X-ray surveys (e.g., Mason et al. 2001; LaRoque et al. 2003), and can be assumed to be orientation-independent provided that the selection is at X-ray fluxes far above the sensitivity limit of the survey. The combination of good X-ray and Sunyaev-Zel'dovich effect data is leading to reliable results for baryon fractions and the Hubble constant.

The next improvement will clearly be to obtain samples of clusters selected entirely in the Sunyaev-Zel'dovich effect, on the basis of blind surveys. Such surveys require sensitive, arcminute-scale, instruments capable of covering many deg^2 of sky in a reasonable length of time—a good first target would be to cover 10 deg^2 to a sensitivity limit $\Delta T_{\mathrm{RJ}} \approx 100 \ \mu$K, yielding a sample of tens of clusters selected only through their Sunyaev-Zel'dovich effects. Such cluster samples would be ideal for cosmological purposes, since they would be almost mass limited. Since the observable thermal Sunyaev-Zel'dovich effect depends linearly on the properties of clusters, and on the angular diameter distance, which is a slow function of redshift for $z \approx 1$, a cluster of a given mass can be detected with almost the same efficiency at any redshift > 0.5. This is illustrated for the OCRA instrument (Browne et al. 2000) in Figure 10.4.

Many of the new projects are aimed at blind surveys of the microwave background sky to find Sunyaev-Zel'dovich effects, and hence complete cluster samples. Figure 10.5 compares the survey speeds of these projects, ranging from radiometer and bolometer arrays (OCRA, BOLOCAM) to interferometers (AMiBA, SZA, AMI). AMiBA and OCRA are discussed in more detail in the next section.

10.4.2 Two New Projects: AMiBA and OCRA

AMiBA, the Array for Microwave Background Anisotropy, is described by Lo et al. (2001). It is a platform-based system, operating at 95 GHz with a bandwidth of 20 GHz, with up to 19 1.2-m antennas, providing an 11$'$ field of view and approximately 2$'$ angular resolution. Its flux density sensitivity, 1.3 mJy per beam in 1 hour, makes it a highly effective interferometer for surveys: a 3-hour observation suffices to detect a $z = 0.5$ cluster with mass 2.7×10^{14} M$_\odot$ at 5σ, and clusters with about a third of this mass could be detected well above the confusion limit in longer integrations.

A trial platform carrying two smaller antennas has been operating on Mauna Loa for some months, and various aspects of the system are being tried out and debugged. AMiBA should be operational in the 2003/4 winter season for studies of primary structure in the microwave background radiation, and full Sunyaev-Zel'dovich effect operations should start about 1 year later.

The high density of short antenna-antenna spacings and the 3-mm operating wavelength make AMiBA the most sensitive of the planned interferometers. However, AMI at 15 GHz (Kneissl et al. 2001), the SZA at 30 GHz (Mohr et al. 2002), and AMiBA at 95 GHz have usefully complementary angular resolutions and operating frequencies. Although the spread of frequencies provided by these three systems is inadequate for useful spectral work, to separate primary anisotropies and kinematic effects from the thermal signal, it is sufficient to provide a control against nonthermal radio sources, and an important cross-check on cluster counts.

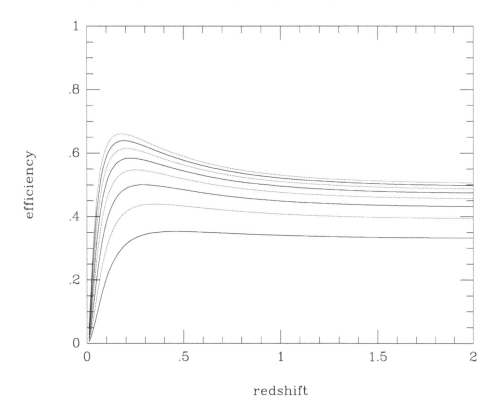

Fig. 10.4. The relative sensitivity of observations of the thermal Sunyaev-Zel'dovich effect in a cluster of galaxies as a function of differencing angle (from $1\overset{\prime}{.}5$ to $5\overset{\prime}{.}0$ from the bottom to the top curve) and redshift for the One Centimeter Radiometer Array (OCRA; Browne et al. 2000). Note the flatness of the sensitivity function at $z > 0.5$ for any differencing angle.

Several thousand individual pointings with AMiBA would be needed for a shallow survey of the entire 64 deg^2 *XMM-Newton* survey field (Pierre et al. 2001). While most of the $\sim 10^3$ Sunyaev-Zel'dovich effect detections expected will correspond to X-ray clusters, a comparison of the *XMM-Newton* and AMiBA detection functions shows that AMiBA will be better at detecting clusters beyond $z = 0.7$, and could find clusters with X-ray luminosities $> 4 \times 10^{43}$ erg s^{-1} (in $0.5 - 10$ keV) to $z > 2$ for later, deeper, X-ray and infrared follow-up.

While AMiBA has the interferometric advantages of rejecting contaminating environmental signals, high sensitivity and fast survey speed are better achieved by radiometer arrays. OCRA, the One Centimeter Radiometer Array (Browne et al. 2000), has been designed to make use of recent advances in radiometer technology to perform wide-field surveys for Sunyaev-Zel'dovich effects and radio sources.

OCRA is an array of ~ 100 30-GHz radiometers, based on *Planck* technology, housed in a single cryostat at the secondary (or tertiary) focal plane of a large antenna, and providing $1'$ beams with $\sim 3'$ beam separations. Each radiometer has excellent flux density sensitivity at 30 GHz. This is a large and expensive system, and two preliminary (but scientifically useful) versions of OCRA will be used before OCRA is built.

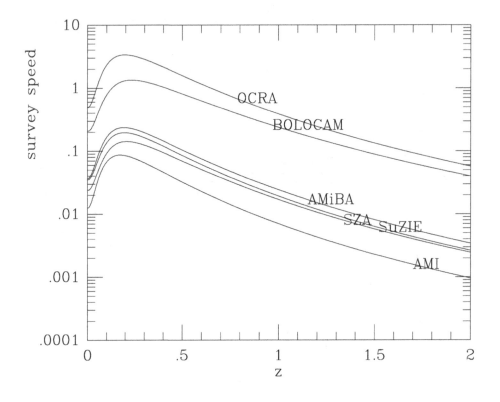

Fig. 10.5. The relative speeds of several future Sunyaev-Zel'dovich effect instruments, as a function of redshift, for a model in which the cluster gas evolves in a non self-similar fashion. The experiments shown are OCRA (Browne et al. 2000; § 10.4.2), BOLOCAM (Glenn et al. 1998), AMiBA (Lo et al. 2001; § 10.4.2), SZA (Mohr et al. 2002), SuZIE (Holzapfel et al. 1997b), and AMI (Kneissl et al. 2001).

The prototype, OCRA-p, is a two-beam system now being tested on the Torun 32-m radio telescope. With a system temperature below 40 K, we expect to achieve a flux density sensitivity of 5 mJy in 10 s provided that the atmospheric noise is well controlled. OCRA-p will be followed by a FARADAY project receiver with eight beams, which uses MMICs on InP substrates, and which is funded by a European Union grant. This receiver is currently under construction and will be in use in about a year. It will achieve a significant improvement in mapping speed over OCRA-p, though still be a factor ~ 10 slower than OCRA itself.

OCRA will be an extremely fast instrument for finding high-redshift clusters. A comparison of its mapping speed with some other systems is shown in Figure 10.5. While OCRA is more susceptible to radio source confusion than AMiBA (indeed, the study of the radio source population at 30 GHz is one aim of OCRA), recent work on the probability of detecting Sunyaev-Zel'dovich effects from clusters at 18.5 GHz (Birkinshaw, in preparation) suggests that the level of source confusion will not prevent OCRA from detecting most rich, distant clusters. Even the FARADAY receiver will be a highly competitive survey instrument.

10.5   Summary

The thermal Sunyaev-Zel'dovich effects of clusters of galaxies, expressed in brightness temperature terms, are redshift-independent measures of the thermal energy content of clusters: they are, effectively, calorimeters of energy releases in clusters after account is taken of the energy radiated by line and bremsstrahlung emission (principally in the X-ray band). With the assistance of X-ray and/or gravitational lensing data, the thermal Sunyaev-Zel'dovich effect measures a wide range of cluster properties, including distance.

Thermal Sunyaev-Zel'dovich effects are mass finders, with strong associations with rich clusters of galaxies, and should be good probes of the most massive structures that exist in the Universe at any redshift. Deep Sunyaev-Zel'dovich effect studies are therefore an excellent way of discovering the degree of cluster formation in the Universe to high redshift, and counts of clusters from forthcoming surveys should place strong constraints on the processes of cluster formation.

Two weaker Sunyaev-Zel'dovich effects, the kinematic and polarization effects, are also of interest, but are harder to measure. Both can give information about the speeds of clusters of galaxies, and hence the evolving dynamics of gravitating objects in the Universe, but both are subject to considerable confusion from primary structures in the microwave background radiation and are therefore likely to be detectable only in a statistical sense for populations of clusters.

Over the past 20 years, the thermal Sunyaev-Zel'dovich effect has gone from being a curiosity to a major tool for cosmology and cluster physics. Substantial results based on Sunyaev-Zel'dovich effect work are to be expected within the next 10 years.

References

Battistelli, E. S., et al. 2002, ApJ, 580, L101

Bennett, C. L., et al. 2003, ApJS, 148, 1

Birkinshaw, M. 1999, Phys. Rep., 310, 97

Birkinshaw, M., & Gull, S. F. 1984, MNRAS, 206, 359

Birkinshaw, M., Gull, S. F., & Northover, K. J. E. 1978, Nature, 185, 245

Browne, I. W. A., Mao, S., Wilkinson, P. N., Kus, A. J., Marecki, A., & Birkinshaw, M. 2000, Proc. SPIE., 4015, 299

Carlstrom, J. E., Holder, G. P., & Reese, E. D. 2002, ARA&A, 40, 643

Carlstrom, J. E., Joy, M., & Grego, L. 1996, ApJ, 456, L75 (erratum: 461, L9)

Cotter, G., Buttery, H. J., Das, R., Jones, M. E., Grainge, K., Pooley, G. G., & Saunders, R. 2002, MNRAS, 334, 323

da Silva, A. C., Barbosa, D., Liddle, A. R., & Thomas, P. A. 2000, MNRAS, 317, 37

de Bernardis, P., et al. 2000, Nature, 404, 955

Dicke, R. H., Peebles, P. J. E., Roll, P. G., & Wilkinson, D. T. 1965, ApJ, 142, 414

Doré, O., Bouchet, F. R., Mellier, Y., & Teyssier, R. 2001, A&A, 375, 14

Fabricant, D., Lecar, M., & Gorenstein, P. 1980, ApJ, 241, 552

Fan, Z., & Chiueh, T. 2001, ApJ, 550, 547

Fixsen, D. J., Cheng, E. S., Gales, J. M., Mather, J. C., Shafer, R. A., & Wright, E. L. 1996, ApJ, 473, 576

Freedman, W. L., et al. 2001, ApJ, 553, 47

Glenn, J., et al. 1998, Proc. SPIE., 3357, 326

Gorski, K. M., Banday, A. J., Bennett, C. L., Hinshaw, G., Kogut, A., Smoot, G. F., & Wright, E. L. 1996, ApJ, 464, L11

Grainge, K., Jones, M. E., Pooley, G., Saunders, R., Edge, A., Grainger, W. F., & Kneissl, R. 2002, MNRAS, 333, 318

Grego, L., Carlstrom, J. E., Reese, E. D., Holder, G. P., Holzapfel, W. L., Joy, M. K., Mohr, J. J., & Patel, S. 2001, ApJ, 552, 2

Hanany, S., et al. 2000, ApJ, 545, L5

Hardcastle, M. J., Birkinshaw, M., Cameron, R. A., Harris, D. E., Looney, L. W., & Worrall, D. M. 2002, ApJ, 581, 948

Hinshaw, G., Banday, A. J., Bennett, C. L., Gorski, K. M., Kogut, A., Smoot, G. F., & Wright, E. L. 1996, ApJ, 464, L17

Holzapfel, W. L., Ade, P. A. R., Church, S. E., Mauskopf, P. D., Rephaeli, Y., Wilbanks, T. M., & Lange, A. E. 1997a, ApJ, 481, 35

Holzapfel, W. L., Wilbanks, T. M., Ade, P. A. R., Church, S. E., Fischer, M. L., Mauskopf, P. D., Osgood, D. E., & Lange, A. E. 1997b, ApJ, 479, 19

Hughes, J. P., & Birkinshaw, M. 1998, ApJ, 501, 1

Jaffe, A. H., et al. 2001, Phys. Rev. Lett., 86, 3475

Jones, M., et al. 1993, Nature, 365, 320

Joy, M., et al. 2001, ApJ, 551, L1

Kneissl, R., Jones, M. E., Saunders, R., Eke, V. R., Lasenby, A. N., Grainge, K., & Cotter, G. 2001, MNRAS, 328, 783

Lapi, A., Cavaliere, A., & De Zotti, G. 2004, in Carnegie Observatories Astrophysics Series, Vol. 3: Clusters of Galaxies: Probes of Cosmological Structure and Galaxy Evolution, ed. J. S. Mulchaey, A. Dressler, & A. Oemler (Pasadena: Carnegie Observatories, http://www.ociw.edu/ociw/symposia/series/symposium3/proceedings.html)

LaRoque, S. J., et al. 2003, ApJ, 583, 559

LaRoque, S. J., Carlstrom, J. E., Reese, E. D., Holder, G. P., Holzapfel, W. L., Joy, M., & Grego, L. 2004, ApJ, in press (astro-ph/0204134)

Lo, K. Y., Chiueh, T., Liang, H., Ma, C. P., Martin, R., Ng, K.-W., Pen, U. L., & Subramanyan, R. 2001, in IAU Symp. 201, ed. A. N. Lasenby, A. W. Jones & A. Wilkinson (San Francisco: ASP), 31

Markevitch, M., et al. 2000, ApJ, 541, 542

Mason, B. S., Myers, S. T., & Readhead, A. C. S. 2001, ApJ, 555, L11

McKinnon, M. M., Owen, F. N., & Eilek, J. A. 1990, AJ, 101, 2026

Mohr, J. J., et al. 2002, in AMiBA 2001: High-z Clusters, Missing Baryons, and CMB Polarization, ed. L. W. Chen et al. (San Francisco: ASP), 43

Molnar, S., & Birkinshaw, M. 2000, ApJ, 537, 542

Molnar, S., Birkinshaw, M., & Mushotzky, R. F. 2002, ApJ, 570, 1

Myers, S. T., Baker, J. E., Readhead, A. C. S., Leitch, E. M., & Herbig, T. 1997, ApJ, 485, 1

Ostriker, J. P., & Vishniac, E. T. 1986, Nature, 322, 804

Penzias, A. A., & Wilson, R. W. 1965, ApJ, 142, 419

Pierre, M., et al. 2001, ESO Messenger, 105, 32

Pyne, T., & Birkinshaw, M. 1993, ApJ, 415, 459

Rees, M. J., & Sciama, D. W. 1968, Nature, 217, 511

Reese, E. D., Carlstrom, J. E., Joy, M., Mohr, J. J., Grego, L., & Holzapfel, W. L. 2002, ApJ, 581, 53

Rephaeli, Y. 1995a, ARA&A, 33, 541

——. 1995b, ApJ, 445, 33

Rephaeli, Y., & Lahav, O. 1991, ApJ, 372, 21

Romer, A. K., et al. 2001, BAAS, 199, 1420

Smail, I., Ellis, R. S., Dressler, A., Couch, W. J., Oemler, A., Sharples, R. M., & Butcher, H. 1997, ApJ, 479, 70

Spergel, D. N., et al. 2003, ApJS, 148, 175

Springel, V., White, M., & Hernquist, L. 2000, ApJ, 549, 681 (erratum: 562, 1086)

Sunyaev, R. A., & Zel'dovich, Ya. B. 1972, Comments Astrophys. Space Phys., 4, 173

Trauber, J. 2001, in IAU Symp. 204, The Extragalactic Infrared Background and its Cosmological Implications, ed. M. Hauser & M. Harwit (ASP: San Francisco), 40

Worrall, D. M., & Birkinshaw, M. 2004, MNRAS, in press (astro-ph/0301123)

Wright, E. L., Bennett, C. L., Gorski, K., Hinshaw, G., & Smoot, G. F. 1996, ApJ, 464, L21

Zaroubi, S., Squires, G., de Gasperis, G., Evrard, A. E., Hoffman, Y., & Silk, J. 2001, ApJ, 561, 600

11

The formation of early-type galaxies: observations to $z \approx 1$

TOMMASO TREU
California Institute of Technology

Abstract

How does the number density of early-type (E+S0) galaxies evolve with redshift? What are their star formation histories? Do their mass density profile and other structural properties evolve with redshift? Answering these questions is key to understanding how E+S0s form and evolve. I review the observational evidence on these issues, focusing on the redshift range $z \approx 0.1 - 1$, and compare it to the predictions of current models of galaxy formation.

11.1   Introduction

Understanding the formation and evolution of early-type galaxies (E+S0, i.e. ellipticals and lenticulars) is not only crucial to unveil the origin of the Hubble sequence, but is also a focal point connecting several unanswered, major astrophysical questions. The hypothesis that E+S0s form by mergers of disks at relatively recent times is one of the pillars of the cold dark matter (CDM) hierarchical scenario. At galactic scales, since they are the most massive galaxies, E+S0s are the key to understanding how and when dark and luminous mass are assembled in galaxies, and to test the universal form and ubiquity of dark matter halos predicted by the CDM paradigm (Navarro, Frenk, & White 1997, hereafter NFW; Moore et al. 1998). On subgalactic scales, the existence of a correlation between black hole mass and spheroid velocity dispersion suggests that the growth of black holes and the activity cycles in active galactic nuclei are somehow intimately connected with the formation of spheroids. Therefore, a unified formation scenario must ultimately be conceived (e.g., Kauffmann & Haehnelt 2000; Monaco, Salucci, & Danese 2000; Volonteri, Haardt, & Madau 2003; see Ho 2004).

Theoretical formation scenarios are often grouped into two categories, broadly referred to as "monolithic collapse" and "hierarchical formation"[*].

In the traditional picture of the monolithic collapse, E+S0s assembled their mass and formed their stars in a rapid event, of much shorter duration than their average age (Eggen, Lynden-Bell, & Sandage 1962; Larson 1975; van Albada 1982). The formation process happened at high redshifts, and proto-early-type galaxies would be star-forming and dust-enshrouded systems. These kinds of models are consistent with a variety of features (see Matteucci 2003, and references therein), including the homogeneity of the present-day stel-

---

[*] A complete review of the theoretical background is beyond the aims of this observational review. For more information the reader is referred to, e.g., the reviews by de Zeeuw & Franx (1991), Bertin & Stiavelli (1993), Merritt (1999), Peebles (2002), de Freitas Pacheco, Michard, & Mohayaee (2003), and references therein.

lar populations (Sandage & Visvanathan 1978), the existence of metallicity gradients (Sandage 1972) and the characteristic $R^{1/4}$ surface brightness profile (de Vaucouleurs 1948).

By contrast, in the hierarchical scenario, hereafter the "standard model," early-type galaxies form by mergers of disks at relatively recent times (Toomre & Toomre 1972; Toomre 1977; White & Rees 1978; Blumenthal et al. 1984). The formation process is continuous: mass is accreted over time, and both major and minor mergers can induce star formation, thus rejuvenating at times the stellar populations. Furthermore, environmental processes, such as galaxy interactions, can be built into the models and predictions made of the properties of E+S0s as a function of environment (Kauffmann 1996; Benson, Ellis, & Menanteau 2002). Examples that can be tested against observations include the global properties of E+S0s, such as the color-magnitude relation or the age of the integrated stellar populations. Increasingly sophisticated numerical cosmological simulations are being developed: it has recently become possible to simulate in detail* the formation of individual E+S0s in a fully cosmological context (Meza et al. 2003). This opens up the possibility of using observations of the internal structure (e.g., the mass density profile) of E+S0s as a test of the standard model.

A common and practical tool are the so-called pure luminosity evolution (PLE) models. In these phenomenological models, E+S0s form at a given redshift of formation (z_f) and evolve only through the evolution of their stellar populations. Typically, the star formation history is assumed independent of present-day luminosity. For a given star formation history, stellar evolution models, and present-day luminosity function (LF), it is straightforward to compute observable properties, such as number counts and observed color distribution. PLE models are often used as toy realizations of the monolithic collapse models, and their predictions are contrasted to the standard model predictions. However, it should be kept in mind that PLE models are only a phenomenological tool, not coincident with monolithic collapse.

For decades, the only way to test and improve our understanding of the formation process was through observations of the local Universe. The only accessible pieces of information were observables such as color or spectra of local E+S0s.

This has dramatically changed in the last few years. The sharp images taken by the *Hubble Space Telescope (HST)*, together with the high-quality ground-based data collected by large-aperture telescopes equipped with modern instruments, have opened up the cosmic time domain. Now E+S0s can be identified, counted, and studied as a function of redshift (i.e. cosmic time) out to look-back times that are a significant fraction of the lifetime of the Universe. Increasingly detailed information (luminosity, color, redshift, internal kinematics, mass estimates from dynamics and lensing) for distant E+S0s can now be obtained, allowing for increasingly stringent tests of the cosmological model.

Clearly, the combination of both pieces of information—local and high-redshift data—is what delivers the most stringent observational tests. Since other speakers at this meeting have covered the local Universe (e.g., Davies 2004), I will focus specifically on the study of distant E+S0s. In particular, I will cover the redshift range $0.1 < z < 1$, corresponding in the currently favored ΛCDM cosmology† to look-back times of 1 to 8 Gyr, thus approximately the second half of the life of the Universe. I will discuss E+S0s in general, irrespective of

---

* Note, however, that crucial mechanisms such as star formation can only be treated in a simplified way by means of semi-analytical recipes.

† When needed, I assume the Hubble constant to be $H_0 = 65\ h_{65}$ km s^{-1} Mpc^{-1} = $100\ h$ km s^{-1} Mpc^{-1}, $h_{65} = 1$. The matter density and cosmological constant in critical units are $\Omega_m = 0.3$ and $\Omega_\Lambda = 0.7$, respectively.

their environment. When needed, I will distinguish between field and cluster E+S0s to contrast different evolutionary histories. A final caveat is that I will mostly consider the broad class of spheroids (E+S0), without distinguishing between pure ellipticals and lenticulars. This simplification should be kept in mind when interpreting observational results, given that Es and S0s might have significantly different formation histories (e.g., Dressler et al. 1997; Trager 2004). When possible I will discuss the results in terms of pure Es or S0s.

I will concentrate on three key questions:

(1) How does the number density of E+S0s evolve with time?
(2) What is the star formation history of E+S0s?
(3) What is the distribution of mass in E+S0s, and how does it evolve with time?

In the next sections, I will review observational work on each of these questions, discuss comparison with model predictions, and briefly comment on future perspective.

11.2   Evolution of the Number Density

How many E+S0s are there at any given redshift? Ideally, in order to compare directly with models of structure formation, we would like observations to deliver the volume density of E+S0s as a function of mass and redshift. The closest available observable to the mass function is the LF $\phi(L,z)$. If luminosity evolution is understood, $\phi(L,z)$ can be used to derive the evolution of the number density at a fixed present-day equivalent luminosity.

Before we proceed, it is useful to introduce a simple parametrization of $\phi(L,z)$ that can be used to express observational results in a synthetic form. Assuming PLE and indicating stellar mass with M_*, we can express luminosity evolution as $\log L(z) = \log L(0) - \left[d \left(\log M_*/L \right) / dz \right] z$, to first order in z. Similarly, assuming that the shape of the LF is time invariant, the evolution of the overall normalization ϕ_* can be parameterized as $\phi_*(z) = \phi_*(0)(1+z)^p$.

11.2.1   *The Luminosity Function of E+S0 Galaxies at $z < 1$*

Before considering distant galaxies, let us briefly summarize our knowledge of the local LF of E+S0s. To this aim, a compilation is shown in Figure 11.1 (heavy lines; see caption for details and references). The compilation includes LFs in various photometric bands, from the blue to the near-infrared. For ease of comparison, the best-fit Schechter (1976) LFs have been transformed to an intermediate wavelength, using the average colors of E+S0s[*] to obtain L_* in the I band. Note that a simple shift in color is only an approximate transformation, because of the existence of the color-magnitude relation and of different definitions of magnitudes, photometric system, and morphological classes adopted by various authors (see, e.g., Kochanek, Keeton, & McLeod 2001). The agreement among the most recent determinations is rather encouraging. Nevertheless, as we will see, the uncertainty in the local LF (the fossil evidence) contributes significantly to the error budget in the measurement of $\phi(L,z)$.

The most extensive study of the evolution of the LF of morphologically selected early-type galaxies is based on the 145 E+S0s with 16.5 mag $< I <$ 22 mag in the Groth Strip Survey (Im et al. 2002; other samples are given in Franceschini et al. 1998; Schade et al. 1999; Benson et al. 2002; given the morphological definition adopted here, I will not review

---

[*] $B - I = 2.1$ mag, $r' - I = 0.24$ mag (Fukugita, Shimasaku, & Ichikawa 1995); $B_Z - I = 2.4$ mag (Im et al. 1996); $I - K = 2.1$ mag (Bower, Lucey, & Ellis 1992; Fugukita et al. 1995).

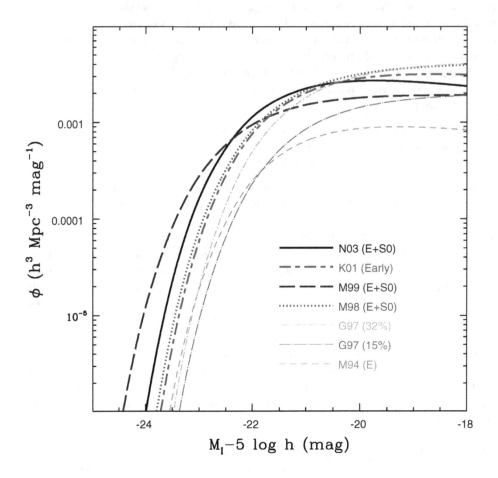

Fig. 11.1. Local LFs of E+S0s transformed to the *I* band; see references for details (N03 = Nakamura et al. 2003, K01 = Kochanek et al. 2001, M99 = Marinoni et al. 1999, M98 = Marzke et al. 1998). Other local LF adopted to construct PLE models are shown for comparison as thin lines (G97 = Gardner et al. 1997 total *K*-band LF scaled by 0.32 and 0.15; M94 = Marzke et al. 1994 LF for pure ellipticals).

similar studies based on spectral classification, such as Cohen 2002 and Willis et al. 2002). The authors use *HST* photometry, together with spectroscopic (45/145) and photometric redshifts (100/145), to compute the rest-frame *B*-band luminosity, and then proceed to measure the evolution of the LF.

Based on their sample of 145 objects in the redshift range 0.1–1.2 (median redshift 0.6), Im et al. (2002) find $d\left(\log M_*/L_B\right)/dz = -0.76 \pm 0.32$ (i.e. 1.89 ± 0.81 mag of brightening to $z = 1$) and $p = -0.86 \pm 0.68$ (i.e. the number density at $z = 1$ is $0.55^{+0.33}_{-0.21}$ of the local value). The large uncertainties arise mostly from the limited size of the sample, but also from the fact that an apparent magnitude-limited sample probes different volumes and absolute magnitude ranges in the local and distant Universe. Therefore, bright E+S0s, dominant at large z, will have very few counterparts in the local Universe, and, *vice versa*, the faint galaxies dominating the counts in the local Universe will go undetected at large z. Im et al. (2002) try to remove this source of uncertainty by fixing the characteristic luminosity of the local LF to some external measurement based on a larger volume. Unfortunately, Im et al. (2002) find very different results according to the local LF they adopt. The Marzke et al. (1998) LF yields $d \left(\log M_* / L_B \right) / dz = -0.79 \pm 0.09$ and $p = -0.95 \pm 0.48$, while the Marinoni et al. (1999) LF yields $d \left(\log M_* / L_B \right) / dz = -0.41 \pm 0.09$ and $p = 0.12 \pm 0.54$ (the two LF are shown in Fig. 11.1, labeled as M98 and M99, respectively). In conclusion, systematic uncertainties in the local LF hinder substantial progress.

From these results it is clear that a two-pronged strategy must be followed in order to improve on the current factor of 3 uncertainty in the number density evolution to $z \approx 1$. On the one hand, it is necessary to increase the size of the high-redshift sample, possibly to several thousands objects with a larger fraction of spectroscopic redshifts. Also a sample collected along multiple independent lines of sight will be desirable to minimize the effects of cosmic variance and clustering of E+S0s (see discussion in Im et al. 2002 and in § 11.2.2). The future prospects appear promising, due to the recent improvements of observational capabilities, both on the imaging side with the Advanced Camera for Survey on *HST* and on the spectroscopic side with the new generation of wide-field, high-multiplexing spectrographs. On the other hand, it is necessary to reduce the uncertainty on the local LF of E+S0s. As illustrated in Figure 11.2, prospects look good and hopefully the optical LF per morphological type will be know with higher accuracy once the morphological classification of large numbers of galaxies in the Sloan Digitized Sky Survey (SDSS) and 2dF is completed.

11.2.2 Extremely Red Objects and the Luminosity Function of E+S0s at $z \approx 1$

Old stellar populations are characterized by a sharp break in their spectral energy distribution around 4000 Å, with larger flux at longer wavelengths. Hence, an old stellar population at $z \approx 1$ appears as an object with extremely red optical to infrared colors, i.e. an "extremely red object" (ERO). Therefore, a census of EROs would provide directly the number density of *old* E+S0s at $z \gtrsim 1$ without the need for spectroscopic redshifts, provided contaminants such as cold stars or dust-enshrouded galaxies could be removed.

Several optical-infrared surveys over the past five years have been conducted with the goal of measuring the number density evolution of E+S0s (Moustakas et al. 1997; Zepf 1997; Barger et al. 1999; Benítez et al. 1999; Menanteau et al. 1999; Thompson et al. 1999; Treu & Stiavelli 1999; Corbin et al. 2000; Daddi et al. 2000a; Daddi, Cimatti, & Renzini 2000b; McCracken et al. 2000; Yan et al. 2000; Martini 2001; McCarthy et al. 2001; Chen et al. 2002; Firth et al. 2002; Roche et al. 2002; Smith et al. 2002). A selection of the results is shown in Figure 11.2. In spite of the slightly different color cuts adopted by various groups, it is clear that those surveys conducted over sufficiently wide areas and/or along multiple lines of sight (where clustering bias and cosmic variance are unimportant) are in good agreement.

A comparison with the local abundance of E+S0s can be done by considering simple PLE models. Given a spectral evolution model and the selection criteria of an ERO survey, we can obtain the set of redshifts $\mathcal{Z}(L)$ at which a galaxy of present-day luminosity L would be included in the sample (typically an interval limited on the low-redshift side by the red color criterion and on the high-redshift side by the detection limit). Then, the density of EROs

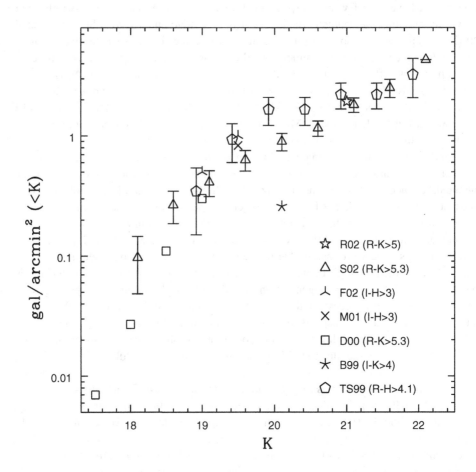

Fig. 11.2. Cumulative number density of EROs for similar color definitions (see references for details: R02 = Roche et al. 2002, S02 = Smith et al. 2002, F02 = Firth et al. 2002, M01 = McCarthy et al. 2001, D00 = Daddi et al. 2000a, B99 = Barger et al. 1999, TS99 = Treu & Stiavelli 1999). The outlier B99 is measured from the Hubble Deep Field alone, which is thought to be deficient in high-redshift E+S0s. This is a clear example of the effects of cosmic variance and an illustration of the need for a wide field of view and multiple lines of sight.

is obtained by integrating the local LF times the cosmic volume per unit solid angle dV/dz over the appropriate range in luminosity and redshift:

$$\int_{L_{\min}}^{L_{\max}} \int_{z(L)} \phi(L) \frac{dV}{dz} dz dL. \tag{11.1}$$

For a local Schechter LF with a flat faint-end slope (cf. Fig. 11.1), the model number density depends linearly on the local characteristic density, while the dependence on the characteristic luminosity is a rapidly varying function of the depth of the survey. At L_*, where the

Table 11.1. *Number Density of EROs.*

Observed/Predicted (%)	Model LF	Area (arcmin^2)	Depth (mag)	Reference
70	0.15×K (G97)	2200	$H < 21.0$	McCarthy et al. (2001)
100	E (M94)	701[a]	$K < 19.2$	Daddi et al. (2000b)
33	E+S0 (K01)	81.5	$K < 21.0$	Roche et al. (2002)
100	E (M94)[b]	49[a]	$K < 21.6$	Smith et al. (2002)
25	0.32×K (G97)	13.8[a]	$H < 23.2$	Treu & Stiavelli (1999)

The first column lists the fraction galaxies observed with respect to the prediction of PLE models. The second column lists the local LF assumed in the PLE models (notation as in Fig. 11.1). [a] Area is a function of depth; I report here the maximum area and depth (see references for details). [b] PLE models from Daddi et al. (2000b).

LF is steep, an uncertainty of 0.3 (0.5) mag in the assumed L_* affects the predicted counts at the 50% (80%) level. At $L_*/10$ the same uncertainties affect the counts only at the 10% (20%) level. The same calculation can be used to estimate the effects of the uncertainties in luminosity evolution. It is clear that in order to perform a reliable comparison with the local LF it is necessary to go significantly fainter than L_* (corresponding to $K = 18 - 18.5$ mag adopting the LF of Nakamura et al. (2003) and a reasonable range of evolutionary models). It is thus necessary to reach beyond $K = 20.5 - 21$ mag to make the uncertainty on L_* and luminosity evolution smaller than the observational errors on the number counts.

The uncertainties related to modeling the star formation history are more dramatic. It is sufficient to have a small amount of recent star formation to make the optical to infrared color significantly bluer and therefore change dramatically $\mathcal{Z}(L)$, and hence the predicted number counts (see discussion in Jimenez et al. 1999). Without more information on the star formation history of E+S0s (see § 11.3), it is convenient to adopt the following approach. Models with no delayed star formation (single burst) will predict the maximum density of red E+S0s at high z. A comparison with these models defines the fraction of local E+S0s already "old" at $z \approx 1$ (Treu & Stiavelli 1999).

How do the counts of EROs compare with PLE models? I summarize some of the observational results in Table 11.1 (the deepest and widest for which I could find a PLE comparison). At first glance the results seem highly discrepant. However, some of the fractions in Table 11.1 are with respect to the LF of pure ellipticals (M94 and 0.15K G97; see Fig. 11.1) and not of ellipticals *and* lenticulars. If we assume for simplicity that E and S0 galaxies are present approximately in equal numbers at these luminosities, we have to halve the fractions of McCarthy et al. (2001), Daddi et al. (2000b), and Smith et al. (2002) to obtain the ratio between the density of EROs and that of local E+S0s (35%, 50%, 50%, respectively). The corrected fractions are in much better mutual agreement and range between 25% and 50%. This range can be readily explained in terms of measurement errors and different star formation histories and local LFs adopted in the PLE models. A further correction is required, because we have to take into account the presence of possible contaminants, such as highly dust-enshrouded starbursts or reddened AGNs (e.g., Smail et al. 2002). To address this, researchers have relied on *HST* images to determine what fraction of EROs are *morphologically* E+S0s. A list of the determinations of morphological fractions among EROs to date is shown in Table 11.2. It is sufficient to multiply the corrected fractions from Table 11.1

Table 11.2. *Morphology of EROs*

Fraction of E+S0	Instrument	Number of EROs	Reference
25%–35%	WFPC2	115	Yan & Thompson (2003)
20%–50%	WFPC2	60	Smith et al. (2002)
50%–80%	NICMOS/WFPC2	41	Moriondo et al. (2000)
55%–75%	NICMOS	30	Stiavelli & Treu (2001)

Fraction of EROs with E+S0 morphology found by various surveys at infrared (NICMOS) or optical (WFPC2) wavelengths.

by the fractions in Table 11.2 to obtain the density of red E+S0s at $z \approx 1$. The density of red E+S0s at $z \approx 1$ is 8%-40% of the local value, with most of the range coming from the uncertainties in Table 11.2. Even allowing for some extra uncertainty related to the range of local LF, it seems clear that the number of E+S0s already "old" at $z \approx 1$ is less than in the local Universe.

Where are the rest of them? Either they are not yet assembled, or they are simply not recognizable. The latter alternative would be, for example, the case in a "frosting" (Trager et al. 2000) scenario, where most of the stellar mass is assembled at early times, while low levels of star formation contribute the rest of the stellar mass a later times. In this scenario, some E+S0s would be too blue at $z \approx 1$ to make it into EROs samples (cf. the range in rest-frame UV colors reported by Moustakas et al. 1997 and McCarthy et al. 2001 for EROs). Star formation can also alter morphology. For example, if the secondary star formation activity is concentrated in the disk of an S0, the disk would become more prominent and active at $z \approx 1$, transforming the S0 into an Sa. This mechanism, of course, would be effective only on lenticulars and not on pure ellipticals. Therefore, it could perhaps provide an explanation of the deficit of "old" spheroids in terms of a demise of lenticulars together with an almost constant number density of ellipticals (similarly to what is seen in clusters: Dressler et al. 1997; Fasano et al. 2000). Deeper, multicolor, high-resolution observations are needed to test whether the number density evolution of Es and S0s is different.

Let us now turn our attention to semi-analytic hierarchical models. How do their predictions compare the surface density of EROs? Generally, models significantly underpredict the surface density of EROs. For example, Firth et al. (2002) and Smith et al. (2002) find 4.5 and 10 times more EROs than expected in the models, a statistically significant disagreement. Again, we face two possible solutions for the disagreement. Either hierarchical models do not produce enough massive systems at $z \approx 1$, or simply their colors are wrong. This latter possibility can occur as a result of an excess of delayed star formation, or as a result of inappropriate treatment of star formation and dust extinction in semi-analytic models.

To summarize, it seems that measurements of the basic observable (surface density of EROs; Fig. 11.2) is reaching a reasonable level of mutual agreement. A further improvement would be to gather redshifts for a large number of EROs in order to pin down the effective redshift selection functions and the three-dimensional space density. Significant efforts are ongoing (Cimatti et al. 2002b; Ellis et al. 2004) and are expected to provide this piece of information soon, at least at the bright end ($K < 19 - 20$ mag). In spite of these achievements, the interpretation of the observations is still open. It seems clear that E+S0s in the local Universe are not all "old" (as in a single burst of star formation at high redshift) and hierarchical models underpredict the density of EROs. It is still disputed whether this discrepancy can be accounted for by improving the treatment of star formation, or it is rather pointing to fundamental problems in the standard model.

Observations can help address this issue in at least two ways. On the one hand, they can provide new constraints less critically dependent on star formation history (such as the distribution of redshifts for K-band selected objects described by Kauffmann & Charlot 1998 and Cimatti et al. 2002a, or the approach focused on mass described in § 11.3). On the other hand, observations can help by providing independent and detailed information on the star formation history of E+S0s.

11.3   Star Formation History

We now turn to observational constraints on the star formation history of E+S0s, particularly those obtained from the redshift evolution of the color-magnitude relation (§11.3.1) and the fundamental plane (§11.3.2). Where possible, I will discuss both cluster and field observations. Contrasting and connecting the trends across environments is not only crucial to obtain a complete empirical picture, but also to test if the formation of E+S0s is delayed in low-density environments as predicted by the standard model (Kauffmann 1996).

11.3.1   The Colors of Distant E+S0 Galaxies

In the local Universe E+S0s obey a color-magnitude relation: brighter E+S0s are redder than less luminous ones. At a given absolute magnitude, the scatter in optical and infrared colors, at least in clusters, is minimal (< 0.05 mag; Bower et al. 1992). The widely accepted interpretation is that brighter E+S0s are more metal rich and that star formation in cluster E+S0s happened and ceased early enough in cosmic time that the effects of possible spread in formation epoch are not detectable through broad-band colors (see Bower, Kodama, & Terlevich 1998 for caveats). The most convincing evidence in support of this interpretation is the redshift evolution of the color-magnitude relation. The almost constant slope of the color-magnitude relation with redshift shows directly that it is not an age-mass sequence (Ellis et al. 1997; Kodama et al. 1998). Similarly, the small scatter found in high-z clusters indicates that the stellar populations of E+S0s in massive clusters are uniformly old ($z_f > 2$) and quiescent (Stanford, Eisenhardt, & Dickinson 1995, 1998; Ellis et al. 1997), with the possible exception of S0s at large cluster radii (van Dokkum et al. 1998b). What prevents this from being a simple and well-defined picture is that E+S0s in high-redshift clusters are not the only possible progenitors of present-day cluster E+S0. Some of the progenitors at $z \approx 1$ might not have yet been accreted onto the cluster, or might not be morphologically recognizable. Therefore, the tightness of the color-magnitude relation in high-z clusters could, in part, be due to a selection effect ("progenitor bias"; van Dokkum & Franx 2001). However, the observed evolution of the morphology-density relation (Dressler et al. 1997; van Dokkum et al. 2001; Treu et al. 2003) can be used to quantify the bias and rules out the most dramatic scenarios.

Less well studied is the color-magnitude relation in the general field. The few studies available seem to indicate that there is a color-magnitude relation in the field out to $z \approx 1$, although with considerably more scatter than in clusters (Franceschini et al. 1998; Kodama, Bower, & Bell 1999; Schade et al. 1999). Similarly, the Hubble Deep Fields show strong variations in internal color, often associated with blue cores (Menanteau, Abraham, & Ellis

2001), at variance with the homogeneous population and red color gradients (Saglia et al. 2000) observed in clusters at similar redshifts.

11.3.2   The Fundamental Plane of Distant E+S0 Galaxies

Can we now measure the star formation history of E+S0s at a given mass? A promising way to do this is by studying the evolution with redshift of the fundamental plane (Djorgovski & Davis 1987; Dressler et al. 1987; Jørgensen, Franx, & Kjærgaard 1996; hereafter FP). The FP is a tight empirical correlation between the effective radius R_e, velocity dispersion σ, and effective surface brightness μ_e

$$\log R_e = \alpha \log \sigma + \beta \mu_e + \gamma, \tag{11.2}$$

where α and β are called the slopes, while γ is called the intercept. The very existence of the FP is a remarkable fact. Any theory of galaxy formation and evolution must be able to account for its tightness (0.08 rms in $\log R_e$). For discussion of possible physical explanations of the FP relation, see references in Treu et al. (2001).

Independent of its origin, the evolution of the FP with redshift can be linked to the evolution of the stellar mass-to-light ratio (M/L) of E+S0s, and hence of their star formation history, in the following way. Let us define an effective mass $M \equiv \sigma^2 R_e$. If homology holds, i.e. early-type galaxies are structurally similar, the total mass $\mathcal{M}$ (including dark matter if present) is proportional to M, and the effective mass can be interpreted in terms of the virial theorem (e.g., Bertin, Ciotti, & Del Principe 2002). Similarly, an effective luminosity can be defined as $\log L = -0.4\mu_e + 2\log R_e + \log 2\pi$. Based on these definitions, the M/L (effective mass-to-light ratio) of a galaxy is readily obtained in terms of the FP observables: $M/L \propto 10^{0.4\mu_e} \sigma^2 R_e^{-1}$. Using the FP relation to eliminate μ_e yields $M/L \propto 10^{-\frac{\gamma}{2.5}} \sigma^{\frac{10\beta-2\alpha}{5\beta}} R_e^{\frac{2-5\beta}{5\beta}}$.

Consider a sample of galaxies at $z > 0$ identified by a running index i. The offset of M/L from the local value can be computed as

$$\Delta \log(M/L)^i = \Delta\left(\frac{10\beta-2\alpha}{5\beta}\right)\log\sigma^i + \Delta\left(\frac{2-5\beta}{5\beta}\right)\log R_e^i - \Delta\left(\frac{\gamma^i}{2.5\beta}\right), \tag{11.3}$$

where the symbol Δ indicates the difference of the quantity at two redshifts, and γ^i is defined as $\log R_e^i - \alpha \log \sigma^i - \beta \mu_e^i$. For the analysis presented here, I will assume that α and β are constant (see Treu et al. 2001a,b for discussion). This assumption is consistent with the observations and makes the interpretation of the results straightforward. If α and β are constant, and there is no structural evolution (so that R_e and σ are constant), then $\Delta \log\left(M/L^i\right) = -\frac{\Delta\gamma^i}{2.5\beta}$, i.e. the evolution of $\log(M/L^i)$ depends only on the evolution of γ^i. Measuring γ^i for a sample of galaxies at intermediate redshift, and comparing it to the value of the intercept found in the local Universe, measures the average evolution of $\log(M/L)$ with cosmic time as $\langle\Delta \log(M/L)\rangle = -\frac{\langle\Delta\gamma\rangle}{2.5\beta}$. If the evolution of the effective mass-to-light ratio measures the evolution of the stellar mass-to-light ratio, then the FP becomes a powerful diagnostic of stellar populations. Not only does this diagnostic connect stellar populations to a dynamical mass measurement, but it is also intrinsically tight, and thus selection effects are small and can be corrected (Treu et al. 2001; Bernardi et al. 2003).

Several groups have applied this technique at $z > 0.1$, both in clusters (Franx 1993; van Dokkum & Franx 1996; Kelson et al. 1997; Bender et al. 1998; van Dokkum et al. 1998a; Jørgensen et al. 1999; Kelson et al. 2000; Ziegler et al. 2001; van Dokkum & Stanford 2003; Fritz et al. 2004) and in the field (Treu et al. 1999, 2001a,b, 2002; van Dokkum et

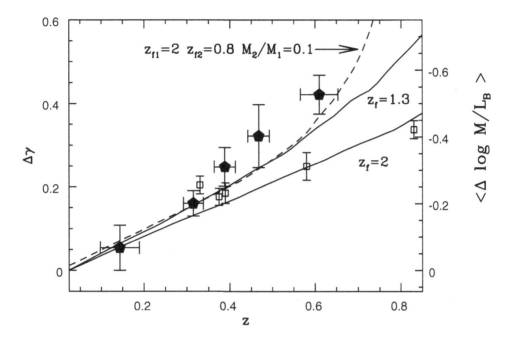

Fig. 11.3. FP in the rest-frame B band. The average offset of the intercept of field galaxies (Treu et al. 2002) from the local FP relation as a function of redshift (large filled pentagons) is compared to the offset observed in clusters (open squares; van Dokkum & Franx 1996; Kelson et al. 1997; Bender et al. 1998; van Dokkum et al. 1998a; Kelson et al. 2000). The solid lines represent the evolution predicted for passively evolving stellar populations formed in a single burst at $z = 1.3$ and 2, computed using the 1996 version of the Bruzual & Charlot (1993) models. The evolution predicted by a double-burst model (dashed line) is also shown for comparison. See Treu et al. (2002) for details.

al. 2001; Bernardi et al. 2003). A representative selection of results is shown in Figure 11.3. The main results of these studies are: (1) E+S0s obey a FP relation out to al least $z \approx 0.8$ with scatter similar to local samples (in the field the scatter in $\log R$ is less than 0.15 at $z \approx 0.8$; Treu et al. 2002); (2) field E+S0s (solid pentagons) evolve faster than cluster ones (open squares). In quantitative terms, Treu et al. (2002) obtain $d \left(\log M_* / L_B \right) / dz = -0.72^{+0.11}_{-0.16}$ for the field sample, while van Dokkum et al. (1998a) obtain -0.49 ± 0.05 for clusters. Note the good agreement with the results obtained by Im et al. (2002) from the evolution of the LF (§11.2.1).

In terms of evolution of stellar populations, the cluster data are consistent with passive evolution of an old stellar population ($z_f \approx 2$). A more recent "epoch of formation," $z_f \approx 1.3$, appears to be needed to explain the field E+S0s evolution in terms of single burst stellar populations. However, as for the EROs PLE models, it is sufficient to rejuvenate an old stellar population with a small amount of recent star formation to obtain an evolution consistent with the data. For example, the data are well described by a model where 90% of the stellar mass is formed at $z_{f1} \approx 2$ and secondary bursts at $z_{f2} \approx 1$ contribute the residual 10%.

11.3.3   Discussion

Both the evolution of colors and the FP to $z \approx 1$ appear to be consistent with the following picture. Massive cluster E+S0s are old and quiescent, while field examples show some relatively recent star formation activity. This picture is also in qualitative agreement with the fact that *at any given morphological type* star formation activity decreases monotonically with (local) galaxy density (see, e.g., Poggianti et al. 1999; Dressler 2004; Nichol 2004; Poggianti 2004), and observations of high-redshift E+S0s based on other spectroscopic diagnostic features (Schade et al. 1999; Kelson et al. 2001; Treu et al. 2002). Further support for this picture comes from the fossil evidence (Bernardi et al. 1998; Trager et al. 2000; Kuntschner et al. 2002) , although interpreting the observations in the local Universe is more difficult, because possible differences could have been quenched by time to a level where uncertainties on dust extinction and absolute distances (Pahre, Djorgovski, & de Carvalho 1998), and the age/metallicity degeneracy (Kuntschner et al. 2002) are significant.

How does this picture compare with CDM predictions? *Qualitatively*, the observational picture is similar to theoretical predictions (Diaferio et al. 2001; Benson et al. 2002). However, *quantitatively*, the observed differences between the star formation history of field and cluster E+S0s are smaller than predicted by models. Whereas observations indicate at most minor departures from a single old stellar populations, hierarchical models predict dramatic differences already at $z < 0.5$ (see Kauffmann 1996 and van Dokkum et al. 2001). As was the case for EROs, improvements in the treatment of star formation or of environmental effects might reconcile the model with the data. Alternatively, this might prove a major problem for the standard model, especially when more precise measurements will be available. From on observational point of view, it has to be noticed that current studies are based on a few tens of objects at most. It is now crucial to collect high-quality data on larger numbers of distant E+S0s to overcome small-sample statistics and cosmic variance.

11.4   The Mass Density Profile of Distant E+S0 Galaxies

So far, in this review, I have interpreted observations in terms of PLE. For example, when expressing the evolution of the FP in terms of evolution of stellar mass-to-light ratio, I have assumed PLE. Is there any way we can relax this assumption and measure directly and simultaneously the internal structure and stellar populations properties of distant E+S0s? If we could, not only we could test if the results obtained under a PLE hypothesis are correct, but also, and most importantly, we could gain new and fundamentally different insight into the evolution of E+S0s. For example, measuring the mass density profile of luminous and dark matter in E+S0s as a function of redshift not only yields an independent determination of the evolution of the stellar mass-to-light ratio, but also tests the existence of the universal dark matter profile predicted by the standard model. Furthermore, theoretical predictions related to the mass density profile and orbital structure might not be as dramatically sensitive to the details of the treatment of star formation as, for example, the number density fo EROs. Therefore, testing these predictions might be a more robust way to test the standard model.

Measuring the mass density profile of E+S0s is already challenging in the local Universe (e.g., Bertin et al. 1994), and traditional methods are inapplicable at high redshift (for example surface brightness dimming prevents the measurement of very extended velocity dispersion profiles). Nevertheless, mass density profile measurements at high redshifts are possible because distant E+S0s are efficient gravitational lenses. The next two sections describe recent results on the mass density profile of distant E+S0s from weak lensing (§11.4.1), and joint strong lensing and dynamical analysis (§11.4.2).

11.4.1 Galaxy-galaxy Lensing

The distortion of background galaxies lensed by an individual E+S0 is not detectable. However, if several (at least hundreds) E+S0s are considered, and the signal from all the background objects is coadded, a statistical measurement of mass density profile of the average galaxy can be derived (Brainerd, Blandford, & Smail 1996). This technique is known as galaxy-galaxy lensing and has proved viable to study the outer regions of the dark matter halos of E+S0s (Griffiths et al. 1996). For example, dark matter halos around red galaxies have been detected out to several hundreds kpc in SDSS images (McKay et al. 2004). Combining information from galaxy-galaxy lensing with the existence of the FP, Seljak (2002) showed that at large radii the mass density profile of E+S0s declines faster than r^{-2}, consistent with the r^{-3} behavior predicted by CDM numerical simulations. Natarajan & Kneib (1997) and Natarajan et al. (1998) showed that dark matter halos of E+S0s can be detected even within clusters if the cluster potential is appropriately modeled. Natarajan, Kneib, & Smail (2002) applied this technique to WFPC2 images of a sample of intermediate-redshift clusters and showed that the dark matter halos of E+S0s are truncated, as expected, from tidal interaction with the cluster.

11.4.2 Strong Lensing and the Lenses Structure and Dynamics Survey

The majority of the almost one hundred galaxian gravitational lenses known are E+S0s. Once the redshifts of the lens and the source are known, the geometry of the multiple images provides a very robust measurement of the mass enclosed by the Einstein radius R_E. The Einstein radius of the typical $z \approx 0.5$ E+S0 lens galaxy is larger than the effective radius. Thus, strong lensing can be used to determine the total mass at large radii for tens of distant E+S0s, independent of the nature and dynamical state of the mass inside R_E.

In some cases, knowledge of the mass enclosed by R_E is already sufficient to show that the average total mass-to-light ratio is larger than expected for reasonable stellar populations, and therefore to prove the existence of dark matter. Unfortunately, not much information is generally provided on how the mass is spatially distributed*.

Nevertheless, assuming a mass density profile, lensing can be used to probe the evolution of the stellar populations. For example, Kochanek et al. (2000) and Rusin et al. (2003) used image separation to estimate the velocity dispersion of lens E+S0s, assuming a singular isothermal total mass density profile (i.e. the total density $\rho_t \propto r^{-2}$). With this assumption they measure the evolution of the FP of lens galaxies and find $d \left(\log M_* / L_B \right) / dz = -0.54 \pm 0.09$, i.e. more similar to the cluster value than the field value [a similar analysis of lens E+S0s by van de Ven, van Dokkum, & Franx (2003) yields -0.62 ± 0.13]. The marginally significant differences with respect to the direct method could be the result of different selection processes (lenses are "mass" selected, while samples used in direct measurements are "light" selected), of different environments (lenses might be preferentially found in groups or small clusters; Fassnacht & Lubin 2002), or of external contributions to the image separation (such as from a nearby group or cluster). Or perhaps, the differences

---

* Except, perhaps, when the lensed source is extended and the detailed geometry can be used to increase the number of constraints (see, e.g., Blandford, Surpi, & Kundic 2001; Kochanek et al. 2001; Saha & Williams 2001).

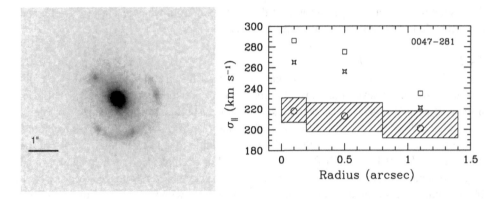

Fig. 11.4. *Left*: *HST* image of 0047–281 at $z = 0.485$. *Right*: velocity dispersion profile of 0047–281 along the major axis. The box height indicates the 68% measurement error, whereas the box width indicates the spectroscopic aperture. The open squares are the corresponding values for an isotropic constant M/L model, which is rejected by the data. See Koopmans & Treu (2003) for details.

could be an indication of small departures from isothermal mass density profiles. However, the ability of a simple singular isothermal mass model to predict with reasonable accuracy the central velocity dispersion is remarkable (as generally confirmed by direct measurement; e.g., Koopmans & Treu 2002). This is an indication of the overall structural homogeneity of E+S0s. The accuracy of the predicted σ is even more remarkable considering that $R_E/R_e \approx 0.5 - 5$, and therefore lensing probes regions dominated by stellar mass as well as regions dominated by dark matter.

More can be learned on the internal mass distribution of distant E+S0s by combining strong lensing constraints with spatially resolved stellar kinematics of the lens galaxy, in a joint lensing and dynamical analysis. The two diagnostics complement each other, reducing the degeneracies inherent to each method. Stellar kinematics constrains the mass distribution within the Einstein radius, while gravitational lensing analysis fixes the mass at the Einstein radius, thus lifting the so-called mass-anisotropy degeneracy (Treu & Koopmans 2002; Koopmans & Treu 2003).

Combining the two diagnostics is the goal of the Lenses Structure and Dynamics (LSD) Survey (Koopmans & Treu 2002, 2003; Treu & Koopmans 2002a, b; hereafter collectively KT). In eight clear nights at the Keck II Telescope we have collected data to measure accurate and spatially resolved stellar kinematics for a sample of 11 gravitational lenses out to $z \approx 1$ with available *HST* images. An example of the data is shown in Figure 11.4.

A family of two-component spherical mass models is used in the joint lensing and dynamical analysis (see KT for details). One component is the stellar component, assumed to follow the surface brightness profile as measured from *HST* images, scaled by a constant stellar mass-to-light ratio (M_*/L_B). The other component is the dark matter halo, modeled as a generalized NFW profile, where the dark matter density goes as r^{-3} at large radii and $r^{-\gamma}$ at small radii. Comparison with the data yields best-fitting models and likelihood contours on the relevant parameters (M_*/L_B and γ).

The best-fitting mass model for MG2016+112 at $z = 1.004$ (Lawrence et al. 1984) is shown in the left panel of Figure 11.5. Luminous mass dominates in the inner ~ 10 kpc,

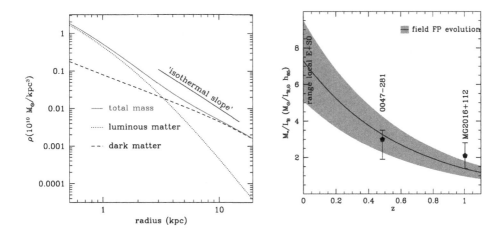

Fig. 11.5. *Left*: Best-fitting mass model of MG2016+112 at $z = 1$ (see Treu & Koopmans 2002a for details). *Right*: Comparison between the evolution of the stellar mass-to-light ratio measured via the FP evolution and via a joint lensing and dynamical analysis by the LSD Survey (see text for detail).

while a flatter dark matter halo contributes most of the mass at larger radii. No dark matter models, or constant total mass-to-light ratio models, are rejected at high confidence level. Remarkably, although neither of the two components is a simple power-law density profile, the *total* mass density profile follows very closely an r^{-2} singular "isothermal" profile (equivalent to a flat rotation curve for spiral galaxies). The same result is recovered by modeling the lens with a simple power law mass density profile $\rho_t \propto r^{-\gamma'}$. Comparison with the data yields $\gamma = 2.0 \pm 0.1 \pm 0.1$. Similar results are found for the other object analyzed so far, 0047−281 (Warren et al. 1996) at $z = 0.485$. There is strong evidence for a dark matter halo more diffuse than the luminous component, and the total mass density profile is close to a singular isothermal profile, well described by a power law with effective slope $\gamma' = 1.90^{+0.05}_{-0.23} \pm 0.1$.

The joint lensing and dynamical analysis also yields a measurement of M_*/L_B, which can be used to measure the evolution of stellar population independently of the FP analysis (§11.3.2). In the right panel of Figure 11.5 the LSD Survey results are compared with the FP results. Since the FP analysis only yields $d\left(\log M_*/L_B\right)/dz$, I adopted for the comparison 7.8 ± 2.7 in solar units for the range of local values (KT). The agreement is very good, consistent with the expectations of a PLE scenario.

Finally, observational limits on γ can be used to test the cuspy dark matter halos predicted by CDM scenarios ($\gamma = 1$ NFW; $\gamma = 1.5$ Moore et al. 1998). For the first two objects we find upper limits $\gamma < 1.4$ and $\gamma < 1.5$ (68% confidence level), consistent with the results of numerical simulations only if the collapse of baryons to form stars (which dominates in the inner regions) did not steepen significantly the dark matter halos. The analysis of the complete sample will hopefully provide more stringent limits.

Although the results so far have to be considered preliminary, since they are based on the first two objects, three facts appear to stand out: (1) E+S0s at high redshift have diffuse dark matter halos; (2) luminous and dark matter, although spatially segregated, "conspire" to follow an almost isothermal total mass density profile, similar to what happens in local E+S0 and spiral galaxies (van Albada & Sancisi 1986; Rix et al. 1997); (3) the agreement between the evolution of the mass-to-light ratio measured by the FP and the direct measurements is consistent with no structural evolution of E+S0s in the past 8 Gyr.

The first point, direct evidence of extended dark matter halos around E+S0s out to $z \approx$ 1, is probably not surprising, but it is a confirmation of the CDM scenario. The second point, the "conspiracy" between luminous and dark matter to produce a total mass density profile r^{-2}, which appears to be a consistent feature of early-type and spiral galaxies out to $z \approx 1$, is something that should be explained by a satisfactory cosmological model. It is not clear if this is the case in the standard model, since simulations do not generally include baryons. Analytic approximations of baryonic collapse (Blumenthal et al. 1986) do not explain naturally this result. An alternate explanation might be that the r^{-2} profile, the limit of (incomplete) violent relaxation (Lynden-Bell 1967), is a *dynamical attractor*. If baryons are transformed into stars early enough (as in the monolithic collapse scenario by van Albada 1982 or as recently proposed by Loeb & Peebles 2003), then they behave as dissipationless particles and could interact with dark matter so as to tend to a *total* mass density profile that is close to the dynamical attractor (Loeb & Peebles 2003), while preserving spatial segregation as a result of different initial conditions. Finally, the third point, the lack of dynamical evolution out to $z \approx 1$, together with other evidence for homogeneity of E+S0s described in the previous sections, appears to be another challenge for the standard model. If E+S0s are formed by mergers, then either mergers have to occur very early in cosmic time, or some sort of fine tuning of the merging process appears to be required in order to produce such homogeneous end products.

Acknowledgements. I am grateful to Giuseppe Bertin and Richard Ellis for their comments on an earlier version of this manuscript, and to my collaborators Stefano Casertano, Léon Koopmans, Palle Møller, and Massimo Stiavelli for innumerous stimulating discussions. I acknowledge useful conversations with Masataka Fukugita, Myungshin Im, Pat McCarthy, Alvio Renzini, David Sand, and Graham Smith. I would like to thank the organizers for this exciting meeting, and the referee, Alan Dressler, for insightful comments.

References

Barger, A. J., Cowie, L. L., Trentham, N., Fulton, E., Hu, E. M., Songaila, A., & Hall, D. 1999, AJ, 117, 102
Bender, R., Saglia, R. P., Ziegler, B., Belloni, P., Greggio, L., Hopp, U., & Bruzual A., G. 1998, ApJ, 493, 529
Benítez, N., Broadhurst, T., Bouwens, R. J., Silk, J., & Rosati, P. 1999, ApJ, 515, L65
Benson, A. J., Ellis, R. S., & Menanteau, F. 2002, MNRAS, 336, 564
Bernardi, M., et al. 2003, AJ, 125, 1866
Bernardi, M., Renzini, A., da Costa, L. N., Wegner, G., Alonso, M. V., Pellegrini, P. S., Rité, C., & Willmer, C. N. A. 1998, ApJ, 508, L143
Bertin, G., et al. 1994, A&A, 292, 381
Bertin, G., Ciotti, L., & Del Principe, M. 2002, A&A, 386, 149
Bertin, G., & Stiavelli, M. 1993, Rep. Prog. Phys., 56, 493
Blandford, R. D., Surpi, G., & Kundic, T. 2001, in Gravitational Lensing: Recent Progress and Future Goals, ed. T. G. Brainerd & C. S. Kochanek (San Francisco: ASP), 65
Blumenthal, G. R., Faber, S. M., Flores, R., & Primack, J. R. 1986, ApJ, 301, 27
Blumenthal, G. R., Faber, S. M., Primack, J. R., & Rees, M. J. 1984, Nature, 311, 517
Bower, R. G., Kodama, T., & Terlevich, A. 1998, MNRAS, 299, 1193
Bower, R. G., Lucey, J. R., & Ellis, R. S. 1992, MNRAS, 254, 601

Brainerd, T. G., Blandford, R. D., & Smail, I. 1996, ApJ, 466, 623

Bruzual A., G., & Charlot, S. 1993, ApJ, 405, 538

Chen, H.-W., et al. 2002, ApJ, 570, 54

Cimatti, A., et al. 2002a, A&A, 391, L1

———. 2002b, A&A, 392, 395

Cohen, J. G. 2002, ApJ, 567, 672

Corbin, M. R., O'Neil, E., Thompson, R. I., Rieke, M. J., & Schneider, G. 2000, AJ, 120, 1209

Daddi, E., Cimatti, A., Pozzetti, L., Hoekstra, H., Röttgering, H. J. A., Renzini, A., Zamorani, G., & Mannucci, F. 2000a, A&A, 361, 535

Daddi, E., Cimatti, A., & Renzini, A. 2000b, A&A, 362, L45

Davies, R. L. 2004, in Carnegie Observatories Astrophysics Series, Vol. 3: Clusters of Galaxies: Probes of Cosmological Structure and Galaxy Evolution, ed. J. S. Mulchaey, A. Dressler, & A. Oemler (Cambridge: Cambridge Univ. Press), in press

de Freitas Pacheco, J. A., Michard, R., & Mohayaee, R. 2003, Recent Research Developments in Astronomy and Astrophysics, in press (astro-ph/0301248)

de Vaucouleurs, G. 1948, Ann. d'Ap., 11, 247

de Zeeuw, T. P., & Franx, M., 1991, ARA&A, 29, 239

Diaferio, A., Kauffmann, G., Balogh, M. L., White, S. D. M., Schade, D., & Ellingson, E. 2001, MNRAS, 323, 999

Djorgovski, S. G., & Davis, M. 1987, ApJ, 313, 59

Dressler, A., et al. 1997, ApJ, 490, 577

Dressler, A. 2004, in Carnegie Observatories Astrophysics Series, Vol. 3: Clusters of Galaxies: Probes of Cosmological Structure and Galaxy Evolution, ed. J. S. Mulchaey, A. Dressler, & A. Oemler (Cambridge: Cambridge Univ. Press), in press

Dressler, A., Lynden-Bell, D., Burstein, D., Davies, R. L., Faber, S. M., Terlevich, R., & Wegner, G. 1987, ApJ, 313, 42

Eggen, O. J., Lynden-Bell, D., & Sandage, A. R. 1962, ApJ, 136, 748

Ellis, R. S., et al. 2004, in preparation

Ellis, R. S., Smail, I., Dressler, A., Couch, W. J., Oemler, A., Jr., Butcher, H., & Sharples, R. M. 1997, ApJ, 483, 582

Fasano, G., Poggianti, B. M., Couch, W. J., Bettoni, D., Kjærgaard, P., & Moles, M. 2000, ApJ, 542, 673

Fassnacht, C. D., & Lubin, L. L. 2002, AJ, 123, 627

Firth, A. E., et al. 2002, MNRAS, 332, 617

Franceschini, A., Silva, L., Fasano, G., Granato, G. L., Bressan, A., Arnouts, S., & Danese, L. 1998, ApJ, 506, 600

Franx, M. 1993, ApJ, 407, L5

Fritz, A., Ziegler, B. L., Bower, R. G., Smail, I., & Davies, R. L. 2004, in Carnegie Observatories Astrophysics Series, Vol. 3: Clusters of Galaxies: Probes of Cosmological Structure and Galaxy Evolution, ed. J. S. Mulchaey, A. Dressler, & A. Oemler (Pasadena: Carnegie Observatories, http://www.ociw.edu/ociw/symposia/series/symposium3/proceedings.html)

Fukugita, M., Shimasaku, K., & Ichikawa, T. 1995, PASP, 107, 945

Gardner, J. P., Sharples, R. M., Frenk, C. S., & Carrasco, B. E. 1997, ApJ, 480, L99

Griffiths, R. E., Casertano, S., Im, M., & Ratnatunga, K. U. 1996, MNRAS, 282, 1159

Ho, L. C. 2004, ed., Carnegie Observatories Astrophysics Series, Vol. 1: Coevolution of Black Holes and Galaxies (Cambridge: Cambridge Univ. Press)

Im, M., Faber, S. M., Koo, D. C., Phillips, A. C., Schiavon, R. P., Simard, L., & Willmer, C. N. A., 2002, ApJ, 571, 136

Im, M., Griffiths, R. E., Ratnatunga, K. U., & Sarajedini, V. L. 1996, ApJ, 461, 79

Jimenez, R., Friaça, A. C. S., Dunlop, J., Terlevich, R. J., Peacock, J., & Nolan, L. A. 1999, MNRAS, 304, L16

Jørgensen, I., Franx, M., Hjorth, J., & van Dokkum, P. G., MNRAS, 1999, 308, 833

Jørgensen, I., Franx, M., & Kjærgaard, P. 1996, MNRAS, 280, 167

Kauffmann, G. 1996, MNRAS, 281, 487

Kauffmann, G., & Charlot, S. 1998, MNRAS, 297, L23

Kauffmann, G., & Haehnelt, M. G. 2000, MNRAS, 311, 576

Kelson, D. D., Illingworth, G. D., van Dokkum, P. G., & Franx, M. 2000, ApJ, 531, 184

Kelson, D. D., van Dokkum, P. G., Franx, M., Illingworth, G. D., & Fabricant, D. 1997, ApJ, 478, L13

Kochanek, C. S., et al. 2000, ApJ, 543, 131

———. 2001, ApJ, 560, 566

Kochanek, C. S., Keeton, C. R., & McLeod, B. A. 2001, ApJ, 547, 50

Kodama, T., Arimoto, N., Barger, A. J., & Aragón-Salamanca, A. 1998, A&A, 334, 99

Kodama, T., Bower, R. G., & Bell, E. F. 1999, MNRAS, 306, 561

Koopmans, L. V. E., & Treu, T., 2002, ApJ, 568, L5

——. 2003, ApJ, 583, 606

Kuntschner, H., Smith, R. J., Colless, M., Davies, R. L., Kaldare, R., & Vazdekis, A. 2002, MNRAS, 337, 172

Larson, R. B. 1975, MNRAS, 173, 671

Lawrence, C. R., Schneider, D. P., Schmidet, M., Bennett, C. L., Hewitt, J. N., Burke, B. F., Turner, E. L., & Gunn, J. E. 1984, Science, 223, 46

Loeb, A., & Peebles, P. J. E. 2003, ApJ, 589, 29

Lynden-Bell, D. 1967, MNRAS, 136, 101

Marinoni, C., Monaco, P., Giuricin, G., & Costantini, B. 1999, ApJ, 521, 50

Martini, P. 2001, AJ, 121, 598

Marzke, R. O., da Costa, L. N., Pellegrini, P. S., Willmer, C. N. A., & Geller, M. J. 1998, ApJ, 503, 617

Marzke, R. O., Geller, M. J., Huchra, J. P., & Corwin, H. G. 1994, AJ, 108, 437

Matteucci, F. 2003, ApS&S, 284, 539

McCarthy, P. J., et al. 2001, ApJ, 560, L131

McCracken, H. J., Metcalfe, N., Shanks, T., Campos, A., Gardner, J. P., & Fong, R. 2000, MNRAS, 311, 707

McKay, T. A., et al. 2004, ApJ, submitted (astro-ph/0108013)

Menanteau, F., Abraham, R. G., & Ellis, R. S. 2001, MNRAS, 322, 1

Menanteau, F., Ellis, R. S., Abraham, R. G., Barger, A. J., Cowie, L. L. 1999, MNRAS, 309, 208

Merritt, D. 1999, PASP, 111, 129

Meza, A., Navarro, J. F., Steinmetz, M., & Eke, V. R. 2003, ApJ, 590, 619

Monaco, P., Salucci, P., & Danese L, 2000, MNRAS, 311, 279

Moore, B., Governato, F., Quinn, T., Stadel, J., & Lake, G., 1998, ApJ, 499, L5

Moriondo, G., Cimatti, A., & Daddi, E. 2000, A&A, 364, 26

Moustakas, L. A., Davis, M., Graham, J. R., Silk, J., Peterson, B. A., & Yoshii, Y. 1997, ApJ, 475, 445

Nakamura, O., Fukugita, M., Yasuda, N., Loveday, J., Brinkmann, J., Schneider, D. P., Shimasaku, K., & SubbaRao, M. 2003, AJ, 125, 1682

Natarajan, P., & Kneib, J.-P. 1997, MNRAS, 287, 833

Natarajan, P., Kneib, J.-P., & Smail, I. 2002, ApJ, 580, L11

Natarajan, P., Kneib, J.-P., Smail, I., & Ellis, R. S. 1998, ApJ, 499, 600

Navarro, J. F., Frenk, C. S., & White, S. D. M. 1997, ApJ, 490, 493

Nichol, R. C. 2004, in Carnegie Observatories Astrophysics Series, Vol. 3: Clusters of Galaxies: Probes of Cosmological Structure and Galaxy Evolution, ed. J. S. Mulchaey, A. Dressler, & A. Oemler (Cambridge: Cambridge Univ. Press), in press

Pahre, M. A., Djorgovski, S. G., & de Carvalho, R. R. 1998, AJ, 116, 1591

Peebles, P. J. E. 2002, in A New Era in Cosmology, ed. N. Metcalfe & T. Shanks (San Francisco: ASP), 351

Poggianti, B. M. 2004, in Carnegie Observatories Astrophysics Series, Vol. 3: Clusters of Galaxies: Probes of Cosmological Structure and Galaxy Evolution, ed. J. S. Mulchaey, A. Dressler, & A. Oemler (Cambridge: Cambridge Univ. Press), in press

Poggianti, B. M., Smail, I., Dressler, A., Couch, W. J., Barger, A. J., Butcher, H., Ellis, R. S., & Oemler, A., Jr. 1999, ApJ, 518, 576

Rix, H.-W., de Zeeuw, P. T., Cretton, N., van der Marel, R. P., & Carollo, C. M. 1997, ApJ, 488, 702

Roche, N. D., Almaini, O., Dunlop, J., Ivison, R. J., & Willott, C. J. 2002, MNRAS, 337, 128

Rusin, D., et al. 2003, ApJ, 587, 143

Saglia, R. P., Maraston, C., Greggio, L., Bender, R., & Ziegler, B. 2000, A&A, 360, 911

Saha, P., & Williams, L. L. R. 2001, AJ, 122, 585

Sandage, A. 1972, 176, 21

Sandage, A., & Visvanathan, N., 1978, ApJ, 225, 742

Schade, D. J., et al., 1999, ApJ, 525, 31

Schechter, P. L. 1976, ApJ, 203, 297

Seljak, U. 2002, MNRAS, 334, 797

Smail, I., Owen, F. N., Morrison, G., Keel, W. C., Ivison, R. J., & Ledlow, M. J. 2002, ApJ, 581, 844

Smith, G. P., et al. 2002, MNRAS, 330, 1

Stanford, S. A., Eisenhardt, P. R. M., & Dickinson, M. 1995, ApJ, 450, 512

——. 1998, ApJ, 492, 461

Stiavelli, M., & Treu, T. 2001, in Galaxy Disks and Disk Galaxies, ed. J. G. Funes & E. M. Corsini (San Francisco: ASP), 603

Thompson, D., et al. 1999, ApJ, 523, 100

Toomre, A. 1977, ARA&A, 15, 437

Toomre, A., & Toomre, J. 1972, ApJ, 178, 623

Trager, S. C. 2004, in Carnegie Observatories Astrophysics Series, Vol. 4: Origin and Evolution of the Elements, ed. A. McWilliam & M. Rauch (Cambridge: Cambridge Univ. Press), in press

Trager, S. C., Faber, S. M., Worthey, G., & González, J. J. 2000, AJ, 120, 165

Treu, T., & Koopmans, L. V. E. 2002a, ApJ, 575, 87

——. 2002b, MNRAS, 337, L6

Treu, T., Ellis, R. S., Kneib, J.-P., Dressler, A., Smail, I., Czoske, O., Oemler, A., & Natarajan, P. 2003, ApJ, 591, 53

Treu, T., & Stiavelli, M. 1999, ApJ, 524, L27

Treu, T., Stiavelli, M., Bertin G., Casertano, C., & Møller, P. 2001a, MNRAS, 326, 237

Treu, T., Stiavelli, M., Casertano, C., Møller, P., & Bertin G. 1999, MNRAS, 308, 1037

——. 2002, ApJ, 564, L13

Treu, T., Stiavelli, M., Møller, P., Casertano, S., & Bertin, G. 2001b, MNRAS, 326, 221

van Albada, T. S. 1982, MNRAS, 201, 939

van Albada, T. S., & Sancisi, R. 1986, Phil. Trans. Royal Soc., Series A, 320, 447

van de Ven, G., van Dokkum, P. G., & Franx, M. 2003, MNRAS, 344, 924

van Dokkum P. G., & Franx M., 1996, MNRAS, 281, 985

——. 2001, ApJ, 553, 90

van Dokkum, P. G., Franx, M., Kelson, D. D., & Illingworth, G. D. 1998, ApJ, 504, L17

——., 2001, ApJ, 553, L39

van Dokkum, P. G., Franx, M., Kelson D. D., & Illingworth G. D., Fisher, D., Fabricant, D., 1998b, ApJ, 500, 714

van Dokkum, P. G., & Stanford, S. A. 2003, ApJ, 585, 78

van Dokkum, P. G., Stanford, S. A., Holden, B. P., Eisenhardt, P. R., Dickinson, M., & Elston, R. 2001, ApJ, 552, L101

Volonteri, M., Haardt, F., & Madau, P. 2003, 582, 559

Warren, S. J., Hewett, P. C., Lewis, G. F., Møller, P., Iovino, A., & Shaver P. A., 1996, MNRAS, 278, 139

White, S. D. M., & Rees, M. J., 1978, MNRAS, 183, 341

Willis, J. P., Hewett, P. C., Warren, S. J., & Lewis, G. F. 2002, MNRAS, 337, 953

Yan, L., McCarthy, P. J., Weymann, R. J., Malkan, M. A., Teplitz, H. I., Storrie-Lombardi, L. J., Smith, M., & Dressler, A. 2000, AJ, 120, 575

Yan, L., & Thompson, D. 2003, ApJ, 586, 765

Zepf, S. E. 1997, Nature, 390, 377

Ziegler, B. L, Bower, R. G, Smail, I., Davies, R. L., & Lee, D., 2001, MNRAS, 325, 1571

12

Evolution of early-type galaxies in clusters

MARIJN FRANX
Leiden Observatory

Abstract

Early-type galaxies dominate rich clusters, and their formation and evolution are intimately related to the formation of the clusters themselves. They form a very homogeneous class at low redshift, and they evolve slowly to higher redshifts. Here we review the evolution of these galaxies, and compare them to early-type galaxies in the field. Furthermore, we discuss recent results on galaxies at $z > 2$, which are likely the progenitors of early-type galaxies at low redshift.

12.1 A Working Definition

Early-type galaxies encompass both elliptical galaxies and S0 galaxies. The distinction between those classes is not very sharp. Whereas ellipticals were originally thought to be axisymmetric or triaxial systems without disks or other substructure, it became clear after the first high signal-to-noise ratio imaging with CCDs that elliptical galaxies contain substantial substructure (e.g., Jedrzejewski 1987; Bender, Döbereiner, & Möllenhoff 1988; Franx, Illingworth, & Heckman 1989). The most convenient observable is the distortion of the isophotes, measured by c_4 (or a_4). If positive, the deviation is "disky" (i.e., similar to the distortion caused by a disk); if negative, it is "boxy." Many L_* ellipticals have disky isophotes, with c_4 at a level of a few percent. The first modeling of these results showed that the disky ellipticals contain substantial disks, and can be thought to form one class with S0s (Rix & White 1990). If these galaxies are viewed edge-on, they are classified as S0s; if they are viewed face-on, they are classified as ellipticals. Hence, the relative contribution of the disks in disky ellipticals is much higher than the values of c_4, and can be on the order of 50%.

Modeling of the isophotal shapes of early-type galaxies in Coma indicated that the L_* ellipticals and S0s were consistent with forming a single "class" with bulge-to-total light ratios homogeneously distributed between 0 and 1 (Jørgensen & Franx 1994). No "diskless" class was necessary to reproduce the observations, except at the high-luminosity end of the luminosity function.

Given these results, we have to conclude that the classes of L_* ellipticals and S0s are overlapping, a galaxy classified as S0 can be classified as elliptical at another viewing angle, and *vice versa*. To avoid the complications, we do not treat them as separate classes below.

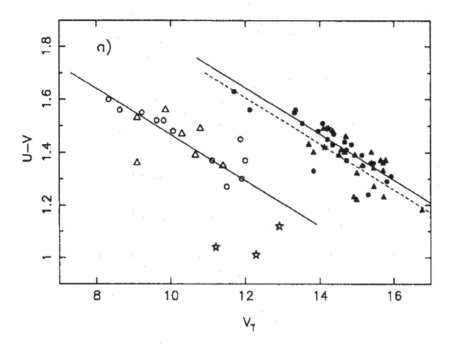

Fig. 12.1. The color magnitude relation of Virgo and Coma, measured by Bower et al. (1992). Open symbols are Virgo galaxies, closed symbols are Coma galaxies. The solid lines show fits to the relation. Open circles and triangles are ellipticals and S0s, and stars are later types. The scatter around the fit is very low for the ellipticals and S0s: 0.05 mag. (From Bower et al. 1992.)

12.2 Homogeneity at Low Redshift

Early-type galaxies in low-redshift clusters form an extremely homogeneous class. It has been known for a long time that early-type galaxies form a tight sequence in the color-magnitude diagram (e.g., Sandage & Visvanathan 1978). Since the colors of stellar populations evolve with age, the width of the color-magnitude relation puts a strong constraint on the age scatter at a given magnitude. Bower, Lucey, & Ellis (1992) measured the color-magnitude diagrams of Coma and Virgo early-type galaxies (Fig. 12.1) and derived that the rms scatter in age was less than 15%. They used this result to argue that these galaxies have likely formed at $z > 2$.

Similarly, the mass-to-light ratios of early-type galaxies are very well behaved (Faber et al. 1987). This is best expressed by the fundamental plane (Dressler et al. 1987, Djorgovski & Davis 1987), which is a relation between velocity dispersion (σ), effective radius (r_e), and surface brightness μ_e: $r_e = \sigma^{1.24}/\mu_e^{0.82}$ (taken from Jørgensen, Franx, & Kjærgaard 1996). Under the assumption of homology, the relation implies that

$$M/L \propto M^{0.24},\tag{12.1}$$

with a small scatter (20%). The small scatter implies a small age scatter at a given mass, as the mass-to-light ratio evolves with time as $M/L \propto t^{0.8}$.

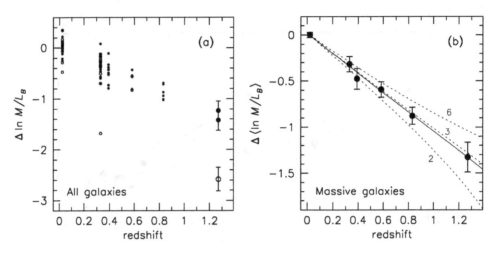

Fig. 12.2. The evolution of the mass-to-light ratios of cluster galaxies out to $z = 1.27$ (van Dokkum & Stanford 2003). The *left* figure shows the mass-to-light ratios of all galaxies; the *right* figure shows the average mass-to-light ratio for each cluster for galaxies more massive than $10^{11} M_\odot$. Dashed lines are model predictions assuming a single burst. The solid line is a model including morphological evolution. The fits imply a mean star formation redshift of 2 or higher. (From van Dokkum & Stanford 2003.)

The picture that emerges from these results is that the early-type galaxies, which dominate the high-density regions of clusters (Dressler 1980), formed at high redshift.

12.3   Evolution to $z = 1$

The most straightforward method to test this is by determining the properties of early-type galaxies at higher redshifts. One expects that the colors evolve slowly, that the scatter in the color-magnitude diagram remains low, but increases slightly, and that the mass-to-light ratios evolve slowly.

With the advent of the *Hubble Space Telescope* and large optical telescopes, it possible to perform such studies beyond a redshift of 1. Many of the results are consistent with the following simple picture.

(1) The evolution of the colors of early-type galaxies is generally slow (Stanford, Eisenhardt, & Dickinson 1998).
(2) Ellis et al. (1997), Stanford et al. (1998), and van Dokkum et al. (2001) found that the scatter in the color-magnitude relation remains very low out to high redshifts, indicating that the relative age spread remains very low, even at substantial look-back times.
(3) van Dokkum & Stanford (2003) have measured the fundamental plane relation for early-type galaxies out to $z = 1.23$, and found a low evolution in the mass-to-light ratio. As can be seen in Figure 12.2, this rules out models with young stellar populations. Best-fit models gave a formation redshift $z_{form} = 2 - 3$.

12.3.1   Evidence for young ages

Whereas all this evidence points to high formation redshifts for cluster early-type galaxies, morphological studies indicate that many early-type galaxies are young. The most

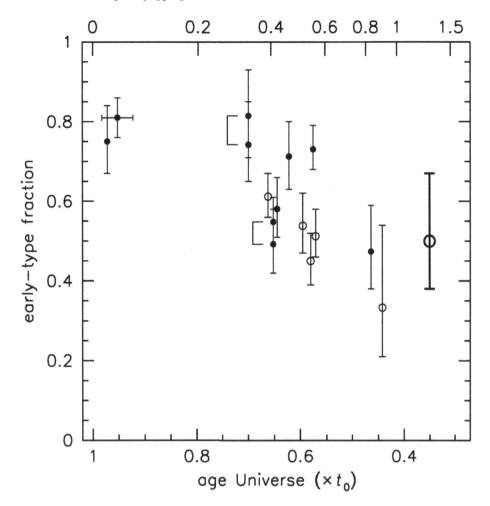

Fig. 12.3. The evolution of the fraction of early-type galaxies in rich clusters, from $z = 1.27$ to $z = 0$. The fraction decreases from 0.8 at $z = 0$ to 0.5 at $z = 1$. X-ray bright clusters, which are indicated by filled symbols, follow this trend. (From van Dokkum et al. 2001.)

comprehensive study is that of Dressler et al. (1997), who found that the fraction of S0 galaxies in concentrated and regular clusters decreases rapidly with increasing redshift. At a redshift of 0.55, the fraction of S0s is as low as 15%–20%, a decrease of a factor of 2–3 compared to the fraction of S0s in similar clusters at low redshift. The fraction of spirals increases correspondingly at high redshift. These results were interpreted by the authors as evidence that the spiral galaxies in the intermediate-redshift clusters were transformed into S0 galaxies.

A concern is the difficulty in distinguishing between elliptical and S0 galaxies, as discussed above. This can be avoided by considering ellipticals and S0s together in the single class of early-type galaxies. Figure 12.3 shows the evolution of the fraction of early-type galaxies in rich clusters, taken from van Dokkum et al. (2001). The fraction decreases from 0.8 at $z = 0$ to ~ 0.5 at $z = 1$, with considerable scatter. The fraction for a given cluster is fairly stable when different groups have classified the galaxies. Hence, the separation between early-type galaxies and spiral galaxies is repeatable.

In summary, the evidence is compelling that a significant fraction of early-type galaxies have transformed at late times from spirals into S0s and/or ellipticals. This may have been caused by gas stripping, merging, or other processes. This appears to be at odds with the evidence that the early-type galaxies are old. Below we discuss models that address this issue.

12.4   Complex Models of Galaxy Evolution

The morphological evolution implies that the formation history of the early-type galaxies is more complex than assumed in the simple "burst" models. Obviously, some of the early-type galaxies at $z = 0$ were spirals at redshifts $z = 0.5 - 1$. Although we lack accurate determinations of the star formation rates in these spirals, some star formation was obviously ongoing, and the models need to incorporate this.

The effect of late star formation on the mass-to-light ratio and late conversion to early-types is shown in Figure 12.4. The figure shows the mass-to-light ratios of a population of early-type galaxies that is enhanced by a steady conversion of star-forming galaxies into "dead" galaxies. As a result, the set of early-type galaxies at $z = 1$ is a very special subset of the galaxies at $z = 0$, and the evolutionary diagnostics such as the evolution of the mass-to-light ratio and color evolution do not give the correct constraints on the ages if they are modeled by single-burst populations. This effect is called "progenitor bias."

A full description of the models can be found in van Dokkum & Franx (2001). The models are very simple, and lack the physical basis of earlier models by Baugh, Cole, & Frenk (1996) and Kauffmann & Charlot (1998). Nevertheless, they demonstrate that the effects are fairly small, because the scatter in the mass-to-light ratios and in the colors are so small. To first order, the correction for progenitor bias is proportional to the scatter in the observed quantity at $z = 0$. van Dokkum & Franx (2001) find that the mean formation redshift of the early-type galaxies decreases from 2.6 to 2 if the maximum correction is applied.

One of the main uncertainties in this modeling is the fact that one needs a comprehensive study of all high-redshift progenitors of low-redshift early-types galaxies in clusters. Many of these galaxies may be outside of the clusters at $z = 1$, and the rich clusters studied at $z = 1$ may be an exceptional, rare environment, atypical for early-type galaxies that end up in nearby rich clusters. Hence, a proper understanding of the evolution of the early-type galaxies at $z = 0$ requires a determination of the environment of their progenitors at $z = 1$, and higher, and this is a very challenging project. It requires an accurate determination of the cluster mass function at $z = 1$, and a study of the morphological types as a function cluster properties.

12.5   Comparison of Field and Cluster Early-type Galaxies

Early-type galaxies in rich clusters have traditionally been studied because of the ease with which one can study many simultaneously using wide-field imagers and spectrographs. We should realize, however, that most early-type galaxies are found outside rich

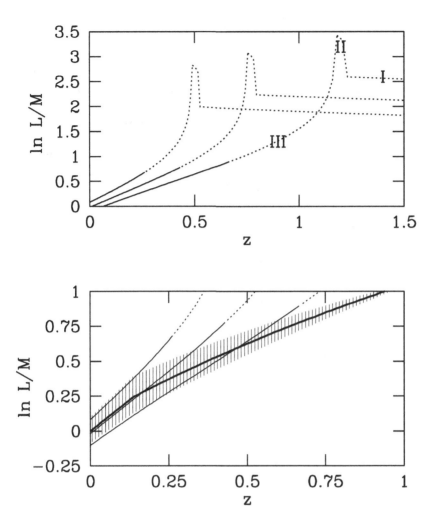

Fig. 12.4. The effects of morphological evolution on the measured mass-to-light ratios. The *top* panel shows the mass-to-light ratios of individual galaxies, which first form stars (indicated by "I"), undergo a burst (phase "II"), become post-starburst galaxies (phase "III"), and finally evolve into early-type systems (solid curve). If galaxies undergo these transformations at different times, the fraction of early-types evolves with redshift, and the sample of early-type galaxies at low redshift is different from that at high redshift (van Dokkum & Franx 2001). The *bottom* panel shows the effect on the measured evolution of the mass-to-light ratios: the measured evolution is slow, as only the oldest early-type galaxies are included in the sample at high redshift. The best-fit curve to the observed evolution out to $z = 1$ underestimates the evolution of all progenitors of early-type galaxies at $z = 0$.

clusters, typically in groups. Hence, whereas the cluster galaxies can provide us reference samples, a proper understanding of early-type galaxies requires field galaxy studies.

Early studies of the color-magnitude relation have established that field early-type galaxies are less homogeneous than cluster early-type galaxies (e.g., Larson, Tinsley, & Caldwell 1980). Similarly, within clusters evidence for environmental effects has been found (e.g., van Dokkum et al. 1998; Terlevich, Caldwell, & Bower 2002): early-type galaxies in the outskirts can have slightly younger populations as derived from their colors.

Similarly, the mass-to-light ratios of early-type galaxies in poor groups is slightly lower than the mass-to-light ratios in rich clusters. This can be derived from the data of Faber et al. (1989), and was also found in a recent study by Bernardi et al. (2003).

The evolution of the mass-to-light ratios of field early-type galaxies has been studied by several groups, with slightly contradictory results. Van Dokkum et al. (2001) and Treu et al. (2001) determined the evolution to $z = 0.4 - 0.6$ and found evolution very similar to that of cluster early-type galaxies. At higher redshifts, however, Treu et al. (2002) found significantly faster evolution, whereas Kochanek et al. (2000), Rusin et al. (2003), and van de Ven, van Dokkum, & Franx (2003) found slow evolution based on lensed galaxies, and van Dokkum & Ellis (2003) found slow evolution to $z = 0.9$ based on an optically selected sample. The last authors found a small offset between field galaxies and cluster galaxies, on the order of 0.14 ± 0.13 mag at $z = 0.88$.

The photometric study of Bell et al. (2004) indicates that the fraction of red galaxies in the field increases significantly from $z = 1$ to $z = 0$. Although a color-selected sample is not the same as a morphologically selected sample, it shows that the effect of morphological evolution and progenitor bias needs to be studies too for the field samples. It may actually be much easier to do this for the field sample than for the cluster sample, as no special selection criteria need to be applied to define the environments.

To conclude, in order to study this topic comprehensively, wide-area studies are needed to measure the evolution of the fraction of early-type galaxies as a function of redshift and to study all progenitors.

The similarity of cluster early-type galaxies and field early-type galaxies cannot be reproduced by models of galaxy formation. The comparisons done by Kochanek et al. (2000) and van Dokkum et al. (2001) show that the semi-analytical models by Diaferio et al. (2001) predict differences that are very large, on the order of 0.6 mag for the mass-to-light difference. The observed differences are much smaller.

12.6   Redshifts $\gg 1$

The high formation redshifts derived for early-type galaxies with $0 < z < 1$ implies that it should be possible to find their star-forming counterparts at $z > 2$. The study of star-forming galaxies at high redshift has progressed rather dramatically in the past years with the discovery of the large population of Lyman-break galaxies (Steidel et al. 1996a,b, 1999). The integrated star formation rate of these galaxies much higher than the star formation rate in the nearby Universe (e.g., Madau et al. 1996; Steidel et al. 1999), and presumably they are the progenitors of early-type galaxies in groups and clusters in the nearby Universe.

Given the rarity of rich clusters at low redshifts, studies of large volumes at high redshifts are needed to determine what the progenitors are of the early-type galaxies in rich clusters.

The masses of the Lyman-break galaxies are thought to be fairly low, on the order of $10^{10} M_\odot$ (Papovich, Dickinson, & Ferguson 2001; Shapley et al. 2001), and hence recurrent merging is necessary to build up massive early-type galaxies in clusters.

On the other hand, the Lyman-break galaxies form a fairly special population of relatively dust-free galaxies, and many other galaxies may have been missed by selection in the rest-

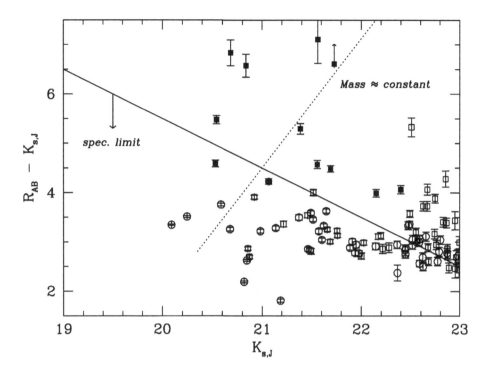

Fig. 12.5. The $R-K$ colors of high-redshift galaxies in the HDF-South (Franx et al. 2003). The galaxies have been selected to have photometric redshifts between 2 and 3.5. The range in $R-K$ colors is very large, more than 5 magnitudes. The filled symbols are galaxies selected to have $J-K > 2.3$, which is an effective criterion to select red galaxies with $z > 2$. Lyman-break galaxies are indicated with round symbols. The two classes have very little overlap. The red galaxies contribute significantly to the integrated K-band flux, and may contribute a comparable fraction to the stellar mass density as the Lyman-break galaxies, down to $K = 22.5$ mag. Larger area studies are needed to verify the density of these red sources. They are possibly the oldest galaxies at high redshift and the progenitors of early-type galaxies at low redshift. (From Franx et al. 2003.)

frame UV. For example, strong starburst galaxies have been identified based on their sub-mm emission (e.g., Smail et al. 2002). Their contribution to the integrated star formation rate is significant.

It is also possible to select high-redshift galaxies based on their near-infrared fluxes. Since a typical L_* galaxy at redshift $z = 0$ can be as faint as $K = 23$ mag at $z = 3$, very deep imaging is required to do this. Recent studies have shown that significant populations of "red" galaxies exist at $z > 2$, which would not be selected by the Lyman-break criterion because they are too faint in the observed optical window (e.g., Franx et al. 2003; van Dokkum et al. 2003)

Figure 12.5 shows the $R-K$ colors of galaxies in the Hubble Deep Field South, with photometric redshifts $2 < z < 3.5$. The Lyman-break galaxies are indicated by closed symbols; the squares are galaxies that satisfy $J-K > 2.3$ mag, which is a very simple criterion to select "red galaxies" with redshifts larger than 2. Although the number of these galaxies is much lower than the number of Lyman-break objects, their contribution to the K-band light is 24%, and their contribution to the stellar mass density may be comparable to that of Lyman-break galaxies.

These red galaxies could be the more evolved counterparts of Lyman-break galaxies, for example those that experienced a burst of star formation at higher redshift. Wide-area studies are needed to determine the true density of these infrared-selected galaxies and their relevance to the formation of early-type galaxies in clusters. Since early-type galaxies in clusters are thought to be the oldest galaxies in the nearby Universe, it might be expected that their counterparts at $z = 2 - 3$ are among the oldest, and hence reddest, galaxies at that redshift.

12.7   Conclusion

Early-type galaxies in clusters form a very homogeneous population, and evolve slowly to $z = 1$. Their formation redshifts are likely higher than 2. Nevertheless, significant morphological evolution occurs in rich clusters: the fraction of early-type galaxies decreases to high redshift, showing that their formation histories are more complex than a single burst. Detailed wide-field studies are necessary to determine the evolution of the progenitors of early-type galaxies at $z = 0.5 - 1$.

Field early-type galaxies are quite comparable to cluster early-type galaxies, but probably slightly younger. Direct evidence for morphological transformations is absent, but the evolution of the color-magnitude diagram suggests a strong evolution of the fraction of red galaxies. The similarity of early-type galaxies in the field and in clusters cannot be reproduced by the semi-analytical models of galaxy formation: in these models, the field early-type galaxies form too late.

These results imply that early-type galaxies formed at $z = 2$ and higher. It has become possible to study these galaxies directly with current telescopes. A large population of UV-bright galaxies has been found by using the Lyman-break technique (Steidel et al. 1996a,b). If these galaxies are the progenitors of early-type galaxies, they need recurrent merging, as their masses are fairly low. On the other hand, sub-mm sources have been found with very large star formation rates, and the role that these galaxies play needs to be elucidated further. Finally, a new population of galaxies has been identified based on very deep near-infrared imaging. These galaxies, selected by $J - K$ color, are generally too faint in the optical to be selected by the Lyman-break technique. They may represent the oldest galaxies at $z > 2$ and may be most typical of progenitors of cluster early-type galaxies.

Acknowledgements. It is a pleasure to thank the organizers for a lively meeting, and the Leids Kerkhoven-Bosscha foundation for financial support.

References

Baugh, C. M., Cole, S., & Frenk, C. S. 1996, MNRAS, 283, 1361
Bell, E. F., Wolf, C., Meisenheimer, K., Rix, H.-W., Borch, A., Dye, S., Kleinheinrich, M., & McIntosh, D. H. 2004, ApJ, submitted
Bender, R., Döbereiner, S., & Möllenhoff, C. 1988, A&AS, 74, 385
Bernardi, M., et al. 2003, AJ, 125, 1866
Bower, R. G., Lucey, J. R., & Ellis, R. S. 1992, MNRAS, 254, 601
Diaferio, A., Kauffmann, G., Balogh, M. L., White, S. D. M., Schade, D., & Ellingson, E. 2001, MNRAS, 323, 999
Djorgovski, S. G., & Davis, M. 1987, ApJ, 313, 59
Dressler, A. 1980, ApJ, 236, 351

Dressler, A., et al. 1997, ApJ, 490, 577

Dressler, A., Lynden-Bell, D., Burstein, D., Davies, R. L., Faber, S. M., Terlevich, R., & Wegner, G. 1987, ApJ, 313, 42

Ellis, R. S., Smail, I., Dressler, A., Couch, W. J., Oemler, A., Jr., Butcher, H., & Sharples, R. M. 1997, ApJ, 483, 582

Faber, S. M., Dressler, A., Davies, R. L., Burstein, D., Lynden-Bell, D., Terlevich, R., & Wegner, G. 1987, in Nearly Normal Galaxies, ed. S. M. Faber (New York: Springer), 175

Faber, S. M., Wegner, G., Burstein, D., Davies, R. L., Dressler, A., Lynden-Bell, D., & Terlevich, R. J. 1989, ApJS, 69, 763

Franx, M. et al. 2003, ApJ, 587, L79

Franx, M., Illingworth, G., & Heckman, T. 1989, AJ, 98, 538

Jedrzejewski, R. I. 1987, MNRAS, 226, 747

Jørgensen, I., & Franx, M. 1994, ApJ, 433, 553

Jørgensen, I., Franx, M., & Kjærgaard, P. 1996, MNRAS, 280, 167

Kauffmann, G., & Charlot, S. 1998, MNRAS, 297, L23

Kochanek, C. S., et al. 2000, ApJ, 543, 131

Labbé, I., et al. 2003, AJ, 125, 1107

Larson, R. B., Tinsley, B. M., & Caldwell, C. N. 1980, ApJ, 237, 692

Madau, P., Ferguson, H. C., Dickinson, M. E., Giavalisco, M., Steidel, C. C., & Fruchter, A. 1996, MNRAS, 283, 1388

Papovich, C., Dickinson, M., & Ferguson, H. C. 2001, ApJ, 559, 620

Rix, H.-W., & White, S. D. M. 1990, ApJ, 362, 52

Rusin, D., et al. 2003, ApJ, 587, 143

Sandage, A., & Visvanathan, N. 1978, ApJ, 225, 742

Shapley, A. E., Steidel, C. C., Adelberger, K. L., Dickinson, M., Giavalisco, M., & Pettini, M. 2001, ApJ, 562, 95

Smail, I., Ivison, R. J., Blain, A. W., & Kneib, J.-P., 2002, MNRAS, 331, 495

Stanford, S. A., Eisenhardt, P. R., & Dickinson, M. 1998, ApJ, 492, 461

Steidel, C. C., Adelberger, K. L., Giavalisco, M., Dickinson, M., & Pettini, M. 1999, ApJ, 519, 1

Steidel, C. C., Giavalisco, M., Dickinson, M., & Adelberger, K. L. 1996a, AJ, 112, 352

Steidel, C. C., Giavalisco, M., Pettini, M., Dickinson, M., & Adelberger, K. L. 1996b, ApJ, 462, L17

Terlevich, A. I., Caldwell, N., & Bower, R. G. 2001, MNRAS, 326, 1547

Treu, T., Stiavelli, M., Bertin, G., Casertano, S., & Møller, P. 2001, MNRAS, 326, 237

Treu, T., Stiavelli, M., Casertano, S., Møller, P., & Bertin, G. 2002, ApJ, 564, L13

van de Ven, G., van Dokkum, P. G., & Franx, M., 2003, MNRAS, 344, 924

van Dokkum, P. G., et al. 2003, ApJ, 587, L83

van Dokkum, P. G., & Ellis, R. S. 2003, ApJ, 592, L53

van Dokkum, P. G., & Franx, M. 2001, ApJ, 553, 90

van Dokkum, P. G., Franx, M., Kelson, D. D., & Illingworth, G. D. 2001, ApJ, 553, L39

van Dokkum, P. G., Franx, M., Kelson, D. D., Illingworth, G. D., Fisher, D., & Fabricant, D. 1998, ApJ, 500, 714

van Dokkum, P. G., & Stanford, S. A. 2003, ApJ, 585, 78

van Dokkum, P. G., Stanford, S. A., Holden, B. P., Eisenhardt, P. R., Dickinson, M., & Elston, R. 2001, ApJ, 552, L101

13

Star-forming galaxies in clusters

ALAN DRESSLER
The Observatories of the Carnegie Institution of Washington

Abstract

Clusters are laboratories for studying the influence of environment on the evolution of the stellar content and structure of galaxies. In this article I shift the emphasis away from present-day clusters, where star formation has nearly ceased, to consider earlier times when the major components of clusters—elliptical and S0 galaxies—were in formation, and when rapidly star-forming disk galaxies were still infalling. During three distinct epochs—the birth of the giant spheroids at $z \gtrsim 2$, the accretion of active spirals into clusters at $z \approx 1$, and the trickle of decaying spirals into present-epoch clusters—the mechanisms of environmental influence on galaxy evolution have been very different.

Both observation and theory seem to be converging on the idea that most of the stars in E and S0 galaxies are old and that—especially in rich clusters—their assembly into mature galaxies occurred early as well, at $z \gtrsim 2$. Because giant spirals are the last such systems to form, this chronology suggests that elliptical galaxies and giant bulges are not likely to be the result of mergers of mature spiral galaxies—the prevalent picture of the last two decades. Rather, they are products of a time when the conditions for star formation and the dynamical interactions among stellar systems were far different from those that characterize present-day mergers of giant spirals. Likewise, the considerable age of most S0 galaxies, and the fact that most appear in lower-density field environments, belies the notion that S0s are produced by ram pressure stripping in rich clusters, another standard interpretation of recent decades. Rather, new observations of intermediate-redshift clusters point to the importance of weak and strong interactions, mergers, and accretions—manifestations of the group-environment stage in cluster building—as principle drivers of the conversion of spiral to S0 galaxies. Both the quenching of star formation through the removal of gas reservoirs, and the exhaustion of disk gas through bursts of star formation are implicated as important mechanisms.

13.1 Introduction

As one of the organizers of this Symposium, I should not be puzzled by the title I was given for this invited paper. At the time, Gus, John, and I imagined that we were dividing up cluster-land into two areas: elliptical/spheroidal galaxies with the emphasis on age and metallicity—reviewed by Marijn Franx (2004)—and the star-forming galaxies, presumably those disk systems that are responsible for the Butcher-Oemler effect, to be reviewed by me. As the time for this Symposium grew near, however, I became increasingly confused about my topic. Looking for star formation in today's clusters is a little bit like searching for the last cashew in a picked-over nut-cup, while at intermediate redshift there is still insufficient high-quality data to settle some very basic disagreements about what is being observed.

This confusion reached a catastrophic level on day one of our conference when we were shown the spectacular N-body simulation by Springel & Hernquist (2003a) of a Coma-like cluster being assembled at high redshift. Watching the clock run as gigayears flew by, with a semi-analytic formulation providing an educated guess of how star formation might track the assembly of dark matter halos (Springel & Hernquist 2003b), I awoke to the obvious—how utterly epoch-dependent the question of "star formation in clusters" actually is. What we see in clusters today is a faint echo of what once was, and even at intermediate redshifts what we observe is often a frosting upon a well-aged, massive spheroidal component. Clearly, what distinguishes a cluster population from the galaxies in groups and the field is the preponderance of old spheroidal systems—ellipticals and bulges. Observations have long indicated that the heyday for their assembly was $z \gtrsim 2$, and now theory seems to be catching up, as the fidelity of the simulations and the precision of the cosmological model are substantially improved. Again, I refer to the results of Springel & Hernquist's N-body simulations that show that not only the birth of the stars, but also their assembly into large spheroidal systems, occurred—for the cluster environment—at a medium epoch of $z = 3$. (Indeed, in these simulations half of *all* star formation, over all environments, occurred before $z = 2$.)

Ultimately, then, the topic of "star-forming galaxies in clusters" has most meaning when we imagine a fireworks of star formation compressed in time and in space unlike anything we see today—when the ellipticals and the big bulged disk systems were assembled. From that time on, the history of star-forming galaxies in clusters has been, in comparison, a tame accretion of infalling star-forming galaxies growing the cluster, a pale glow compared to what went before, and, by today, barely a glow at all.

13.2   Connecting Star Formation History and Morphology in Cluster Galaxies

The history of star formation cannot be separated from that of structure and morphology. From the very first, in the classification scheme proposed by Hubble (1936; see also Sandage 1961, 1981; cf. de Vaucouleurs, de Vaucouleurs, & Corwin 1976; van den Bergh 1976), there was an implied relationship: Hubble believed that spheroidal systems were an early stage of evolution and that as galaxies aged they would acquire arms, turning last into spirals. This is the origin of the "early-type" and "late-type" labels that often accompany a discussion of the Hubble sequence.

Clusters of galaxies are often cited as laboratories in which to study the formation of the Hubble sequence (Dressler 1984), and much of the discussion about cluster populations focuses on star formation rates and the effect of environment. In what follows I will consider this connection through three distinct epochs: (1) $z \gtrsim 2$, the time of elliptical and large bulge formation; (2) $z \approx 0.5 - 1.0$, the time of rapid cluster building through the infall of disk galaxies; and (3) $z \approx 0$, the present epoch when clusters are, in comparison with galaxies in loose groups or the field, nearly empty of significant star formation. Many lines of evidence point to the fact that elliptical galaxies, and to some extent the spheroids of large-bulge disk galaxies, were formed in this earliest period (see Trager 2004; Treu 2004), and that this time of intense star formation is in fact the place to start when describing star formation in clusters, rather than the present epoch, when activity has been slowed to a trickle, or even 5 Gyr ago, when rich clusters still contained substantial populations of star-forming galaxies.

13.2.1   *A Blend of Old and New Paradigms of Elliptical Formation*

Eggen, Lynden-Bell, & Sandage's (1962) notion of a monolithic collapse of the Milky Way halo long predated any more general notions of how structure formed in the Universe. Rather, the idea followed closely from observations that the majority of Galactic halo stars are old, including those in globular clusters, and that they exhibit particular correlations between kinematics and metal abundance. Metal-poor stars were found to have highly elliptical orbits that would characterize a system forming in freefall, while more metal-rich systems, presumably formed later, had progressively more disk-like distributions and kinematics, suggesting increasing levels of dissipation. In a subsequent, more detailed study of the abundances and kinematics of halo objects, Searle & Zinn (1978) suggested that the collapse had been more lumpy than monolithic, with the evolution of individual clumps of stars and gas relatively isolated until late in the game. Although no elliptical galaxies are nearby enough to perform the kind of star-by-star analysis made for the Milky Way *, spectrophotometric observations in integrated light of their metal abundance and kinematics had been (through the 1980s) reasonably interpreted as suggesting that ellipticals are analogous to the Galaxy's spheroid and halo, that is, formed from an early and more-or-less coherent collapse of cooling gas leading to a rapid phase of star formation. Nevertheless, even rather large bulges such as the one of the Sombrero galaxy are much more likely to have sustained some star formation into the recent past (Trager 2004).

With the 1980s came the popularity of hierarchical model of structure formation (White & Negroponte 1982; Primack et al. 1984; Davis et al. 1985; Frenk et al. 1988). The success of the cold dark matter (CDM) model in accounting for a variety of features in the large-scale structure—on scales of groups of galaxies and larger—was strong inducement to apply its methodology to the much smaller scale of galaxy formation itself. From the beginning, however, numerical simulations faced two difficult problems that clearly limited the fidelity of the models: (1) insufficient dynamic range to cover sub-galaxy scales within an evolving larger-scale structure, and (2) lack of a convincing model of star formation that could be linked to the behavior of the dark matter. Today's simulations are finally overcoming the first limitation, with perhaps hundreds of individual mass points representing a galaxy in supercluster or larger simulations (see, e.g., Jenkins et al. 1998; Pearce et al. 2001; Mihos 2004). For the second—a more challenging problem—semi-analytic models (Lacey & Cole 1993, 1994; Kauffmann, White, & Guiderdoni 1993; Cole et al. 2000) appear at present to be the most successful approach, but even these are not without problems (Navarro, Frenk, & White 1995), and most would concede that there is a long way to go before such simulations have a true predictive power. (Indeed, the necessity of "tuning" the many adjustable parameters in order to match some characteristics of real galaxies may be more revealing of star formation process than galaxy formation.)

Perhaps the most striking "prediction" of numerical simulations, as early as the 1980s but much more developed in the 1990s, related to the birth of elliptical galaxies. Going all the way back before Baade's time, and certainly since Larson & Tinsley's (1978) use of Sandage & Visvanathan's (1978) photometry, observers had concluded that elliptical galaxies are the oldest galactic systems, much like globulars are the oldest star clusters. However, in 1972 Toomre & Toomre suggested that major mergers were a likely mechanism of constructing spheroidal systems out of disk galaxies, and the idea became very popular, one might say

---

* It is, however, becoming possible with the new generation of large telescopes to make detailed observations of globular clusters in elliptical galaxies in the Local Supercluster out to the Virgo cluster.

the paradigm, ever since (see, e.g., Barnes 1988, 1992; Schweizer & Seitzer 1992; Mihos & Hernquist 1994, 1996; Schweizer 2000; see also Mihos 2004). Since merging is fundamental to hierarchical clustering, it was natural for those performing CDM N-body simulations to embrace this idea (Negroponte & White 1983): through the 1990s such simulations consistently showed that large spheroidal systems such as elliptical galaxies were relatively late arrivals in the galactic zoo, and that they achieved both spheroidal structure and large masses through hierarchical merging. Although details depend on the cosmology and resolution of the model, the prediction was universal that most ellipticals had formed since $z < 2$, with a sizeable fraction since $z < 1$. This topic is also discussed in Treu's (2004) review.

This was especially troubling to observational astronomers studying the populations of rich clusters of galaxies, our topic here. Colors, spectra, and number density in clusters with increasing look-back time all seemed to point to an advanced age for ellipticals; besides, two-body mergers should be strongly suppressed in the high-speed encounters occurring in the rich cluster environment.* Regardless of expectations, the various attempts to place better constraints on the ages of cluster ellipticals consistently yielded a typical formation epoch of $z = 2-3$, if not for the structures of these ellipticals, at least for the stars now contained within them (Bower, Lucey, & Ellis 1992; Ellis et al. 1997; Bower, Kodama, & Terlevich 1998; Kuntschner & Davies 1998; Stanford, Eisenhardt, & Dickinson 1998; Trager et al. 2000; Kelson et al. 2001; Treu et al. 2002, and references therein for a discussion of constraints based on the fundamental plane; Davies 2004; Trager 2004).

Further evidence comes from the analysis of the chemical composition of elliptical galaxies, which show marked differences from what would be expected of the late merger of two typical (Sc) spirals (Trager et al. 2000; Kelson et al. 2004). In addition, the general similarity of field and cluster ellipticals discourages the notion that only cluster ellipticals are old, although it is clear that star formation has continued at some level in many field ellipticals to a relatively recent epoch (Faber et al. 1999).

Fortunately for the harmony of theorists and observers, this disagreement seems to be coming to an end, at least for cluster ellipticals. The new high-resolution N-body simulations, particularly those with the now-standard Λ cosmology, no longer predict that ellipticals in rich clusters are late-comers. Indeed, the simulations by Springel & Hernquist that we were shown in this Symposium predict an average stellar age of $z \approx 3$ for ellipticals in rich clusters, with an average "assembly age" of $z \approx 2$. To my knowledge, these predictions are consistent with the considerable body of observational data that have been collected. Most ellipticals are old, in composition *and* assembly, especially in clusters.

Quite at odds with this is the heavily cited case of MS 1054–03 at $z = 0.83$ (van Dokkum et al. 2000), which shows ~ 10 major mergers of early-type, luminous galaxies. Particularly puzzling about this case is that most of the mergers are between systems that are already early-type rather than late-type, star-forming galaxies. I am not aware of any other high-redshift clusters that show this behavior; until there are more examples, it is not sensible to take this one case as confirmation of a picture that has so much other evidence stacked against it.

I believe that it is important to draw an even stronger conclusion from the evolution of

---

* It is now recognized, however, that many clusters show strong subclustering; as these subclusters merge to make a richer cluster, their lower velocity dispersions would be more favorable to mergers. Still, in the rich environments of these subgroups (velocity dispersions ~ 500 km s^{-1}), mergers are much less likely than in poor groups or the field.

N-body simulations—the demise of the paradigm of elliptical formation by mergers of spiral galaxies, at least for cluster ellipticals. At the time the cluster ellipticals were formed in rich clusters, there were simply few if any spirals to merge! It is clear from *N*-body simulations, as well as observations of the high-*z* Universe (Driver et al. 1998; Dickinson 2000; Rodighiero et al. 2000), that large spiral galaxies as we know them today were very rare at *z* > 2, although recent work suggests that some do exist (Dawson et al. 2003; Labbe et al. 2003; Trujillo et al. 2004).

This would be especially true in the dense environment of a proto-rich-cluster, where galaxies had too little time to form large disks from the accretion of high-angular momentum material, and likely experienced a too-violent bumper-car ride as they traversed the cluster. True spirals, with their remarkedly thin disks, require a much longer formation time with few or no major disturbances (Toth & Ostriker 1992). Support for this is evident in the images from *HST*'s Medium Deep Survey and Deep Fields: beyond *z* = 1 the number density of large, well-formed spirals begins to rapidly diminish in favor a smaller, chaotically arranged systems. Spiral galaxies are the true late-comers in the galactic club.

So, if cluster ellipticals should not be thought of as the result of mergers of spiral galaxies, which did not even exist at the right time and place, how were cluster ellipticals born? No one can give a unambiguous answer to this yet, but the *N*-body simulations, particularly those that try to add the evolution of the gas component explicitly, should become better and better at developing the picture. My impression from looking at the Springel & Hernquist simulations is that, at the $z = 2 - 3$ epoch, in the densest regions where the cores of rich clusters were born, the component parts for making an elliptical were more globule-like than disk-like, i.e., high-density transition objects with masses between giant H II regions such as 30 Doradus (or M33's NGC 504) and the Large Magellanic Cloud itself. Their typical masses were $10^{7-9} M_\odot$, with velocity dispersions of some tens of kilometers per second. In some ways, I imagine the development of these fragments of elliptical assembly to be a giant version of the Searle-Zinn components of the formation of our own halo. Perhaps this is why the monolithic collapse model, even in its naive form, was so successful in accounting for the Galactic data—the true situation might have been much more like a spherical collapse than a model where the pieces are large disks that have undergone considerable dissipation.

By focusing on cluster ellipticals, I have skidded past an important puzzle concerning this picture of elliptical formation. That is the existence of present-day mergers that first caught the Toomres' attention and have been the worthy subjects of so many studies since, none more notable or refined than those of my Carnegie colleague Francois Schweizer. Francois argues persuasively that most (all?) of the features of ellipticals, including the induced formation of a increased number of globular clusters, are generated in the merger of two late-type galaxies (Whitmore & Schweizer 1995; Schweizer et al. 1996). If this is true, does it mean that these two different modes of elliptical formation lead to the same animal? This is possible, of course, but uncomfortable.

Put another way, the characteristics of field ellipticals are somewhat different than their rich-cluster counterparts (somewhat younger with episodes of recent stars formation, for example, as discussed in Trager's 2004 review), but unless you look closely, your first impression is that field ellipticals and cluster ellipticals are the same beast. For this reason, it is less than satisfying to propose that cluster ellipticals formed in one way and field ellipticals in another. I have for a long time suspected that even field ellipticals—the vast majority of ellipticals altogether—formed in nearly the same way as cluster ellipticals, that is, they are the remnants of small-mass but high-density regions in the early Universe: the elliptical represents the final state of all the available baryonic matter in that region, just as in the cluster case, but with only enough matter to form only a few large galaxies. We will learn the answers to such questions when we know the space density, masses, and stellar compositions of ellipticals all the way back to this putative time of formation, $z = 2-3$. Already the results are intriguing, but they allow a wide range of interpretations (see, e.g., McCarthy et al. 2001; Daddi et al. 2002; Roche et al. 2002; Treu 2004). Our understanding of these issues should rapidly improve.

I have taken the position, then, that the paradigm for elliptical formation should not be the merger of typical disk galaxies, since for the most part these did not exist at the time most ellipticals were formed. The weak form of this proposition applies only to cluster ellipticals; the strong form applies to all. What, then, do I say to my colleague Francois, who presents me with persuasive evidence of spirals in the act of merging to make a spheroid? I do not know—maybe most of these will become early-type disk galaxies, S0 or Sa, or maybe the elliptical population includes both types, and they are difficult to distinguish. We might expect, however, that if this is true, the new generation of large telescopes with their integral field spectrographs and polychromatic imagers might be able to find differences between ellipticals formed at very different epochs and under very different conditions.

13.2.2 Star Formation and Morphology

If the previous discussion contains any truth about the mechanism by which ellipticals are formed, then it is striking how much of a role the initial conditions, rather than the later influence on a galaxy of its environment, play in the development of the Hubble sequence. Again, through the popularity of CDM models of structure formation, the emphasis has been on the influence of later environment. Consider three general descriptions of how the Hubble sequence came about:

(1) Galaxy type is mutable and transient, so that a galaxy could well appear as a spiral as one part of its life, become part of an elliptical sometime later, perhaps returning to a spiral after that (see, e.g., Baugh, Cole, & Frenk 1996). This view, which became popular during the 1990s as CDM *N*-body simulations became larger and more sophisticated, emphasizes the role of happenstance—chance encounters with other galaxies—in determining the Hubble type during the lifetime of the galaxy.

(2) Galaxy type evolves steadily in one direction or the other, as Hubble originally proposed. A modern-day example of this is the first attempt to link hierarchical clustering with galaxy morphology that I describe above, where it is thought that all galaxies started out as spirals but, over time, mergers built larger spheroidal components in many of them, in some destroying their disks altogether, evolution in the opposite direction than Hubble had imagined. Another example is the proposal by Zhang (1999), who has developed the idea that secular evolution, the disturbance of disks by gravitational encounters with other galaxies (not requiring actual mergers) forces a spiral galaxy to evolve steadily toward a more spheroidal structure with steadily declining star formation rates (as the disk becomes more gas depleted and more puffed up—less conducive to continued star formation as perturbations in the disk are damped.

(3) Galaxy type is destiny. The idea here is that the environment *at birth* and soon after will be the main determinant of Hubble type. Sandage, Freeman, & Stokes (1970) and Faber & Gallagher (1976) were the first to suggest that initial conditions alone might be largely

responsible for the Hubble type each galaxy would eventually assume. The distribution of angular momentum of halos and its transfer and dissipation by baryons could have been the key factor. In relatively dense environments such as protoclusters, nondissipative mergers of subgalactic units would produce spheroidal systems at an early epoch (Larson 1975, 1976; Fall 1979; Fall & Efstathiou 1980). Large spirals, particularly those with modest spheroids—take the Milky Way (the "modal type" of giant galaxies)—would evolve in regions of low density where they can accrete high-angular momentum material from large distances. Of course, a certain amount of serendipity may influence a galaxy's morphological type; for example, the processes of (2) are capable of moving the Hubble type, perhaps substantially, but in (3) these are seen as second-order processes.

I believe there is much merit in (2) and (3), and little or none in (1). It is difficult to point to any standard Hubble types that look like they have been through back-and-forth transformations. Indeed, when a spheroidal system tries to recapture a disk, the result looks odd and forced—like a dog wearing shoes—and not likely to settle down into a normal spiral (see the Hubble Heritage image of NGC 4650A). For my tastes, (3) provides a better foundation for describing how the Hubble sequence has formed. This preference began with my 1980 study of the correlation of local density with morphology in rich clusters, a relation that has been shown to extend down to the very modest enhancements in density that are found in loose groups and the field (Postman & Geller 1984; Giovanelli, Haynes, & Chincarini 1986). You might think at first that a correlation with local density is direct evidence for late environmental effects; indeed, the studies of present-epoch clusters have always focused on the transformation of spiral to S0 galaxies through ram pressure stripping, a mechanism exclusively connected with rich clusters. What made me conclude the opposite—that the morphology-density relationship was evidence that the environment at and soon after the galaxy's birth played the major role—was the extremely weak dependence of type on local density. The fraction of spiral galaxies changes by a factor of 2 over a change of 100 in projected density, equivalent to a factor of 1000 in space density! How could the correlation of Hubble type be such a weak function of density? This would, however, make perfect sense if the probability for Hubble type was set early on when these density differences were much more subtle, before the larger-scale perturbations grew and amplified the density contrasts.

What put people off from this explanation, indeed, from any idea that local density (past or present) could influence Hubble type, was the notion that galaxies could not "know" what environments they would enhabit in the future, so that any gradients would be very hard to explain—that is, the effects would be washed out as galaxies left their birth environment or the local density peaks they enhabit today as compared to 5 Gyr ago. N-body simulations have clearly shown that this is not a concern: those particles that are born in overdense regions spend all of their lives in more-or-less dense environments, and those born in lower-density regions remain so, by and large. (This is in contrast to the case of violent relaxation of a "top hat" density distribution where the positions of infalling bodies are truly scrambled—but this is not the way the Universe is put together.) The environment and later-encounter experiences of a 10^{11} $M_\odot$ bound mass will be very different depending on where it is born, and this, I think, is probably to be the major contributor to its future form and star formation history—not the only, but the major one. In a sense, it *is* the initial conditions that determine galaxy morphology, by predetermining the influences that galaxy will experience for the rest of its life.

13.3 Today: The Cold Ashes of Today's Clusters

In the context of the idea that the heyday of galaxy clusters is long past, I would like to briefly review some of the aspects of star formation in rich clusters in today's Universe. There is quite an interesting history of this [see Biviano's (2000) informative and entertaining review], which tells us that, before there were observations of any clusters at earlier epochs, many astronomers believed that spirals—star-forming galaxies—were completely absent from rich clusters. Abell (1962) was skeptical that there were any spirals in rich clusters—I remember a plot he produced of the distribution of spirals galaxies in the field of Coma that showed no obvious concentration at all toward the center as defined by early-type galaxies. In response, Rood et al. (1972) went to some trouble to attribute cluster membership, by means of radial velocities, to a number of spirals in the field of Coma. By 1974, Gus Oemler had made a convincing case for spiral membership in many clusters in his seminal study of cluster properties. Contrasting with Oemler's categorization of "spiral rich" and "spiral poor" clusters, Melnick & Sargent (1977) parameterized the spiral population of X-ray luminous clusters as a smoothly increasing function of distance from the cluster core.

Of course, as less extreme clusters came under study, for example, Hercules and Virgo, and with the recognition that rich clusters are embedded in spiral-rich superclusters (Gregory & Thompson 1978), it became apparent that spiral galaxies *are* members of rich clusters, albeit at a much reduced fraction than the field. Realizing that spiral galaxies must be infalling into rich clusters, and mindful of Baade & Spitzer's (1951) suggestion that spirals might be stripped of their gas—becoming S0s—by galaxy-galaxy collisions, Gunn & Gott (1972) proposed that ram pressure stripping would be an effective means of promoting conversion of spiral to S0 galaxies (see Balsara, Livio, & O'Dea 1994 and Quilis, Moore, & Bower 2000 for contemporary treatments). Removing disk gas would presumably end star formation, provided sufficient gas did not reaccumulate through further infall and the return of gas through stellar evolution. In the same year, substantial arguments were made by Rood et al. that most S0 galaxies in the Coma cluster, let alone the S0 galaxies in the field, were not likely to be the result of ram pressure stripping of typical spirals. Nevertheless, although many additional objections have been raised since (see Dressler 1984 for a review), the ram pressure stripping paradigm is still today the most-cited environmental mechanism for turning spiral galaxies into S0s.

In their classic work on the subject of H I deficiency in cluster spirals using the Arecibo telescope, Giovanelli & Haynes (1983) were careful not to draw "sweeping" conclusions about the ability of the cluster environment—specifically the hot, X-ray-emitting gas first seen so clearly in *Einstein* X-ray images (see the review by Forman & Jones 1982)—to affect transformation of galaxy morphological type. Nevertheless, they did find clear evidence that cluster spirals of given Hubble type are deficient in H I gas in comparison to their field counterparts, and that this deficiency correlates with cluster X-ray luminosity, which is in turn a sensitive probe of gas density. Their work (see Haynes, Giovanelli, & Chincarini 1984 for a review) confirmed the notion of "anemia" in star formation that van den Bergh had noted earlier (1976). I myself found a correlation of orbits with H I deficiency that bolstered the idea (Dressler 1986). These studies had no purchase on the colder molecular gas; subsequent work by Young & Scoville 1991) has shown that the higher-density gas of molecular clouds is usually confined to the inner few kiloparsecs of a spiral galaxy. Theoretical work suggests that this component is more difficult to strip (Kundić, Spergel, & Hernquist 1993; Quilis et al. 2000).

In recent years, Jacqueline van Gorkom (1996, 2004) has made beautifully detailed, spatially resolved H I maps with the VLA that show how H I disks in spirals galaxies shrink systematically with proximity to the cluster center—clearly, the effect on the galaxy's interstellar medium can be substantial. If these galaxies had survived as normal, vigorous spirals until their plunge into the center, this might be evidence that at least *some* spirals were doomed to S0-dom by the stripping of their gas. My colleagues and I—the Morphs—believe that our observations of intermediate-redshift galaxies suggest that most of the damage has already been done on the way in, that the ram pressure stripping that occurs is a shot to the heart of a galaxy that was about beaten to death—already.

The modern work by Solanes et al. (2001) has further strengthened the case that the high-density environment, plausibly the hot plasma itself, has extracted a penalty from all spirals that make it close to the center of a cluster. The trends are clear, but remarkably weak, and the effect is really only seen in the inner 1 Mpc (see, e.g., Fig. 4 of Solanes et al.). Given that gas density rises so rapidly in cluster centers, and the fact that the orbits of these infalling galaxies, which originate far from the cluster core, cannot be purely radial, there seems to be little chance that many spiral galaxies in full flower could suffer a violent stripping of their gas that could turn them into S0 galaxies.

Adding to this circumstantial evidence of H I deficiency is "smoking gun" evidence that ram pressure stripping certainly is at work (see, e.g., Irwin et al. 1987; Kenney & Koopmann 1999); however, there is no real evidence that this has much to do with the prevalence of S0 galaxies in clusters (Abadi, Moore, & Bower 1999). Although Quilis et al. claim to demonstrate that ram pressure stripping is sufficient to completely remove the H I gas from a cluster spiral, they do not consider how often spirals reach the high-density core where such thorough stripping can occur. Furthermore, as I discuss in the next section, it is clear from the work on intermediate-redshift clusters that the effect of environment on an infalling spirals manifests itself far out in the cluster, at much lower gas densities than those at which ram pressure stripping is effective. Furthermore, ram pressure stripping is far less effective in poorer clusters such as Virgo, yet S0s are common even in these clusters. Finally, as Quilis et al. concede, the similarity of cluster and field S0 galaxies must be taken as a coincidence in this picture, since ram pressure stripping is of no consequence outside of rich cluster cores. Rather, it seems to me that the contribution to ram pressure stripping in producing S0 galaxies might be to hasten spirals to become and remain gas free in rich clusters, a nonunique, possibly redundant role. The power of the model was that it was supposed to explain the dearth of spirals and prevalence of S0 galaxies in rich clusters: there seems to me to be no evidence that is even persuasive, let alone compelling, that ram pressure stripping is the principle explanation of this phenomenon.

Star formation in rich clusters *today* is, then, a pretty sad affair. Infall is low and rapidly diminishing, if you believe we live in a Λ-dominated Universe. Spirals themselves are "running down" compared to half-a-Hubble time ago. The spirals that will be drawn into rich clusters in the future will die the death of a thousand cuts—in the rich group environment into which they have for so long been entrained, they are likely already to have had their fates sealed long ago.

13.4   5 Gyr Ago: Flameouts of Dying Galaxies

Against this bleak backdrop, it is perhaps easier to appreciate the surprise (and disbelief) that accompanied Butcher & Oemler's (1978a, b) discovery of a population of blue galaxies in rich clusters at $z \approx 0.5$. Presciently, they suggested that these were actively star-forming *spiral* galaxies (without benefit of morphological information), which were joining an already ancient population of elliptical and S0 galaxies. If spirals are so effectively extinguished in clusters today, why were they not in clusters 5 Gyr ago? This puzzle was itself projected against a dominant view in the late 1970s that the basic characteristics of galaxies were set long ago, well before this "late" 5 Gyr-ago epoch.

Indeed, there is general agreement that most clusters show a marked increase in the fraction of blue, star-forming galaxies by $z = 0.5$, but the question of "why" remains unanswered. An opposing view has been advanced by De Propris et al. (2003), who argue that K-band limited samples of distant clusters show little color evolution. We should not find that surprising, since selecting in the infrared is closer to a mass-limited sample, and, of course, giant elliptical and S0 galaxies dominate the bright end of cluster luminosity functions at $z = 0.5$ and today. But, this does not contradict observations that show clearly that—in contrast to today's clusters—L^* galaxies with strong star formation do appear in $z = 0.5$ clusters. But, are these M^* galaxies or, as De Propris et al. suggest, more dwarfish galaxies much inflamed. Mass-to-light ratios that come from optical colors suggest that this cannot be the case, but dust remains a troubling caveat. This question will be resolved soon with studies such as those of Metevier, Koo, & Simard (2002) and Ziegler et al. (2002), which use rotation curves to determine the mass of these systems—so far, the evidence suggests that these are massive systems. Undoubtedly related to this is Caldwell & Rose's (1997) finding of significant populations of star-forming *dwarf* galaxies in the outskirts of the Coma cluster; this suggests that the Butcher-Oemler effect can be seen as an evolution that depends on mass, in the sense that at higher redshift many luminous disk galaxies are still forming stars vigorously, while today this activity persists only for low-mass systems. Perhaps none of this seems surprising in light of hierarchical clustering, a different situation than in 1974. Nevertheless, the fact that does still need explaining is why this evolution is so rapid over the last 5 Gyr. I will come back to this.

There are five major groups working on intermediate-redshift cluster populations; their studies all include morphology with *HST*, photometry and spectroscopy with ground-based telescopes. Oemler, Poggianti, and I are the active members of the Morphs group at the moment, which has included Smail, Ellis, Couch, Butcher, and Sharples. A well-known study called CNOC1 was done by Yee, Ellingson, Morris, Carlberg, Balogh, and collaborators, and this group is continuing with new samples. The group of Franx, Illingworth, van Dokkum, Kelson, and collaborators has done only five clusters, but these cover the widest redshift range; the ACSCS group, Postman, Lubin, and Oke, has published results for two high-redshift clusters, with many more scheduled to be observed with *HST+ACS*. The newest entry, the EDisCS team of White, Aragón-Salamanca, Saglia, Dalcanton, Poggianti, and Kauffmann has a large amount of VLT time to study an optically selected sample of 10 clusters at $z \approx 0.4$ and 10 at $z \approx 0.8$. Put together, these account for a sample of perhaps 40 clusters of $0.3 < z < 1.0$, but for many of these the data are just being gathered. Except for the Morphs and EDisCS, X-ray luminosity has been the principal selection criteria of these samples (see Donahue et al. 2002 and Rosati 2004)—this results in a well-controlled, but strongly biased sample.

What have been the main results of morphological studies using *HST* images? For one, the Morphs group claims that the fractional population of S0 galaxies grow with time, from the epoch of $z = 0.5$ until today (Dressler et al. 1999). Fabricant, Franx, & van Dokkum

(2000) are not so sure, and Andreon (1998) is *certain* the Morphs are wrong, but no more certain than I am that he is in error. Rather than restate the Morphs case (see Dressler et al. 1997), I will leave this subject by saying that with the new generation of large telescopes with multislit spectrographs, it will be straightforward to measure rotations of early-type galaxies to distinguish between ellipticals and S0s, the main point of contention and concern.

The second result is that those star-forming galaxies that are found in intermediate-redshift clusters are noted by all to be more irregular, distorted, or contorted versions of disk galaxies today (Oemler, Dressler, & Butcher 1997; Smail et al. 1998). It is very likely that environment has played a role in this: some may be mergers or accretion events, others may have been rocked by strong but temporary gravitational interactions. A complication of deciphering what exactly can be blamed on the cluster environment is that, in general, disk galaxies in the field at intermediate redshift also look more disheveled than today's examples (Griffiths et al. 1994; Driver, Windhorst, & Griffiths 1995), although it is difficult to be quantitative. Perhaps future kinematical studies with integral-field spectrographs on big telescopes will be able to distinguish true two-body mergers from other events.

As I have said, photometry leads all these groups to conclude that there is a Butcher-Oemler effect, with some disagreement about whether some clusters at intermediate redshift are immune. Spectroscopic results are more varied and interesting. The Franx et al. group has concentrated on spectroscopy of early types, both for studying the evolution of the fundamental plane (Kelson et al. 2000; see also Jørgensen et al. 1999), and—a related issue—evidence from spectral lines that stellar populations are different in early-type galaxies in intermediate-redshift clusters (Kelson et al. 2001).

Concerning the star-forming galaxies, CNOC1 and the Morphs have weighed in most heavily up until now, with somewhat different conclusions and emphasis. The Morphs group reports high fractions, $\sim 30\%$, of Balmer-strong galaxies, which they associate with starbursting or post-starburst systems (Dressler et al. 1999; Poggianti et al. 1999). Other groups see some of these but generally with lower frequency, sometimes significantly so. If the difference is real, there are many possible causes—X-ray versus optical selection, magnitude limits, quality of spectra—that have not been adequately investigated. Perhaps the dispersion in the frequency of Balmer-strong galaxies is larger than the Morphs group has found for their sample, and modulated by stages in cluster evolution.

I now return to the question of why these star-forming galaxies are seen in $z = 0.5$ clusters but rarely today. We now know that infall into clusters has begun to slow dramatically due to the effects of the acceleration of the Universe. We also know that the star formation rates of field spirals show a factor of several decline since $z = 1$ (Lilly et al. 1996). Do the combination of these two effects alone account for the factor of 10 decrease in star-forming galaxies in rich clusters in only 5 billion years?

To answer this question it will be necessary to finally determine which are the dominant mechanisms through which the cluster environment influences the evolution of spiral galaxies. Despite decades of work, there is no consensus on the importance of different effects. I have already discussed ram pressure stripping and made some arguments that suggest to me that this plays only a secondary role. Evaporative gas removal (Cowie & Songaila 1976; Nulsen 1982) is not usually cited as a possible mechanism for significantly altering the evolution of a spiral galaxy, but is probably an important cause of a variety of more subtle differences of galaxies in dense environments (Burstein & Blumenthal 2002). The hydrodynamics of cold gas clouds bathed in a much hotter gas is complex: contemporary studies benefit from the increased power of numerical simulations (see, e.g., Henser & Vieser 2002). A full understanding of the effects of the intracluster medium on infalling galaxies will certainly include both ram pressure stripping and the evaporative process.

In denser environments there are obviously more opportunities for mergers, accretions, and strong interactions, and therefore reason to suspect that these could be driving galaxy evolution in groups and clusters. If N-body simulations evolve to accurately track star formation, we might be in a good position to judge how important these events are. As I mentioned earlier, such processes should be particularly effective in the subclusters we expect to see in higher-redshift clusters, due to the lower *local* velocity dispersions in these yet-unrelaxed clusters. The "harassment" mechanism of Moore et al. (1996), which combines both global (galaxy-cluster) and local (galaxy-galaxy) tidal effects, and particularly includes the contribution of *rapid* encounters, predicts that infalling galaxies will be substantially whittled down—unfortunately, this does not seem to throw much light on the star-forming disk galaxies that are observed.

The "mechanism *du jour*" is starvation (or strangulation, if you prefer, which I do not), of Bekki, Couch, & Shioya (2002). This derives from Larson, Tinsley, & Caldwell's (1980) idea that, if galaxies are continually refueled by extended gas reservoirs, these would be easily stripped as the galaxy enters a dense environment, resulting in the choking off of subsequent star formation. I digress to mention that this mechanism, originally to account for the Butcher-Oemler effect, was also offered as a way around the observation that most spirals should run out of gas in less than a Hubble time at present gas-consumption rates of their disk gas. From today's perspective, this seems like *a fact* rather than something to find a way out of. Imagine the papers that must have been written 6 billion years ago when some creature realized that the (now) S0 galaxies could not sustain star formation at their present rates. Although no gas reservoirs of the kind proposed are known to exist, they might in fact be difficult to detect (Benson et al. 2000). The starvation mechanism might have particular importance in intermediate-redshift clusters where galaxies still have a "full tank" compared to today's infalling spirals.

The starvation mechanism, much in vogue, is particularly touted by the CNOC1 group, who believe that the decline in star formation rates (measured from [O II] line strengths) as star-forming galaxies fall into clusters can be completely understood as a result of such a starvation mechanism. Ellingson et al. (2001) have produced detailed models of the structural evolution of clusters that match the CNOC1 observations well. By considering the population fractions as a continuous function of radius, they are able to predict the star formation rates of succeeding generations of field galaxies as they "turn around" and migrate toward the cluster center. This is an important new direction for cluster studies in the future—to fully map the range of star formation histories with epoch and radial distance from the cluster center.

Returning, then, to the question of the rapid decay of star-forming galaxies over the last 5 billion years, if these two mechanisms—galaxy-galaxy interactions and starvation—are the ones most influencing the evolution of spiral galaxies, it is interesting to ask if either would be more effective in the past than it is today. As far as removing gas reservoirs, it is not obvious why this would be less effective in intermediate-redshift clusters than today. However, the difference might be that the group environment in rich clusters has changed substantially over this epoch, as pointed out by Kauffmann (1995): the epoch $z \approx 0.5$ may represent a peak in the importance of groups in the building of clusters. For the infalling galaxies that are building today's clusters the protracted time they have spent in the group phase may have already succeeded in exhausting their gas through star formation and perhaps through starvation. Combining this with the normal decline in star formation seen for field galaxies over this epoch, and the effect of Λ in shutting down infall, the different histories of galaxies in groups could account for the rapid decline in star-forming galaxies that is the Butcher-Oemler effect.

Back to the past, do these spiral galaxies go quietly as they enter the apparently hostile environment of a rich cluster at $z \approx 0.5$? Here is where the Morphs and CNOC1 stories diverge: the CNOC1 group sees infalling spirals as victims of starvation and other mechanisms that simply truncate star formation; the Morphs sees these galaxies jostled and shocked into great bursts of star formation that hasten their demise, in dramatic fashion. The Morphs group emphasizes that the CNOC1 survey, for reasons yet to be determined, does not have a representative sample of galaxies with strong Balmer lines—*starbursts*. Despite the large number of CNOC1 clusters, only four overlap the redshift range covered by the 10 Morph's clusters (most are below $z = 0.3$), so this may be just small number statistics. Balogh et al.'s (1999) search for Balmer-strong galaxies in the CNOC1 sample turns up only a few percent of galaxies with strong Balmer lines, but even their analysis of the Morphs data shows a strong enhancement over the intermediate-redshift field (see Balogh et al., Fig. 29). New, deeper, better data on more clusters will soon answer the question of why the two groups get different results (the other groups seem to report intermediate results—higher fractions than CNOC1, but less than the Morphs). Rather than embarking a lengthy and ultimately inconclusive discussion, I will instead show what we believe is incontrovertible evidence that the Morphs data do show a strong starburst/post-starburst signal, i.e., this is not a matter of data quality or analysis technique. Our group has been experimenting with forming *composite spectra* in each cluster by coadding luminosity-weighted spectra. This avoids a major problem in comparing different studies by producing high-S/N spectra for which Balmer lines can be measured robustly: no complex statistical analysis of the kind performed by Balogh et al. is required. In Dressler et al. (2004) we present different combinations of these composite spectra, including simply adding all cluster members together: this produces the diagram Figure 13.1, which shows the trend of [O II] versus Hδ *cluster by cluster* with epoch.

Here I also show as Figure 13.2 the "recipe" for a intermediate-redshift (Morphs) cluster: by adding together all the spectra of all our clusters along categories of our spectral classifications, template spectra are produced that show clearly the different kinds of galaxies found in these clusters, including the large population of those with strong Balmer lines. These spectra demonstrate that the Morphs spectra are correctly interpreted as having strong Balmer lines—the whole of the Balmer sequence is now clearly seen. I hope that forming such composite spectra will become a common practice in future studies, since comparing one or a few excellent composite spectra per cluster is far better than comparing variously defined categories of hundreds of spectra of mixed quality. Integrated spectra will also provide a straightforward way to make Madau-like diagrams of star formation rates as a function of redshift for clusters (see Fig. 15 of Postman et al. 2001) and eventually for the full range of environments.

As Poggianti (2004) discusses, Dressler & Gunn (1983) and later Couch & Sharples (1987) correctly associated the presence of strong Balmer lines with starbursts, past or present. The Morphs interpretation of the distribution of star-forming galaxies in intermediate-

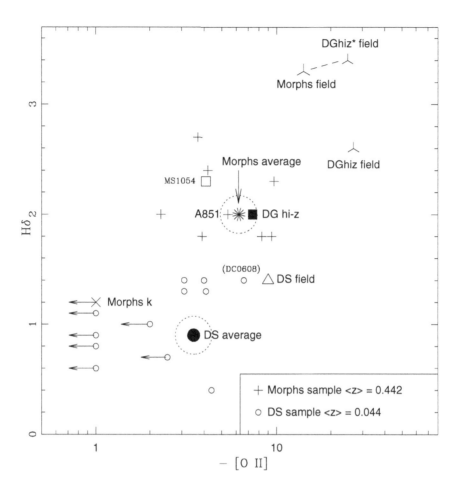

Fig. 13.1. Measurements of [O II] and Hδ for the various field and cluster samples of the Morphs study and the Dressler/Shectman low-redshift sample. There is a clear separation of the "mean" values for the Morphs and Dressler/Shectman samples, but there is also no overlap in the individual measurements for the clusters involved, i.e., there is a clean separation. Even higher-redshift clusters share the properties of the Morphs sample, with no obvious further increase in activity from $z \approx 0.5$ to $z \approx 0.8$. The field relation parallels that for the cluster galaxies, but is shifted toward stronger [O II] and Hδ at each redshift.

redshift clusters is consistent with the picture advocated by the CNOC1 group in the sense that both attribute such changes to the effects of the denser cluster environment. However, the Morphs data lead us to add a strong emphasis on understanding why *starbursts* are prevalent in (at least) many of these clusters, but not in present-epoch clusters. If the field population itself contained a sufficiently large fraction of starbursts, perhaps the simple infall/starvation picture of Ellingson et al. would suffice—this seems to be true for the CNOC1 sample with its lower fraction of starbursts—but there are a factor of 2–3 too few field starbursts to explain the populations of the Morphs clusters, and this is even without

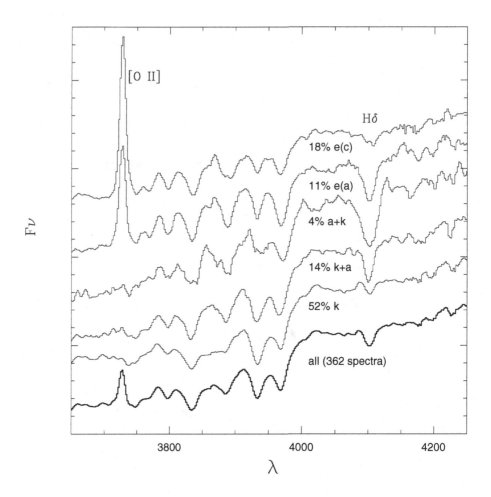

Fig. 13.2. The composite spectra for galaxies in eight clusters of the Morphs sample at $z \approx 0.5$ showing a "recipe" for a typical intermediate-redshift cluster. For the spectra shown above, the individual spectra were sorted according to their spectral classifications (Poggianti et al. 1999) and then coadded, weighted by luminosity. Galaxies with strong Hδ make up 28% of the Morphs sample. For spectra that were classified as k+a, a+k, and e(a), not only Hδ used in this classification, but also the higher-order Balmer lines, are clearly seen. This demonstrates the robustness of these classifications despite the lower S/N and greater variation of quality of the individual spectra.

explaining how starbursts with moderate-to-low duty cycles would fall in at the proper time (see Poggianti et al. 1999). This leads the Morphs to conclude that the cluster environment does more than truncate star formation, but also *stimulates* star formation in many infalling disk galaxies, often with spectacular results. In a new study Milvang-Hensen et al. (2003) come to the same conclusion about stimulated star formation for infalling spirals for a cluster at $z = 0.83$. Galaxy-galaxy interactions in infalling groups seems to be a likely mechanism, as discussed below.

Why is it important to see if stimulated starbursts are an important part of galaxy cluster evolution? The Morphs think that this phenomenon might be representative of the even higher-redshift Universe in a general way. Perhaps the dominant mode of star formation for spiral galaxies at $z \gtrsim 1$ is the starburst, even for massive, giant systems. If true, this is clearly an important feature of how galaxies evolve; the association with the cluster environment is, once again, an important clue as to the mechanism and how it works. As a final thought about this, once again it appears that what we see may be an evolution through mass with epoch. Consider the growing body of work on local dwarf galaxies where the fossil record of star formation can be read, for example, in the study by Gallart et al. (1999). Many dwarf galaxies appear to have evolved by bursts, unlike the modulated but generally continuous star formation of present-epoch giant spirals. The presence of large numbers of bursting galaxies in both cluster and field at $z > 0.4$ could signal a sensitivity to this kind of behavior for giant galaxies that dominated over their lifetimes, but now has abated—an important insight if true.

Finally, to further pursue this possibility that starbursts are an important result of the cluster environment on their galaxies, the Morphs are beginning to study galaxies much further out in these clusters. The idea is to separate the effects of global processes, such as harassment or ram pressure stripping, both of which are only effective within the inner Mpc (Treu et al. 2003), from galaxy-galaxy processes such as tidal interactions, which may dominate because of the way clusters are built from infalling groups in the supercluster environment. In Abell 851, in particular, we have obtained spectra of 50 additional cluster members in the range 2 Mpc $< R <$ 3.5 Mpc (Oemler et al. 2004). More than 50% of these are starbursts according to the definitions of Poggianti et al. Even if this remarkably high fraction is not common in intermediate-redshift clusters (if, for example, Abell 851 is caught at an unusually favorable epoch of assembly), the ability of the cluster environment to trigger such a display is *prima facia* evidence that local processes, most effective in modest-density enhancement, and not global processes are governing the last stages of star formation of these galaxies. Treu et al. (2003) have reached a similar conclusion about the dominance of local processes over the global influences of the high-density cluster core for the cluster CL 0024 + 16, based not on spectra but morphological considerations.

SIRTF will play a crucial role in this area, because, as Poggianti et al. have shown, the starbursts in these clusters are very dusty (see also Poggiant & Wu 2000)—substantial amounts of ongoing star formation hidden in optical (rest frame) observations. VLA observations by Morrison et al. (2002) and *ISO* observations by Fadda et al. (2000) demonstrate the importance of pursuing this issue with wavelengths that penetrate the dusty regions of starbursting galaxies.

As with the Ellingson et al. (2001) study, we believe that the tracking of star formation history with epoch and radius is a crucial next step in this kind of research. Our intention is to use the wide field of IMACS on Magellan to follow the star formation history of infalling galaxies in representative rich clusters of the Red Galaxy Survey (Gladders & Yee 2001) over different epochs out into the cluster periphery.

The idea that the evolution of star-forming galaxies is affected at modest density enhancements over the field, as opposed to the extreme environment of cluster cores, seems to be gaining support. We have heard from others at this conference other evidence for a "critical local density" that seems to mark changes in galaxy properties. Kodama et al. (2001) actually have used the same cluster discussed above, Abell 851, in their study of the colors of galaxies out into the supercluster environment. They report an abrupt change in the colors of *low-luminosity galaxies* at a projected density of 100 galaxies Mpc^{-2}, 2–3 orders of magnitude below that of the dense cluster core. In related studies, the first data release from the Sloan Digital Sky Survey analyzed by Gómez et al. (2003) shows a dramatic decrease in strongly star-forming Sc galaxies at a projected density of 1 Mpc^{-2}, which they identify as the boundary of the infall regions of clusters, and Bower & Balogh (2004) also find a similar effect in the CNOC2 survey. However, it is important to remember that these latter studies deal with present-day, luminous field galaxies and so are not directly comparable with the Kodama et al. results. Nevertheless, both results point to the effectiveness of some environmental process, probably associated with galaxy-galaxy interacts and mergers, that occurs at relatively low density.

13.5   One Billion Years ABB: A Cluster is Born

As I began, I conclude that the subject of "star formation in clusters" is a very epoch-dependent question. The work in this subject started, of course, with the study of present-epoch clusters, which show very weak ongoing star formation, most of it in low-luminosity galaxies. Despite early attempts to associate environmental influences such as ram pressure stripping with fundamental aspects of galaxy morphology or star formation, the preponderance of the evidence is that these characteristics derive from earlier epochs and very different conditions in clusters than we see today. In particular, the large population of elliptical galaxies that anchor today's clusters are very old—theory and observation both seem to be on the same side of this now—and this means that a spectacular epoch of star formation at $z \approx 2-3$ must have marked the beginning of cluster evolution. Probably because they are rare, no such protocluster of the mass of Coma has yet been found, but studies of Lyman-break galaxies or SCUBA sources, to give two examples, will probably turn up a few as the volume of surveyed space rapidly increases.

The properties of galaxies in intermediate-redshift clusters, particularly those back at $z \approx 1$ that are as yet little explored, will of course link the two epochs, telling us how the mostly old S0 galaxies evolved and how and when giant spirals were formed and, in some cases, destroyed. Connecting the galaxies in these clusters to their counterparts in the field at that epoch will be, as it has been at low redshift, essential to understanding the role of environment in shaping the galaxies of the modern Universe.

References

Abadi, M. G., Moore, B., & Bower, R. G. 1999, MNRAS, 308, 947
Abell, G. O. 1958, ApJS, 3, 211
——. 1962, in IAU Symp. 15, Problems of Extra-Galactic Research, ed. G. C. McVittie (New York: Macmillan Press), 213
Andreon, S. 1998, ApJ, 501, 533
Balogh, M. L., Morris, S. L., Yee, H. K. C., Carlberg, R. G., & Ellingson, E. 1999 ApJ, 527, 54
Balogh, M. L., Schade, D., Morris, S. L., Yee, H. K. C., Carlberg, R. G., & Ellingson, E. 1998, ApJ, 504, L75
Balsara, D., Livio, M., & O'Dea, C. P. 1994, ApJ, 437, 83
Barnes, J. E. 1988, ApJ, 335, 699
——. 1992, ApJ, 393, 484
Baugh, C. M., Cole, S., & Frenk, C. S. 1996, MNRAS, 283, 1361
Bekki, K., Couch, W. J., & Shioya, Y. 2002, ApJ, 577, 651
Benson, A. J., Bower, R. G., Frenk, C. S., & White, S. D. M. 2000, MNRAS, 314, 557
Biviano, A. 2000, in Constructing the Universe with Clusters of Galaxies, ed. F. Durret & D. Gerbal, http://nedwww.ipac.caltech.edu/level5/Biviano2/frames.html

Blumenthal, G. R., Faber, S. M., Primack, J. R., & Rees, M. J. 1984, Nature, 311, 517

Bothun, G. D., & Sullivan, W. T., III 1980, ApJ, 242, 903

Bower, R. G., & Balogh, M. L. 2004, in Carnegie Observatories Astrophysics Series, Vol. 3: Clusters of Galaxies: Probes of Cosmological Structure and Galaxy Evolution, ed. J. S. Mulchaey, A. Dressler, & A. Oemler (Cambridge: Cambridge Univ. Press), in press

Bower, R. G., Kodama, T., & Terlevich, A. 1998, MNRAS, 299, 1193

Bower, R. G., Lucey, J. R., & Ellis, R. S. 1992, MNRAS, 254, 589

Burstein, D., & Blumenthal, G. 2002, ApJ, 574, 17

Butcher, H., & Oemler, A., Jr. 1978a, ApJ, 219, 18

——. 1978b, ApJ, 226, 559

Caldwell, N., & Rose, J. A. 1997, AJ, 113, 492

Caldwell, N., Rose, J. A., Sharples, R. M., Ellis, R. S., & Bower, R. G. 1993, ApJ, 106, 473

Cole, S., Lacey, C. G., Baugh, B. M., & Frenk, C. S. 2000, MNRAS, 319, 168

Couch, W. J., & Sharples, R. M. 1987, MNRAS, 229, 423

Cowie, L. L., & Songaila, A. 1977, Nature, 266, 501

Daddi, E., et al. 2002, A&A, 384, L1

Davies, R. L. 2004, in Carnegie Observatories Astrophysics Series, Vol. 3: Clusters of Galaxies: Probes of Cosmological Structure and Galaxy Evolution, ed. J. S. Mulchaey, A. Dressler, & A. Oemler (Cambridge: Cambridge Univ. Press), in press

Davis, M., Efstathiou, G., Frenk, C. S., & White, S. D. M. 1985, ApJ, 292, 371

Dawson, S., McCrady, N., Stern, D., Eckart, M. E., Spinrad, H., Liu, M. C., & Graham, J. R. 2003, AJ, 125, 1236

de Vaucouleurs, G., de Vaucouleurs, A., & Corwin, H. G. 1976, Second Reference Catalog of Bright Galaxies (Austin: Univ. Texas Press)

De Propris, R., Stanford, S. A., Eisenhardt, P., & Dickinson, M. 2003, Ap&SS, 285, 43

Dickinson, M. 2000, Phil. Trans. R. Soc. Lond. A, 358, 200

Donahue, M., et al. 2002, ApJ, 569, 689

Dressler, A. 1980, ApJ, 236, 351

——. 1984, ARA&A, 22, 185

——. 1986, ApJ, 301, 35

Dressler, A., et al. 1997, ApJ, 490, 577

Dressler, A., & Gunn, J. E. 1983, ApJ, 270, 7

Dressler, A., Oemler, A., Jr., Poggianti, B. M., Smail, I., Trager, S., Shectman, S. A., Ellis, R. S., & Couch, W. J. 2004, in preparation

Dressler, A., Smail, I., Poggianti, B. M., Butcher, H., Couch, W. J., Ellis, R. S., & Oemler, A., Jr. 1999, ApJS, 122, 51

Driver, S. P., Fernandez-Soto, A., Couch, W. J., Odewahn, S. C., Windhorst, R. A., Phillips, S., Lanzetta, K., & Yahil, A. 1998, ApJ, 496, L93

Driver, S. P., Windhorst, R. A., & Griffiths, R. E. 1995, ApJ, 453, 48

Eggen, O. J., Lynden-Bell, D., & Sandage, A. 1962, ApJ, 136, 748

Ellingson, E., Lin, H., Yee, H. K. C., & Carlberg, R. G. 2001, ApJ, 547, 609

Ellis, R. S., Smail, I., Dressler, A., Couch, W. J., Oemler, A., Jr., Butcher, H., & Sharples, R. M. 1997, ApJ, 483, 582

Faber, S. M., & Gallagher, J. S. III 1976, ApJ, 204, 365

Faber, S. M., Trager, S. C., González, J. J., & Worthey, G. 1999, Ap&SS, 267, 273

Fabricant, D., Franx, M., & van Dokkum, P. 2000, ApJ, 539, 577

Fadda, D., Elbaz, D., Duc, P.-A., Flores, H., Franceschini, A., Cesarsky, C. J., & Moorwood, A. F. M. 2000, A&A, 361, 827

Fall, S. M. 1979, Nature, 281, 200

Fall, S. M., & Efstathiou, G. 1980, MNRAS, 193, 189

Forman, W., & Jones, C. 1982, ARA&A, 20, 547

Franx, M. 2004, in Carnegie Observatories Astrophysics Series, Vol. 3: Clusters of Galaxies: Probes of Cosmological Structure and Galaxy Evolution, ed. J. S. Mulchaey, A. Dressler, & A. Oemler (Cambridge: Cambridge Univ. Press), in press

Frenk, C. S., et al. 1999, ApJ, 525, 554

Frenk, C. S., White, S. D. M., Davis, M., & Efstathiou, G. 1988, ApJ, 327, 507

Gallart, C., Freedman, W. L., Aparicio, A., Bertelli, G., & Chiosi, C. 1999, AJ, 118, 2245

Giovanelli, R., & Haynes, M. P. 1983, AJ, 88, 881

Giovanelli, R., Haynes, M. P., & Chincarini, G. L. 1986, ApJ, 300, 77

Gladders, M. D., & Yee, H. K. C. 2001 in The New Era of Wide Field Astronomy, ed. R. Clowes, A. Adamson, & G. Bromage (San Francisco: ASP), 126

Gméz, P. L., et al. 2003, 584, 210

Gregory, S. A., & Thompson, L. A. 1978, ApJ, 222, 784

Griffiths, R. E., et al. 1994, ApJ, 435, L19

Gunn, J. E., & Gott, J. R. 1972, ApJ, 176, 1

Haynes, M. P., Giovanelli, R., & Chincarini, G. L. 1984, ARA&A, 22, 445

Hensler, G., & Vieser, W. 2002, Ap&SS, 281, 275

Hubble, E. 1936, The Realm of the Nebula (New Haven: Yale Univ. Press)

Hubble, E., & Humason, M. L. 1931, ApJ, 74, 43

Hubble, E. P. 1936, The Realm of the Nebulae (New Haven: Yale Univ. Press)

Irwin, J. A., Seaquist, E. R., Taylor, A. R., & Duric, N. 1987, ApJ, 313, L91

Jenkins, A., et al. 1998, ApJ, 499, 20

Jørgensen, I., Franx, M., Hjorth, J., & van Dokkum, P. G. 1999, MNRAS, 308, 833

Kauffmann, G. 1995, MNRAS, 274, 153.

Kauffmann, G., White, S. D. M., & Guiderdoni, B. 1993, MNRAS, 261, 921

Kelson, D. D., et al. 2004, in preparation

Kelson, D. D., Illingworth, G. D., Franx, M., & van Dokkum, P. G. 2001, ApJ, 552, 17

Kelson, D. D., Illingworth, G. D., van Dokkum, P. G., & Franx, M. 2000, ApJ, 531, 184

Kenney, J. P. D., & Koopmann, R. A. 1999, AJ, 117, 181

——. 2001, in Gas and Galaxy Evolution, ed. J. E. Hibbard, M. Rupen, & J. H. van Gorkom (San Francisco: ASP), 577

Kodama, T., Smail, I., Nakata, F., Okamura, S., & Bower, R. G. 2001, ApJ, 562, L9

Kundić, T., Spergel, D. N., & Hernquist, L. 1993, BAAS, 183, 8708

Kuntschner, H., & Davies, R. L. 1998, MNRAS, 295, L29

Labbe, I., et al. 2003, ApJ, 591, L95

Lacey, C. G., & Cole, S. 1993, MNRAS, 262, 627

——. 1994, MNRAS, 271, 676

Larson, R. B. 1975, MNRAS, 173, 671

——. 1976, MNRAS, 176, 31

Larson, R. B., & Tinsley, B. M. 1978, ApJ, 219, 46

Larson, R. B., Tinsley, B. M., & Caldwell, C. N. 1980, ApJ, 237, 692

Lilly, S. J., Le Fèvre, O., Hammer, F., & Crampton, D. 1996, ApJ, 460, L1

McCarthy, P. J., et al. 2001, ApJ, 560, L131

Melnick, J., & Sargent, W. L. W. 1977, ApJ, 215, 401

Metevier, A. J., Koo, D. C., & Simard, L. 2002, in Tracing Cosmic Evolution with Galaxy Clusters, ed. S. Borgani, M. Mezzetti, & R. Valdarnin (San Francisco: ASP), 173

Mihos, J. C. 2004, in Carnegie Observatories Astrophysics Series, Vol. 3: Clusters of Galaxies: Probes of Cosmological Structure and Galaxy Evolution, ed. J. S. Mulchaey, A. Dressler, & A. Oemler (Cambridge: Cambridge Univ. Press), in press

Mihos, J. C., & Hernquist, L. 1994, ApJ, 425, L13

——. 1996, ApJ, 464, 641

Milvang-Hensen, B., Aragón-Salamanca, A., Hau, G. K. T., Jørgensen, I., & Hjorth, J. 2003, MNRAS, 339, L1

Moore, B., Katz, N., Lake, G., Dressler, A., & Oemler, A., Jr. 1996, Nature, 379, 613

Morgan, W. W. 1958, PASP, 70, 364

Morrison, G. E., Owen, F. N., Ledlow, M. J., Keel, W. C., Hill, J. M., & Voges, W. 2002, in Tracing Cosmic Evolution with Galaxy Clusters, ed. S. Borgani, M. Mezzetti, & R. Valdarnin (San Francisco: ASP), 419

Navarro, J. F., Frenk, C. S., & White, S. D. M. 1995, MNRAS, 275, 56

Negroponte, J., & White, S. D. M. 1983, MNRAS, 205, 1009

Nulsen, P. E. J. 1982, MNRAS, 198, 1007

Oemler, A., Jr. 1974, ApJ, 194, 10

Oemler, A., Jr., et al. 2004, in preparation

Oemler, A., Jr., Dressler, A., & Butcher, H. R. 1997, ApJ, 474, 561

Pearce, F. R., Jenkins, A., Frenk, C. S., White, S. D. M., Thomas, P. A., Couchman, H. M. P., Peacock, J. A., & Efstathiou, G. 2001, MNRAS, 326, 649

Poggianti, B. M. 2004, in Carnegie Observatories Astrophysics Series, Vol. 3: Clusters of Galaxies: Probes of

Cosmological Structure and Galaxy Evolution, ed. J. S. Mulchaey, A. Dressler, & A. Oemler (Cambridge: Cambridge Univ. Press), in press

Poggianti, B. M., Smail, I., Dressler, A., Couch, W. J., Barger, A. J., Butcher, H., Ellis, R. S., & Oemler, A., Jr. 1999, ApJ, 518, 576

Poggianti, B. M., & Wu, H. 2000, ApJ, 529, 157

Postman, M., & Geller, M. J. 1984, ApJ, 281, 95

Postman, M., Lubin, L. M., & Oke, J. B. 2001, AJ, 122, 1125

Quilis, V., Moore, B., & Bower, R. 2000, Science, 288, 1617

Roche, N. D., Almaini, O., Dunlop, J., Ivison, R. J., & Willott, C. J. 2002, MNRAS, 337, 1282

Rodighiero, G., Granato, G. L., Franceschini, A., Fasano, G., & Silva, L. 2000, A&A, 364, 517

Rood, H. J., Page, T. L., Kintner, E. C., & King, I. R. 1972, ApJ, 175, 627

Rosati, P. 2004, in Carnegie Observatories Astrophysics Series, Vol. 3: Clusters of Galaxies: Probes of Cosmological Structure and Galaxy Evolution, ed. J. S. Mulchaey, A. Dressler, & A. Oemler (Cambridge: Cambridge Univ. Press), in press

Sandage, A. 1961, The Hubble Atlas of Galaxies (Washington, DC: Carnegie Inst. Washington)

Sandage, A., Freeman, K. C., & Stokes, N. R. 1970, ApJ, 160, 831

Sandage, A., & Tammann, G. A. 1981, A Revised Shapely-Ames Catalog of Bright Galaxies (Washington, DC: Carnegie Inst. Washington)

Sandage, A., & Visvanathan, N. 1978, ApJ, 225, 742

Schweizer, F. 2000, Phil. Trans. R. Soc. Lond. A, 358, 2063

Schweizer, F., Miller, B. W., Whitmore, B. C., & Fall, S. M. 1996, AJ, 112, 1839

Schweizer, F., & Seitzer, P. 1992, AJ, 104, 1039

Searle, L., & Zinn, R. 1978, ApJ, 225, 357

Smail, I., Dressler, A., Couch, W. J., Ellis, R. S., Oemler, A., Jr., Butcher, H., & Sharples, R. M. 1997, ApJS, 110, 213

Solanes, J. M., Manrique, A., García-Gómez, C., González-Casado, G., Giovanelli, R., & Haynes, M. P. 2001, ApJ, 548, 97

Spitzer, L., & Baade, W. 1951, ApJ, 113, 413

Springel, V., & Hernquist, L. 2003a, MNRAS, 339, 312

———. 2003b, MNRAS, 339, 289

Stanford, S. A., Eisenhardt, P. R., & Dickinson, M. 1998, ApJ, 492, 461

White, S. D. M., & Negroponte, J. 1982, MNRAS, 201, 401

Whitmore, B. C., & Schweizer, F. 1995, AJ, 109, 960

Toomre, A. 1977, in The Evolution of Galaxies and Stellar Populations, ed. B. M. Tinsley & R. B. Larson (New Haven: Yale Univ. Press), 401

Toomre, A., & Toomre, J. 1972, ApJ, 178, 623

Toth, G., & Ostriker, J. P. 1992, ApJ, 389, 4

Trager, S. C. 2004, in Carnegie Observatories Astrophysics Series, Vol. 4: Origin and Evolution of the Elements, ed. A. McWilliam & M. Rauch (Cambridge: Cambridge Univ. Press), in press

Trager, S. C., Faber, S. M., Worthey, G., & Gonzáles, J. J. 2000, AJ, 120, 165

Treu, T. 2004, in Carnegie Observatories Astrophysics Series, Vol. 3: Clusters of Galaxies: Probes of Cosmological Structure and Galaxy Evolution, ed. J. S. Mulchaey, A. Dressler, & A. Oemler (Cambridge: Cambridge Univ. Press), in press

Treu, T., Ellis, R. S., Kneib, J.-P., Dressler, A., Smail, I., Czoske, O., Oemler, A., Jr., & Natarajan, P. 2003, ApJ, 591, 53

Treu, T., Stiavelli, M., Casertano, S., Møller, P., & Bertin, G. 2002, ApJ, 564, L13

Trujillo, I., et al. 2004, ApJ, submitted (astro-ph/0307015)

van den Bergh, S. 1976, ApJ, 206, 883

van Dokkum, P. G., Franx, M., Fabricant, D., Illingworth, G. D., & Kelson, D. D. 2000, ApJ, 541, 95

van Dokkum, P. G., Franx, M., Kelson, D. D., & Illingworth, G. D. 2001, ApJ, 553, L39

van Gorkom, J. 1996, in Cold Gas at High Redshift, ed. M. N. Bremer & N. Malcolm (Dordrecht: Kluwer), 145

———. 2004, in Carnegie Observatories Astrophysics Series, Vol. 3: Clusters of Galaxies: Probes of Cosmological Structure and Galaxy Evolution, ed. J. S. Mulchaey, A. Dressler, & A. Oemler (Cambridge: Cambridge Univ. Press), in press

van Gorkom, J. H., et al. 2003, Ap&SS, 285, 219

Young, J. S., & Scoville, N. Z. 1991, ARA&A, 29, 581

Zhang, X. 1999, ApJ, 518, 613

Ziegler, B. L., et al. 2002, ApJ, 564, L69

14

The stellar content of galaxy clusters

ROGER L. DAVIES

Astrophysics, University of Oxford

Abstract

Clusters of galaxies are well-defined laboratories in which to study the evolution of the stellar content of galaxies. Direct detections of individual stars in intracluster space, as well as measurements of the diffuse light between galaxies, indicate that the gravitational interactions experienced by galaxies in clusters, with each other and with the cluster potential, result in 10%–20% of the stars being expelled into the intracluster space. I consider the properties of cluster galaxies as a function of position and local density within clusters and as a function of redshift: the color-magnitude relation, star formation rate, fundamental plane, age, and metallicity. These data strongly suggest that the population of luminous, high-velocity dispersion galaxies in rich clusters are old and form a metallicity sequence. Furthermore the increasing enhancement of magnesium abundance in high-dispersion galaxies implies a short star formation time scale. These observations suggest that massive, early-type galaxies in clusters formed at high redshift by monolithic collapse and appear counter to the expectations of hierarchical models of galaxy assembly. In lower luminosity ($\sigma < 150$ km s^{-1}) cluster galaxies, galaxies on the periphery of clusters, and galaxies in lower density regions, secondary star formation played an important role in galaxy evolution, consistent with hierarchical assembly. I illustrate how we can attempt to order the events in galaxy assembly using data from the SAURON integral-field spectrograph.

14.1 Introduction

Galaxy clusters have been used for decades in studies of the evolution of galaxies and their stellar populations because they contain large numbers of galaxies at a common distance that inhabit a common environment. Exploring the variation in the systematic properties of galaxies as a function of cluster type, radius within the cluster, and cluster redshift has proved to be a fruitful avenue of research (Schneider, Gunn, & Hoessel 1983; Dressler, Gunn, & Schneider 1985; Binggeli, Tammann, & Sandage 1987). Here I will confine my remarks to galaxies that are luminous (in the brightest two decades of the luminosity function) and that are of early morphological type. The morphology-density relation (Dressler 1980) ensures that this includes the vast majority of luminous galaxies in clusters of appreciable richness. I will not consider the population of dwarf galaxies, although there have been new results in this area recently (e.g., Conselice, Gallagher, & Wyse 2003). I will also not consider the globular cluster populations of cluster galaxies; recent work on the Coma system is reported in Harris et al. (2000).

The current state of knowledge of the stellar content of galaxies leads us to pose the following questions. How and when did cluster galaxies assemble? What physical mechanisms give rise to the global scaling relations found among cluster galaxies? What mechanisms are responsible for the variation in galaxy properties with environment? Can we develop a consistent view of galaxy evolution in clusters from $z = 0.5$ to the present day? In § 14.2 I will describe recent advances in the study of intracluster stars. I will discuss the integrated properties of cluster galaxies and their scaling relations, most of which have been established in clusters, in § 14.3. These relationships span structural properties such as luminosity, radius, and surface brightness, dynamical properties such as velocity dispersion and rotation and stellar population diagnostics such as colors, absorption- and emission-line strengths. Much of our knowledge of cluster galaxies arises from comparisons of these relationships within and among clusters, both locally and at high redshifts. I cannot consider all of these relationships but start by reviewing the trends in the color-magnitude relation (CMR) and star formation rate of galaxies as a function of cluster radius and as a function of redshift. I follow that with a summary of the results on the "fundamental plane," connecting size, dispersion, and surface brightness. Section 14.3 concludes with a discussion of the controversial topic of estimating the age and metalicity of the integrated stellar populations of galaxies using line indices. I will review these, illustrating the discussion with respect to the Fornax cluster and by considering the relative abundances of iron and magnesium. Before summarizing, I illustrate in § 14.4 how integral-field spectroscopy can give us a much clearer picture of the evolution of individual galaxies.

14.2   The Stars between Galaxies

Models of gravitational interactions of cluster galaxies with each other and with the potential of the cluster lead us to expect that stars will be ejected from galaxies into the intracluster space (e.g., Moore et al. 1996). Individual stars have now been detected. Diffuse emission in clusters, with a morphology indicative of tidal interactions between galaxies, has also been detected.

14.2.1   *Planetary Nebulae*

Using narrow-band imaging centered on the [O III] emission line, Theuns & Warren (1997) identified 10 candidate intergalactic planetary nebulae in the Fornax cluster in three fields chosen well away from any galaxy. Much of the follow-up work has been carried out by groups measuring the distance to clusters using the planetary nebula luminosity function. For example, in the Virgo cluster the current state-of-the-art has been reported by Arnaboldi et al. (2003). They identified intracluster planetary nebulae by imaging with the Subaru Suprime-Cam through two narrow-band filters centered at the redshifted wavelengths of [O III] and $H\alpha$. Spectroscopic observations confirmed their success in identifying true planetary nebulae. Based on their identification of 36 intracluster planetary nebulae, of which 25% were assumed to be contaminating background objects, they derived a lower limit of 10% and an upper limit of 40% for the fraction of the B-band Virgo cluster light contributed by intracluster stars.

14.2.2   *Red Giant Branch Stars*

In 1998, Ferguson, Tanvir, & von Hippel used deep imaging on the *Hubble Space Telescope (HST)* to discover red giant branch (RGB) stars in a blank field in the Virgo cluster. They suggested that the intracluster stars account for roughly one-tenth of the total stellar

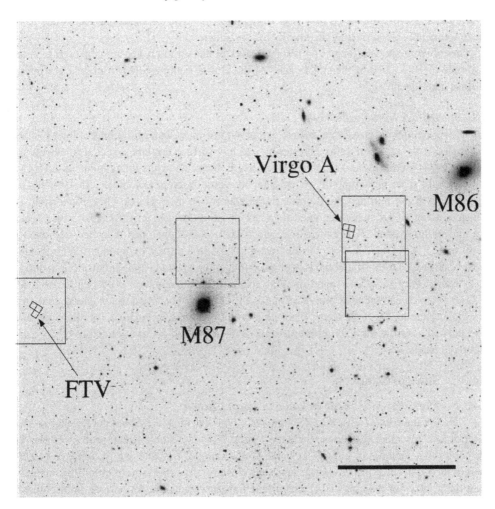

Fig. 14.1. Digitized Sky Survey image of the central region of the Virgo cluster, labeled with the location of the *HST* WFPC2 field used by Durrell et al. (2002) to search for RGB stars (labeled "Virgo A") and that used by Ferguson, Tanvir, & von Hippel (1998). The image is 2° on a side, with North at the top and East to the left. Also shown are the four fields used by Feldmeier et al. (2002) to search for intracluster planetary nebulae. (From Durrell et al. 2002.)

mass of the cluster. More recent studies (e.g., Durrell et al. 2002) have used deep F814W images to estimate the number of RGB stars in two fields in Virgo using the two Hubble Deep Fields as a control. At magnitudes fainter than $I = 27$ they found an excess of point sources in the Virgo fields. Modeling the luminosity function, they concluded that the tip of the RGB in the intracluster stars is at $I = 27.3$ mag and that the metalicity of the stars spans $-0.8 <$ [Fe/H] < -0.2, consistent with them originating in luminous galaxies. An analysis of the surface density of RGB stars led them to estimate that the intracluster population accounts for 10%–20% of the *I*-band luminosity of the Virgo cluster. In Figure 14.1, reproduced from

Durrell et al. (2002), the fields in which Virgo cluster planetary nebulae and RGB stars have been found are shown on a Digitized Sky Survey image.

These two methods of detecting resolved stars in the Virgo cluster are converging on an estimate of 10%–20% of the stellar mass of the cluster being ejected from galaxies into a diffuse component.

14.2.3   Diffuse Intracluster Light

Attempts to measure a diffuse component of intracluster light have a long history entangled with studies of galaxy clustering and low-surface brightness galaxies. Shectman (1973, 1974) predicted, and attempted to measure, the scale of the fluctuations in the background light from photographic plates scanned with a microdensitometer. In 1991 Uson, Boughn, & Kuhn reported the first measurements of the intergalactic background light in clusters using CCDs, which enabled them to measure down to 2×10^{-4} of the sky background in the R band. Their investigation of Abell 2029 concluded that, between cluster radii of 125–425 kpc, 5%–10% of the cluster light is in a diffuse component. With the advent of large CCDs this method has been used to identify specific components of tidal debris in clusters. For example, V-band imaging to $\mu_V = 26.5$ mag arcsec^{-2} by Feldmeier et al. (2002) revealed an extended plume of material in MKW7 but found no extended tidal arcs (Feldmeier et al. 2004). It seems likely that with the advent of large CCDs this work will expand and confront the varying predictions of the scale and distribution expected for tidal debris in clusters.

14.3   Stars in Galaxies

14.3.1   The Color-Magnitude Relation in Local Clusters

We have known for decades that early-type galaxies in clusters get slighter redder as they become more luminous (Sandage 1972). This is a small effect, seen most clearly in colors that include an ultraviolet band. For example, Visvanathan & Sandage (1977) determined that galaxies in nine groups and clusters obey the same relationship, namely that the $u - V$ color reddens by 0.1 magnitude for each V magnitude increase in brightness. Furthermore, they established that the relationship has a scatter of 0.1 magnitude in color and is common to elliptical and lenticular galaxies alike.

Bower, Lucey, & Ellis (1992) measured U, V, J, and K magnitudes for 94 early-type galaxies in Coma and Virgo. The rms scatter in the measurements was between 1.5%–3.0%. They found no difference between the intrinsic colors of galaxies in the two clusters to a precision of a few percent, checking their result using the distance-independent color-velocity dispersion relation. By carefully estimating the observational error they concluded that the *intrinsic* scatter in colors of galaxies was 3%–4%. This remarkable uniformity of galaxy properties between and within the two clusters, when analyzed using stellar population models (Bruzual 1983), led Bower et al. to conclude that the spheroidal components of early-type galaxies in clusters are unlikely to have formed below a redshift of 2.

The Bower et al. (1992) data extended to no fainter than $B = 16$ mag and therefore included only the first decade of the galaxy luminosity function in the Coma cluster. The fainter Virgo galaxies showed some signs of increased scatter in the CMR but the sample is not numerous enough to draw a definitive conclusion. Terlevich et al. (1999) extended the Coma $U - V$ vs. V relation by almost two magnitudes and discovered a substantially increased scatter among the fainter population (see Fig. 14.2). Furthermore, they used metal-

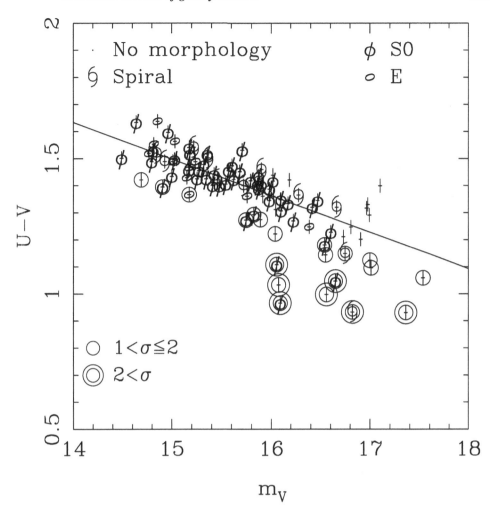

Fig. 14.2. The $U-V$ CMR relation for 101 Coma galaxies. Colors and magnitudes are within an 8.8-diameter aperture. The standard deviation about the best fit (solid line) is 0.07±0.01 mag. The symbols are coded by the morphological type of each galaxy. Galaxies that lie between 1 and 2 standard deviations blueward of the CMR best fit have an additional circle drawn around the symbol. Galaxies that lie more than 2 standard deviations blueward of the CMR best fit have two extra circles. (Taken from Terlevich et al. 1999.)

and Balmer-line index measurements, together with broad-band colors, to interpret the CMR in terms of changes in age and metalicity, using the stellar population models of Worthey (1994) and Worthey & Ottaviani (1997). They showed that the change in color over the most luminous decade of the CMR can be accounted for by roughly a factor of 2 increase in metalicity; a similar result was found by Vazdekis et al. (2001) for the Virgo cluster. To account for the blueward scatter of the fainter galaxy population, however, they needed to add a population of young stars. They demonstrated that the stronger Balmer lines and bluer colors could be accounted for by a population of 1 Gyr old stars containing 15% of the mass.

Terlevich, Caldwell, & Bower (2001) explored variations in the CMR as a function of

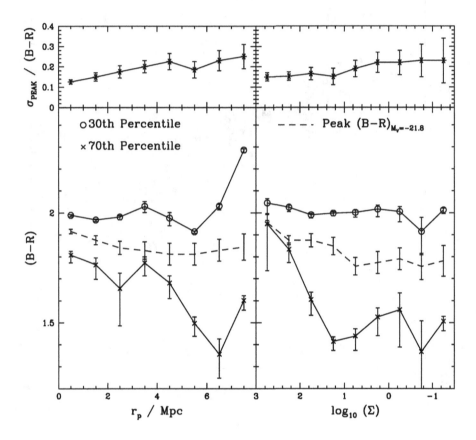

Fig. 14.3. $B-R$ color as a function of cluster radius and local density (lower panels). The three lines represent the peak, 30th, and 70th percentiles of the color histograms measured relative to the red end of the distribution. The 30th percentile of the distributions is nearly constant over at least the inner 6 Mpc of the combined cluster, while the peak shifts to the blue. The upper panel displays the width of the Gaussian fit to the color histograms against radius and local galaxy density. (Taken from Pimbblet et al. 2002).

radius in the Coma cluster and found that the slope remains constant as a function of radius, but that at fixed luminosity the colors get bluer toward the outskirts. This is a small but significant effect, which they were not able to account for simply on the basis of dust in the cluster, as this would produce too much scatter in the CMR compared to the data. They concluded that the galaxies up to 1 Mpc from the cluster center have younger luminosity-weighted ages by at most 2 Gyr compared to those in the core. Pimbblet et al. (2002), reporting the results of the LARCS project, extended the study of the CMR as a function of radius to a homogeneous sample of 11 luminous X-ray clusters at $z \approx 0.1$. By combining the galaxy populations of the clusters, they were able to extend the investigation to much larger radii, out to 6 Mpc. They found that, although the position of the red envelope of the CMR does not vary as a function of projected radius (or local galaxy density), the modal colors of galaxies lying on the CMR become bluer with increasing projected radius or decreasing density (see Fig. 14.3). They inferred that as projected radius increases the red envelope remains unchanged but the fraction of bluer galaxies increases. They concluded that toward the periphery of clusters there is an increasing fraction of galaxies that exhibit signatures of star formation in the recent past and that this characteristic is not present in the evolved galaxies in the highest density regions of the clusters.

14.3.2 The Star Formation Rate in Local Clusters

The local CMR results have been extended to measurements of the star formation rate through the work of the 2dF and SDSS redshift survey teams (Lewis et al. 2002; Gomez et al. 2003). Lewis et al. (2002) made Hα emission-line equivalent width measurements for 11,006 galaxies brighter than $M_B = -19$ mag in 17 known rich clusters with $0.05 < z < 0.1$. They deduced the star formation rate, normalized to L^*, as a function of distance from the cluster center and found that it is skewed toward larger values with increasing radius, reaching the value for the field population at distances greater than three times the virial radius. They found a correlation between normalized star formation rate and local projected density (Fig. 14.4), which saturates at projected densities below about 1 galaxy Mpc^{-2}, for galaxies brighter than $M_B = -19$, corresponding roughly to the mean density at the cluster virial radius. Importantly, they were able to establish that this relation between star formation rate and local density is valid in density enhancements that are more than two virial radii from the cluster center. Therefore, they concluded that environmental influences on galaxy properties are not restricted to cluster cores, but are effective in all groups where the density exceeds a critical value.

14.3.3 The Color-Magnitude Relation in Clusters to $z = 1$

The remarkable uniformity of the CMR in local early-type galaxies has made it an attractive diagnostic with which to directly probe the evolution of galaxies in clusters. The availability of *HST* imaging has enabled several groups to address this. The MORPHS collaboration (Ellis et al. 1997) explored the CMR in three $z \approx 0.5$ clusters. They used *HST* images to measure the rest-frame UV–optical colors of galaxies selected on the basis of their morphology. They found a small scatter in color (<0.1 mag rms) for galaxies classified as E and E/S0, both within each cluster and from cluster to cluster. The found no trend for the scatter to increase with decreasing luminosity or any evidence for a distinction between the scatter observed for galaxies classified as ellipticals and S0. This uniformity drove them to conclude that most of the star formation in the elliptical galaxies in dense clusters was completed before $z = 3$.

Kodama et al. (1998) compiled data from *HST* on the CMR for early-type galaxies in 17 distant clusters in the redshift range 0.31–1.27 to attempt to distinguish between models in which the CMR is a sequence of increasing metalicity or age. They used the Kodama & Arimoto (1997) evolutionary model in which elliptical galaxies formed rapidly in a monolithic collapse followed by a galactic wind. They were able to reproduce the CMR for Coma ellipticals by assuming they formed either a metalicity sequence or an age sequence. They then used the model to predict the slope and zeropoint of the CMR as a function of redshift under the two hypotheses. By assuming all elliptical galaxies are coeval with a mean stellar

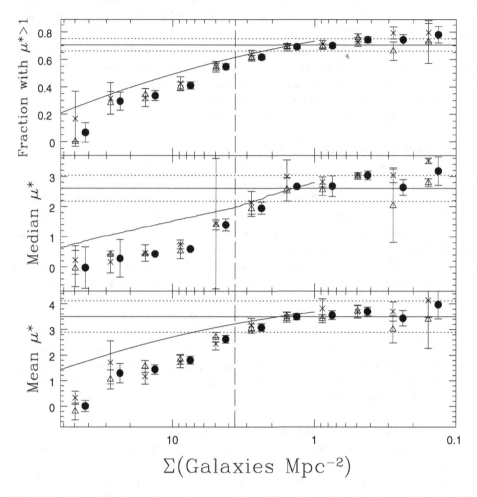

Fig. 14.4. Normalized star formation rate, μ^* as a function of local projected density. The bottom and middle panels show the mean and median value of μ^* in the cluster sample. The top panel shows the fraction of galaxies with $\mu^* > 1 M_\odot \mathrm{yr}^{-1}$. Solid points represent the full galaxy sample, while the triangles and crosses are from clusters with velocity dispersion greater or less than 800 km s^{-1}, respectively (offset in density for clarity). The horizontal, solid line represents the value of each statistic in the field sample; the dotted lines are an estimate of the field-to-field standard deviation. The vertical, dashed line represents the mean projected density of galaxies within the virial radius of the cluster. The solid curves are the expected trends due to the morphology-density relation of Dressler (1980), assuming the field population is composed of 18% E, 23% S0, and 59% spiral galaxies (Whitmore, Gilmore, & Jones 1993). (Taken from Lewis et al. 2002.)

metalicity that increases with luminosity, they found no significant differences between the CMR of the clusters and the models. They used the zeropoint of the CMRs to constrain the epoch of major star formation in early-type galaxies to $z \approx 4$.

In a complementary study, Gladders et al. (1998), measured the slope of the red sequence of galaxies in 44 Abell clusters at $z < 0.15$ and used these as a control sample for comparison with six clusters with *HST* imaging in the range $0.2 < z < 0.75$. The Kodama & Arimoto (1997) models predict a change in the observed slope of the red sequence of elliptical galaxies as a function of redshift because as galaxies age the metal-rich systems become redder faster than the more metal-poor ones. Specifically, we expect the red sequence to be significantly flatter up to 4 Gyr after formation than at later times. Thus Gladders et al. used the *slope* of the red sequence to constrain the formation epoch of cluster galaxies and did not need to rely on precise absolute photometry to determine the zeropoints. They detected the expected steepening with increasing redshift and concluded that the galaxies could not have formed more recently than $z = 2$.

Both the local CMR studies and those of clusters to $z = 1$ conclude that the luminous early-type galaxies in cluster cores formed at early epochs and that secondary star formation is rare. There is strong evidence that an increase in metalicity with luminosity is responsible for the CMR. However, it is also clear that for lower luminosity galaxies and galaxies away from the cluster core secondary star formation is more common and results in a larger fraction of blue galaxies and more scatter in the CMR. The results of emission-line determinations of star formation rate confirm these results and have shown that there is a threshold density above which star formation is quenched. This density is remarkably low, suggesting that physical mechanisms that work *only* in the dense regions of cluster cores, such as ram presure stripping, cannot be entirely responsible for the observed changes.

14.3.4 The Fundamental Plane

The fundamental plane (FP) links the velocity dispersion of the stars to the size and surface brightness of elliptical galaxies. It was discovered through studying the early-type galaxy population of clusters in the late 1980s (Djorgovski & Davis 1987; Dressler et al. 1987), and it has been used as a distance indicator to measure cluster peculiar motions (e.g., Faber et al. 1989; Colless et al. 1999; Hudson et al. 2001). Colless et al. (2001) report the following relationship based on their 29 best-observed clusters:

$$\log R_e = 1.22 \log \sigma + 0.33 \langle \mu_e \rangle - 8.66,$$

where R_e is the effective radius, $\langle \mu_e \rangle$ is the mean R-band surface brightness inside R_e, and σ is the central velocity dispersion. The zeropoint is essentially that which gives the Coma cluster zero peculiar velocity. Elliptical galaxies fall on this plane with very little scatter, 0.064 in $\log R_e$. The FP can be used as an indicator of the homogeneity of the stellar population of cluster galaxies. Its small scatter was used by Ciotti & Renzini (1993) to estimate that the range in mass-to-light ratio (M/L) at any given location in the plane is only 12%. In their analysis of the κ-space distribution, Bender, Burstein, & Faber (1992) estimated an increase in blue M/L, from 5 to 10, over the absolute magnitude range from -20.5 to -23.5. They argued that the scatter in age and/or metalicity at fixed σ must be less than 15%. Colless et al. (1999) similarly demonstrated that the small scatter in the FP constrains the range of ages of early-type galaxies in clusters to $\pm 30\%$ (as long as there is no strong anticorrelation between age and metalicity). These remarkably uniform properties stimulated work to measure the evolution of galaxies directly by measuring the FP for clusters at redshifts up to $z \approx 1$. Such work relies on *HST* imaging to make measurements of R_e and large ground-based telescopes to measure the stellar velocity dispersions. Jørgensen et al.

(1999) compared the FP for six clusters out to $z = 0.58$. They find that the scatter around the FP is similar for all six clusters and that the mass-to-light ratio decreases with redshift: $\log (M/L)_r = -0.26 \pm 0.06\Delta z$, consistent with passive evolution of a stellar population that formed at a redshift larger than 5.

Franx et al. (2000) and van Dokkum et al. (2000) extended this work to $z = 0.83$, finding that early-type galaxies in clusters follow a well-defined relation out to the highest redshift and confirming that the mass-to-light ratio evolution is slow. They considered this result in the context of current galaxy evolution models involving mergers, starbursts, and the resulting changes in morphology and luminosity. They speculated that the youngest early-type galaxies at low redshift have different (pre-merger) morphological types at high redshift and therefore do not appear in high-redshift samples of early-type galaxies; they dubbed this a "progenitor bias." They noted that there is a high fraction of close companions or merging galaxies in the periphery of their highest redshift clusters, a number comparable to the fraction of ellipticals. These merger candidates fall on the same CMR as the ellipticals but have roughly twice the scatter. These results suggest that the present-day population of early-type galaxies in clusters has not been constant, but is evolving, and that ellipticals are created as galaxy pairs or groups fall into clusters. This picture is qualitatively consistent with the low-redshift result that the star formation rate increases monotonically with radius in clusters and suggests that roughly half the population of cluster ellipticals has assembled since $z = 1$. In a follow-up study, Kelson et al. (2001) measured the velocity dispersions and Balmer absorption-line strengths of galaxies in four clusters with $0.06 < z < 0.83$ and found that the evolution of the Balmer-line strength at fixed σ was consistent with passive evolution of single stellar population models and similar to that derived from the FP, once again requiring an early redshift of formation.

14.3.5   *Line Strengths: The Age/Metalicity Dilemma*

Models of the integrated spectral energy distributions of old stellar populations exhibit a degeneracy between increasing the age and increasing the metalicity of the population, such that it has been difficult to distinguish between changes in these on the basis of, for example, broad-band colors alone. This problem provided the motivation for studies of the CMR and FP to make measurements at high redshift where these two factors can be separated. In the 1990s models became available that identified index combinations, such as Hβ and an average of iron and magnesium features ([MgFe]), which together have some leverage between age and metalicity (Worthey 1994). This enabled age and metalicity to be determined in local galaxies using high-precision spectroscopy. These models were expanded and refined throughout the last decade (Worthey & Ottaviani 1997; Vazdekis 1999) so that we now have a powerful set of tools to address questions of the age and metalicity of local galaxies. Some of these models have even been developed to include relative abundances that are not in the solar ratios, to address some anomalies in the standard model fits (e.g., Thomas, Maraston, & Bender 2003).

The first application of the Worthey models was by González (1993) who analyzed long-slit spectra of 41 elliptical galaxies taken with a CCD system at Lick Observatory. The sample of galaxies in González's thesis exhibited a factor of 10 range in age, while spanning only a factor of 2–3 in metalicity. These galaxies were much closer to an age sequence than a metalicity sequence, in contradiction to the view developed from cluster CMR and FP studies, although the sample was drawn from the full range of environments in which

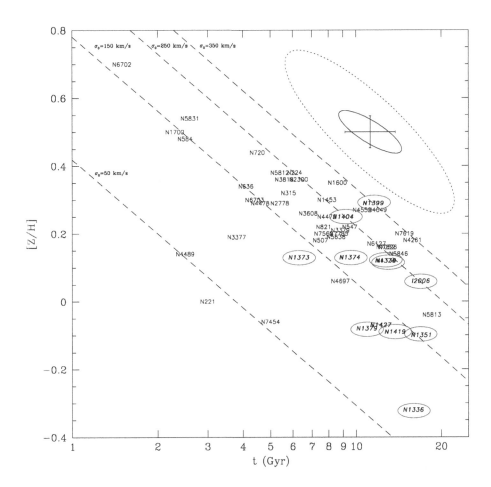

Fig. 14.5. A face-on view of the "Z-plane" from Trager et al. (2000), adapted to highlight the Fornax elliptical galaxies from Kuntschner & Davies (1998). At fixed velocity dispersion (dashed lines), younger galaxies have higher metalicities than older galaxies. The solid error ellipse in the top right-hand corner is typical of the González (1993) sample; the dotted ellipse is typical of the highest quality data in the Lick/IDS galaxy sample (Trager et al. 1998). The slope of the error ellipses is nearly identical to the of lines of constant velocity dispersion, indicating that poor data can masquerade as a real trend. The Fornax ellipticals are ringed and can be seen to be coeval within the uncertainty of the best data.

ellipticals are found. In contrast, Kuntschner & Davies (1998) found that in the Fornax cluster the elliptical galaxies fall on a tight metalicity sequence of constant age, whereas the lenticular galaxies scatter to luminosity-weighted ages as young as 2 Gyr. González's work was re-visited by Trager et al. (2000), who included galaxies from the Kuntschner & Davies sample. As illustrated in Figure 14.5, they found that at fixed σ there is an anticorrelation between age and metalicity. They demonstrated that this anticorrelation can account for the tightness of the Mg-σ relation because if galaxies experience secondary star forma-

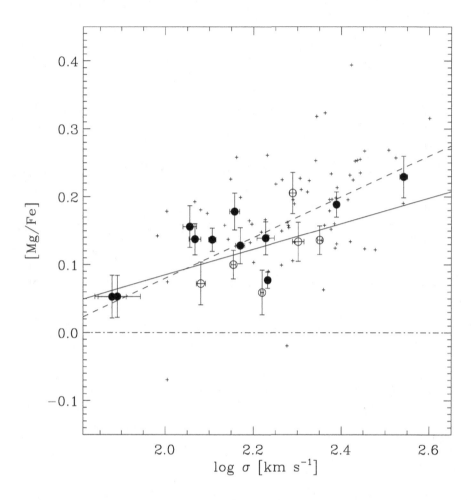

Fig. 14.6. The relationship between [Mg/Fe] and log σ for galaxies in the Fornax cluster (large symbols: filled circles are elliptical galaxies, open circles are S0 galaxies); the best fit to these data is the solid line. The sample compiled by Kuntschner et al. (2001), which is dominated by galaxies in the Coma and Virgo clusters, is shown as crosses; the best fit to these data is the dashed line. This figure is adapted from Figure 16 of Kuntschner et al. (2002).

tion (changing their mean age) they move along the relation rather increasing the scatter. I have therefore not included this relationship among those used to argue that luminous cluster ellipticals are old. Trager et al., however, agreed that the ellipticals in clusters are predominantly old and that it is the low-luminosity galaxies and galaxies in lower density regions that have experienced secondary star formation, which they referred to as a "frosting" of younger stars. Recently, Kuntschner et al. (2002) have carried out a similar analysis on a sample of ellipticals drawn from extreme low-density environments and shown that indeed they have younger luminosity-weighted ages than cluster galaxies.

One failing of the single stellar population models is that they are not able to account

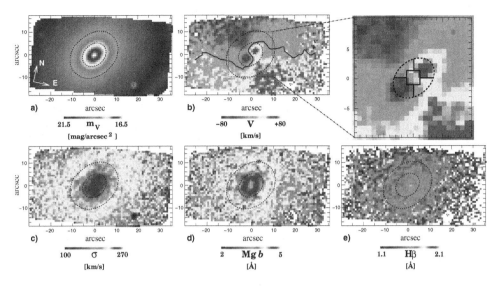

Fig. 14.7. Maps of (a) reconstructed surface brightness; (b) mean streaming velocity V, (c) velocity dispersion σ, (d) Mg b line strength, and (e) Hβ line strength for NGC 4365. Isophotal contours of $\mu_V = 18$ and 19 mag arcsec^{-2} are overplotted (dotted lines). A zero-velocity contour line (solid black line) is shown in (b). The enlarged core region of the velocity map indicates the regions used for line-strength analysis; see Figure 14.8. The central point is indicated by the square, the decoupled core by the sectors along the major axis, and the main body at the same radius by the sectors along the minor axis. (Taken from Davies et al. 2001.)

for the strength of the magnesium absorption features in metal-rich populations. This is why most of the above analysis is done using a combination of magnesium and iron lines. Over the last 15 years we have come to realize that the abundance patterns in early-type populations is not solar and that the magnesium-to-iron ratio is enhanced compared to solar values (Peletier 1989; Worthey, Faber, & Gonzáles 1992; Davies, Sadler, & Peletier 1993). A number of possible origins for this abundance pattern were considered, and the consensus view is that the overabundance arises from magnesium production in Type II supernovae that occurred in a burst of star formation lasting less than 1–2 Gyr, perhaps considerably less. The iron, which is produced primarily in Type Ia supernovae at later times, does not enrich the population because by then star formation has ceased, for example by the onset of a galactic wind. Trager et al. used models that include a parameterization of the overabundance [E/Fe], based on corrections to the Worthey models (Tripicco & Bell 1995), to estimate the level of enhanced abundances and concluded that elliptical galaxies fall on a plane in the four-space of velocity dispersion (log σ), age (log t), metallicity ([Z/H]), and "enhancement" ratio ([E/Fe]). They decomposed this plane into the "Z-plane" (Fig. 14.5) and the [E/Fe] vs. log σ plane. Figure 14.6, adapted from Kuntschner et al. (2002), shows [Mg/Fe] as a function of log σ for the Fornax galaxies and those compiled in Kuntschner et al. (2001), a sample dominated by galaxies in the Coma and Virgo clusters. This relationship shows that the high-velocity dispersion galaxies in clusters have a systematic overabundance of magnesium, which implies a short time scale for star formation in these systems (Bender 1996).

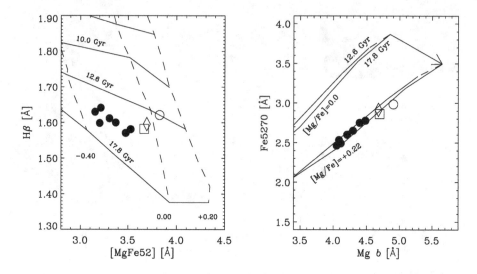

Fig. 14.8. *Left:* [MgFe52] vs. Hβ equivalent width diagram. The open circle represents the average line strength of the very central data points ($r < 1.''6$), the open diamond represents the region of the decoupled core, and the open square is the mean of the data in the main body of the galaxy at the same radii as the decoupled core but along the minor axis (see also Fig. 14.7, enlarged core region). For larger radii, the data were averaged in elliptical annuli centered on the photometric nucleus (filled circles) from $5.''6$ to $26.''1$. Overplotted are the predictions of stellar population models from Vazdekis (1999). The solid lines are lines of constant age, and the dashed lines are lines of constant metalicity. *Right:* Mg b vs. Fe5270 equivalent width diagram. Overplotted are stellar population models from Vazdekis (1999) with solar abundance ratios for ages of 12.6 and 17.8 Gyr, which are degenerate. In addition, a correction of the models for nonsolar abundance ratios, of [Mg/Fe] = 0.22 dex, is shown (Trager et al. 2000). The error bars on the mean line strength in a given zone are omitted for clarity, as they are similar to the size of the symbols in the diagrams. (Adapted from Davies et al. 2001.)

14.4   The Decoupled Core in NGC 4365

To address the issue of how and when galaxies assembled, we would ideally need to estimate the age of each stellar component of a galaxy to determine, for example, whether the core formed earlier or later than the bulk of the galaxy, or whether disks form stars over a longer period than bulges. To understand how galaxies assemble, we need to order the formation events. The results I have discussed so far arise from measurements of central or integrated values of color, dispersion, line strength, etc., and so are not able to address the formation epoch of different stellar components. Integral-field spectrographs, however, provide maps of these quantities at each point across the face of a galaxy. The SAURON spectrograph (Bacon et al. 2001) was built to carry out these studies and has been used in an extensive survey designed to characterize the local early-type galaxy population in terms of luminosity, environment, and morphology (de Zeeuw et al. 2002). The potential of such measurements is well illustrated by their maps of NGC 4365, an elliptical galaxy with a decoupled core (Davies et al. 2001). In Figure 14.7, we see that the velocity maps show two independent kinematic subsystems: the central 300 pc × 700 pc rotates about the projected minor axis, and the main body of the galaxy (3 kpc × 4 kpc) rotates almost at right angles to this. The misalignment between the photometric and kinematic axes of the main body is unambiguous evidence of triaxiality. Remarkably, the decoupled core does not show a signature in the line-strength maps. Figure 14.8*a* shows that the metallicity of the stellar population decreases from a central value greater than solar to one-half solar at a radius of 2 kpc. The Vazdekis (1999) models imply that the decoupled core and main body of the galaxy have the same luminosity-weighted age, 14 Gyr. In Figure 14.8*b* we see that the whole galaxy possesses the same elevated magnesium-to-iron ratio. The two kinematically distinct components have thus shared a common star formation history. We infer that the galaxy underwent a sequence of mergers associated with dissipative star formation that ended roughly 12 Gyr ago. The similarity of the stellar populations in the two components suggests that the observed kinematic structure has not changed substantially in 12 Gyr.

14.5 Summary

(1) Direct measurements of intracluster stars show that they account for 10%–20% of the total stellar luminosity of clusters.

(2) There is strong evidence that the population of luminous, high-dispersion galaxies in rich clusters is old and forms a metallicity sequence (the CMR and FP, both locally and to $z \approx 1$, the Hα-inferred star formation rate, the line-strength analysis of age and metallicity). Furthermore, the increasing enhancement of magnesium abundance in high-dispersion galaxies implies short star formation time scales in these galaxies. These observations are counter to the hierarchical picture of galaxy assembly in which the most massive objects assemble latest. They suggest that massive early-type galaxies in clusters formed by an alternative route; the short time scale for formation suggests a scheme similar to the monolithic collapse modeled by Larson (1975) and Carlberg (1984). One caveat is that the very central, brightest cluster galaxies frequently show emission lines, nuclear activity, and evidence for secondary star formation, which are associated with their special location at the center of the cluster potential. With this exception, luminous cluster ellipticals appear to be the quiescent remnants of an early starburst. This early phase has recently been invoked to account for the high star formation rates implied for the SCUBA sources (e.g., Smail et al. 2003).

(3) For lower luminosity galaxies ($\sigma < 150$ km s^{-1}) and galaxies found in the outer reaches (lower density environments) of clusters, the same evidence shows that secondary star formation has played an important role in their evolution. It seems likely that these galaxies have assembled hierarchically. The low star formation rates found at large cluster radius (low density) rule out physical processes that require high gas densities to operate, such as ram pressure stripping of disk gas, as being completely responsible for the variations in galaxy properties with environment.

(4) The availability of integral-field spectrographs that can map the kinematics and stellar populations of galaxy components offers the potential for directly measuring the assembly history of galaxies.

Acknowledgements. I would like to thank the PPARC for the award of a Senior Fellowship and the University of Oxford for research support. I am pleased to thank Harald Kuntschner and Marc Sarzi for assistance with the production of Figures 14.6 and 14.8. I am grateful to Carol Longbone and North Lincolnshire Libraries and Information Services who provided Figure 14.9.

Fig. 14.9. The "Ashby Free Library" in Ashby, North Lincolnshire. This photograph was taken shortly after it opened in 1904. The library was built using a £1500 donation from Andrew Carnegie. The author spent many childhood hours in this library, including those spent on his first astronomy project. Thanks to Carol Longbone of North Lincolnshire Libraries for providing the photograph.

References

Arnaboldi, M., et al. 2003 AJ, 125, 514
Bacon, R., et al. 2001 MNRAS, 326, 23
Bender, R. 1996, in IAU Symp. 171, New Light on Galaxy Evolution, ed. R. L. Davies & R. Bender (Dordrecht:

Kluwer), 181

Bender, R., Burstein, D., & Faber, S. M. 1992, ApJ, 399, 462

Binggeli, B., Sandage, A. R., & Tammann, G. A. 1987, AJ, 94, 251

Bower, R. G., Lucey, J. R., & Ellis, R. S. 1992, MNRAS, 254, 589

Bruzual A., G. 1983, ApJ, 273, 105

Carlberg, R. G. 1984, ApJ, 286, 403

Ciotti, L., & Renzini, A. 1993, ApJ, 416, L49

Colless, M., Burstein, D., Davies, R. L., McMahan, R. K., Jr., Saglia, R. P., & Wegner, G. 1999, MNRAS, 303, 813

Colless, M., Saglia, R. P., Burstein, D., Davies, R. L., McMahan, R. K., Jr., & Wegner, G. 2001 MNRAS, 321, 277

Conselice, C. J., Gallagher, J. S., III, & Wyse, R. F. G. 2003, AJ, 125, 66

Davies, R. L., et al. 2001, ApJ, 548, L33

Davies, R. L., Sadler, E. M., & Peletier, R. F. 1993, MNRAS, 262, 650

de Zeeuw, P. T., et al. 2002 MNRAS, 329, 513

Djorgovski, S. G., & Davis, M. 1987, ApJ, 313, 59

Dressler, A. 1980, ApJ, 236, 351

Dressler, A., Gunn, J. E., & Schneider, D. P. 1985, ApJ, 294, 70

Dressler, A., Lynden-Bell, D., Burstein, D., Davies, R. L., Faber, S. M., Terlevich, R., & Wegner, G. 1987, ApJ, 313, 42

Durrell, P. R., Ciardullo, R., Feldmeier, J. J., Jacoby, G. H., & Sigurdsson, S. 2002, ApJ, 570, 119

Ellis, R. S., Smail, I., Dressler, A., Couch, W. J., Oemler, A., Jr., Butcher, H., & Sharples, R. M. 1997, ApJ, 483, 582

Faber, S. M., Wegner, G., Burstein, D., Davies, R. L., Dressler, A., Lynden-Bell, D., & Terlevich, R. J. 1989, ApJS, 69, 763

Feldmeier, J. J., Mihos, J. C., Morrison, H. L., Harding, P., & Kaib, N. 2004, in Carnegie Observatories Astrophysics Series, Vol. 3: Clusters of Galaxies: Probes of Cosmological Structure and Galaxy Evolution, ed. J. S. Mulchaey, A. Dressler, & A. Oemler (Pasadena: Carnegie Observatories, http://www.ociw.edu/ociw/symposia/series/symposium3/proceedings.html)

Feldmeier, J. J., Mihos, J. C., Morrison, H. L., Rodney, S. A., & Harding, P. 2002, ApJ, 575, 779

Ferguson, A. M. N., Tanvir, N. R., & von Hippel, T. 1998, Nature, 391, 461

Franx, M., van Dokkum, P. G., Kelson, D., Fabricant D. G., & Illingworth, G. D. 2000, Philosophical Transactions: Mathematical, Physical & Engineering Sciences, 358, 2109

Gladders, M. D., Lopez-Cruz, O., Yee, H. K. C., & Kodama, T. 1998, ApJ, 501, 571

Gomez, P. L., et al. 2003, ApJ, 584, 210

González, J. J. 1993, Ph.D. Thesis, University of California, Santa Cruz

Harris, W. E., Kavelaars, J. J., Hanes, D. A., Hesser, J. E., & Pritchet, C. J. 2000, ApJ, 533, 137

Hudson, M. J., Lucey, J. R., Smith, R. J., Schlegel, D. J., & Davies, R. L. 2001, MNRAS, 327, 265

Jørgensen, I., Franx, M., Hjorth, J., & van Dokkum, P. G. 1999, MNRAS, 308, 833

Kelson, D. D., Illingworth, G. D., Franx, M., & van Dokkum, P. G. 2001, ApJ, 552, L17

Kodama, T., & Arimoto, N. 1997, A&A, 320, 41

Kodama, T., Arimoto, N., Barger, A. J., & Aragón-Salamanca, A. 1998, A&A, 334, 99

Kuntschner, H., & Davies, R. L. 1998, MNRAS, 295, L29

Kuntschner, H., Lucey, J. R., Smith, R. J., Hudson, M. J., & Davies, R. L. 2001, MNRAS, 323, 615

Kuntschner, H., Smith, R. J., Colless, M., Davies, R. L., Kaldare, R., & Vazdekis, A. 2002, MNRAS, 337, 172

Larson, R. B. 1975, MNRAS, 173, 671

Lewis, I. J., et al. 2002, MNRAS, 334, 673

Moore, B., Katz, N., Lake, G., Dressler, A., & Oemler, A. 1996, Nature, 379, 613

Peletier, R. F. 1989, Ph.D. Thesis, University of Groningen

Pimbblet, K. A., Smail, I., Kodama, T., Couch, W. J., Edge, A. C., Zabludoff, A. I., & O'Hely, E. 2002, MNRAS, 331, 333

Sandage, A. 1972, ApJ, 176, 21

Schneider, D. P., Gunn, J. E., & Hoessel, J. G. 1983, ApJ, 264, 337

Shectman, S. 1973, ApJ, 179, 681

———. 1974, ApJ, 188, 233

Smail, I., Ivison, R. J., Gilbank, D. G., Dunlop, J. S., Keel, W. C., Motohara, K., & Stevens, J. A. 2003, ApJ, 583, 551

Terlevich, A. I., Caldwell, N., & Bower, R. G. 2001, MNRAS, 326, 1547

Terlevich, A. I., Kuntschner, H., Bower, R. G., Caldwell, N., & Sharples, R. M. 1999, MNRAS, 310, 445

Theuns, T., & Warren, S. J. 1997, MNRAS, 284, L11

Thomas, D., Maraston, C., & Bender, R. 2003, MNRAS, 339, 897

Trager, S. C., Faber, S. M., Worthey, G., & González, J. J. 2000, AJ, 120, 165

Trager, S. C., Worthey, G., Faber, S. M., Burstein, D., & González, J. J. 1998, ApJS, 116, 1

Tripicco, M. J., & Bell, R. A. 1995, AJ, 110, 3035

Uson, J. M., Boughn, S. P., & Kuhn, J. R. 1991, ApJ, 369, 46

Visvanathan, N., & Sandage, A. 1977, ApJ, 216, 214

Whitmore, B. C., Gilmore, D. M., & Jones, C. 1993, ApJ, 407, 489

Worthey, G. 1994, ApJS, 95, 107

Worthey, G., Faber, S. M., & González, J. J. 1992, ApJ, 398, 69

Worthey, G., & Ottaviani, D. L. 1997, ApJS, 111, 377

van Dokkum, P. G., Franx, M., Fabricant, D., Illingworth, G. D., & Kelson, D. D. 2000, ApJ, 541, 95

Vazdekis, A. 1999, ApJ, 513, 224

Vazdekis, A., Kuntschner, H., Davies, R. L., Arimoto, N., Nakamura, O., & Peletier, R. F. 2001, ApJ, 551, L127

15

Modeling stellar populations in cluster galaxies

BIANCA M. POGGIANTI

Astronomical Observatory, Padova, Italy

Abstract

In this review I highlight the role played by spectrophotometric models in the study of galaxy evolution in distant clusters. I summarize the main achievements of the modeling of k+a spectra, the derivation of the star formation rate in emission-line galaxies, and the stellar ages of red cluster galaxies. The current knowledge of the dependence of the star formation histories on the galaxy morphology and luminosity is also presented.

15.1 Introduction

The goal of any spectrophotometric modeling is to reconstruct the star formation history of galaxies, to learn what has been the star formation rate (SFR) at each epoch and, as a consequence, the enrichment history of metals and the evolution of the galactic luminosity and mass. In this review I focus on galaxies in mostly distant clusters ($z > 0.2$), and present the main results obtained using spectrophotometric modeling as a tool to interpret the data.

Quite a few of the spectrophotometric models available today have the ability to include the three main components that produce the spectral energy distribution of a galaxy: (1) the stellar contribution; (2) the emission lines and emission continuum from gas ionized by young massive stars, and (3) the extinction due to dust. Nowadays it is common to include stellar populations of different metallicities, and models can have sufficient wavelength resolution ($3-4$ Å or better) to study stellar absorption features in detail.

In this context it is useful to think of three types of spectra: those *with emission lines* and, among those without emission lines, *k+a* and *passive* spectra. For each one of these types of spectra there is a stellar time scale associated: $< 5 \times 10^7$ yr for emission-line spectra (the lifetime of the massive stars able to ionize the gas); between 5×10^7 and 1.5×10^9 yr for k+a spectra (the period during which stars with strong Balmer lines dominate the integrated spectrum of a stellar system); and $> 1.5 \times 10^9$ yr for passive spectra. Each one of these spectral types is described and discussed below.

Whenever my own work will be mentioned in the following, credit should be given to my collaborators of the MORPHS group (H. Butcher, W. Couch, A. Dressler, R. Ellis, A. Oemler, R. Sharples, I. Smail) for distant cluster work, and of the Coma collaboration (T. Bridges, D. Carter, B. Mobasher, Y. Komiyama, S. Okamura, N. Kashikawa, et al.) for results concerning the Coma cluster.

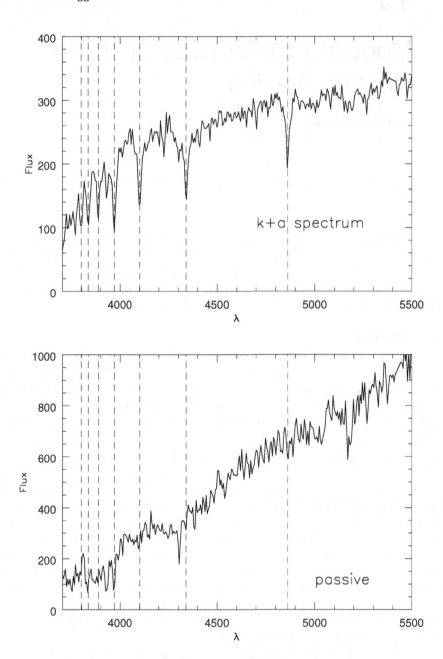

Fig. 15.1. Example of a k+a spectrum (*top*) and of a passive spectrum (*bottom*). The lines of the Balmer series (from left to right Hθ, Hη, Hζ, Hε, Hδ, Hγ, Hβ) are highlighted by dashed lines.

15.2   k+a Spectra

Historically, k+a spectra (originally named "E+A's", also known as Hδ-strong galaxies, etc.) were the first type of spectra to be noticed and modeled in distant cluster studies.

They are spectra with no emission lines and strong Balmer lines in absorption [EW(Hδ) > 3 Å]. A strong-lined example of a k+a galaxy is shown in Figure 15.1 and contrasted with a passive spectrum.

The second paper that appeared presenting spectra of distant cluster galaxies already pointed out what is still the current interpretation of (the strongest) k+a spectra, that they "indicate a large burst of star formation 10^9 years before the light left the galaxy" (Dressler & Gunn 1983). A large number of successive papers have discussed k+a spectra in distant clusters (e.g., Henry & Lavery 1987; Fabricant, McClintock, & Bautz 1991; Fabricant, Bautz, & McClintock 1994; Dressler & Gunn 1992; Belloni et al. 1995; Belloni & Roeser 1996; Fisher et al. 1998; Couch et al. 1998; Balogh et al. 1999; Dressler et al. 1999; Poggianti et al. 1999; Bartholomew et al. 2001; Ellingson et al. 2001; Tran et al. 2004); those that have performed spectrophotometric modeling to interpret them include Dressler & Gunn (1983), Couch & Sharples (1987), Newberry, Boroson, & Kirshner (1990), Charlot & Silk (1994), Jablonka & Alloin (1995), Abraham et al. (1996), Barger et al. (1996), Poggianti & Barbaro (1996, 1997), Morris et al. (1998), Bekki, Shioya, & Couch (2001), Shioya, Bekki, & Couch (2001), and Shioya et al. (2002). The main conclusions of k+a modeling can be summarized as follows:

(1) In k+a galaxies, star formation stopped typically between 5×10^7 and 1.5×10^9 yr before the epoch at which the galaxy is observed. The evolution of the equivalent width of Hδ versus color is shown in Figure 15.2. This diagram reproduces a plot from the influential paper of Couch & Sharples (1987). Soon after the moment when star formation is interrupted, the galaxy moves in this diagram toward higher EW(Hδ) and progressively redder colors, reaching a maximum in EW(Hδ) after $(3-5) \times 10^8$ yr. Then, the strength of the line starts to decline until the locus occupied by typical ellipticals is reached, in the right bottom corner of the plot.

(2) Spectra with very strong EW(Hδ) (above 5 Å) require a starburst prior to the truncation of the star formation, with high galactic mass fractions involved in the burst (10%–20% or higher). k+a spectra with a moderate Hδ line, on the other hand, may be post-starburst galaxies in a late stage of evolution, but may also be reproduced by simply interrupting "normal" star formation activity. The 5 Å limit mentioned above should only be considered indicative: given the strong dependence of the EW value on the method adopted to measure it, the strength of the lines should always be measured in the same way on data and models.

(3) A combination of EW(Hδ) and color, such as that shown in Figure 15.2, can help in roughly estimating the time elapsed since the halting of star formation. A more sophisticated method to age date the starburst using Balmer-line indices can be found in Leonardi & Rose (1996).

(4) A slowly declining SFR, with a time scale comparable to, or longer than, that of the k+a phase (1 Gyr or longer), is not able to produce a k+a spectrum, which is more easily obtained if the star formation is halted quickly. The reason for such behavior is that while the SFR is slowly declining emission lines are still present in the galaxy spectrum, thus the galaxy will not be classified as a k+a. When finally the star formation terminates, the contribution of A-type stars to the integrated spectrum is low, due to the low average level of star formation during the previous Gyr.

A great advantage of the k+a spectral classification is the fact that, in principle, it requires a single measurement (the Hδ line), though this must be of sufficiently high signal-to-noise ratio and must be coupled with a complete inspection of the spectrum to verify the absence of any emission line. Analysis of the other higher-order lines of the Balmer series usually

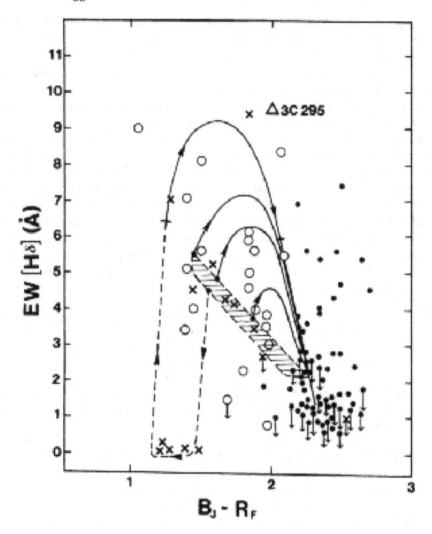

Fig. 15.2. Color versus rest-frame equivalent width of Hδ of galaxies in three clusters at $z = 0.3$. Emission-line galaxies are indicated by crosses and non-emission-line galaxies by filled (red galaxies) and empty (blue galaxies) circles. The dashed region shows the locus of typical spirals today, which generally display emission lines and therefore are comparable to the crosses in this diagram. Ellipticals occupy the bottom right region of this diagram, at $B - R > 2$ mag and EW(Hδ) < 1 Å. The lines are evolutionary sequences of galaxies in which star formation is halted with or without a starburst preceding the quenching of star formation. The former case is shown as a dotted+continuous line leading downward from the spiral sequence, while examples of truncated models with no starbursts are the solid tracks leading upward from the spiral sequence. (From Couch & Sharples 1987.)

greatly helps in confirming the strength of Hδ. Moreover, modeling of k+a spectra is relatively simple: the age-metallicity degeneracy that affects the interpretation of passive spectra is not an issue here, because the effect in the strength of the Balmer lines is so large that only age can be the cause for it. The strong Balmer lines arise in A- to F-type stars, which are stars of ~ 2 solar masses on the main sequence, hence in a well known and easily modeled evolutionary phase.

The importance of k+a modeling depends obviously on the importance of the k+a phenomenon in cluster galaxy data. Many works (see list above) have found distant clusters to host a significant number of k+a galaxies. If k+a spectra occur proportionally more in distant clusters than in the field at similar redshifts (as found, e.g., by Dressler et al. 1999; but see Balogh et al. 1999 for a different view), then they provide strong evidence that star formation is rapidly quenched in clusters, and they could be used as tracers both of the star formation history of galaxies infalling into clusters and of the infall history itself. The strongest k+a spectra are a solid signature of a recent starburst; whether this burst was or was not induced by the cluster environment is still an open issue and a subject of ongoing investigation.

Finally, it is not a mere intellectual exercise to note that if a galaxy formed all of its stars on a short time scale, as ellipticals are expected to do in a monolithic-collapse scenario of galaxy formation, then there must have been a phase/epoch when it displayed a spectacular k+a spectrum of an almost pure single stellar population dominated by A stars. In an $\Omega_\Lambda = 0.7$, $H_0 = 70$ km s^{-1} Mpc^{-1} cosmology, a galaxy forming all stars within 0.5 Gyr between $z = 3$ and $z = 2.5$ would appear as a k+a until $z = 1.5$. If it stopped forming stars at $z = 2$, it would still be recognizable as a k+a at $z = 1.3$. The k+a phase of massive ellipticals might be within reach, and possibly we have begun observing it (van Dokkum & Stanford 2003).

15.3   Emission-line Spectra and Dust

The luminosities of some emission lines in an integrated spectrum can be used to infer the current SFR in a galaxy because, as schematically described below for Hα, the luminosity of the line is proportional to the number of ionizing photons. In a star-forming galaxy the number of ionizing photons is proportional to the number of massive young stars and, for a given stellar initial mass function, to the total current SFR:

$$L_{\mathrm{H}\alpha} \propto N_{ion-photons} \propto N_{massive-stars} \propto SFR. \tag{15.1}$$

In distant galaxy studies, the [O II] $\lambda 3727$ line is often used instead of Hα because the latter is redshifted out of the optical window. Spectrophotometric models are used to determine the proportionality coefficient between $N_{ion-photons}$ and $N_{massive-stars}$, and they provide the standard relations between the SFR and the luminosity of the lines (Kennicutt 1992, 1998):

$$SFR = 0.9 \times 10^{-41} L_{\mathrm{H}\alpha} E_{\mathrm{H}\alpha} \ M_\odot \, \mathrm{yr}^{-1}, \tag{15.2}$$

$$SFR = 2.0 \times 10^{-41} L_{[\mathrm{O\ II}]} E_{\mathrm{H}\alpha} \ M_\odot \, \mathrm{yr}^{-1}, \tag{15.3}$$

where the line luminosity is in erg s^{-1} and $E_{\mathrm{H}\alpha}$ is the extinction correction factor at Hα. A major uncertainty in these estimates is, therefore, the extinction by dust.

Modeling has focused on starburst and dusty galaxies, while little theoretical work has been done on galaxies forming stars in a continuous, regular fashion, which are another sizable component present in distant clusters. In the following I will not treat the quiescent star-forming galaxies, but will summarize the modeling results for starburst and dusty galaxies.

Several lines of evidence suggest that dust can play an important role in some distant cluster galaxies. In the optical, spectra with weak to moderate [O II] emission and unusually strong, higher-order Balmer lines were noted to represent a nonnegligible fraction of both cluster and field spectra (Dressler et al. 1999). It was realized that in the local Universe such spectra are rare among normal spirals, while they are common in infrared-luminous starburst galaxies, which are known to have large dust extinction (Poggianti et al. 1999; Poggianti & Wu 2000; see also Liu & Kennicutt 1995). Hence, regardless of the weakness of their emission lines, these galaxies were suggested to be in a starburst phase in which the dust extinction works selectively: the youngest stellar generations are more affected by dust obscuration than older stellar populations, which have had time to free themselves or drift away from their dust cocoons.

Models with selective extinction are empirically motivated by observations of star-forming regions in nearby galaxies, and they explain why different values of $E(B-V)$ are usually measured within the same spectrum when using different spectral features, for example why the extinction measured from the emission lines is usually stronger than that measured from the continuum. Selective extinction is consistent with the fact that in the nearby Universe galaxies with the highest current SFR are generally *not* those with the strongest emission lines, which tend to be low-mass, very late-type galaxies of moderate total SFR.

Quantitative modeling of these dusty spectra have been done by Shioya & Bekki (2000), Shioya et al. 2001, Bekki et al. (2001), and Poggianti, Bressan, & Franceschini (2001c). While a dust-screen model, with any extinction law, preserves the equivalent width of the lines (because it affects both the line and the underlying continuum by a proportional amount), an age-selective extinction can produce this peculiar spectral combination with weak emission lines (originating in regions with highly extincted, young massive stars) and exceptionally strong, higher-order Balmer absorption lines (from exposed intermediate-age stars).

While a spectrum with moderate emission and unusually strong Balmer lines in absorption is therefore a good candidate for a starburst galaxy, it has been shown that this modeling is highly degenerate, that this peculiar type of spectra does not guarantee SFRs above a certain limit, and that the total SFR remains unknown without dust-free SFR estimators such as far-infrared or radio continuum fluxes. Radio continuum observations of a cluster at $z = 0.4$, for example, surprisingly have detected some of the strongest (i.e. youngest) k+a galaxies in the cluster (Smail et al. 1999), suggesting that star formation might still be ongoing and be totally obscured at [O II] by dust in some of the k+a's (see also the Hα detection in some k+a's by Balogh & Morris 2000 and Miller & Owen 2002). Mid-infrared 15 μm data obtained with ISOCAM have been used to estimate the far-infrared flux and, therefore, the SFR in distant clusters (Duc et al. 2002). As described in Duc et al. (2004), ISOCAM observations of a cluster at $z = 0.55$ have yielded an unexpectedly high number of cluster members that are luminous infrared galaxies with exceptionally high SFRs.

15.4   Passive Galaxies and Evolutionary Links

A large fraction of galaxies in clusters up to $z = 1$ have passive (non-k+a, non-emission-line) spectra. The evolutionary histories of the passive galaxies are the subject of other reviews and contributed talks in these proceedings (see Franx 2004 and Treu 2004); thus, I will limit the following discussion to a description of the most fundamental contributions given to this area of research by spectrophotometric models.

The ages of the stellar populations in luminous, early-type galaxies in clusters are known to be old. Solid evidence for this comes from the analysis and evolution of the red color-magnitude (CM) sequence in clusters, whose slope, scatter, and zeropoint indicate a passive evolution of stars formed at $z > 2-3$ (Bower, Lucey, & Ellis 1992; Aragón-Salamanca, Ellis, & Sharples 1993; Rakos & Schombert 1995; Stanford, Eisenhardt, & Dickinson 1995, 1998; Stanford et al. 1997; Schade et al. 1996; Ellis et al. 1997; Schade, Barrientos, & Lopez-Cruz 1997; Barger et al. 1998; Gladders et al. 1998; Kodama et al. 1998; van Dokkum et al. 1998, 1999, 2000; De Propris et al. 1999; Terlevich et al. 1999; Terlevich, Caldwell, & Bower 2001; van Dokkum & Franx 2001; Vazdekis et al. 2001). Studies of the fundamental plane, mass-to-light ratios, and the magnesium-velocity dispersion relation agree with these findings (Bender, Ziegler, & Bruzual 1996; van Dokkum & Franx 1996; Kelson et al. 1997, 2000, 2001; Ziegler & Bender 1997; Bender et al. 1998; van Dokkum et al. 1998; Ziegler et al. 2001), though they are necessarily limited to the brightest subset of galaxies.

How can these results be reconciled with the presence of numerous blue galaxies in distant clusters (the Butcher-Oemler effect), given that these blue, star-forming galaxies have largely "disappeared" (i.e. become red) by $z = 0$? This has been investigated by works that have modeled the evolution of galaxy colors and magnitudes (Bower, Kodama, & Terlevich 1998; Smail et al. 1998; Kodama & Bower 2001). Figure 15.3 presents the results of Kodama & Bower (2001), who have shown that by evolving the observed CM diagram of intermediate-redshift clusters it is possible to obtain a diagram at $z = 0$ that is mostly composed of red galaxies, similar to the observed CM diagram of the Coma cluster. Moreover, it is now clear that the CM red sequence of clusters is comprised not only of early-type galaxies, but also contains morphologically classified spirals (Poggianti et al. 1999; Couch et al. 2001; Terlevich et al. 2001; Balogh et al. 2002; Goto et al. 2003, 2004).

These results imply that the CM sequence today is composed of a varied population of galaxies with different star formation histories. As shown by several works, including the one by Kodama & Bower (2001), although the principal driver of the CM sequence is the correlation between galaxy luminosity and metallicity, there is still room for a relatively recent epoch of star formation activity in a significant fraction of the (today) red galaxies. When contrasting this with the homogeneity and old ages of red sequence galaxies derived from CM and fundamental plane studies, it is important to take into account two aspects: the morphological transformations in clusters and the evolution of the galaxy luminosities.

15.4.1 Morphological Evolution

The morphological mix of galaxies in clusters at various redshifts strongly suggests that a significant fraction of the spirals in distant clusters have evolved into the S0s or, more generally, the early-type galaxies that dominate clusters today (Dressler et al. 1997; Fasano et al. 2000; van Dokkum et al. 2000; Lubin, Oke, & Postman 2002; see Dressler 2004). This morphological evolution is likely to lead to a "progenitor bias": in distant clusters we would be observing as early-type galaxies only the "oldest" subset of the present-day early-types, those that were already assembled and stopped forming stars at high redshift (van Dokkum & Franx 1996, 2001; Stanford et al. 1998).

The signature of relatively recent star formation activity in some of the early-type galaxies in clusters has been searched for in several ways. Recently, significant differences between the ages of the stellar populations of ellipticals and a fraction of the S0 galaxies have been detected, supporting the scenario of spirals evolving into S0s (Kuntschner & Davies 1998; van Dokkum et al. 1998; Terlevich et al. 1999; Poggianti et al. 2001b; Smail et al. 2001; Thomas 2002). Many of these works derive luminosity-weighted ages and metallicities by comparing a metallicity-sensitive and an age-sensitive spectral index with a grid of spec-

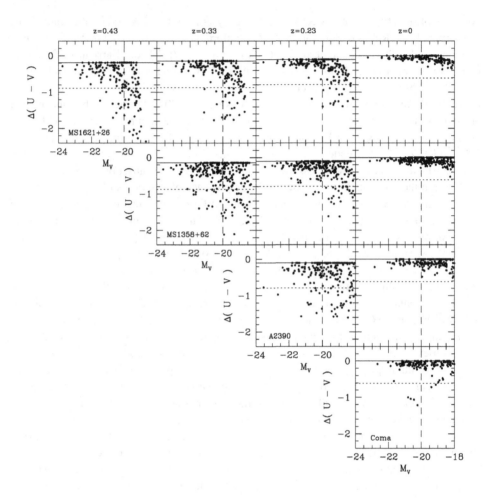

Fig. 15.3. Evolution of the CM diagram. From left to right, redshift is decreasing as shown on the top of the plots. The leftmost panels are observed CM diagrams of clusters, while the subsequent evolution are examples of Monte-Carlo simulations in which star formation is truncated after accretion of galaxies from the field. The CM diagram of Coma is shown for comparison in the bottom right panel. (From Kodama & Bower 2001.)

trophotometric models. Similarly, an age-sensitive and a metallicity-sensitive *color index* can be used, as shown in Figure 15.4 (Smail et al. 2001). The plot shows the color-color diagram of galaxies in Abell 2218 at $z = 0.17$, where a group of faint S0 galaxies is seen to lie at younger luminosity-weighted ages (lower $V - I$ color) than all the ellipticals and the rest of the S0s. The possibility of a spiral-to-S0 transformation has been quantitatively investigated in terms of galaxy numbers and morphologies by Kodama & Smail (2001) and from the spectrophotometric point of view by Bicker, Fritz-v. Alvensleben, & Fricke (2002), providing interesting constraints on the possible evolutionary paths.

However, not all studies find differences between S0s and ellipticals. Ellis et al. (1997), Jørgensen (1999, and references therein), Jones, Smail, & Couch (2000), and Ziegler et al.

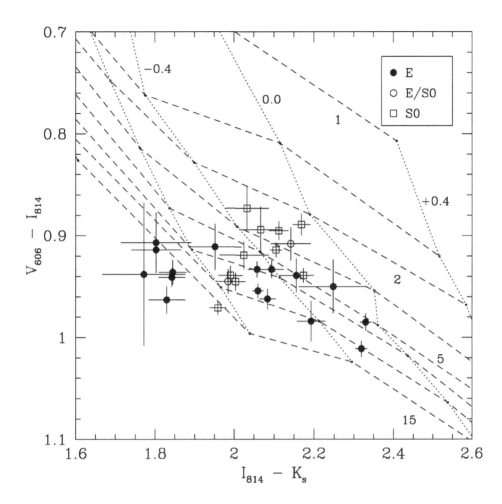

Fig. 15.4. Color-color diagram of galaxies in Abell 2218. The grid of lines shows models of single stellar populations of constant age (pseudo-horizontal lines) and constant metallicity (pseudo-vertical lines). As shown in the legend of the plot, the group of galaxies with younger ages (lower $V - I$ colors) than the rest are all S0 galaxies. (From Smail et al. 2001.)

(2001) do not detect a significant morphological dependence of the stellar population ages and metallicities in early-type cluster galaxies. It is important to stress that the presence of recent star formation in S0s has never been detected in the *brightest* S0s but becomes a prominent effect when probing fainter down the luminosity function. For example, Coma S0s with recent star formation are fainter than $M^* + 1.3$, as expected given the luminosities of their possible spiral progenitors at intermediate redshift (Poggianti et al. 2001b). When studying the dependence of the stellar population properties on the morphological type of galaxies, it is therefore important to simultaneously disentangle the dependence on the galaxy luminosity and to consider the effects of luminosity evolution. This is further discussed in the following section.

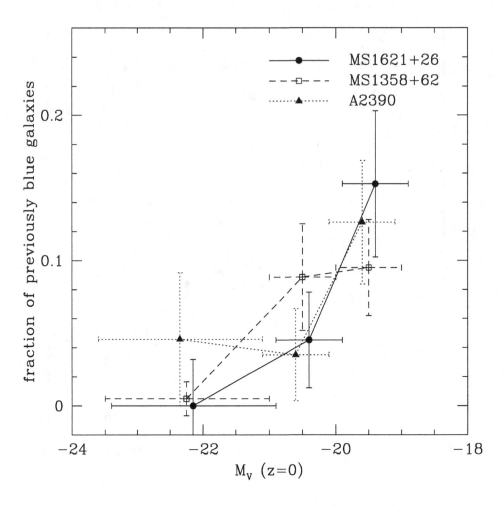

Fig. 15.5. Fraction of galaxies observed as blue at intermediate redshifts (three clusters at $z = 0.2$, 0.3, and 0.4) as a function of the present-day ($z = 0$) absolute V magnitude. The $z = 0$ magnitude is computed assuming a decline of star formation as a consequence of the accretion onto the cluster. (From Kodama & Bower 2001.)

15.4.2   *Luminosity Evolution*

There are two aspects of the evolution of galaxy luminosities that are important to stress. The first aspect, already mentioned above, is the *fading*. As discussed in §15.2, the interruption of star formation seems to be a phenomenon involving a significant fraction of the cluster galaxies. As a consequence, galaxies not only become redder but also fainter: this evolution in luminosity is obviously stronger than the evolution occurring both in the case of passive evolution of an old population and in the case of a galaxy continuously forming stars with no large variation in SFR until $z = 0$. Kodama & Bower (2001) computed the expected fraction of previously blue galaxies as a function of the present-day absolute magnitude, as shown in Figure 15.5, starting from the CM diagram of three intermediate-redshift clusters and computing the luminosity evolution using models with truncated star formation. The figure shows how the fraction of present-day red galaxies that were blue at higher redshift increases progressively as one goes fainter. The relevance of the fading and the evolution as a function of galaxy luminosity are discussed also in many other works (Postman, Lubin, & Oke 1998; Smail et al. 1998, 2001; van Dokkum et al. 1998; Terlevich et al. 1999; Ferreras & Silk 2000; Poggianti et al. 2001b; Shioya et al. 2002; Merluzzi et al. 2003; Tran et al. 2004).

Related to this, there is a second interesting point related to luminosity, known as the "downsizing effect": the fact that, as one goes to lower redshifts, the maximum luminosity/mass of galaxies with significant star formation activity seems to be progressively decreasing, possibly both in clusters (Bower et al. 1999) and in the field (Cowie et al. 1996).

An attempt to derive the distribution of luminosity-weighted ages as a function of galaxy luminosity over a range of almost 7 magnitudes has been done for Coma in a new magnitude limited spectroscopic survey (Poggianti et al. 2001a). In this study spectrophotometric models were also used to derive luminosity-weighted ages for spectra without emission lines (the great majority in Coma) from index-index diagrams. These ages give a rough estimate of the time elapsed since the latest star formation activity stopped. This time has been transformed into a cosmological epoch (i.e. redshift) adopting an $\Omega_\Lambda = 0.7$, $H_0 = 70$ km s^{-1} Mpc^{-1} cosmology. About half of all non-emission-line galaxies of any magnitude in Coma show no detectable star formation activity during the last 9 Gyr, i.e. since $z \approx 1.5$. The other half *do* show signs of some star formation below this redshift and, interestingly, display a trend of epoch of latest star formation with galaxy magnitude: recent ($z < 0.25$) star formation is detected in 25%–30% of the dwarf galaxies, while only in 5%–10% of the giant galaxies. In contrast, 30%–40% of the giants reveal some star formation at intermediate redshifts ($0.25 < z < 1.5$), compared to 15%–20% for the dwarfs.

Overall, including all galaxies with M_B between -21 and -14 mag, the fraction of galaxies with signs of star formation at the various epochs is presented in Figure 15.6 (left). Going a step forward, one can try to quantify the stellar mass involved, transforming the observed luminosities into a mass by adopting the mass-to-light ratio of a single stellar population with an age equal to the luminosity-weighted age derived from the spectral indices. As shown in the right panel of Figure 15.6, galaxies with some star formation below $z = 1$ account for only about 20% of the mass, and those with some star formation at $z < 0.25$ for only 3% of the mass, being mostly low-mass dwarf galaxies. Note that this is *not* the mass fraction formed at each epoch, but it is the *total* mass in those galaxies that had some (unquantified) star formation. An example of an attempt to estimate the total mass fraction formed at each redshift using population synthesis models of cluster galaxy colors and magnitudes can be found in Merluzzi et al. (2003), who find, for quiescent, non-starbursting star formation histories, that about 20% of the mass was formed below redshift 1.

Other evidence of a downsizing effect comes from k+a galaxies in clusters at low redshift. A number of works have investigated the presence of Balmer-strong absorption-line galaxies in nearby clusters, especially in Coma (Caldwell et al. 1993, 1996; Caldwell & Rose 1997, 1998; Caldwell, Rose, & Dendy 1999; Castander et al. 2001; Rose et al. 2001). Some of these works have highlighted the tendency of these Balmer-strong galaxies to be intrinsically fainter than the k+a galaxies observed at high redshift. In our Coma survey (Mobasher et al. 2001; Poggianti et al. 2001a) we have found no luminous k+a galaxy down to a magnitude-limit comparable to the MORPHS limit at $z = 0.5$, while there are strong and frequent cases

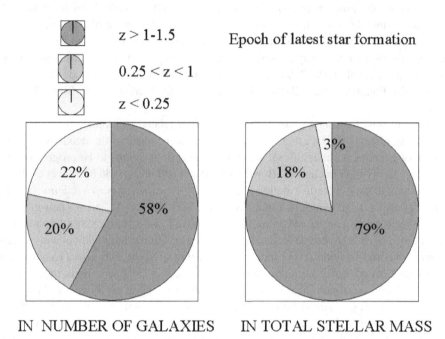

Fig. 15.6. *Left*: Fractions of non-emission-line Coma galaxies with signs of star formation in three cosmological intervals. These have been computed from the luminosity-weighted ages derived from spectral index diagrams in Poggianti et al. (2001a). *Right*: The number of galaxies have been transformed into fractions of stellar mass in galaxies with signs of star formation during the three epochs (see text).

among faint Coma galaxies at $M_V > -18.5$ mag. At these magnitudes 10% to 15% of the cluster galaxies are k+a's, and many of these dwarf spectra are exceptionally strong, post-starburst galaxies (Poggianti et al. 2004).

15.5 Summary

In this contribution I have summarized the main achievements of spectrophotometric modeling of galaxy evolution in distant clusters. Five main points have been mentioned:

(1) k+a spectra unequivocally indicate post-star-forming or post-starburst galaxies, depending on the strength of the Balmer lines.
(2) A galaxy with weak to moderate emission lines *and* strong Balmer lines in absorption is a good candidate for an ongoing starburst, or a galaxy currently forming stars at a vigorous rate.
(3) There are observed differences in the stellar population ages between ellipticals and S0 galaxies that are consistent with a significant fraction of the latter having evolved from star-forming spirals at intermediate redshifts. This result is not universal, though, and probing sufficiently faint magnitudes seems to be crucial for detecting any difference, at least at low redshift.

(4) The CM sequence in clusters today appears to be a mixed bag. The slope is dominated by metallicity, but there is room for relatively recent star formation in a significant fraction of the galaxies. The bright end of the CM sequence seems to be populated by homogeneously old galaxies, while more and more remnants of galaxies with star formation at intermediate redshifts ($z \approx 0.5$) can be found as one goes fainter along the sequence.

(5) Related to the previous point, a "downsizing effect" is observed in clusters, probably in parallel to a similar effect in the field: toward lower redshifts, the maximum luminosity/mass of galaxies with significant star formation activity progressively decreases. The downsizing effect could be responsible of the fact that k+a cluster spectra appear to occur in luminous galaxies at $z = 0.5$ and in faint (mostly dwarf) galaxies at $z = 0$.

In the next few years the efforts of spectrophotometric modeling will likely concentrate on modeling at higher redshifts than it has been done so far, in order to aid the interpretation of the large photometric and spectroscopic data sets that are being acquired in clusters at $0.5 < z < 1.5$. We can expect further progress also at low redshift, where the model improvements, especially regarding the metal abundance ratios, coupled with the new large surveys of nearby clusters and groups, are expected to reveal detailed characteristics of cluster galaxies that can be helpful in constraining their formation histories. As more and better data are obtained, the most important contribution of the modeling will be the endeavor to reach a quantitative general picture that can simultaneously account for the observations at all redshifts.

Acknowledgments. The author wishes to thank the organizers for a very pleasant and interesting Symposium, A. Oemler for reviewing this paper and suggesting numerous improvements to the text, and L. Ho for careful and patient copy-editing.

References

Abraham, R. G., et al. 1996, ApJ, 471, 694
Aragón-Salamanca, A., Ellis, R. S., & Sharples, R. M. 1991, MNRAS, 248, 128
Balogh, M. L., Bower, R. G., Smail, I., Ziegler, B. L., Davies, R. L., Gaztelu, A., & Fritz, A. 2002, MNRAS, 337, 256
Balogh, M. L.,& Morris, S. L. 2000, MNRAS, 318, 703
Balogh, M. L., Morris, S. L., Yee, H. K. C., Carlberg, R. G., & Ellingson, E. 1999, ApJ, 527, 54
Barger, A. J., et al. 1998, ApJ, 501, 522
Barger, A. J., Aragón-Salamanca, A., Ellis, R. S., Couch, W. J., Smail, I., & Sharples, R. M. 1996, MNRAS, 279, 1
Bartholomew, L. J., Rose, J. A., Gaba, A. E., & Caldwell, N. 2001, AJ, 122, 2913
Bekki, K., Shioya, Y., & Couch, W. J. 2001, ApJ, 547, L17
Belloni, P. Bruzual A., G., Thimm, G. J., & Roeser, H.-J. 1995, A&A, 297, 61
Belloni, P., & Roeser, H.-J. 1996, A&AS, 118, 65
Bender, R., Saglia, R. P., Ziegler, B., Belloni, P., Greggio, L., Hopp, U., & Bruzual A., G. 1998, ApJ, 493, 529
Bender, R., Ziegler, B., & Bruzual A., G. 1996, ApJ, 463, L51
Bicker, J., Fritze-v. Alvensleben, U., & Fricke, K. J. 2002, A&A, 387, 412
Bower, R. G., Kodama, T., & Terlevich, A. 1998, MNRAS, 299, 1193
Bower, R. G., Lucey, J. R., & Ellis, R. S. 1992, MNRAS, 254, 601
Bower, R. G., Terlevich, A., Kodama, T., & Caldwell, N. 1999, in The Formation History of Early-Type Galaxies, ed. P. Carral & J. Cepa (San Francisco: ASP), 211
Caldwell, N., & Rose, J. A. 1997, AJ, 113, 492
———. 1998, AJ, 115, 1423
Caldwell, N., Rose, J. A., & Dendy, K. 1999, AJ, 117, 140
Caldwell, N., Rose, J. A., Franx, M., & Leonardi, A. J. 1996, AJ, 111, 78
Caldwell, N., Rose, J. A., Sharples, R. M., Ellis, R. S., & Bower, R. G. 1993, AJ, 106, 473
Castander, F. J., et al. 2001, AJ, 121, 2331

Charlot, S., & Silk, J. 1994, ApJ, 432, 453

Couch, W. J., Balogh, M. L., Bower, R. G., Smail, I., Glazebrook, K., & Taylor, M. 2001, ApJ, 549, 820

Couch, W. J., Barger, A. J., Smail, I., Ellis, R. S., & Sharples, R. M. 1998, ApJ, 497, 188

Couch, W. J., & Sharples, R. M. 1987, MNRAS, 229, 423

Cowie, L. L., Songalia, A., Hu, E. M., & Cohen, J. G. 1996, AJ, 112, 839

De Propris, R., Stanford, S. A., Eisenhardt, P. R., Dickinson, M., & Elston, R. 1999, AJ, 118, 719

Dressler, A., et al. 1997, ApJ, 490, 577

Dressler, A. 2004, in Carnegie Observatories Astrophysics Series, Vol. 3: Clusters of Galaxies: Probes of
    Cosmological Structure and Galaxy Evolution, ed. J. S. Mulchaey, A. Dressler, & A. Oemler (Cambridge:
    Cambridge Univ. Press), in press

Dressler, A., & Gunn, J. E. 1983, ApJ, 270, 7

——. 1992, ApJS, 78, 1

Dressler, A., Smail, I., Poggianti, B. M., Butcher, H., Couch, W. J., Ellis, R. S., & Oemler, A. 1999, ApJS, 122, 51

Duc, P.-A., et al. 2002, A&A, 382, 60

——. 2004, in Carnegie Observatories Astrophysics Series, Vol. 3: Clusters of Galaxies: Probes of Cosmological
    Structure and Galaxy Evolution, ed. J. S. Mulchaey, A. Dressler, & A. Oemler (Pasadena: Carnegie
    Observatories, http://www.ociw.edu/ociw/symposia/series/symposium3/proceedings.html)

Ellingson, E., Lin, H., Yee, H. K. C., & Carlberg, R. G. 2001, ApJ, 547, 609

Ellis, R. S., Smail, I., Dressler, A., Couch, W. J., Oemler, A., Jr., Butcher, H., & Sharples, R. M. 1997, ApJ, 483,
    582

Fabricant, D. G., Bautz, M. W., & McClintock, J. E., 1994, AJ, 107, 8

Fabricant, D. G., McClintock, J. E., & Bautz, M. W., 1991, ApJ, 381, 33

Fasano, G., Poggianti, B. M., Couch, W. J., Bettoni, D., Kjærgaard, P., & Moles, M. 2000, ApJ, 542, 673

Ferreras, I., & Silk, J. 2000, ApJ, 541, L37

Fisher, D., Fabricant, D., Franx, M., & van Dokkum, P. 1998, ApJ, 498, 195

Franx, M. 2004, in Carnegie Observatories Astrophysics Series, Vol. 3: Clusters of Galaxies: Probes of
    Cosmological Structure and Galaxy Evolution, ed. J. S. Mulchaey, A. Dressler, & A. Oemler (Cambridge:
    Cambridge Univ. Press), in press

Gladders, M. D., Lopez-Cruz, O., Yee, H. K. C., & Kodama, T. 1998, ApJ, 501, 571

Goto, T., et al. 2003, PASJ, 55, 757

——. 2004, in Carnegie Observatories Astrophysics Series, Vol. 3: Clusters of Galaxies: Probes of Cosmological
    Structure and Galaxy Evolution, ed. J. S. Mulchaey, A. Dressler, & A. Oemler (Pasadena: Carnegie
    Observatories, http://www.ociw.edu/ociw/symposia/series/symposium3/proceedings.html)

Henry, J. P, & Lavery, R. J. 1987, ApJ, 323, 473

Jablonka, P., & Alloin, D. 1995, A&A, 298, 361

Jones, L., Smail, I., & Couch, W. J. 2000, ApJ, 528, 118

Jørgensen, I. 1999, MNRAS, 306, 607

Kelson, D. D., Illingworth, G. D., Franx, M., & van Dokkum, P. G. 2001, ApJ, 552, L17

Kelson, D. D., Illingworth, G. D., van Dokkum, P. G., & Franx, M. 2000, ApJ, 531, 184

Kelson, D. D., van Dokkum, P. G., Franx, M., Illingworth, G. D., & Fabricant, D. 1997, ApJ, 478, L13

Kennicutt, R. C. 1992, ApJ, 388, 310

——. 1998, ARA&A, 36, 189

Kodama, T., Arimoto, N., Barger, A. J. & Aragón-Salamanca, A. 1998, A&A, 334, 99

Kodama, T., & Bower, R. G. 2001, MNRAS, 321, 18

Kodama, T., & Smail, I. 2001, MNRAS, 326, 637

Kuntschner, H., & Davies, R. L. 1998, MNRAS, 295, L29

Leonardi, A. J., & Rose, J. A. 1996, AJ, 111, 182

Liu, C. T., & Kennicutt, R. C. 1995, ApJ, 450, 547

Lubin, L. M., Oke, J. B., & Postman, M. 2002, AJ, 124, 1905

Merluzzi, P., La Barbera, F., Massarotti, M., Busarello, G., & Capaccioli, M. 2003, ApJ, 589, 147

Miller, N. A, & Owen, F. N. 2002, AJ, 124, 2453

Mobasher, B., et al. 2001, ApJS, 137, 279

Morris, S. L., Hutchings, J. B., Carlberg, R. G., Yee, H. K. C., Ellingson, E., Balogh, M. L., Abraham, R. G., &
    Smecker-Hane, T. A. 1998, ApJ, 507, 84

Newberry, M. V., Boroson, T. A., & Kirshner, R. P. 1990, ApJ, 350, 585

Poggianti, B. M., & Wu, H. 2000, ApJ, 529, 157

Poggianti, B. M., et al. 2001a, ApJ, 562, 689

———. 2001b, ApJ, 563, 118

———. 2004, in preparation (astro-ph/0208181)

Poggianti, B. M., & Barbaro, G. 1996, A&A, 314, 379

———. 1997, A&A, 325, 1025

Poggianti, B. M., Bressan, A., & Franceschini, A. 2001c, ApJ, 550, 195

Poggianti, B. M., Smail, I., Dressler, A., Couch, W. J., Barger, A. J., Butcher, H., Ellis, R. S., & Oemler, A. 1999, ApJ, 518, 576

Poggianti, B. M., & Wu, H. 2000, ApJ, 529, 157

Postman, M., Lubin, L. M., & Oke, J. B. 1998, AJ, 116, 560

Rakos, K. D., & Schombert, J. M. 1995, ApJ, 439, 47

Rose, J. A., Gaba, A. E., Caldwell, N., & Chaboyer, B. 2001, AJ, 121, 793

Schade, D., Barrientos, L. F., & Lopez-Cruz, O. 1997, ApJ, 477, L17

Schade, D., Carlberg, R. G., Yee, H. K. C., Lopez-Cruz, O., & Ellingson, E. 1996, ApJ, 464, L63

Shioya, Y., & Bekki, K. 2000, ApJ, 539, L29

Shioya, Y., Bekki, K., & Couch, W. J. 2001, ApJ, 558, 42

Shioya, Y., Bekki, K., Couch, W. J., & De Propris, R. 2002, ApJ, 565, 223

Smail, I., Edge, A. C., Ellis, R. S., Blandford, R. D. 1998, MNRAS, 293, 124

Smail, I., Kuntschner, H., Kodama, T., Smith, G. P., Packham, C., Fruchter, A. S., & Hook, R. N. 2001, MNRAS, 323, 839

Smail, I., Morrison, G., Gray, M. E., Owen, F. N., Ivison, R. J., Kneib, J.-P., & Ellis, R. S. 1999, ApJ, 525, 609

Stanford, S. A., Eisenhardt, P. R. M., & Dickinson, M. 1995, ApJ, 450, 512

———. 1998, ApJ, 492, 461

Stanford, S. A., Elston, R., Eisenhardt, P. R., Spinrad, H., Stern, D., & Dey, A. 1997, AJ, 114, 2232

Terlevich, A. I., Caldwell, N., & Bower, R. G. 2001, MNRAS, 326, 1547

Terlevich, A. I., Kuntschner, H., Bower, R. G., Caldwell, N., & Sharples, R. M. 1999, MNRAS, 310, 445

Thomas, T. 2002, Ph.D. Thesis, Univ. Leiden

Tran, K., Franx, M., Kelson, D. D., Illingworth, G. D., van Dokkum, P. G., & Kelson, D. D. 2004, in Carnegie Observatories Astrophysics Series, Vol. 3: Clusters of Galaxies: Probes of Cosmological Structure and Galaxy Evolution, ed. J. S. Mulchaey, A. Dressler, & A. Oemler (Pasadena: Carnegie Observatories, http://www.ociw.edu/ociw/symposia/series/symposium3/proceedings.html)

Treu, M. 2004, in Carnegie Observatories Astrophysics Series, Vol. 3: Clusters of Galaxies: Probes of Cosmological Structure and Galaxy Evolution, ed. J. S. Mulchaey, A. Dressler, & A. Oemler (Cambridge: Cambridge Univ. Press), in press

van Dokkum, P. G., & Franx, M. 1996, MNRAS, 281, 985

———. 2001, ApJ, 553, 90

van Dokkum, P. G., Franx, M., Fabricant, D., Illingworth, G. D., & Kelson, D. D. 2000, ApJ, 541, 95

van Dokkum, P. G., Franx, M., Fabricant, D., Kelson, D. D., & Illingworth, G. D. 1999, ApJ, 520, L95

van Dokkum, P. G., Franx, M., Kelson, D. D., Illingworth, G. D., Fisher, D., & Fabricant, D. 1998, ApJ, 500, 714

van Dokkum, P. G., & Stanford, S. A. 2003, 585, 78

van Dokkum, P. G., Stanford, S. A., Holden, B. P., Eisenhardt, P. R., Dickinson, M., & Elston, R. 2001, ApJ, 552, L101

Vazdekis, A., Kuntschner, H., Davies, R. L., Arimoto, N., Nakamura, O., & Peletier, R. F. 2001, ApJ, 551, L127

Ziegler, B., & Bender, R. 1997, MNRAS, 291, 527

Ziegler, B. L., Bower, R. G., Smail, I., Davies, R. L., & Lee, D. 2001, MNRAS, 325, 1571

16

The chemistry of galaxy clusters

ALVIO RENZINI
European Southern Observatory

Abstract

From X-ray observations of galaxy clusters one derives the mass of the intracluster medium along with its chemical composition. Optical/infrared observations are used to estimate the mass of the stellar components of galaxies, along with their chemical composition and age. This review shows that when combining all this information, several interesting inferences can be drawn, including: (1) galaxies lose more metals than they retain; (2) clusters and the general field have converted the same fraction of baryons into stars, hence the metallicity of the $z = 0$ Universe as a whole has to be nearly the same we see in clusters, $\sim 1/3$ solar; (3) for the same reason, the thermal content of the intergalactic medium is expected to be nearly the same as the preheating energy of clusters; (4) a strong increase of the Type Ia supernova (SN) rate with lookback time is predicted if SNe Ia produce a major fraction of cosmic iron; (5) the global metallicity of the $z \approx 3$ Universe was already $\sim 1/10$ solar; and (6) the Milky Way disk formed out of material that was pre-enriched to $\sim 1/10$ solar by the bulge stellar population.

16.1 Introduction

In one of the rare cases in which theory anticipates observations, the existence of large amounts of heavy elements in the intracluster medium (ICM) was predicted shortly before it was actually observed (Larson & Dinerstein 1975). This came from (now old-fashioned) so-called *monolithic* models of elliptical galaxy formation, in which the observed color-magnitude relation is reproduced in terms of a metallicity trend. In turn, this trend is established by SN-driven galactic winds being more effective in less massive, fainter galaxies with shallow potential wells, compared to more massive galaxies harbored in deep potential wells. While these models may now be inadequate in quite many respects, their prediction was confirmed the following year by the discovery of the strong iron-K line in the X-ray spectrum of galaxy clusters (Mitchell et al. 1976).

This Carnegie Symposium on clusters of galaxies covers all the manifold aspects of these largest bound systems in the Universe. This review is meant to focus on one specific topic, the metal content of the clusters. I show that from this one can infer quite a number of intriguing consequences on galaxy formation and evolution on a wide scale, as well as on the evolution of some global properties of the baryonic component of the Universe. The following will be a *broad-brush* picture about facts and inferences, and is meant to stimulate a deeper look at each of the issues that will be cursorily touched upon here, and which include the following main topics:

- The metal content of clusters: ICM and galaxies
- The composition of the ICM "metallicity" (elemental ratios)
- The ICM/galaxies iron share in clusters
- Metal production: Type Ia and Type II SNe, and the cosmic evolution of their rate
- Metal transfer from galaxies to ICM: ejection vs. extraction
- Metals as tracers of the "ICM preheating"
- Clusters vs. field at $z = 0$
- The major epoch of metal production in clusters
- The metallicity of the Universe at $z = 3$
- The early chemical evolution of the Milky Way

The production and circulation of iron and other heavy elements on galaxy cluster scale has been widely discussed since their early discovery (e.g., Vigroux 1977; Matteucci & Vettolani 1988; Ciotti et al. 1991; Arnaud et al. 1992; Renzini et al. 1993; Loewenstein & Mushotzky 1996; Ishimaru & Arimoto 1997; Renzini 1997, 2000; Chiosi 2000; Aguirre et al. 2001; Pipino et al. 2002).

16.2   The Heavy Elements in Clusters: ICM and Galaxies

16.2.1   Iron

Iron is the best studied element in clusters of galaxies, as ICM iron emission lines are present in all clusters and groups, either warm or hot. Figure 16.1 shows the iron abundance in the ICM of clusters and groups as a function of ICM temperature from an earlier compilation (Renzini 2000). For $kT \gtrsim 3$ keV the ICM iron abundance is constant at $Z^{\mathrm{Fe}} \simeq 0.3 Z_\odot^{\mathrm{Fe}}$, independent of cluster temperature. Abundances for clusters in this *horizontal* sequence come from the iron-K complex at ~ 7 keV, whose emission is due to transitions to the K level of H-like and He-like iron ions. At lower temperatures the situation is much less simple. Figure 16.1 shows data from Buote (2000), with the iron abundance having been derived with both one-temperature and two-temperature fits. The one-temperature fits give iron abundances for those cool groups that are more or less in line with those of the hotter clusters. The abundances of the two-temperature fits, instead, form an almost vertical sequence, with a great deal of dispersion around a mean value of ~ 0.75 solar. Earlier estimates gave extremely low values for cooler groups, $kT \lesssim 1$ keV (Mulchaey et al. 1996). Compiling values from the literature, a strong dependence of the abundance on ICM temperature is apparent, being very low at low temperatures, steeply increasing to a maximum around $kT \approx 2$ keV, then decreasing to reach ~ 0.3 solar by $kT \approx 3$ keV (Renzini 1997; see also Mushotzky 2002).

Is this strong temperature dependence real? Perhaps some caution is in order. Besides the ambiguity as to whether one- or two-temperature fits are preferable, additional uncertainties for the iron abundances at $kT \lesssim 2$ keV come from their being derived from the iron-L complex at ~ 1 keV, whose emission lines are due to transitions to the L level of iron ions with three or more electrons. In these cooler groups/clusters iron is indeed in such lower ionization stages, and the iron-K emission disappears. The atomic configurations of these more complex ions are not as simple as those giving rise to the iron-K emission, and their (calculated) collisional excitation probabilities may be more uncertain. In summary, iron abundances derived from the iron-L emission should be regarded with a little more caution compared to those from the iron-K emission.

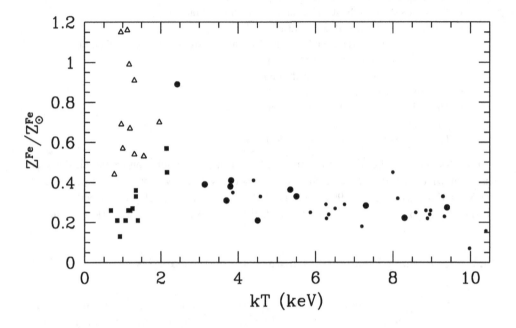

Fig. 16.1. A compilation of the iron abundance in the ICM as a function of the ICM temperature for a sample of clusters and groups, including several clusters at moderately high redshift with $\langle z \rangle \simeq 0.35$, represented by small filled circles. For temperatures less than about 2 keV, 11 groups are shown from Buote (2000), with temperatures and abundances determined from one- and two-temperature fits (filled squares and open triangles, respectively). (Adapted from Renzini 2000.)

Abundances shown in Figure 16.1 refer to the cluster central regions. However, radial gradients in the iron abundance have been reported for several clusters, starting with *ASCA* and then *ROSAT* data (e.g., Fukazawa et al. 1994; Dupke & White 2000; Finoguenov, David, & Ponman 2000; White 2000; Finoguenov, Arnaud, & David 2001). From *Beppo-SAX* data, De Grandi & Molendi (2001) have conducted a systematic study of the radial distribution of iron (metals) in many clusters. Figure 16.2 shows that clusters break up into two distinct groups: so-called cool core clusters (formerly known as "cooling flow" clusters before the failure of the cooling flow model was generally acknowledged) are characterized by a steep metallicity (mostly iron) gradient in the core, reaching ~ 0.6 solar near the center, and non-cold core clusters (where no temperature gradient is found), which show no metallicity gradient. The origin of the dichotomy remains to be understood. The fact that metallicity gradients are found to be associated with large temperature gradients in the central regions may look suspicious, as noted for the strong dependence of Z^{Fe} on ICM temperature, but it appears to be well established.

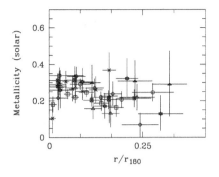 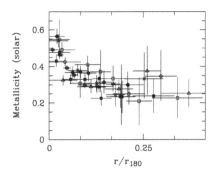

Fig. 16.2. Projected metallicity distributions for non-cold core clusters (left panel) and cold-core clusters (right panel), from *Beppo-SAX* data. The radial coordinate is normalized to the radius with an overdensity factor of 180. (From De Grandi & Molendi 2001.)

16.2.2 Elemental Ratios

X-ray observatories (especially *ASCA*, *Beppo-SAX*, and *XMM-Newton*) have such high spectral resolution that besides those of iron the emission lines of many other elements can be detected and measured. These include oxygen, neon, magnesium, calcium, silicon, sulfur, argon, and nickel. Most of these are α elements, predominantly synthesized in massive stars exploding as Type II SNe. As is well known, iron-peak elements are mainly produced by Type Ia SNe, and 50% – 75% of the iron in the Sun may come from them.

Early estimates from *ASCA* suggested a sizable α-element enhancement, $\langle[\alpha/Fe]\rangle \simeq +0.4$ (Mushotzky (1994), later reduced to +0.2 (Mushotzky et al. 1996), and eventually found consistent with solar proportions $\langle[\alpha/Fe]\rangle \simeq 0.0$ (Ishimaru & Arimoto 1997). More recently, Finoguenov et al. (2000) report near-solar Ne/Fe, slightly enhanced Si/Fe, and slightly depleted S/Fe, but with rather large error bars. From a systematic reanalysis of the *ASCA* archival data, Mushotzky (2002) reports a systematic increase of Si and Ni and a decrease of Ca with ICM temperature. Note that both silicon and calcium are α elements, and apparently they do not follow the same trend! No simple interpretation has so far emerged of these trends in terms of the relative role of the two SN types (Gibson & Matteucci 1997; Loewenstein 2001; Finoguenov et al. 2002).

I would conclude that no compelling evidence exist for other than near-solar $[\alpha/Fe]$ ratios in the ICM, when all α elements are lumped together. This argues for stellar nucleosynthesis having proceeded in much the same way in the solar neighborhood as well as at the galaxy cluster scale. In turn, this demands a similar ratio of the number of Type Ia to Type II SNe, as well as a similar stellar initial mass function (IMF), suggesting that the star formation process (IMF, binary fraction, etc.) is universal, with little or no dependence on the global characteristics of the parent galaxies (and their large-scale structure environment) in which molecular clouds are turned into stars. Alternatively, one can take at face value the variations of the abundance ratios with cluster temperature, as well as the overabundance of some α elements and the underabundance of others. One can then be forced to rather contrived conclusions, such as the mix of the two SN types, and perhaps even the nucleosynthesis of massive stars, depends on what the temperature of the ICM will be billions of years *after* star formation has ceased. On the other hand, one may argue that rich galaxy clusters are "special" places in many senses, and that ICM abundances reflect not only SN nucleosynthesis yields, but also how efficiently these are ejected, mixed into, and retained in the ICM. However, no simple understanding of the apparent empirical trends has yet emerged.

In summary, in the following I will assume that clusters, on a global scale, have solar elemental ratios and the total heavy element abundance is 0.3 solar, or 0.006 by mass.

16.2.3 The Iron Mass-to-Light Ratio

One useful quantity is the iron-mass-to-light-ratio (M^{Fe}/L) of the ICM, the ratio M^{Fe}_{ICM}/L_B of the total iron mass in the ICM over the total B-band luminosity of the galaxies in the cluster. In turn, the total iron mass in the ICM is given by the product of the iron abundance times the mass of the ICM, $M^{Fe}_{ICM} = M_{ICM} Z^{Fe}_{ICM}$. Figure 16.3 shows the resulting M^{Fe}/L from an earlier compilation (Renzini 1997). The drop of the M^{Fe}/L in poor clusters and groups (i.e., for $kT \lesssim 2$ keV) can be traced back to a drop in both the iron abundance (which, however, may not be real; see above) *and* in the ICM mass. Such groups appear to be gas poor compared to clusters, which suggests (1) that they may have been subject to baryon and metal losses due to strong galactic winds driving much of the ICM out of them (Renzini et al. 1993; Renzini 1997; Davis, Mulchaey, & Mushotzky 1999), (2) that such winds have *preheated* the gas around galaxies, thus preventing it to fall inside groups, or (3) that they have *inflated* the gas distribution. In one way or another, the *break* seen in Figure 16.3 is likely to be related to the break of self-similarity in the X-ray-temperature relation (see later).

For the rest of this paper I will mainly deal with clusters with $kT \gtrsim 2-3$ keV, for which the interpretation of the data appears more secure. Yet, several cautionary remarks are in order. The first is that the iron abundances used to construct Figure 16.3 did not take into account that some clusters have sizable iron gradients. In principle, X-ray observations can give both the run of gas density and abundance with radius, so to make possible to integrate their product over the cluster volume to get M^{Fe}_{ICM}. To my knowledge, so far this has been attempted only for one cluster (Pratt & Arnaud 2003). However, the extra iron contained within the core's iron gradient seems to be the product of the central cD galaxy, and may represent only a small fraction of the whole M^{Fe}_{ICM} (De Grandi & Molendi 2002).

Another concern is that two of the three ingredients entering into the calculation of the M^{Fe}/L values shown in Figure 16.3 (namely M_{ICM} and L_B) may not be measured in precisely the same way in the various sources used in the compilation. Both quantities come from a radial integration up to an ill-defined cluster boundary, such as the Abell radius, the virial radius, or to a radius of some fixed overdensity. Sometimes it is quite difficult to ascertain what definition has been used by one author or another, with the complication that in general X-ray and optical data have been collected by different groups using different assumptions. There is certainly room for improvement here. Finally, estimated total luminosities (L_B) refer to the sum over all cluster galaxies and do not include the population of stars that is diffusely distributed throughout the cluster, which may account for at least $\sim 10\%$ of the total cluster light (Ferguson, Tanvir, & von Hippel 1998; Arnaboldi et al. 2003).

While keeping these cautions in mind, we see from Figure 16.3 that M^{Fe}/L runs remarkably flat with increasing cluster temperature, for $kT \gtrsim 2-3$ keV. This constancy of the M^{Fe}/L comes from both Z^{Fe}_{ICM} and M_{ICM}/L_B, showing very little trend with cluster temperature [see Fig. 16.1 and Fig. 4 in Renzini (1997), where $M_{ICM}/L_B \simeq 25 h_{70}^{-1/2}$ $(M_\odot/L_\odot)$.] The resulting M^{Fe}/L is therefore

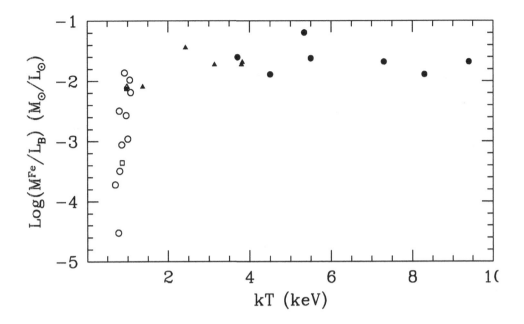

Fig. 16.3. The iron mass-to-light ratio of the ICM of clusters and groups as a function of the ICM temperature from an earlier compilation (Renzini 1997). Data are taken from the following sources: filled circles, Arnaud et al. (1992); filled triangles, Tsuru (1992); open triangle, David et al. (1994); open square, Mulchaey et al. (1993); filled square, Ponman et al. (1994); and open circles, Mulchaey et al. (1996). (Adapted from Renzini 1997.)

$$\frac{M_{ICM}^{Fe}}{L} = Z_{ICM}^{Fe}\frac{M_{ICM}}{L_B} \simeq 0.3 \times Z_{\odot}^{Fe} \times 25 h_{70}^{-1/2} \simeq 0.01\, h_{70}^{-1/2}(M_{\odot}/L_{\odot}). \tag{16.1}$$

That is, in the ICM there is about 0.01 solar masses of iron for each solar luminosity of the cluster galaxies. This value is $\sim 30\%$ lower than adopted in Renzini (1997) and shown in Figure 16.3, having consistently adopted here for $Z_{\odot}^{Fe}$ the recommended *meteoritic* iron abundance (Anders & Grevesse 1989), $Z_{\odot}^{Fe} = 0.0013$. Assuming solar elemental proportions for the ICM, the ICM *metal mass-to-light ratio* is therefore $\sim 0.3 \times 0.02 \times 25 h_{70}^{-1/2} = 0.15\,(M_{\odot}/L_{\odot})$, having adopted $Z_{\odot} = 0.02$ and for $h_{70} = 1$.

A very accurate analysis was performed recently for the A1983 cluster (Pratt & Arnaud 2003), paying attention to measure M_{ICM} and L_B within the same radius. The result is $M^{Fe}/L = (7.5 \pm 1.5) \times 10^{-3} h_{70}^{-1/2}(M_{\odot}/L_{\odot})$, in fair agreement with the estimate above.

The most straightforward interpretation of the constant M^{Fe}/L is that clusters did not lose iron (hence baryons), nor differentially acquired pristine baryonic material, and that the conversion of baryonic gas into stars and galaxies has proceeded with the same efficiency and the same stellar IMF in all clusters (Renzini 1997). Otherwise, there should be cluster-to-cluster variations of Z_{ICM}^{Fe} and M^{Fe}/L. All this is true insofar as the baryon-to-dark matter ratio is the same in all $kT \gtrsim 2$ keV clusters (White at al. 1993), and the ICM mass-to-light ratio and the gas fraction are constant. Nevertheless, there may be hints for some of these quantities showing (small) cluster-to-cluster variations (Arnaud & Evrard 1999; Mohr, Mathiesen, & Evrard 1999; Pratt & Arnaud 2003), but no firm conclusion has been reached yet.

16.2.4   The Iron Share Between ICM and Cluster Galaxies

The metal abundance of the stellar component of cluster galaxies is derived from integrated spectra coupled to synthetic stellar populations. Much of the stellar mass in clusters is confined to passively evolving spheroids (ellipticals and bulges), for which the iron abundance Z_*^{Fe} may range from $\sim 1/3$ solar to a few times solar. For example, among ellipticals metal-sensitive spectral features such as the magnesium index Mg_2 range from values slightly lower than in the most metal-rich globular clusters of the Milky Way bulge (which are nearly solar), to values for which models indicate a metallicity a few times solar (e.g., Maraston et al. 2003). The M^{Fe}/L of cluster galaxies is then given by:

$$\frac{M_{\mathrm{gal}}^{\mathrm{Fe}}}{L} = Z_*^{\mathrm{Fe}}\frac{M_*}{L_{\mathrm{B}}} \simeq 0.0046\, h_{70} \quad (M_{\odot}/L_{\odot}), \tag{16.2}$$

where we have adopted $M_*/L_{\mathrm{B}} = 3.5\, h_{70}$ (White et al. 1993) and $Z_*^{\mathrm{Fe}} = Z_{\odot}^{\mathrm{Fe}}$. The *total* cluster M^{Fe}/L (ICM+galaxies) is therefore $\sim 0.015 \quad (M_{\odot}/L_{\odot})$, for $h_{70} = 1$, and the ratio of the iron mass in the ICM to the iron mass locked into stars and galaxies is

$$\frac{Z_{\mathrm{ICM}}^{\mathrm{Fe}}M_{\mathrm{ICM}}}{Z_*^{\mathrm{Fe}}M_*} \simeq 2.2 h_{70}^{-3/2}, \tag{16.3}$$

having adopted $Z_{\mathrm{ICM}}^{\mathrm{Fe}} = 0.3\, Z_{\odot}^{\mathrm{Fe}}$, $Z_*^{\mathrm{Fe}} = 1\, Z_{\odot}^{\mathrm{Fe}}$, and $M_{\mathrm{ICM}}/M_* = 9.3 h_{70}^{-3/2}$ as for the Coma cluster (White et al. 1993). So, it appears that there is ~ 2 times more iron mass in the ICM than locked into cluster stars (galaxies), perhaps even more if Z_*^{Fe} is subsolar due to an abundance gradient within clusters (Arimoto et al. 1997). In turn, this empirical iron share (ICM vs. galaxies) sets a strong constraint on models of the chemical evolution of galaxies. Under the same assumptions as above, the total metal mass-to-light ratio (ICM + galaxies) is therefore $\sim 0.15 h_{70}^{-1/2} + 0.07 h_{70} \simeq 0.2\, (M_{\odot}/L_{\odot})$. This can be regarded as a fully empirical determination of the metal yield of (now) old stellar populations.

16.3   Metal Production: The Parent Stellar Population

The constant M^{Fe}/L of clusters means that the total mass of iron in the ICM is proportional to the total optical luminosity of the cluster galaxies (Songaila, Cowie, & Lilly 1990; Ciotti et al. 1991; Arnaud et al. 1992; Renzini et al. 1993). The simplest interpretation is that the iron and all the metals now in the ICM have been produced by the (massive) stars of the same stellar generation to which belong the low-mass stars now radiating the bulk of the cluster optical light. Since much of the cluster light comes from old spheroids (ellipticals and bulges), one can conclude that *the bulk of cluster metals were produced by the stars destined to make up the old spheroids that we see today in clusters.*

It is also interesting to ask which galaxies have produced the bulk of the iron and the other heavy elements, i.e. the relative contribution as a function of the present-day luminosity of cluster galaxies. From their luminosity function it is easy to realize that the bright galaxies (those with $L \gtrsim L^*$) produce the bulk of the cluster light, while the dwarfs contribute a negligible amount of light, in spite of their dominating the galaxy counts by a large margin

(Thomas 1999). In practice, most galaxies do not do much, while only the brightest $\sim 3\%$ of all galaxies contribute $\sim 97\%$ of the whole cluster light. Giants dominate the scene, while dwarfs do not count much. Following the simplest interpretation, according to which the metals were produced by the same stellar population that now shines, one can also conclude that the bulk of the cluster metals have been produced by the giant galaxies that contain most of the stellar mass. The relative contribution of dwarfs to ICM metals may have been somewhat larger than their small relative contribution to the cluster light, since metals can more easily escape from their shallower potential wells (Thomas 1999). Yet, this is unlikely to alter the conclusion that the giants dominate metal production by a very large margin.

16.4   Metal Production: Type Ia vs. Type II Supernovae

As is well known, clusters are now dominated by E/S0 galaxies, which produce only Type Ia SNe at a rate of $\sim (0.16 \pm 0.06) h_{70}^2$ SNU (Cappellaro, Evans, & Turatto 1999), with 1 SNU corresponding to 10^{-12} SNe yr$^{-1} L_{B\odot}^{-1}$. Assuming such rate to have been constant through cosmological times (~ 13 Gyr), the number of SNe Ia exploded in a cluster of present-day luminosity L_B is therefore $\sim 1.6 \times 10^{-13} \times 1.3 \times 10^{10} L_B h_{70}^2 \simeq 2 \times 10^{-3} L_B$. With each SN Ia producing $\sim 0.7 M_\odot$ of iron, the resulting $M^{\rm Fe}/L$ of clusters would be:

$$\left(\frac{M^{\rm Fe}}{L_B}\right)_{\rm SN\ Ia} \simeq 1.4 \times 10^{-3} h_{70}^2, \tag{16.4}$$

which falls short by a factor ~ 10 compared to the observed cluster $M^{\rm Fe}/L$ (0.015 for $h_{70} = 1$). The straightforward conclusion is that either SNe Ia did not play any significant role in manufacturing iron in clusters, or their rate in what are now E/S0 galaxies had to be much higher in the past. This argues for a strong evolution of the SN Ia rate in E/S0 galaxies and bulges, with the past average being $\sim 5-10$ times higher than the present rate (Ciotti et al. 1991). This may soon be tested directly by observations.

In the case of SNe Ia we believe to have a fairly precise knowledge of the amount of iron released by each event, while the ambiguities affecting the progenitors make theory unable to predict the evolution of the SN Ia rate past a burst of star formation (e.g., Greggio 1996). The case of Type II SN's is quite the opposite: one believes to have unambiguously identified the progenitors (stars more massive than $\sim 8 M_\odot$), while a great uncertainty affects the amount of iron produced by each SN II event as a function of progenitor's mass, $M_{\rm II}^{\rm Fe}(M)$. The SN luminosity at late times can be used to infer the amount of radioactive Ni-Co (hence eventually iron) that was ejected, and an early study indicated small variations from one event to another ($0.04 - 0.10 M_\odot$; Patat et al. 1994). This led Renzini et al. (1993) to assume $M_{\rm II}^{\rm Fe}$ to be a weak function of initial mass, with an average yield of $0.07 M_\odot$ of iron per event (as in SN 1987A). More recent studies based on a larger sample of SN II events have actually detected very large differences from one event to another (ranging from $\sim 0.002 M_\odot$ to $\sim 0.3 M_\odot$; Turatto 2003). However, an average over 16 well-studied SNe II gives $\langle M^{\rm Ni} \rangle = 0.062 M_\odot$ (Hamuy 2004), close to the adopted value.

The total number of SNe II, $N_{\rm II}$, is obtained by integrating the stellar IMF from, for example, 8 to 100 $M_\odot$, with the IMF being $\psi(M) = 3.0 L_B M^{-(1+x)}$, where L_B is the luminosity of the stellar population when it ages to $\gtrsim 10^{10}$ yr (Renzini 1998b). Clearly, the flatter the IMF slope the larger the number of massive stars per unit present luminosity, the larger the number of SNe II, and the larger the implied $M^{\rm Fe}/L$. Thus, integrating the IMF one gets

$$\left(\frac{M^{\mathrm{Fe}}}{L_{\mathrm{B}}}\right)_{\mathrm{SN\ II}} = \frac{M_{\mathrm{II}}^{\mathrm{Fe}} N_{\mathrm{II}}}{L_{\mathrm{B}}} \simeq \begin{cases} 0.003 & \text{for } x = 1.7 \\ 0.009 & \text{for } x = 1.35 \\ 0.035 & \text{for } x = 0.9. \end{cases} \tag{16.5}$$

Hence, if the Galactic IMF slope ($x = 1.7$; Scalo 1986) applies also to cluster ellipticals, then SNe II underproduce iron by about a factor of ~ 5. Instead, making all the observed iron by SNe II would require an IMF flatter than Salpeter's $x = 1.35$ (Renzini et al. 1993).

In summary, with an IMF with $1.35 \lesssim x \lesssim 1.7$ and a past average rate of SNe Ia in ellipticals $\gtrsim 5$ times the present rate, the iron content of clusters and the global ICM [α/Fe] ratio are grossly accounted for, with SNe Ia then having produced $\sim 1/2$–$3/4$ of the total cluster iron, not unlike in standard chemical models of the Milky Way galaxy. This is not to say that this has been firmly proved, but it seems to me to be premature to abandon the attractive simplicity of a universal nucleosynthesis process (i.e., IMF and SN Ia/SN II ratio) for embarking toward more complex, multi-parametric scenarios.

16.5   Metals from Galaxies to the ICM: Ejection vs. Extraction

Having established that most metals in clusters are out of the parent galaxies, it remains to be understood how they were transferred from galaxies to the ICM. There are two possibilities: extraction by ram pressure stripping as galaxies plow through the ICM, and ejection by galactic winds powered from inside galaxies themselves. In the latter case the power can be supplied by SNe (the so-called star formation feedback) and/or by AGN activity. Three arguments favor ejection over extraction:

- There appears to be no trend of either $Z_{\mathrm{ICM}}^{\mathrm{Fe}}$ or the M^{Fe}/L with cluster temperature or cluster velocity dispersion (σ_{v}), while the efficiency of ram pressure stripping should increase steeply with increasing σ_{v}.
- Field ellipticals appear to be virtually identical to cluster ellipticals. They follow basically the same $Mg_2 - \sigma$ relation (Bernardi et al. 1998, 2003), which does not show any appreciable trend with the local density of galaxies. If stripping was responsible for extracting metals from galaxies one would expect galaxies in low-density environments to have retained more metals, hence showing higher metal indices for a given σ, which is not seen.
- Nongravitational energy injection of the ICM seems to be required to account for the break of the self-similar X-ray luminosity-temperature relation for groups and clusters (Ponman, Cannon, & Navarro 1999; see below). While galactic winds are an obvious vehicle for preheating, no preheating would be associated with metal transfer by ram pressure.

One can quite safely conclude that metals in the ICM have been *ejected* from galaxies by SN (or AGN) driven winds, rather than stripped by ram pressure (Renzini et al. 1993; Dupke & White 1999). Two kinds of galactic winds are likely to operate: *early winds* driven by the starburst forming much of the galaxy's stellar mass itself, and *late winds* or outflows where the gas comes from the cumulative stellar mass loss as the stellar populations passively age. Direct observational evidence for early winds exists for Lyman-break galaxies (Pettini et al. 2001), as well as for local massive starbursts (Heckman et al. 2000). Late winds are also likely to operate, as the stellar mass loss from the aging population flows out of spheroids, being either continuously driven by a declining SN Ia rate (Ciotti et al. 1991), or intermittently by recurrent AGN activity (Ciotti & Ostriker 2001).

16.6 Metals as Tracers of ICM Preheating

The total amount of iron in clusters represents a record of the overall past SN activity as well as of the past mass and energy ejected from cluster galaxies. The empirical values of M^{Fe}/L can be used to set a constraint on the energy injection into the ICM by SN-driven galactic winds (Renzini 1994). The total SN heating is given by the kinetic energy released by one SN ($\sim 10^{51}$ erg) times the number of SNe that have exploded. It is convenient to express this energy per unit present optical light L_B,

$$\frac{E_{SN}}{L_B} = 10^{51} \frac{N_{SN}}{L_B} = 10^{51} \left(\frac{M^{Fe}}{L_B}\right)_{tot} \frac{1}{\langle M^{Fe}\rangle} \simeq 10^{50} \quad (\text{erg}/L_\odot), \tag{16.6}$$

where the total (ICM+galaxies) $M^{Fe}/L=0.015\,M_\odot/L_\odot$ is adopted, and the average iron release per SN event is assumed to be $0.15\,M_\odot$ (appropriate if SNe Ia and SNe II contribute equally to the iron production). This estimate should be accurate to within a factor of 2 or 3.

The kinetic energy injected into the ICM by galactic winds, again per unit cluster light, is given by 1/2 the ejected mass (M^{Fe}_{ICM}/Z^{Fe}_w) times the square of the typical wind velocity,

$$\frac{E_w}{L_B} = \frac{1}{2}\frac{M^{Fe}_{ICM}}{L_B}\frac{\langle v^2_w\rangle}{\langle Z^{Fe}_w\rangle} \simeq 1.5 \times 10^{49} \frac{Z^{Fe}_\odot}{Z^{Fe}_w} \cdot \left(\frac{v_w}{500\,\text{km s}^{-1}}\right)^2 \simeq 10^{49} \quad (\text{erg}/L_\odot), \tag{16.7}$$

where the empirical M^{Fe}/L for the ICM has been used and the average metallicity of the winds Z^{Fe}_w is assumed to be 2 times solar. As usual in the case of thermal winds, the wind velocity v_w is of the order of the escape velocity from individual galaxies. Again, this estimate may be regarded as accurate to within a factor of 2 or so.

A first inference is that of order $\sim 5\% - 20\%$ of the kinetic energy released by SNe is likely to survive as kinetic energy of galactic winds, thus contributing to the heating of the ICM. A roughly similar amount goes into work to extract the gas from the potential well of individual galaxies, while the rest of the SN energy has to be radiated away locally and does not contribute to the feedback. This estimated energy injection represents a small fraction of the thermal energy of the ICM of rich (hot) clusters and so had only a minor impact on the history of the ICM. However, in groups it represents a nonnegligible fraction of the thermal energy of the ICM, thus affecting its evolution and present structure (Renzini 1994). The necessity of some nongravitational heating (or preheating) was recognized from the observed break of the self-similarity of the X-ray luminosity-temperature relation, especially when groups are included (Ponman et al. 1999).

The estimated $\sim 10^{49}$ erg/$L_\odot$ correspond to a preheating of ~ 0.1 keV per particle, for a typical cluster $M_{ICM}/L_B \simeq 25\,M_\odot/L_\odot$. This is $\gtrsim 10$ times lower than the ~ 1 keV/particle preheating that some models require to fit the cluster $L_X - T$ relation (Wu, Fabian, & Nulsen 2000; Borgani et al. 2001, 2002; Tozzi & Norman 2001; Pipino et al. 2002; Finoguenov et al. 2003). This estimate depends somewhat on the gas density (hence environment and redshift) where/when the energy is injected, because what matters is the entropy change induced by the preheating, $\Delta S = k\Delta T/n_e^{2/3}$ (Kaiser 1991; Cavaliere, Colafrancesco, & Menci 1993); hence, the required energy decreases if it is injected at a lower gas density. Nevertheless, this extreme (1 keV/particle) requirement would be met only if virtually all the SN energy were to go to increase the thermal energy of the ICM. Such extreme preheating requirement points toward an additional energy (entropy) source, such as AGN energy injection (e.g., Valageas & Silk 1999; Wu et al. 2000). Note, however, that in powerful starbursts most SNe explode inside hot bubbles made by previous SNe, thus reducing radiative losses, and the feedback efficiency may approach unity (Heckman 2002). More recently it has been suggested that preheating requirements may be relaxed somewhat if the energy injection takes place at relatively low density, so as to boost the entropy increase with less energy deposition (Ponman, Sanderson, & Finoguenov 2003). For example, preheating could take place within the filaments, prior to their coalescing to form clusters. Numerical simulations are exploring this possibility (see the review by Evrard 2004), which may eventually reduce the energetic requirement to be more in line with a conservative (i.e., $\sim 0.1-0.3$ keV/particle) SN-driven galactic wind scenario.

16.7   Clusters vs. Field at $z = 0$ and the Overall Metallicity of the Universe

To what extent are clusters fair samples of the $z \sim 0$ Universe as a whole? In many respects clusters look much different from the field, for example in the morphological mix of galaxies, or in the star formation activity, which in clusters has almost completely ceased while it is still going on in the field. Yet, when we restrict ourselves to some global properties clusters and field are not so different. For example, the baryon fraction of the Universe is $\Omega_b/\Omega_m \simeq 0.16 \pm 0.02$ (Bennett et al. 2003), which compares to ~ 0.15 as estimated for clusters (White et al. 1993), adopting $h_{70} = 1$. This tells us that no appreciable baryon vs. dark matter segregation has taken place at a cluster scale (White et al. 1993).

Even more interesting may be the case of the stellar mass over baryon mass in clusters and in the field. For the field in the local Universe, Fukugita, Hogan, & Peebles (1998) estimate $\Omega_* = 0.0035 h_{70}^{-1}$ for the stellar contribution to Ω. From the 2dF K-band luminosity function, Cole et al. (2001) estimate $\Omega_* = 0.0041 h_{70}^{-1}$, with a $\sim 15\%$ uncertainty (adopting a Salpeter IMF). The total baryon density is $\Omega_b = 0.039 h_{70}^{-2}$, as derived from standard Big Bang nucleosynthesis (and confirmed by *WMAP*; Bennett et al. 2003). This gives a global baryon to star conversion efficiency $\Omega_*/\Omega_b \simeq 0.10 h_{70}$—that is, over the whole cosmic time $\sim 10\%$ of the baryons have been converted and locked into stars. At the galaxy cluster level, the same efficiency can be measured directly, and following White et al. (1993) one gets

$$\frac{M_*}{M_{\text{ICM}} + M_*} \simeq \frac{1}{9.3 h_{70}^{-3/2} + 1} \simeq 0.1. \tag{16.8}$$

For clusters Fukugita et al. (1998) obtain a $\sim 30\%$ larger value, which, however, is well within the uncertainty affecting these estimates. One can safely conclude that *the efficiency of baryon to galaxies/stars conversion has been $\sim 10\%$, quite the same in the "field" as well as within rich clusters of galaxies*. The environment seems to be irrelevant!

Two interesting inferences can be drawn from this intriguing cluster-field similarity:

(1) *The metallicity of the present Universe is $\sim 1/3$ solar.* The metallicity of the local Universe has to be virtually identical to that measured in clusters ($\sim 1/3$ solar), since star formation, and hence the ensuing metal enrichment, have proceeded at the same level. In analogy to clusters, a majority share of the metals now reside outside galaxies in a warm intergalactic medium (IGM) containing the majority of the baryons. Most baryons as well as most metals in the local Universe remain unaccounted.

(2) *The thermal energy (temperature) of the local Universe is about the same as the preheating energy of clusters.* Similar overall star formation activities most likely result not only in similar metal productions but also in similar energy deposition by galactic winds. Hence, the temperature in the local IGM is likely to be $kT \approx 0.1-1$ keV, whatever the cluster preheating will turn out to be. Attempts are currently under way to detect this warm, metal-rich IGM.

The detection of O VI-absorbing clouds physically located within the Local Group is a first important step in this direction (Nicastro et al. 2003).

16.8 The Major Epoch of Metal Production

Most stars are either in spheroids or in disks. According to Fukugita et al. (1998) ~3/4 of the total mass in stars in the local Universe is now in spheroids, ~1/4 in disks, and less than 1% in irregular galaxies. Other authors give less extreme estimates; Dressler & Gunn (1990) estimate that the stellar mass in spheroids and disks is about the same (see also Benson, Frenk, & Sharples 2002). In clusters the dominance of spheroids is likely to be even stronger than in the general field. The prevalence of spheroids offers an opportunity to estimate the epoch (redshift) at which (most) metals were produced and disseminated, since we now know quite well when most stars in spheroids were formed.

16.8.1 In Clusters

Following the first step in this direction (Bower, Lucey, & Ellis 1992), I believe that the most precise estimates of the age (redshift of formation) of stellar populations in cluster elliptical galaxies come from the tightness of several correlations, such as the color-magnitude, fundamental plane, and the $Mg_2 - \sigma$ relations, and especially by such relations remaining tight all the way to $z \approx 1$ (Stanford, Eisenhardt, & Dickinson 1998; van Dokkum & Franx 2001; see also Renzini 1999 for an extensive review and reference list). This has taught us that the best way of breaking the age-metallicity degeneracy is to look back at high-redshift galaxies. All evidence converges to indicate that most stars in cluster ellipticals formed at $z \gtrsim 3$, while only minor episodes of star formation may have occurred later.

With most of star formation having taken place at such high redshift, most cluster metals should also have been produced and disseminated at $z \gtrsim 3$. Little evolution of the ICM composition is then expected all the way to high redshifts, with the possible exception of iron from SNe Ia, whose rate of release does not follow the star formation rate, but is modulated by the distribution of the delays between formation of the precursor and explosion time. Still, one expects that the SN Ia rate peaks shortly after a burst of star formation and then rapidly declines, with most events taking place within 1–2 Gyr after formation (e.g., Greggio & Renzini 1983). If so, no appreciable evolution of the iron abundance in clusters should be detectable from $z = 0$ to $z \approx 1$. Note, however, that *late winds* will keep enriching the ICM at a decreasing rate approximately $\propto t^{-1.4}$ (Ciotti et al. 1991).

16.8.2 In the Global Universe

As already noted, at $z \approx 0$ field early-type galaxies show very little differences with respect to their cluster analogs. Moreover, bulges appear very similar to ellipticals in integrated properties, such as the $Mg_2 - \sigma$ and fundamental plane relations (Jablonka, Martin, & Arimoto 1996; Falcón-Baroso, Peletier, & Balcells 2002). In the well studied case of the Milky Way bulge, no trace of stars younger than halo-bulge globular clusters could be found (Zoccali et al. 2003). At $z \approx 1$ old early-type galaxies are also found in sizable numbers in the general field, although it appears that star formation may have been a little more extended than in clusters (Cimatti et al. 2002; Treu et al. 2002; Bell et al. 2004).

Therefore, spheroids in the general field appear almost as old as cluster ellipticals, with the bulk of their stellar populations having formed at $z \gtrsim 2 - 3$. In this spirit, Hogg et al. (2002) estimate that at least 65% of the stellar mass is at least 8 Gyr old, or formed at

$z > 1$. With $\sim 50\%$ of the stellar mass in spheroids that formed $\gtrsim 80\%$ of their mass at $z \gtrsim 2-3$, one can conclude that $\gtrsim 30\%$ of the stellar mass we see today was already in place by $z \approx 3$ (Renzini 1998a). This *indirect* estimate is ~ 3 times higher than *directly* measured in the HDF-N (Dickinson et al. 2003). However, this latter result may be subject to cosmic variance, given the small size of the explored field, and indeed HDF-S appears to be much richer in massive galaxies, hence in stellar mass, at high redshift (Franx et al. 2003).

16.8.3   *The Metallicity of the Universe at $z = 3$*

With $\sim 30\%$ of all stars having formed by $z = 3$, $\sim 30\%$ of the metals should also have been formed before such an early epoch. I have argued that the global metallicity of the present-day Universe is $\sim 1/3$ solar; hence, the metallicity of the $z = 3$ Universe should be $\sim 1/10$ solar (Renzini 1998a). This simple argument supports the notion of a *prompt initial enrichment* of the early Universe. While $\sim 10\%$ solar at $z = 3$ is a very straightforward estimate, observational tests are not so easy.

Figure 16.4 (adapted from Pettini 2004) shows that at $z = 3$ the Universe had developed extremely inhomogeneously in chemical composition, with the metallicity ranging from supersolar in the central regions of young/forming spheroids and in QSOs likely hosted by them, down to $\sim 10^{-3}$ solar in the Lyα forest. Making the proper (mass-) average abundance of the heavy elements requires one to know the fractional mass of each baryonic component at $z = 3$, not an easy task. While the Lyα forest may fill most of the volume at $z = 3$ and perhaps contain most of the baryons, it may contain as little as just a few percent of the metals produced by $z = 3$. At this early time most metals are likely to be locked into stars, in metal-rich winds, and in shocked IGM which has already diluted wind materials. This latter medium may have been detected, thanks to its O VI absorption (Simcoe, Sargent, & Rauch 2002).

16.9   The Early Chemical Evolution of the Milky Way

In the K band the Galactic bulge luminosity is $\sim 10^{10} L_{K,\odot}$ (Kent, Dame, & Fazio 1991), and in the B band the bulge luminosity is $L_B^{bulge} \simeq 6 \times 10^9 L_{B,\odot}$. From the empirical yield of metals in clusters ($\sim 0.2 \times L_B M_\odot$), it follows that the Galactic bulge has produced $M_Z \simeq 0.2 L_B^{bulge} = 0.2 \times 6 \times 10^9 \simeq 10^9 M_\odot$ of metals. Where are all these metals? One billion solar masses of metals should not be easy to hide: part of it must be in the stars of the bulge itself; part of it must have been ejected by winds. The stellar mass of the bulge follows from its K-band mass to light ratio, $M_*^{bulge}/L_K = 1$ (Kent 1992; Zoccali et al. 2003), and its luminosity; hence, $M_*^{bulge} \simeq 10^{10} M_\odot$. Its average metallicity is about solar or slightly lower (McWilliam & Rich 1994; Zoccali et al. 2003), i.e. $Z = 0.02$, and therefore the bulge stars altogether contain $\sim 2 \times 10^8 M_\odot$ of metals. Only $\sim 1/5$ of the metals produced when the bulge was actively star forming some 11–13 Gyr ago are still in the bulge! Hence, $\sim 80\%$, or still $\sim 10^9 M_\odot$, wad ejected into the surrounding space by an early wind.

At the time of bulge formation, such $\sim 10^9 M_\odot$ of metals ran into largely pristine ($Z = 0$) material, experienced Rayleigh-Taylor instabilities leading to chaotic mixing, and established a distribution of metallicities in a largely inhomogeneous IGM surrounding the young Milky Way bulge. For example, this enormous amount of metals was able to bring to a metallicity 1/10 solar (i.e., $Z = 0.002$) about $5 \times 10^{11} M_\odot$ of pristine material, several times the mass of the yet-to-be-formed Galactic disk. Therefore, it is likely that the Galactic disk

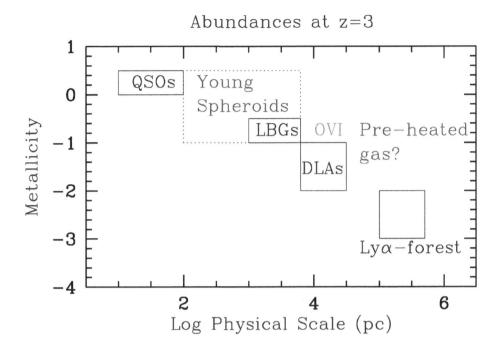

Fig. 16.4. Summary of current knowledge of metal abundances at $z \approx 3$. On the vertical axis the logarithmic abundance relative to solar is reported. The horizontal axis gives the typical linear dimensions of the structures for which direct abundance measurements are available. This figure has been adapted from Pettini (2004) by the inclusion of the box for "young spheroidals," for which the estimate is indirect, as based on the present-day observed metallicity range and on the estimated redshift of formation. The figure also includes the approximate location of the O VI absorbers (Simcoe et al. 2002), and the hypothetical location of the intergalactic medium enriched and preheated by early galactic winds.

formed and grew out of such pre-enriched material, which provides a quite natural solution to the classical "G dwarf problem" (Renzini 2002).

16.10 Summary

A number of interesting inferences are derived starting from a few empirical facts, namely the iron and metal content of the ICM and cluster galaxies, the fraction of the baryons locked into stars in clusters and in the field, and the age and baryon fraction of stellar populations of galactic spheroids. Such inferences include:

- In clusters and in the general field alike there are more metals in the diffused gas (ICM and IGM) than there are locked into stars inside galaxies. The loss of metals to the surrounding media is therefore a major process in the chemical evolution of galaxies.
- In clusters and in the general field alike $\sim 10\%$ of the baryons are now locked into stars inside galaxies. At this global level, the outcome of star formation through cosmic time is largely independent of environment, most likely just because a major fraction of all stars formed before cluster formation.

- Various arguments support the notion that the metals now in the ICM/IGM were ejected by galactic winds, rather then being extracted from galaxies by ram pressure.
- Having processed the same fraction of baryons into stars, the global metallicity of the local Universe has to be nearly the same one can measure in clusters, namely ∼1/3 solar.
- For the same reason, one expects the IGM to have experienced nearly the same amount of preheating as the ICM, and therefore to be at a temperature of ∼ 0.1 − 1 keV, whatever is the amount of preheating that is required for clusters.
- Given the predominance and formation redshift of galactic spheroids, both in clusters as well as globally in the Universe, it is likely that the Universe experienced a prompt metal enrichment, with the global metallicity possibly reaching ∼1/10 solar already at $z \approx 3$. However, most metals remain unaccounted both at low as well as high redshift, and may reside in a warm IGM, the existence of which we have only preliminary observational hints.
- This same scenario may well hold down to the scale of our own Milky Way galaxy, with early winds from the forming Galactic bulge having pre-enriched to ∼1/10 solar a much greater mass of gas, out of which the Galactic disk started to form and grew.

References

Aguirre, A., Hernquist, L., Schaye, J., Katz, N., Weinberg, D. H., & Gardner, J. 2001, ApJ, 561, 521

Anders, E., & Grevesse, N. 1989, Geochimica et Cosmochimica Acta, 53, 197

Arimoto, N., Matsushita, K., Ishimaru, Y., Ohashi, T., & Renzini, A. 1997, ApJ, 477, 128

Arnaboldi, M., et al. 2003, AJ, 125, 514

Arnaud, M., & Evrard, A. E. 1999, MNRAS, 305, 631

Arnaud, M., Rothenflug, R., Boulade, O., Vigroux, R., & Vangioni-Flam, E. 1992, A&A, 254, 49

Bell, E. F., Wolf, C., Meisenheimer, K., Rix, H.-W., Borch, A., Dye, S., Kleinheinrich, M., & McIntosh, D. H. 2004, ApJ, submitted (astro-ph/0303394)

Bennett, C. L., et al. 2003, ApJS, 148, 1

Benson, A. J., Frenk, C. S., & Sharples, R. M. 2002, ApJ, 574, 104

Bernardi, M., et al. 2003, AJ, 125, 1882

Bernardi, M., Renzini, A., da Costa, L. N., Wegner, G., Alonso, M. V., Pellegrini, P. S., Rité, C., & Willmer, C. N. A. 1998, ApJ, 508, L143

Borgani, S., Governato, F., Wadsley, J., Menci, N., Tozzi, P., Lake, G., Quinn, T., & Stadel, J. 2001, ApJ, 559, L71

Borgani, S., Governato, F., Wadsley, J., Menci, N., Tozzi, P., Quinn, T., Stadel, J., & Lake, G. 2002, MNRAS, 336, 424

Bower, R. G., Lucey, J. R., & Ellis, R. S. 1992, MNRAS, 254, 613

Buote, D. A. 2000, MNRAS, 311, 176

Cappellaro, E., Evans, R., & Turatto, M. 1999, A&A, 351, 459

Cavaliere, A., Colafrancesco, S., & Menci, N. 1993, ApJ, 415, 50

Chiosi, C. 2000, A&A, 364, 423

Cimatti, A., et al. 2002, A&A, 381, L68

Ciotti, L., D'Ercole, A., Pellegrini, S., & Renzini, A. 1991, ApJ, 376, 380

Ciotti, L., & Ostriker, J.P. 2001, ApJ, 551, 131

Cole, S., et al. 2001, MNRAS, 326, 255

David, L. P., Jones, C., Forman, W., & Daines, S. 1994, ApJ, 428, 544

Davis, D. S., Mulchaey, J. S., & Mushotzky, R. F. 1999, ApJ, 511, 34

De Grandi, S., & Molendi, S. 2001, ApJ, 551, 153

——. 2002, in Chemical Enrichment of Intracluster and Intergalactic Medium, ed. R. Fusco-Femiano & F. Matteucci (San Francisco: ASP), 3

Dickinson, M., Papovich, C., Ferguson, H. C., & Budavari, T. 2003, ApJ, 587, 25

Dressler, A., & Gunn, J. E. 1990, in Evolution of the Universe of Galaxies, Proceedings of the Edwin Hubble Centennial Symposium, ed. R. G. Kron (San Francisco: ASP), 200

Dupke, R. A., & White, R. E., III 2000, ApJ, 537, 123

Evrard, A. E. 2004, in Carnegie Observatories Astrophysics Series, Vol. 3: Clusters of Galaxies: Probes of

Cosmological Structure and Galaxy Evolution, ed. J. S. Mulchaey, A. Dressler, & A. Oemler (Cambridge: Cambridge Univ. Press), in press

Falcón-Barroso, J., Peletier, R. F., & Balcells, M. 2002, MNRAS, 335, 741

Ferguson, H. C., Tanvir, N. R., & von Hippel, T. 1998, Nature, 391, 461

Finoguenov, A., Arnaud, M., & David, L. P. 2001, ApJ, 555, 191

Finoguenov, A., Borgani, S., Tornatore, L., & Böhringer, H. 2003, A&A, 398, L35

Finoguenov, A., David, L. P., & Ponman, T. J. 2000, ApJ, 544, 188

Finoguenov, A., Jones, C., Böhringer, H., & Ponman, T.J. 2002, ApJ, 578, 74

Franx, M., et al. 2003, ApJ, 587, L79

Fukazawa, Y, Ohashi, T., Fabian, A. C., Canizares, C. R., Ikebe, Y., Makishima, K., Mushotzky, R. F., & Yamashita, K. 1994, PASJ, 46, L55

Fukugita, M., Hogan, C. J., & Peebles, P. J. E. 1998, ApJ, 503, 518

Gibson, B., & Matteucci, F. 1997, MNRAS, 291, L8

Greggio. L. 1996, in The Interplay between Massive Star Formation, the ISM and Galaxy Evolution, ed. D. Kunth et al. (Gif-sur-Yvettes: Edition Frontières), 89

Greggio. L., & Renzini, A. 1983, A&A, 118, 217

Hogg, D. W., et al. 2002, AJ, 124, 646

Hamuy, M. 2004, in Core Collapse of Massive Stars, ed. C. L. Fryer (Dordrecht: Kluwer), in press (astro-ph/0301006)

Heckman, T. M. 2002, in Extragalactic Gas at Low Redshift, ed. J. S. Mulchaey & J. Stocke (San Francisco: ASP), 292

Heckman, T. M., Lehnert, M. D., Strickland, D. K., & Armus, L. 2000, ApJS, 129, 493

Ishimaru, Y., & Arimoto, N. 1997, PASJ, 49, 1

Jablonka, P., Martin, P., & Arimoto, N. 1996, AJ, 112, 1415

Kaiser, N. 1991, ApJ, 383, 104

Kent, S. M. 1992, ApJ, 387, 181

Kent, S. M., Dame, T. M., & Fazio, G. 1991, ApJ, 378, 131

Larson, R. B., & Dinerstein, H. L. 1975, PASP, 87, 511

Loewenstein, M. 2001, ApJ, 557, 573

Loewenstein, M., & Mushotzky, R. F. 1996, ApJ, 466, 695

Maraston, C., Greggio, L., Renzini, A., Ortolani, S., Saglia, R. P., Puzia, T. H., & Kissler-Patig, M. 2003, A&A, 400, 823

Matteucci, F., & Vettolani, G. 1988, A&A, 202, 21

McWilliam, A., & Rich, R. M. 1994, ApJS, 91, 794

Mitchell, R. J., Culhane, J. L., Davison, P. J., & Ives, J. C. 1976, MNRAS, 175, 29P

Mohr, J. J., Mathiesen, B., & Evrard, A. E. 1999, ApJ, 627, 649

Mulchaey, J. S., Davis, D. S., Mushotsky, R. F., Burstein, D. 1993, ApJ, 404, L9

——. 1996, ApJ, 456, 80

Mushotzky, R. F. 1994, in Clusters of Galaxies, ed. F. Durret, A. Mazure, & J. T. Thanh Van (Gyf-sur-Yvette: Editions Frontières), 167

——. 2002, Phil. Trans. R. Soc. Lond. A, 360, 2019

Mushotzky, R. F., Loewenstein, M., Arnaud, K. A., Tamura, T., Fukazawa, Y., Matsushita, K., Kikuchi, K., & Hatsukade, I. 1996, ApJ, 466, 686

Nicastro, F., et al. 2003, Nature, 421, 719

Patat, F., Barbon, R., Cappellaro, E., & Turatto, M. 1994, A&A, 282, 731

Pettini, M. 2004, in Cosmochemistry: The Melting Pot of Elements (Cambridge: Cambridge Univ. Press), in press (astro-ph/0303272)

Pettini, M., Shapley, A. E., Steidel, C. C., Cuby, J.-G., Dickinson, M., Moorwood, A. F. M., Adelberger, K. L., & Giavalisco, M. 2001, ApJ, 554, 981

Pipino, A., Matteucci, F., Borgani, S., & Biviano, A. 2002, NewA, 7, 227

Ponman, T. J., Allan, D. J., Jones, L. R., Merrifield, M., McHardy, I. M., Lehto, H. J., & Luppino, G. A. 1994, Nature, 369, 462

Ponman, T. J., Cannon, D. G., & Navarro, J. F. 1999, Nature, 397, 135

Ponman, T. J., Sanderson, A. J. R., & Finoguenov, A. 2003, MNRAS, 343, 331

Pratt, G. W., & Arnaud, M. 2003, A&A, 408, 1

Renzini, A. 1994, in Clusters of Galaxies, ed. F. Durret, A. Mazure, & J. T. Thanh Van (Gyf-sur-Yvette: Editions Frontières), 221

——. 1997, ApJ, 488, 35

——. 1998a, in The Young Universe: Galaxy Formation and Evolution at Intermediate and High Redshift, ed. S. D'Odorico, A. Fontana, & E. Giallongo (San Francisco: ASP), 298

——. 1998b, AJ, 115, 2459

——. 1999, in The Formation of Galactic Bulges, ed. C. M. Carollo, H. C. Ferguson, & R. F. G. Wyse (Cambridge: Cambridge Univ. Press), 9

——. 2000, in Large Scale Structure in the X-ray Universe, ed. M. Plionis & I. Georgantopoulos (Gif-sur-Yvette: Editions Frontiéres), 103

——. 2002, in Chemical Enrichment of Intracluster and Intergalactic Medium, ed. R. Fusco-Femiano & F. Matteucci (San Francisco: ASP), 331

Renzini, A., Ciotti, L., D'Ercole, A., & Pellegrini, S. 1993, ApJ, 419, 52

Scalo, J. M. 1986, Fund. Cosmic Phys. 11, 1

Simcoe, R. A., Sargent, W. L. W., & Rauch M. 2002, ApJ, 578, 737

Songaila, A., Cowie, L. L., & Lilly, S. J. 1990, ApJ, 348, 371

Stanford, S. A., Eisenhardt, P. R., & Dickinson, M. 1998, ApJ, 492, 461

Thomas, D. 1999, in Chemical Evolution from Zero to High Redshift, ed. J. R. Walsh & M. R. Rosa (Berlin: Springer), 197

Tozzi, P., & Norman, C. 2001, ApJ, 546, 63

Treu, T., Stiavelli, M., Casertano, S., Møller, P., Bertin, G. 2002, Apj, 564, L13

Tsuru, T. 1992, Ph.D. Thesis, Univ. Tokyo

Turatto, M. 2003, in Supernovae and Gamma-Ray Bursters, ed. K. W. Weiler (Berlin: Springer), 21

Valageas, P., & Silk, J. 1999, A&A, 350, 725

van Dokkum, P. G., & Franx, M. 2001, ApJ, 553, 90

Vigroux, L. 1977, A&A, 56, 473

White, D. A. 2000, MNRAS, 312, 663

White, S. D. M., Navarro, J. F., Evrard, A. E., & Frenk, C. S. 1993, Nature, 366, 429

Wu, K. K. S., Fabian, A. C., & Nulsen, P. E. J. 2000, MNRAS, 318, 889

Zoccali, M., et al. 2003, A&A, 399, 931

17

Interactions and mergers of cluster galaxies

J. CHRISTOPHER MIHOS
Case Western Reserve University

Abstract

The high-density environment of galaxy clusters is ripe for collisional encounters of galaxies. While the large velocity dispersion of clusters was originally thought to preclude slow encounters, the infall of smaller groups into the cluster environment provides a mechanism for promoting slow encounters and even mergers within clusters. The dynamical and star-forming response of galaxies to a close encounter depends on both their internal structure and on the collisional encounter speed—fast encounters tend to trigger modest, disk-wide responses in luminous spirals, while slow encounters are more able to drive instabilities that result in strong nuclear activity. While the combined effects of the cluster tidal field and ram pressure stripping make it difficult for individual cluster galaxies to participate in many merger-driven evolutionary scenarios, infalling groups represent a natural site for these evolutionary processes and may represent a "preprocessing" stage in the evolution of cluster galaxies. Meanwhile, the efficiency of tidal stripping also drives the formation of the diffuse intracluster light in galaxy clusters; deep imaging of clusters is beginning to reveal evidence for significant substructure in the intracluster light.

17.1   Interactions of Cluster Galaxies

The importance of collisions in the life of cluster galaxies can be seen through a simple rate argument. A characteristic number of interactions per galaxy can be written as $N \approx n\sigma v t$, where n is the number density of galaxies in a cluster, σ is the cross section for interactions, v is the encounter velocity, and t is the age of the cluster. If $\sigma = \pi r_p^2$, where r_p is the impact parameter, and $v = \sqrt{2}\sigma_v$, then for a cluster like Coma we have

$$ N \approx 4 \left(\frac{n}{250 \text{ Mpc}^{-3}} \right) \left(\frac{r_p}{20 \text{ kpc}} \right)^2 \left(\frac{\sigma_v}{1000 \text{ km s}^{-1}} \right) \left(\frac{t}{10 \text{ Gyr}} \right). $$

While very crude, this calculation shows that it is reasonable to expect that over the course of its lifetime in the cluster, a typical galaxy should experience several close interactions with other cluster members.

While interactions should be common in clusters, they will also be fast. Because the characteristic encounter velocity is much higher than the typical circular velocities of galaxies, these perturbations will be impulsive in nature. Simple analytic arguments suggest that both the energy input and dynamical friction should scale as v^{-2} (Binney & Tremaine 1987), so that a fast encounter does less damage and is much less likely to lead to a merger than are the slow encounters experienced by galaxies in the field. A common view has arisen, therefore, that slow interactions and mergers of galaxies are a rarity in massive clusters (e.g., Ostriker 1980), and that much of the dynamical evolution in cluster galaxy populations is driven by the effects of the global tidal field (e.g., Byrd & Valtonen 1990; Henriksen & Byrd 1996). Combined with the possible effects of ram pressure stripping of the dense interstellar medium (ISM) (Gunn & Gott 1972) or hot gas in galaxy halos ("strangulation"; Larson, Tinsley, & Caldwell 1980), a myriad of processes, aside from galaxy interactions themselves, seemed available to transform cluster galaxies.

However, the abandonment of individual collisions as a mechanism to drive cluster galaxy evolution has proved premature. More recent work on the dynamical evolution of cluster galaxies has emphasized the importance of fast collisions. Moore et al. (1996) and Moore, Lake, & Katz (1998) have shown that *repeated* fast encounters, coupled with the effects of the global tidal field, can drive a very strong response in cluster galaxies. For galaxy-like potentials, the amount of heating during an impulsive encounter scales like $\Delta E/E \sim r_p^{-2}$, such that distant encounters impart less energy. However, this effect is balanced by the fact that under simple geometric weighting the number of encounters scales as r_p^2, so that the total heating, summed over all interactions, can be significant. While this simple argument breaks down when one considers more realistic galaxy potentials and the finite time scales involved, in hindsight it is not surprising that repeated high-speed encounters—"galaxy harassment"—should drive strong evolution. However, the efficacy of harassment is largely limited to low-luminosity hosts, due to their slowly rising rotation curves and low-density cores. In luminous spirals, the effects of harassment are much more limited (Moore et al. 1999). This effect leads to a situation where harassment can effectively describe processes such as the formation of dwarf ellipticals (Moore et al. 1998), the fueling of low-luminosity AGNs (Lake, Katz, & Moore 1998), and the destruction of low-surface brightness galaxies in clusters (Moore et al. 1999), but is less able to explain the evolution of luminous cluster galaxies.

Even as the effects of high-speed collisions are being demonstrated, a new attention is focusing on slow encounters and mergers in clusters. The crucial element that is often overlooked in a classical discussion of cluster interactions is the fact that structure forms *hierarchically*. Galaxy clusters form not by accreting individual galaxies randomly from the field environment, but rather through the infall of less massive groups falling in along the filaments that make up the "cosmic web." Observationally, evidence for this accretion of smaller galaxy groups is well established. Clusters show ample evidence for substructure in X-rays, galaxy populations, and velocity structure (see, e.g., reviews by Buote 2002; Girardi & Biviano 2002). These infalling groups have velocity dispersions that are much smaller than that of the cluster as a whole, permitting the slow, strong interactions normally associated with field galaxies. Imaging of distant clusters show populations of strongly interacting and possibly merging galaxies (e.g., Dressler et al. 1997; van Dokkum et al. 1999), which may contribute to the Butcher-Oemler effect (e.g., Lavery & Henry 1988). Even in nearby (presumably dynamically older) clusters, several notable examples of strongly interacting systems exist (e.g., Schweizer 1998; Dressler, this volume), including the classic Toomre-sequence pair "The Mice" (NGC 4476) located at a projected distance of $\sim$5 Mpc from the center of the Coma cluster. Interactions in the infalling group environment may in effect represent a "preprocessing" step in the evolution of cluster galaxies.

While the observational record contains many examples of group accretion and slow encounters, numerical simulations are also beginning to reveal this evolutionary path for clus-

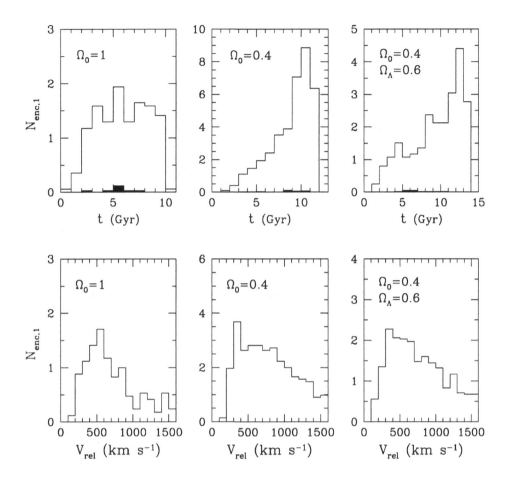

Fig. 17.1. Top: Number of close encounters per galaxy per Gyr in a simulated $\log(M/M_\odot) =$ 14.6 cluster under different cosmologies. Bottom: The distribution of galaxy encounter velocities in the simulated clusters. (From Gnedin 2003.)

ter galaxies. Modern N-body calculations can follow the evolution of individual galaxy-mass dark matter halos in large-scale cosmological simulations, allowing the interaction and merger history of cluster galaxies to be probed. Ghigna et al. (1998) tracked galaxy halos in an $\Omega_m = 1, \log(M/M_\odot) = 14.7$ cluster and showed at late times ($z < 0.5$) that, while no mergers occurred within the inner virialized 1.5 Mpc of the cluster, in the outskirts the merger rate was $\sim$5%–10%. Similar models by Dubinski (1998) confirmed these results, and showed more intense activity between $z = 1$ and $z = 0.4$. Gnedin (2003) expanded on these works by studying the interaction and merger rates in clusters under different cosmological models (Fig. 17.1). In $\Omega_m = 1$ cosmologies, the encounter rates in clusters stays relatively constant with time as the cluster slowly accretes, while under open or Λ-dominated cosmologies, the interaction rate increases significantly once the cluster virializes, as the galaxies experience many high-speed encounters in the cluster core. The distribution of velocities in these encounters shows a large tail to high encounter velocities, but a significant fraction of encounters, largely those in the cluster periphery or those occurring at higher redshift, before the cluster has fully collapsed, occur at low relative velocities ($v_{rel} < 500$ km s^{-1}). Clearly, not *all* interactions are of the high-speed variety!

In summary, clusters are an active dynamical environment, with a multitude of processes available to drive evolution in the galaxy population. Disentangling these different processes continues to prove difficult, in large part because nearly all of them correlate with cluster richness and clustercentric distance. Indeed, seeking to isolate the "dominant" mechanism driving evolution may be ill motivated, as these processes likely work in concert, such as the connection between high-speed collisions and tidal field that describes galaxy harassment. What is clear from both observational and computational studies is that slow encounters and mergers of galaxies can be important over the life of a cluster—at early times when the cluster is first collapsing, and at later times in the outskirts as the cluster accretes groups from the field.

Finally, it is also particularly important to remember two points. First, clusters come in a range of mass and richness: not every cluster is as massive as the archetypal Coma cluster, with its extraordinarily high velocity dispersion of $\sigma_v = 1000$ km s^{-1}. In smaller clusters and groups, the lower velocity dispersion will slow the encounter velocities and make them behave more like field encounters. Second, as we push observations out to higher and higher redshift, we begin to probe the regime of cluster formation, where unvirialized dense environments can host strong encounters.

17.2   Lessons from the Field ...

To understand the effects of interactions and mergers on cluster galaxies, we start with lessons learned from the study of collisions in the field environment. In §17.3, we will then ask how the cluster environment modifies these results. In this discussion, we focus largely on two aspects of encounters: the triggering of starbursts and nuclear activity, and the late-time evolution of tidal debris and possible reformation of gaseous disks.

The role of galaxy interactions in driving activity and evolution of field spirals has been well documented through a myriad of observational and theoretical studies. Models of interactions have demonstrated the basic dynamical response of galaxies to close encounter (e.g., Toomre & Toomre 1972; Negroponte & White 1982; Barnes 1988, 1992; Noguchi 1988; Barnes & Hernquist 1991, 1996; Mihos & Hernquist 1994, 1996). Close interactions can lead to a strong internal dynamical response in the galaxies, driving the formation of spiral arms and, depending on the structural properties of the disks, strong bar modes. These non-axisymmetric structures lead to compression and inflow of gas in the disks, elevating star formation rates and fueling nuclear starburst/AGN activity. If the encounter is sufficiently close, dynamical friction leads to an eventual merging of the galaxies, at which time violent relaxation destroys the dynamically cold disks and produces a kinematically hot merger remnant with many of the properties found in the field elliptical galaxy population (see, e.g., Barnes & Hernquist 1992).

Observational studies support much of this picture. Interacting systems show preferentially elevated star formation rates, enhanced on average by factors of a few over those of isolated spirals (Larson & Tinsley 1978; Condon et al. 1982; Keel et al. 1985; Kennicutt et al. 1987). Nuclear starbursts are common, with typical starburst mass fractions that involve a few percent of the luminous mass (Kennicutt et al. 1987). More dramatically, infrared-selected samples of galaxies reveal a population of interacting "ultralumi-

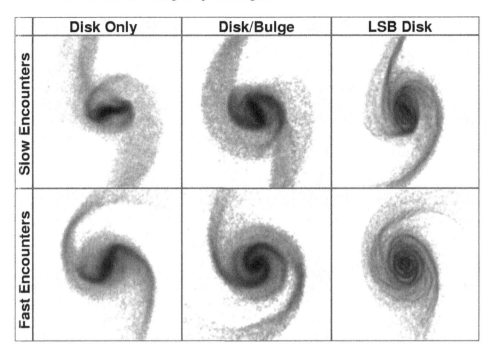

Fig. 17.2. The morphological response of galaxies to a close encounter. The galaxy models are viewed one rotation period after the initial collision. Top panels show the response to a slow, parabolic encounter, while the bottom panels show the response to fast encounters. The left columns show the response of a pure disk system, the middle panels show a disk/bulge system, and the right panels show a low-density, dark matter dominated disk.

nous infrared galaxies," where star formation rates are elevated by 1–2 orders of magnitude and dust-enshrouded nuclear activity is common (Soifer et al. 1984; Lawrence et al. 1989). These systems are preferentially found in late-stage mergers (e.g., Veilleux, Kim, & Sanders 2002) and have been suggested as the precursors of luminous quasars (Sanders et al. 1988). This diversity of properties for interacting systems argues that the response of a galaxy to a close interaction is likely a complicated function of encounter parameters, galaxy type, local environment, and gas fraction.

Can we isolate the different determining factors to understand what drives the strong response in interacting systems? Numerical modeling of interactions has shown that gaseous inflow and central activity in an interacting disk is driven largely by gravitational torques acting on the gas—not from the companion galaxy, but by the developing non-axisymmetric structures (spiral arms and/or central bar) in the host disk (Noguchi 1988; Barnes & Hernquist 1991; Mihos & Hernquist 1996). This result argues that the structural properties of galaxies play a central role in determining the response to interactions. In particular, disks that are stable against the strong growth of disk instabilities will experience a weaker response, exhibiting modestly enhanced, disk-wide star formation (unless and until they ultimately merge). This stability can be provided by the presence of a centrally concentrated bulge (Mihos & Hernquist 1996) or a lowered disk surface density (at fixed rotational speed; Mihos, McGaugh, & de Blok 1997). In contrast, disk-dominated systems are more susceptible to global bar modes and experience the strongest levels of inflow and nuclear activity (see § 17.3).

Interacting and merging galaxies also show a wide variety of tidal features, from long thin tidal tails to plumes, bridges, and other amorphous tidal debris. The evolution of this material was first elegantly described by the computer models of Toomre & Toomre (1972) and Wright (1972). Gravitational tides during a close encounter lead to the stripping of loosely bound material from the galaxies (see Fig. 17.3); rather than being completely liberated, the lion's share of this material (> 95%) remains bound, albeit weakly, to its host galaxy (Hernquist & Spergel 1992; Hibbard & Mihos 1995). Material is sorted in the tidal tails by a combination of energy and angular momentum—the outer portions of the tails contain the least bound material with the highest angular momentum. At any given time, material at the base of the tail has achieved turn-around and is falling back toward the remnant. Further out in the tail, material still expands away, resulting in a rapid drop in the luminosity density of the tidal tails due to this differential stretching. As a result, the detectability of these tidal features is a strong function of age and limiting surface brightness; after a few billion years of dynamical evolution, they will be extremely difficult to detect (Mihos 1995).

In mergers, the gas and stars ejected in the tidal tails fall back onto the remnant in a long-lived "rain" that spans many billions of years (Hernquist & Spergel 1992; Hibbard & Mihos 1995). In the merger simulation shown in Figure 17.3, this fallback manifests itself as loops of tidal debris that form as stars fall back through the gravitational potential of the remnant. Tidal gas will follow a different evolution, as it shocks and dissipates energy as it falls back. The most tightly bound gaseous material returns to the remnant over short time scales and can resettle into a warped disk (Mihos & Hernquist 1996; Naab & Burkert 2001; Barnes 2002). Such warped H I disks have been observed in NGC 4753 (Steiman-Cameron, Kormendy, & Durisen 1992) and in the nearby merger remnant Centaurus A (Nicholson, Bland-Hawthorn, & Taylor 1992). Over longer time scales, the loosely bound, high-angular momentum gas falls back to ever-increasing radii, forming a more extended but less-organized distribution of gas outside several effective radii in the remnant. Many elliptical galaxies show extended neutral hydrogen gas, sometimes in the form of broken rings at large radius, perhaps arising from long-ago merger events (e.g., van Gorkom & Schiminovich 1997).

The ultimate fate of this infalling material is uncertain, but may have important ramifications for interaction-driven galaxy evolution models. If efficient star formation occurs in this gas, such as that observed in the inner disk of the merger remnant NGC 7252 (Hibbard et al. 1994), this may present a mechanism for building disks in elliptical galaxies. If the amount of gas resettling into the disk is significant, in principle the remnant could evolve to become a spheroidal system with a high bulge-to-disk ratio, perhaps forming an S0 or Sa galaxy (e.g., Schweizer 1998).

17.3   ... Applied to Clusters

In clusters, a number of environmental effects may modify the dynamical response and evolution of interacting galaxies described above. First, the relative velocities of interacting systems tend to be higher, although, as argued earlier, many low-velocity encounters still occur within smaller groups falling in from the cluster periphery. Second, the global tidal field of the cluster must also play a role in the evolution of interacting systems, stripping away the loosely bound tidal material and potentially adding energy to bound groups.

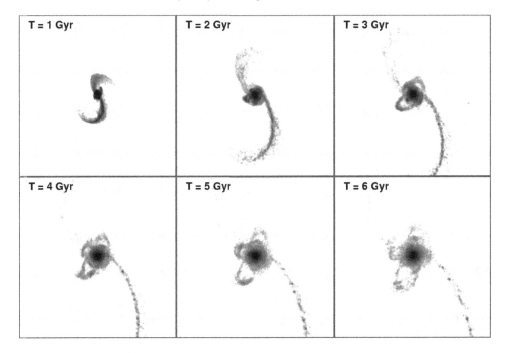

| T = 1 Gyr | T = 2 Gyr | T = 3 Gyr |
| T = 4 Gyr | T = 5 Gyr | T = 6 Gyr |

Fig. 17.3. Evolution of the tidal debris in an equal-mass merger of two disk galaxies occurring in isolation. Each frame is approximately 0.9 Mpc on a side. Note the sharpness of the tidal debris, as well as the loops that form as material falls back into the remnant over long time scales.

The hot intracluster medium (ICM) can act to further strip out low-density ISM in galaxies, particularly the diffuse tidal gas ejected during collisions.

The most obvious difference between interactions in the field and in the cluster environment is the collision speed of the encounter. While slow interactions are able to drive a strong dynamical response in disk galaxies, faster encounters result in a perturbation that is much shorter lived and less resonant with the internal dynamics of the disk. Figure 17.2 shows the different response of galaxies to slow and fast collisions. In each case, the galaxies experience an equal-mass encounter inclined 45° to the orbital plane and with a closest approach of six disk scale lengths. Three structural models are used for the galaxies. The first is a pure disk system where the disk dominates the rotation curve in the inner two scale lengths; the second is identical to the first, but with a central bulge with bulge-to-disk ratio of 1:3. In both these models, the disk-to-halo mass ratio is 1:5.8. The third system is identical · to the first, but with the disk surface density lowered by a factor of 8; this system represents a dark matter dominated, low-surface brightness (LSB) disk galaxy. In the slow collision, the galaxies fall on a parabolic (zero-energy) orbit with a velocity at closest approach that is approximately twice the circular velocity. The fast collision takes place with a hyperbolic orbit with an encounter velocity of twice that of the parabolic encounters.

Notable differences can be seen in the dynamical response of the galaxy models, particularly in the growth of global bar modes. During slow encounters, both the disk and disk/bulge galaxy models develop dramatic bars and spiral arms, which can drive strong inflow and central activity. The lowered surface density of the LSB model results in a weaker self-gravitating response (Mihos et al. 1997); a very small, weak bar is present, but the overall response is one of a persistent oval distortion, which would be much less able to drive gaseous inflow. In contrast, the response of the fast encounters depends more strongly on the structural properties of the galaxy. The pure-disk model develops a relatively strong bar mode, while the disk/bulge system sports a two-arm spiral pattern with no central bar. These results are similar to those shown in Moore et al. (1999), who modeled high-speed encounters of disk galaxies of varying structural properties. The LSB model, on the other hand, lacks the disk self-gravity to amplify the perturbation into any strong internal response. However, the vulnerability of LSBs lies not in their internal response to a *single* encounter, but rather in their response to *repeated* high-speed collisions in the cluster environment (Moore et al. 1998).

Because the star-forming response of a galaxy is intimately linked to its dynamical response, we can use these results to guide our expectations of starburst triggering mechanisms in cluster galaxies. Because of their stability toward high-speed encounters, luminous, early-type spirals should experience modestly enhanced, disk-wide starbursts. Low-luminosity, late-type disks will be more susceptible to stronger inflows, central starbursts, and AGN fueling; even the LSBs will succumb to the effects of repeated encounters and the cluster tides (Moore et al. 1996), which drive a much stronger response. As a result, these high-speed, "harassment-like" encounters are effective at driving evolution in the low-luminosity cluster populations (Moore et al. 1996; Lake et al. 1998), but if harassment is the whole story in driving cluster galaxy evolution, it is hard to explain strong starburst activity in *luminous* cluster spirals at moderate redshift. On the other hand, the slower collisions expected in infalling substructure are able to drive a stronger response regardless of galaxy type.

Aside from driving stronger starbursts in interacting cluster galaxies, slow collisions also heat and strip galaxies more efficiently than do high-speed encounters. They also raise the possibility for mergers among cluster galaxies, and the potential for merger-driven evolutionary scenarios. However, unlike slow collisions in the field, cluster galaxies must also contend with the effects of the overall tidal field of the cluster. How will this affect the evolution of close interactions, in particular the longevity and detectability of tidal debris, and the ability for galaxies to reaccrete tidal material? To address this question, Figure 17.4 shows the evolution of an equal-mass merger (identical to the merger shown in Fig. 17.3) occurring in a cluster tidal field. The cluster potential is given by an Coma-like Navarro, Frenk, & White (1996) profile with total mass $M_{200} = 10^{15} M_{\odot}$, $r_{200} = 2$ Mpc, and $r_s = 300$ kpc. The binary pair travels on an orbit with $r_{peri} = 0.5$ Mpc and $r_{apo} = 2$ Mpc, and passes through periclustercon twice, at T $\approx$ 1 and 4 Gyr.

The tidal field has a number of effects on the evolution of this system. First, the merger time scale has been lengthened—the cluster tidal field imparts energy to the galaxies' orbits, extending the time it takes for the system to merge (by $\sim$50% for this calculation). In this case, the encounter is close enough that the galaxies do still ultimately merge, but it is not hard to envision encounters where the tidal energy input is sufficient to unbind the galaxy pair. This raises the interesting possibility that infalling pairs may experience the close, slow collisions that drive strong activity, yet survive the encounter whole without merging.

The most dramatic difference between field mergers and those in a cluster is in the evolution of the tidal debris. In the field, most of the material ejected into the tidal tails remains loosely bound to the host galaxy, forming a well-defined tracer of the tidal encounter. In the

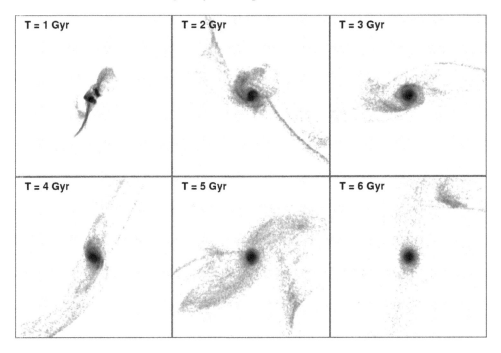

Fig. 17.4. Evolution of an equal-mass merger, identical to that in Fig. 17.3, but occurring as the system orbits through a Coma-like cluster potential (see text). Note the rapid stripping of the tidal tails early in the simulation; the tidal debris seen here is more extended and diffuse than in the field merger, and late infall is shut off due to tidal stripping by the cluster potential.

cluster encounter, the tidal field quickly strips this loosely bound material from the galaxies, dispersing it throughout the cluster and adding to the diffuse intracluster starlight found in galaxy clusters. This rapid stripping of the tidal debris is clear in Figure 17.4: shortly after the first passage the debris becomes very extended and diffuse and is removed entirely from the system after the subsequent passage through the cluster interior. In this encounter, 20% of the stellar mass is stripped to large distances (> 50 kpc), fully twice the amount in a similar field merger.

This rapid stripping has a number of important ramifications. First, these tidal tracers are very short-lived; identifying a galaxy as a victim of a close interaction or merger will be very difficult indeed shortly after it enters the cluster potential. The tidal features we do see in cluster galaxies are likely signatures of a very recent interaction, such that interaction rates derived from the presence of tidal debris may underestimate the true interaction rates in clusters. Second, the rebuilding/resettling of gaseous disks in interaction/merger remnants will be severely inhibited, as both cluster tides and ram pressure stripping act to strip off all but the most bound material in the tidal debris, leaving little material able to return. Finally, this stripping will contribute both to the intracluster light and to the ICM, as gas and stars in the tidal debris are mixed in to the diffuse cluster environment. We discuss these processes in the following sections.

17.4   Galaxy Evolution: Mergers, Elliptical, and S0 Galaxies

In the field environment, interactions and mergers of galaxies are thought to be a prime candidate for driving evolution in galaxy populations. Major mergers of spiral galaxies may lead to the transformation of spirals into ellipticals (e.g., Toomre 1978; Schweizer 1982): the violent relaxation associated with a merger effectively destroys the galactic disks and creates a kinematically hot, $r^{1/4}$-law spheroid, while the concurrent intense burst of star formation may process the cold ISM of a spiral galaxy into the hot X-ray halo of an elliptical. These processes are not totally efficient, however, and leave signatures behind that identify the violence of the merging process: diffuse loops and shells of starlight, extended H I gas, significant rotation in the outskirts of the remnant, and dynamically distinct cores (see, e.g., the review by Schweizer 1998). While it is an open question as to what fraction of ellipticals formed this way, it is clear that at least some nearby field ellipticals—Centaurus A being a notable example (Schiminovich et al. 1994)—have had such violent histories.

Mergers have also been proposed as a mechanism to drive the formation of S0 galaxies. In this case, the scenarios are varied. For S0s with very large bulge-to-disk ratios, reaccretion of gas after a major merger (either from returning tidal material or from the surrounding environment) may rebuild a disk inside a newly formed spheroid. In unequal-mass mergers of disk galaxies, disk destruction is not complete, and the resulting remnant retains a significant amount of rotation (Bendo & Barnes 2000; Cretton et al. 2001) and may be identified with a disky S0, particularly if a significant amount of cold gas is retained by the system to reform a thin disk (Bekki 1998). Finally, minor mergers between spirals and their satellite companions can significantly heat, but not destroy, galactic disks (Toth & Ostriker 1992; Quinn, Hernquist, & Fullagar 1993; Walker, Mihos, & Hernquist 1996), while simultaneously helping to "sweep the disk clean" of cold gas via a gravitationally induced bar driving gas to the nucleus (Hernquist & Mihos 1995). The resulting disks have many similarities to disky S0s (Mihos et al. 1995): thickened disks, little or no spiral structure, cold gas, or ongoing star formation. These different scenarios vary mainly in the proposed strength of the interaction—from major to minor mergers—and it has been proposed that this parameter may, in fact, determine the ultimate morphological classification of galaxies all the way from early-type ellipticals to late-type spirals (e.g., Schweizer 1998; Steinmetz & Navarro 2002).

Applying these arguments to cluster populations, it seems that building cluster ellipticals through a wholesale merging of spirals within the established cluster environment is a difficult proposition. Clusters ellipticals are an old, homogeneous population showing little evolution since at least a redshift of $z \approx 1$ (e.g., Dressler et al. 1997; Ellis et al. 1997). Within the cores of massive clusters, merging has largely shut off due to the high velocity dispersion of the virialized cluster (Ghigna et al. 1998). The accretion of merger-spawned ellipticals from infalling groups may still occur, and these will be hard to identify morphologically as merger remnants—the combination of cluster tides and hot ICM will strip off any telltale tidal debris and sweep clean any diffuse cold gas in the tidal tails (recall Fig. 17.4) or low-density reaccreting disk. However, the small scatter in the color-magnitude relation and weak evolution of the fundamental plane of cluster ellipticals (see, e.g., the review by van Dokkum 2002) argues that such lately formed ellipticals likely do not contribute to the bulk of the cluster elliptical population. This does not mean that mergers have not played a role in the formation of cluster ellipticals. In any hierarchical model for structure formation, galaxies form via the accretion of smaller objects. Luminous cluster ellipticals may well have formed from mergers of galaxies at high redshift, in the previrialized environment of the protocluster. However, at these redshifts ($z \gg 1$), the progenitor galaxies are likely to have looked very different from the present-day spiral population.

Unlike the rather passive evolution observed in cluster ellipticals, much stronger evolution is observed in the population of cluster S0s. The fraction of S0s in rich cluster has increased significantly since a redshift of $z \approx 1$, with a corresponding decrease in the spiral fraction (Dressler et al. 1997). Can the same collisional processes that have been hypothesized to drive S0 formation in the field account for the dramatic evolution in cluster S0 populations? S0 formation scenarios that rely on reaccretion of material after a major merger (disk rebuilding schemes) seem difficult to envision wholly within the cluster environment. While mergers are possible in infalling groups, the combination of tidal and ram pressure stripping will shut down reaccretion and ablate any low-density gaseous disks that have survived the merger process. For example, it is unlikely the H I disk in Centaurus A (Nicholson et al. 1992), likely a product of merger accretion, would survive passage through the hot ICM of a dense cluster. Satellite merger mechanisms trade one dynamical problem for another—because the mergers involve bound satellite populations there is no concern about the efficacy of high-speed mergers, but, instead, the issue is whether or not satellite populations can stay bound to their host galaxy as it moves through the cluster potential. And, of course, this mechanism relies on the very *local* environment of galaxies, which does not explain why S0 formation would be enhanced in clusters.

None of the proposed merger-driven S0 formation mechanisms appear to work well deep inside the cluster potential. On the other hand, these processes should operate efficiently in the group environment, where the encounter velocities are smaller and cluster tides and the hot ICM do not play havoc with tidal reaccretion. The group environment may create S0s and feed them into the accreting cluster, but if there is wholesale transformation of cluster spirals into S0s in the cluster environment, it needs to occur via other mechanisms.

Other cluster-specific methods for making S0 galaxies have been proposed, including collisional heating and ram pressure stripping of the dense ISM (Moore et al. 1999; Quilis, Moore, & Bower 2000) and strangulation, the stripping of hot halo gas from spirals (Larson et al. 1980; Bekki, Couch, & Shioya 2002). While these models, by design, explain the preferential link between clusters and S0 galaxies, they are not without problems themselves. While the effects of ram pressure stripping on the extended neutral hydrogen gas in cluster galaxies is clear (e.g., van Gorkom, this volume), its efficacy on the denser molecular gas is unclear. For example, the H I deficient spirals in the Virgo cluster still contain significant quantities of molecular gas (e.g., Kenney & Young 1989), while studies of the molecular content of cluster spirals show no deficit of CO emission (Casoli et al. 1998). If the molecular ISM survives, it is unclear why star formation should not continue in these disks. Strangulation models suffer less from concerns of the efficacy of ram pressure stripping, since it is much easier to strip low-density halo gas than a dense molecular ISM, although it must be noted that there currently is little observational evidence for hot halos in (non-starbursting) spiral galaxies. In addition, neither of these methods leads to the production of a luminous spheroid—the S0s that might be produced in these ways would have low bulge-to-disk ratios.

Ultimately, S0s are a heterogeneous class, from bulge-dominated S0s to the disky S0s seen in galaxy clusters, and it should not be surprising that a single mechanism cannot fully account for the range of S0 types (e.g., Hinz, Rix, & Bernstein 2001). Whether there is a systematic difference between cluster and field S0s is unclear, an issue fraught with selection and classification uncertainties. What is clear is that, even in clusters, S0s often show evidence for accretion events, similar to that observed in the field S0 population (see, e.g., , the discussion in Schweizer 1998). It is likely that many of these S0s were "processed" via mergers in the group environment before being incorporated into clusters.

17.5   Tidal Stripping and Intracluster Light

As galaxies orbit in the potential well of a galaxy cluster, stars are tidally stripped from their outer regions, mixing over time to form a diffuse "intracluster light" (ICL). First proposed by Zwicky (1951), the ICL has proved very difficult to study—at its *brightest*, it is only $\sim$1% of the brightness of the night sky. Previous attempts to study the ICL have resulted in some heroic detections (Oemler 1973; Thuan & Kormendy 1977; Bernstein et al. 1995; Gregg & West 1998; Gonzalez et al. 2000), verified by observations of intracluster stars and planetary nebulae in Virgo (Feldmeier, Ciardullo, & Jacoby 1998; Ferguson, Tanvir, & von Hippel 1998; Arnaboldi et al. 2002). While the ICL is typically thought of as arising from the stripping of starlight due to the cluster potential, in fact the role of interactions between cluster galaxies in feeding the ICL is quite strong. Galaxy interactions significantly enhance the rate at which material is stripped; as illustrated in Figure 17.4, the strong, local tidal field of a close encounter can strip material from deep within a galaxy's potential well, after which the cluster tidal field can liberate the material completely. Interactions, particularly those in infalling groups, act to "prime the pump" for the creation of the ICL.

The properties of the ICL in clusters, particularly the fractional luminosity, radial light profile, and presence of substructure, may hold important clues about the accretion history and dynamical evolution of galaxy clusters. Material stripped from galaxies falling in the cluster potential is left on orbits that trace the orbital path of the accreted galaxy, creating long, low-surface brightness tidal arcs (e.g., Moore et al. 1996), which have been observed in a few nearby clusters (Trentham & Mobasher 1998; Calcáneo-Roldán et al. 2000). However, these arcs will only survive as discrete structures if the potential is quiet; substructure will dynamically heat these arcs, and the accretion of significant mass (i.e., a cluster merger event) may well destroy these structures. If much of the ICL is formed early in a cluster's dynamical history, before the cluster has been fully assembled, the bulk of the ICL will be morphologically smooth and well mixed by the present day, with a few faint tidal arcs showing the effects of late accretion. In contrast, if the ICL formed largely after cluster virialization, from the stripping of "quietly infalling" galaxies, the ICL should consist of an ensemble of kinematically distinct tidal debris arcs. Clusters that are dynamically younger should also possess an ICL with significant kinematic and morphological substructure.

Early theoretical studies of the formation of the ICL suggested that it might account for anywhere from 10% to 70% of the total cluster luminosity (Richstone 1976; Merritt 1983, 1984; Miller 1983; Richstone & Malumuth 1983; Malumuth & Richstone 1984). These studies were based largely on analytic estimates of tidal stripping, or on simulations of individual galaxies orbiting in a smooth cluster potential well. Such estimates miss the effects of interactions with individual galaxies (e.g., Moore et al. 1996), intermediate-scale substructure (Gnedin 2003), and priming due to interactions in the infalling group environment. As a result, these models underpredict the total amount of ICL as well as the heating of tidal streams in the ICL. Now, however, cosmological simulations can be used to study cluster

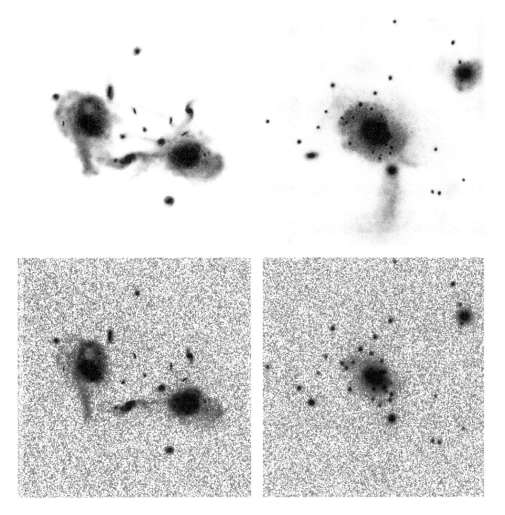

Fig. 17.5. Visualizations of the cluster simulations of Dubinski (1998). Top panels show the distribution of luminous starlight, with the faintest contours corresponding to a surface brightness of $\mu_V \approx 30$ mag arcsec^{-2}. Bottom panels show the effect of adding noise characteristic of current observational limits. Left panels show the cluster early in collapse (at $z = 2$), while right panels show the virialized cluster at $z = 0$.

collapse and tidal stripping at much higher resolution and with a cosmologically motivated cluster accretion history (e.g., Moore et al. 1998; Dubinski, Murali, & Ouyed 2001).

One example of the modeling of ICL is shown in Figure 17.5. These images are derived from the *N*-body simulations of Dubinski (1998), who simulated the collapse of a $\log(M/M_\odot) = 14.0$ cluster in a standard cold dark matter Universe. Starting from a cosmological dark matter simulation, the 100 most massive halos are identified at a redshift of $z = 2.2$ and replaced with composite disk/bulge/halo galaxies, whereafter the simulation is continued to $z = 0$ (see Dubinski 1998 for more details). To quantify the diffuse light in these cluster models, we assign luminosity to the stellar particles based on a mass-to-light ratio of

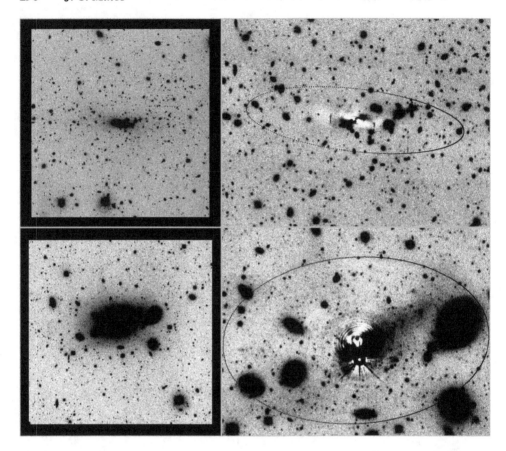

Fig. 17.6. Deep imaging of cD clusters from Feldmeier et al. (2002). Top panels show Abell 1413; bottom panels show MKW 7. The left panels show the full $10' \times 10'$ view of the cluster, while the right panels show a close up of the clusters once a smooth elliptical fit to the cD cluster envelope has been removed. The oval shows the radius inside which the model has been subtracted.

1. The top panels show the cluster at two different times. On the left, the cluster is shown early in the collapse, at $z = 2$, where it consists of two main groups coming together. The right panels show the cluster at $z = 0$, when the cluster has virialized and formed a massive cD galaxy at the center. In each case, the lowest visible contour is at a surface brightness of $\mu_V \approx 30$ mag arcsec^{-2}. The bottom panels show the effects of adding observational noise typical of our ICL imaging data (discussed below) and illustrate the difficulties in detecting this diffuse light.

In the early stages of cluster collapse, material is being stripped out of galaxies and into the growing ICL component. This material has a significant degree of spatial structure in the form of thin streams and more diffuse plumes, much of it at observationally detectable surface brightnesses. At later times this material has become well mixed in the virialized cluster, forming a much smoother distribution of ICL and substructure that is visible only at much fainter surface brightnesses, well below current levels of detectability. Along these

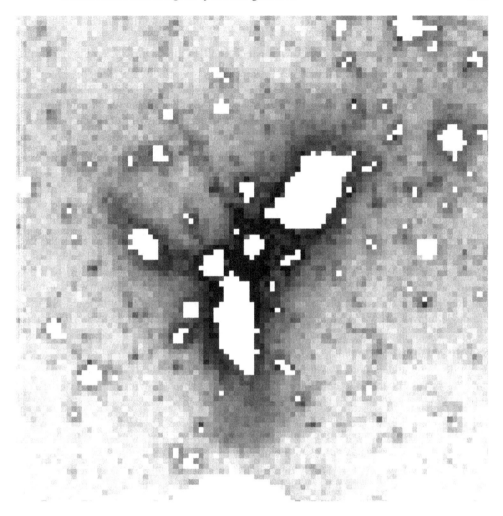

Fig. 17.7. The ICL in Abell 1914 (from Feldmeier et al. 2004).

lines, the degree of ICL substructure may act as a tracer of the dynamical age of galaxy clusters.

Indeed, galaxy clusters do show a range of ICL properties. We (Feldmeier et al. 2002, 2004) have recently begun a deep imaging survey of galaxy clusters, aimed at linking their morphological properties to the structure of their ICL. As the detection of ICL is crucially dependent on reducing systematic effects in the flat fielding, we have taken significant steps to alleviate these issues, including imaging in the Washington M filter to reduce contamination from variable night sky lines, flat fielding from a composite of many night sky flats taken at similar telescope orientations, and aggressive masking of bright stars and background sources (see Feldmeier et al. 2002 for complete details). With this data, we achieve a signal-to-noise ratio of 5 at $\mu_V = 26.5$ mag arcsec^{-2} and a signal-to-noise ratio of 1 at $\mu_V = 28.3$ mag arcsec^{-2}. We have targeted two types of galaxy clusters thus far: cD-dominated Bautz-

Morgan class I clusters (Feldmeier et al. 2002) and irregular Bautz-Morgan class III clusters (Feldmeier et al. 2004).

Figure 17.6 shows results for the cD clusters Abell 1413 and MKW 7. Similar to the clusters studied by Gonzalez et al. (2000; see also Gonzalez, Zabludoff, & Zaritsky 2004), the cD galaxies are well fit by a $r^{1/4}$ law over a large range in radius, with only a slight luminosity excess in the outskirts of each cluster. In each case, we search for ICL substructure by using the STSDAS `ellipse` package to subtract a smooth fit to the cD galaxy extended envelope. In the case of Abell 1413, we see little evidence for any substructure in the ICL; the small-scale arcs we observe are likely to be due to gravitational lensing. MKW 7 shows a broad plume extending from the cD galaxy to a nearby bright elliptical, but little else in the way of substructure.

In contrast, we see evidence for more widespread ICL substructure in our Bautz-Morgan type III clusters. Figure 17.7 shows our image of Abell 1914, binned to a resolution of $3''$ after all stars and galaxies have been masked. Here we see a variety of features: a fan-like plume projecting from the southern clump of galaxies, another diffuse plume extending from the galaxy group to the east of the cluster, and a narrow stream extending to the northeast from the cluster center. The amount of substructure seen here is consistent with an unrelaxed cluster experiencing a merger, similar to the features seen in the unrelaxed phase of the model cluster shown in Figure 17.5. We see similar plumes in other type III clusters, suggesting that the ICL in these types of clusters does reflect a cluster that is dynamically less evolved than the cD-dominated clusters of Feldmeier et al. (2002).

While these studies point toward significant substructure in the ICL of galaxy clusters, imaging surveys continue to be hampered by systematic effects. With so much of the ICL substructure present only at surface brightnesses fainter than $\mu_V > 28$ mag arcsec^{-2}, issues of flat fielding, scattered light, and sky variability become severe. An interesting alternative is to use the significant numbers of intracluster planetary nebulae now being found in emission-line surveys of nearby galaxy clusters (Feldmeier et al. 1998; Arnaboldi et al. 2002). These studies have very different detection biases than deep surface photometry and have the potential to probe the ICL down to much lower surface densities. Planetary nebulae offer an added bonus: as emission-line objects, follow-up spectroscopy can determine the kinematics of the ICL, giving yet another view of the degree to which the ICL is dynamically relaxed (Dubinski et al. 2001). An interesting analogy can be made between the search for kinematic substructure due to tidal stripping in galaxy clusters and the search for kinematic substructure due to tidally destroyed satellites in the Milky Way's halo (e.g., Morrison et al. 2002). In both cases, kinematic substructure can be used to trace the dynamical accretion history of the system. With the advent of multi-object spectrographs on 8-m class telescopes, new and exciting opportunities now exist for studying this substructure in the diffuse starlight of galaxy clusters.

Acknowledgements. I would like to thank all my collaborators for their contributions to this work, and John Feldmeier in particular for providing Figures 17.5 – 17.7. John Dubinski graciously provided the cluster simulations from which Figure 17.5 was made. I also thank the conference organizers for putting together such a wonderful scientific program. This research has been supported in part by an NSF Career Award and a Research Corporation Cottrell Scholarship.

References

Arnaboldi, M., et al. 2002, AJ, 123, 760

Barnes, J. E. 1988, ApJ, 331, 699

——. 1992, ApJ, 393, 484

——. 2002, MNRAS, 333, 481

Barnes, J. E., & Hernquist, L. E. 1991, ApJ, 370, L65

——. 1992, ARA&A, 30, 705

——. 1996, ApJ, 471, 115

Bekki, K. 1998, ApJ, 502, L133

——. 2001, Ap&SS, 276, 847

Bekki, K., Couch, W. J., & Shioya, Y. 2002, ApJ, 577, 651

Bendo, G. J., & Barnes, J. E. 2000, MNRAS, 316, 315

Bernstein, G. M., Nichol, R. C., Tyson, J. A., Ulmer, M. P., & Wittman, D. 1995, AJ, 110, 1507

Binney, J., & Tremaine, S. 1987, Galactic Dynamics (Princeton: Princeton Univ. Press)

Buote, D. A. 2002, in Merging Processes in Galaxy Clusters, ed. L. Feretti, I. M. Gioia, & G. Giovannini (Dordrecht: Kluwer), 79

Byrd, G., & Valtonen, M. 1990, ApJ, 350, 89

Calcáneo-Roldán, C. , Moore, B. , Bland-Hawthorn, J., Malin, D., & Sadler, E. M. 2000, MNRAS, 314, 324

Casoli, F., et al. 1998, A&A, 331, 451

Condon, J. J., Condon, M. A., Gisler, G., & Puschell, J. J. 1982, ApJ, 252, 102

Cretton, N., Naab, T., Rix, H.-W., & Burkert, A. 2001, ApJ, 554, 291

Dressler, A., et al. 1997, ApJ, 490, 577

Dubinski, J. 1998, ApJ, 502, 141

Dubinski, J., Murali, C., & Ouyed, R. 2001, unpublished preprint

Ellis, R. S., Smail, I., Dressler, A., Couch, W. J., Oemler, A., Jr., Butcher, H., & Sharples, R. M. 1997, ApJ, 483, 582

Feldmeier, J. J., Ciardullo, R., & Jacoby, G. H. 1998, ApJ, 503, 109

Feldmeier, J. J., Mihos, J. C., Morrison, H. L., Harding, P., & Kaib, N. 2004, in preparation

Feldmeier, J. J., Mihos, J. C., Morrison, H. L., Rodney, S. A., & Harding, P. 2002, ApJ, 575, 779

Ferguson, H. C., Tanvir, N. R., & von Hippel, T. 1998, Nature, 391, 461

Ghigna, S., Moore, B., Governato, F., Lake, G., Quinn, T., & Stadel, J. 1998, MNRAS, 300, 146

Girardi, M., & Biviano, A. 2002, in Merging Processes in Galaxy Clusters, ed. L. Feretti, I. M. Gioia, & G. Giovannini (Dordrecht: Kluwer), 39

Gnedin, O. Y. 2003, ApJ, 582, 141

Gonzalez, A. H., Zabludoff, A. I., & Zaritsky, D. 2004, Carnegie Observatories Astrophysics Series, Vol. 3: Clusters of Galaxies: Probes of Cosmological Structure and Galaxy Evolution, ed. J. S. Mulchaey, A. Dressler, & A. Oemler (Pasadena: Carnegie Observatories, http://www.ociw.edu/ociw/symposia/series/symposium3/proceedings.html)

Gonzalez, A. H., Zabludoff, A. I., Zaritsky, D., & Dalcanton, J. J. 2000, ApJ, 536, 561

Gregg, M. D., & West, M. J. 1998, Nature, 396, 549

Gunn, J. E., & Gott, J. R. 1972, ApJ, 176, 1

Henriksen, M., & Byrd, G. 1996, ApJ, 459, 82

Hernquist, L., & Mihos, J. C. 1995, ApJ, 448, 41

Hernquist, L., & Spergel, D. N. 1992, ApJ, 399, L117

Hibbard, J. E., Guhathakurta, P., van Gorkom, J. H., & Schweizer, F. 1994, AJ, 107, 67

Hibbard, J. E., & Mihos, J. C. 1995, AJ, 110, 140

Hinz, J. L., Rix, H.-W., & Bernstein, G. M. 2001, AJ, 121, 683

Keel, W. C., Kennicutt, R. C., Hummel, E., & van der Hulst, J. M. 1985, AJ, 90, 708

Kenney, J. D. P., & Young, J. S. 1989, ApJ, 344, 171

Kennicutt, R. C., Keel, W. C., van der Hulst, J. M., Hummel, E., & Roettiger, K. A 1987, AJ, 93, 1011

Lake, G., Katz, N., & Moore, B. 1998, ApJ, 495, 152

Larson, R. B., & Tinsley, B. M. 1978, ApJ, 219, 46

Larson, R. B., Tinsley, B. M., & Caldwell, C. N. 1980, ApJ, 237, 692

Lavery, R. J., & Henry, J. P. 1988, ApJ, 330, 596

Lawrence, A., Rowan-Robinson, M., Leech, K., Jones, D. H. P., & Wall, J. V. 1989, MNRAS, 240, 329

Malumuth, E. M., & Richstone, D. O. 1984, ApJ, 276, 413

Merritt, D. 1983, ApJ, 264, 24

——. 1984, ApJ, 276, 26

Mihos, J. C. 1995, ApJ, 438, L75

Mihos, J. C., & Hernquist, L. 1994a, 425, L13

——. 1994b, ApJ, 431, L9

——. 1996, ApJ, 464, 641

Mihos, J. C., McGaugh, S. S., & de Blok, W. J. G. 1997, ApJ, 477, L79

Mihos, J. C., Walker, I. R., Hernquist, L., Mendes de Oliveira, C., & Bolte, M. 1995, ApJ, 447, L87

Miller, G. E. 1983, ApJ, 268, 495

Moore, B., Katz, N., Lake, G., Dressler, A., & Oemler, A. 1996, Nature, 379, 613

Moore, B., Lake, G., & Katz, N. 1998, ApJ, 495, 139

Moore, B., Lake, G., Quinn, T., & Stadel, J. 1999, MNRAS, 304, 465

Morrison, H., et al. 2002, The Dynamics, Structure, and History of Galaxies: A Workshop in Honour of Professor Ken Freeman, ed. G. S. Da Costa & H. Jerjen. (San Francisco: ASP), 123

Naab, T., & Burkert, A. 2001, in The Central Kpc of Starbursts and AGN: The La Palma Connection, ed. J. H. Knapen et al. (San Francisco: ASP), 735

Navarro, J. F., Frenk, C. S., & White, S. D. M. 1996, ApJ, 462, 563

Negroponte, J., & White, S. D. M. 1983, MNRAS, 205, 1009

Nicholson, R. A., Bland-Hawthorn, J., & Taylor, K. 1992, ApJ, 387, 503

Noguchi, M. 1988, A&A, 203, 259

Oemler, A. 1973, ApJ, 180, 11

Ostriker, J. P. 1980, Comments on Astrophysics, 8, 177

Quilis, V., Moore, B., & Bower, R. 2000, Science, 288, 1617

Quinn, P. J., Hernquist, L., & Fullagar, D. P. 1993, ApJ, 403, 74

Richstone, D. O. 1976, ApJ, 204, 642

Richstone, D. O., & Malumuth, E. M. 1983, ApJ, 268, 30

Sanders, D. B., Soifer, B. T., Elias, J. H., Neugebauer, G., & Matthews, K. 1988, ApJ, 328, L35

Schiminovich, D., van Gorkom, J. H., van der Hulst, J. M., & Kasow, S. 1994, ApJ, 423, L101

Schweizer, F. 1982, ApJ, 252, 455

——. 1998, in Saas-Fee Advanced Course 26, Galaxies: Interactions and Induced Star Formation, ed. R. C. Kennicutt, Jr., et al. (Berlin: Springer-Verlag), 105

Soifer, B. T., et al. 1984, ApJ, 278, L71

Steiman-Cameron, T. Y., Kormendy, J., & Durisen, R. H. 1992, AJ, 104, 1339

Steinmetz, M., & Navarro, J. F. 2002, NewA, 7, 155

Thuan, T. X., & Kormendy, J. 1977, PASP, 89, 466

Toomre, A. 1978, IAU Symp. 79, The Large Scale Structure of the Universe (Dordrecht: Reidel), 109

Toomre, A., & Toomre, J. 1972, ApJ, 178, 623

Toth, G., & Ostriker, J. P. 1992, ApJ, 389, 5

Trentham, N., & Mobasher, B. 1998, MNRAS, 293, 53

van Dokkum, P. G. 2002, in Tracing Cosmic Evolution with Galaxy Clusters, ed. S. Borgani, M. Mezzetti, & R. Valdarnini (San Francisco: ASP), 265

van Dokkum, P. G., Franx, M., Fabricant, D., Kelson, D. D., & Illingworth, G. D. 1999, ApJ, 520, L95

van Gorkom, J., & Schiminovich, D. 1997, in The Nature of Elliptical Galaxies, 2nd Stromlo Symposium, ed. M. Arnaboldi, G. S. Da Costa, & P. Saha (San Francisco: ASP), 310

Veilleux, S., Kim, D.-C., & Sanders, D. B. 2002, ApJS, 143, 315

Walker, I. R., Mihos, J. C., & Hernquist, L. 1996, ApJ, 460, 121

Wright, A. E. 1972, MNRAS, 157, 309

Zwicky, F. 1951, PASP, 63, 61

18

Evolutionary processes in clusters

BEN MOORE

Institute for Theoretical Physics, University of Zürich, Switzerland

Abstract

Are the morphologies of galaxies imprinted during an early and rapid formation epoch, or are they due to environmental processes that subsequently transform galaxies between morphological classes? Recent numerical simulations demonstrate that the cluster environment can change the morphology of galaxies, even at a couple of cluster virial radii. The gravitational and hydrodynamical mechanisms that could perform such transformations were proposed in the 1970s, even before the key observational evidence for environmental dependences—the morphology-density relation and the Butcher-Oemler effect.

18.1   Introduction

Galaxies are observed to have a wide range of morphologies and stellar configurations, classified as disklike, with subclasses depending on the degree of disk instability, gas fraction, and central nucleation, or as spheroidal configurations of varying shapes and concentrations. Within both of these broad sequences we have various combinations of irregularities, subclasses, sizes, luminosities, and star formation histories. Observational studies of clusters and groups have played an important role in helping us to understand the origin of galactic morphologies. This is due to three main reasons: (1) evolutionary processes are accelerated in high-density environments, (2) some classes of galaxies are only found within larger virialized systems, and (3) clusters can be found easily at higher redshifts and therefore can be used to detect evolution directly.

The observational data are consistent with the idea that the visible baryons are concentrated at the center of much larger dark matter halos (Fischer et al. 2000). Thus, the interpretation of galaxy morphologies is closely linked to understanding dark matter clustering on different scales. The baryons are observed to have a scale length that is about 1/10th that of the dark matter, implying that dissipation must have played a key role in galaxy formation. On average, galactic mass halos have accumulated close to the universal baryon fraction, implying that violent feedback that leads to mass ejection of baryons has played a less important role in determining galaxy morphology. If most of the baryons quietly dissipated with little merging between subhalos, then the first galaxies to form are expected to be disks, due to conservation of angular momentum as the gas radiates energy, sinks within the dark matter halos, and spins faster. Indeed, most of the galaxies in quiescent environments (outside of other virialized systems) are disks—nearly all other classes of galaxies are found orbiting inside deeper potentials as satellite galaxies (halos within halos).

Spheroidal stellar configurations span a factor of 10^7 in luminosity, from central cluster cDs to the Local Group dwarf spheroidals, whereas the disks range from the giant low-surface brightness galaxies to the tiny Local Group dwarf irregulars. In general the spheroidals reach higher masses and luminosities. It is not obvious why this is the case. The gas cooling times cale limits the maximum size of cold baryonic systems, but it is a puzzle why we do not observe ~30 kpc disks at the center of clusters where the cooling times are short. Perhaps harassment or local feedback from a central active galactic nucleus may be suppressing central disk formation in massive halos.

The fact that few spiral galaxies are found anywhere within the central regions (~Mpc) of rich clusters (Dressler 1980) could be explained in two ways: (1) cluster galaxies were never disks since they formed in a more merger-prone environment as lenticulars (S0), spheroidals (E), or dwarf ellipticals (dE); or (2) disks formed first and have subsequently been transformed into other morphological classes by virtue of the cluster environment that formed later. The *Hubble Space Telescope* allowed a direct observational test of the morphological change over the past few Gyr within dense and proto-dense environments. High-resolution images of the "Butcher-Oemler" (Butcher & Oemler 1978) clusters at $z \approx 0.5$ showed that the luminous spheroidal population was already in place at this epoch, but that the majority of the other galaxies were indeed disks (Dressler et al. 1997), and even the S0 population is deficient (Smail et al. 1997).

Many of these distant Butcher-Oemler clusters are complex merging systems of groups that will eventually have similar masses and galaxy densities to the nearby rich clusters like Coma. Clusters at high redshift that already have the mass and virial state of a rich cluster may already look similar to Coma (for example, MS 1054–03; van Dokkum et al. 1998). These comparisons are complex due to projection effects and background subtraction, but also the interpretation is difficult since frequently one is comparing systems identified at different epochs that are in different states of virialization.

18.2   The Paradigms for Disk and Spheroid Formation

In order for gas to concentrate and form stars at the centers of dark halos it must first be shock heated so that it can dissipate and cool to high densities. As it cools it must spin faster, as it conserves its primordial angular momentum generated from tidal torques (Hoyle 1951). The natural end state of cooling gas within an isolated dark matter halo is a rotationally supported disk; thus, one might postulate that disks are the initial building block from which the entire morphological sequence is constructed. Once the first disky objects have formed (or even while they are forming), a process of multiple mergers and the associated central gas inflows are expected to create the cD–E sequence that extends to the faintest ellipticals (the high-surface brightness, M32-like systems).

It should be remembered that this is primarily theoretical speculation—the details of this process are far from being worked out, and considerable numerical resources are required to fully investigate these ideas. Firstly, we need to understand how the gas is shock heated and cools to the central disk and how the angular momentum of the gas evolves during this process. Forming a bulgeless, late-type disk galaxy may be one of the most difficult challenges for the cold dark matter (CDM) model. Understanding the formation of spheroids has equally challenging problems. For example, why is there such a narrow spread in the luminosity of cD galaxies, and why do they lie offset brightward from the E galaxy luminosity function? Why are the smallest ellipticals rotating faster than the massive ellipticals? Why are there so few isolated field ellipticals, which are the probable end state of the M_*

groups, where M_* is the characteristic nonlinear mass today? The kinematics, colors, and ages of ellipticals are also challenges for this paradigm, which have yet to be understood. On a more general note, it is still unclear how CDM-type models that predict a steep mass spectrum of halos can produce a flat luminosity function, while at the same time matching the correlation between baryonic mass/luminosity and dynamical mass.

Independent of the model, we can pose the question of whether it is theoretically possible to reproduce the entire sequence of galaxies starting from disky systems. Starting with Holmberg's (1941) N-body experiments with lights and photometers, and later confirmed by Toomre's computer calculations (Toomre 1977), it has been shown that it is possible for galaxies to interact gravitationally and produce spectacular tidal features. Longer numerical calculations performed by Gerhard (1981) showed that the end states of mergers will violently relax into spheroidal configurations, but additional dissipation is required to produce the high phase space densities observed in ellipticals. As is usual with N-body simulations, the detailed comparisons can be quite complex, and the more details one simulates the more discrepancies one finds between data and theory (Naab, Burkert, & Hernquist 1999). More realistically, most ellipticals probably formed from the rapid and multiple mergers of a variety of baryonic systems—not too unlike a clumpy monolithic collapse—and distinguishing between the two standard hypotheses is difficult. However, in this review I will concentrate on "non-merging" mechanisms that can drive evolution and transform galaxies between and across morphological classes in the environments of galactic, group, and cluster halos.

18.3   Mechanisms for Transformation

Early theoretical work predicted that clusters of galaxies are harsh environments for galaxies to inhabit. Hydrodynamical processes (Spitzer & Bade 1951; Gunn & Gott 1972; Cowie & Songalia 1977; Norman & Silk 1979; Nulsen 1982) were proposed as important mechanisms for stripping the interstellar medium from galaxies. The importance of gravitational encounters and tidal forces as mechanisms for forming central cD galaxies, creating diffuse light, and for influencing morphological transformation was proposed by many authors in the 1970s and 1980s (Gallagher & Ostriker 1972; Ostriker & Tremaine 1975; Richstone 1976; White 1976; Hausman & Ostriker 1978; Merritt 1983).

Many of these mechanisms are efficient only in massive galaxy clusters, and they may be invoked to explain the morphology-density and Butcher-Oemler effects (Solanes & Salvador-Sole 1992; Kauffmann 1995). However, the role of evolution in lower-density environments may play an important role. The evolution of bright/massive galaxies within groups and poor clusters with velocity dispersions below 400 km s^{-1} should be dominated by dynamical friction and mergers. An important question remains: Is galaxy evolution in clusters driven by pre-processing of galaxies in groups? Indeed, star formation appears to be truncated within galaxies that lie in lower-density environments (Balogh, Navarro, & Morris 2000). However, it is clear that galaxies in virialized systems have different morphologies than the field. If environment is responsible for transition, then it should be possible to quantify observationally and theoretically where, when, and how these transitions take place.

In his excellent book on clusters, Sarazin (1988) makes six points to argue that the morphologies of galaxies are set at the time of formation rather than by subsequent environmental processes. Many of these points are still open questions. For example, Dressler (1980) claimed that the bulges of S0s are more luminous than those of the spirals, a frequently cited result that was disputed by Solanes, Salvador-Sole, & Sanroma (1989), who analyzed the

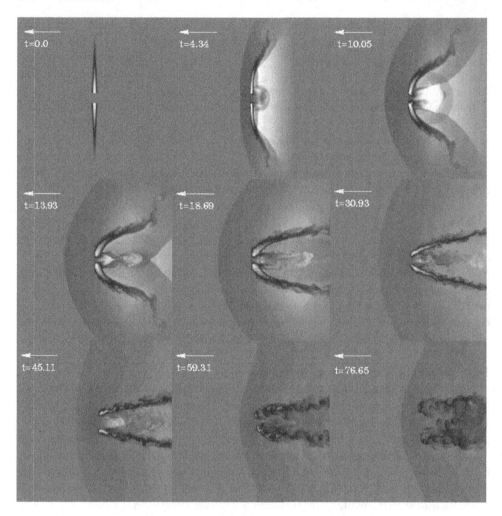

Fig. 18.1. A high-resolution Eulerian simulation of gas stripping of a galaxy falling into a gas density representative of the center of the Coma cluster at 3000 km s^{-1}. The time unit is millions of years, and each frame is 50 kpc on a side. (Adapted from Quilis et al. 2000.)

entire Dressler sample and took into account additional selection effects. Most of Sarazin's arguments were made with the idea that gas processes would be driving a morphological evolution. Sarazin makes the point that gas dynamics would not reproduce the thick disks of S0s, although it is a natural outcome of gravitational interactions.

Indeed, recent numerical work has demonstrated that gravitational and hydrodynamical processes could be responsible for many of the observed aspects of galaxy morphology and evolution in different environments (Byrd & Valtonen 1990; Valluri & Jog 1991; Valluri 1993; Summers, Davis, & Evrard 1995; Moore et al. 1996b; Dubinski 1998; Moore, Lake, & Katz 1998; Abadi, Moore, & Bower 1999; Dubinski, Mihos, & Hernquist 1999; Mihos 1999; Balogh et al. 2000; Quilis, Moore, & Bower 2000; Vollmer et al. 2000; Mayer et al. 2001; Vollmer, Balkowski, & Cayatte 2002; Gnedin 2003a,b). The following sections will

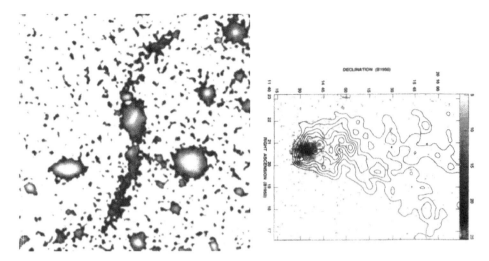

Fig. 18.2. The *left* image shows a symmetric stellar tidal tail from a disk galaxy that is orbiting in a galaxy cluster at $z = 0.5$ (courtesy of I. Smail and the MORPHS group). The *right* image shows the trailing radio emission from the hydrodynamical stripping of a cluster galaxy (courtesy of G. Gavazzi). Direct observational evidence for either gravitational or hydrodynamical effects is hard to find. In general, the tidal debris from harassed disks is too faint to observe, and ram pressure stripping acts on such a short time scale that it is rare to catch a galaxy in the act of being stripped.

discuss these studies in the context of understanding the wide variety of galactic morphologies.

18.4 A New Paradigm for the Formation of S0/dS0/dE/dSph/UCD Galaxies

Once a disky object has formed, then it may enter a denser environment, whether a galactic, group or cluster mass system. If the velocity dispersion of the deeper potential is more than $\sim$5 times the velocity dispersion of the infalling galaxy, then it is unlikely to merge with either the central object (by dynamical friction) or with another satellite. Only external processes will effect its evolution, and we can speculate, with support from numerical simulations, that the entire sequence of remaining galaxies (S0, dS0, dE, dSph...) is created by the transformation of spiral or dIrr systems by impulsive and resonant gravitational interactions, with some additional help from hydrodynamical processes.

Ram pressure stripping is effective at removing the entire gas supply from galaxies that pass through the cores of rich clusters (cf. Fig. 18.1). This will suppress star-formation in cluster galaxies, but will it be effective at large distances from the centers of clusters, where the gas density is low? Gravitational interactions may be important over a larger region of the cluster since more dark matter is bound to galaxies further from the central cluster potential (thus the encounters are stronger), but the number of encounters between halos decreases. The orbits of galaxies in clusters are nearly isotropic, which results in most of the galaxies orbiting through the dense inner region of the cluster (Ghigna et al. 1998). In fact, 10% of orbits will take galaxies through the core and to beyond twice the cluster virial radii ($\sim$6 Mpc for Coma). Therefore, the environment near rich clusters may host galaxies

z= 0

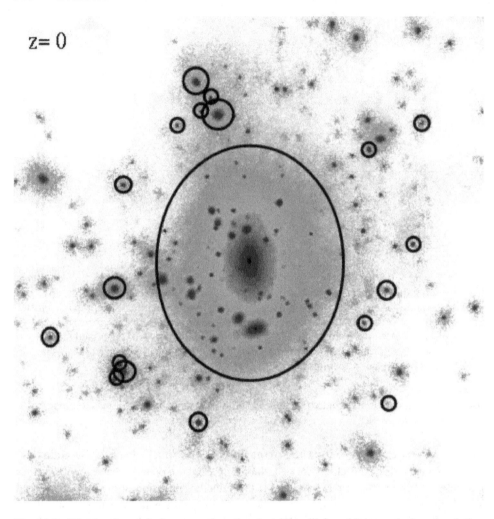

Fig. 18.3. The density of dark matter plotted to 3 virial radii for a high-resolution simulation of a galaxy cluster. The large inner circle is the virial radius of the cluster. The smaller circles highlight halos that are currently outside the cluster but have orbited within $0.25\,r_{200}$. (This is the final frame of the mpeg movie showing the formation history of this cluster and can be downloaded from www.nbody.net.)

that have suffered significant gravitational perturbations and have been partially stripped of gas (cf. Figs. 18.2 and 18.3) by virtue of orbiting through the cluster core.

The definition of an S0 is basically a featureless disky galaxy with little or no interstellar medium, so the simplest way to create an S0 is to remove the interstellar medium from a Sa/Sb disk (Solanes et al. 1989; Solanes & Salvador-Sole 1992). We have two possible methods for accomplishing this. Any galaxy passing through the core of a massive cluster will have its gas supply completely stripped by ram pressure and viscous stripping on a time scale shorter than the core-crossing time. The remaining dense molecular gas may recycle into the interstellar medium and also be removed or consumed in a burst of star formation due to the pressure increase as the galaxy reaches the cluster center. Alternatively, gravitational interactions with the cluster and its substructures will heat disks, raising the Toomre "Q" parameter, which naturally suppresses spiral structure and other instabilities in the disk, further suppressing star-formation as molecular cloud growth is halted (Moore et al. 1998; Gnedin 2003a,b).

18.4.1 Cluster Dwarf Ellipticals (dEs) and Transition Galaxies (dS0s).

Dwarf ellipticals are the most numerous type of galaxy in clusters (Binggeli, Sandage, & Tammann 1985). They are flattened exponential systems, sometimes with a bright, compact nucleus. These systems are in various dynamical states, ranging from pressure-supported spheroids to rotationally flattened disky systems (Geha, Guhathakurta, & van der Marel 2002). Numerical simulations of disks orbiting within cluster potentials have shown that a dramatic transformation to a diffuse spheroidal system is likely to occur (Moore et al. 1996b). The transformation sequence begins with a violent bar instability. Subsequent perturbations cause the bar to lose angular momentum, and it eventually collapses into a spheroidal system through a buckling-type instability. During the early stages of the transformation, which takes roughly a cluster orbital time, the morphology of the galaxy may appear as a rotationally supported dS0 system.

A fundamental prediction from gravitational heating mechanisms is that dEs should be embedded within very low-surface brightness tidal streams of stellar debris, as demonstrated in Figure 18.4, and these streams should trace the orbital path of the progenitor galaxy. Since the relaxation time is only short in the cluster cores, these streams should survive relatively intact at the edge of clusters, but they should become well mixed near the cluster centers.

18.4.2 Intracluster Diffuse Light, Overmerging, and UCDs

The nature of the dark matter is critical to the survival of galaxies in clusters. If galaxies had constant-density cores of just a few kpc (or cuspy potentials with very low concentrations), then they would all be easily disrupted by the cluster potential. This process is directly analogous with the overmerging problem that was responsible for the dissolution of subhalos in early *N*-body simulations (Moore, Katz, & Lake 1996a). On the other hand, if galactic halos are cuspy and concentrated they would all survive in clusters and we would not expect to observe a significant component of diffuse light.

The "ultra-compact dwarfs" (UCDs) recently found in the Fornax cluster are most likely the dense nuclei of nucleated dE galaxies that have been "overmerged" by gravitational interactions (Drinkwater et al. 2000). These tracer cores show that the central concentrations of the progenitor galaxies must have been low enough so that they have been completely disrupted by the present day. Both the fraction of diffuse light stripped from galaxies and the abundance and locations of the UCDs could be used to constrain the structure of galaxies and the efficiency of gravitational interactions in different environments.

18.4.3 Local Group Dwarf Spheroidals

The Local Group dwarf spheroidals are the extreme tip of the galaxy luminosity function, and, as the shallowest potentials and faintest galaxies known, they provide a strong test of our understanding of galaxy formation (Kormendy 1989). The Local Group morphology-density relation is similar to that of rich clusters in that the spheroidal galaxies with no diffuse gas are located close to the host galactic potentials of the Milky Way and

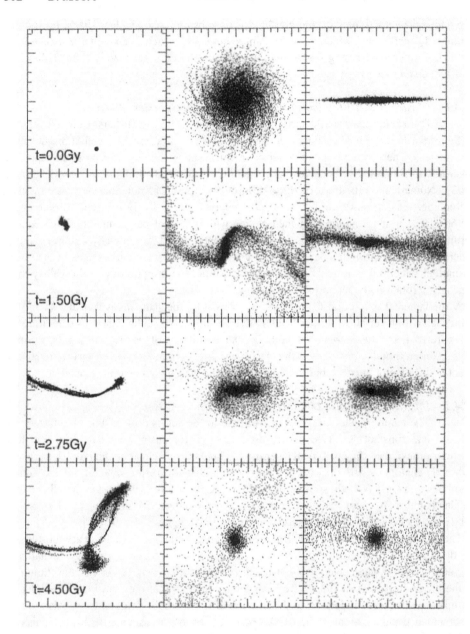

Fig. 18.4. Gravitational tides from a deeper potential are sufficient to drive an evolution from disks to dwarf spheroidals. Here I show the evolution of an Sc spiral galaxy on a 6:1 (apo:peri) orbit, where the pericenter is 0.1 r_{virial} and the ratio of circular velocities of the two systems are 5:1. This system could therefore be rescaled to represent a galaxy like the SMC orbiting within the Milky Way or an L_* low-surface brightness galaxy orbiting in a galaxy cluster. The left frame shows the entire orbit, while the center and right frames are face-on and edge-on views centered on the galaxy. The forced bar instability is so violent that the evolution is better described by a secondary violent relaxation of the stellar disk—most of the galaxy is stripped into symmetric debris streams leaving a small pressure-supported spheroidal galaxy with an exponential light profile (courtesy of C. Calcaneo-Roldan).

M31. Again, we can ask the question: Do disks know not to form near the site of a massive potential, or were they originally disks that have been transformed to spheroidals through interactions? Observationally this is unclear and a difficult question to answer because we cannot observe Local Group progenitors at high redshifts.

However, a huge amount of detailed data exists for these faint Local Group galaxies (e.g., Grebel 2001), which must be reproduced by a successful model for their formation. Simulations of galaxy formation in a cosmological context that can resolve the faintest satellites are some years away; therefore, only the most basic comparisons between theoretical models and data have been made to date. The dynamics of these systems indicate that their spheroidal shapes are due to random stellar motions and that rotational support is small. Many of these systems show evidence for continuous star formation or widely separated bursts, indicating that ram pressure stripping by a hot Galactic halo component has not been completely efficient at removing their fuel. It also indicates that supernovae winds have not been effective at ejecting the bulk of the gas from these systems.

Galaxies are on average some 5 Gyr older than galaxy clusters, which allows time for gravitational interactions to transform disks to dSphs in the potential of a more massive system (Mayer et al. 2001). A high-density, Draco-like dSph must have had a progenitor with a similar high dark matter density, such as the dIrr galaxy GR8. However, one should not compare GR8 directly with Draco since GR8 has evolved for 10 Gyr longer in the field than the system that accreted into the Galaxy to evolve into Draco. The morphological transformation predicted by the numerical simulations is similar to that for cluster galaxies, but on a longer time scale since only the main galactic potential has been considered as the perturber. If the CDM model is in fact correct, then the transformation for the dSphs may be more rapid due to encounters with dark matter substructure predicted to fill galactic halos.

18.5   Conclusions

Numerical simulations have shown that environmental processes are sufficient to reproduce some of the basic properties of a large fraction of the Hubble sequence (S0, dE, dS0, dSph, UCD) by gravitationally interactions acting on disks. Hydrodynamical processes can can also play a role in suppressing star formation on a shorter time scale, but only in regions of high gas density. Dramatic improvements to the modeling will come when algorithms have improved to the extent that we can form realistic disk systems from *ab initio* initial conditions. This will allow the morphological evolution in groups and clusters to be studied in great detail and within a cosmological context, enabling quantitative observational comparisons and predictions to be made. Realistically, this is probably 5–10 years away. For comparison with theory, observers need to quantify the Butcher-Oemler effect using mass-selected samples of clusters at different epochs through a combination of lensing and high-resolution spectrophotometric data. Probing down to dwarf galaxy luminosities and group mass scales will be of great interest to theorists working to constrain evolutionary scenarios and cosmological models.

References

Abadi, M. G., Moore, B., & Bower R. G. 1999, MNRAS, 308, 947
Balogh, M. L., Navarro, J. F., & Morris, S. L. 2000, ApJ, 540, 113
Binggeli, B., Sandage, A. R., & Tammann, G. A. 1985, AJ, 90, 1681
Butcher, H., & Oemler, A., Jr. 1978, ApJ, 219, 18
Byrd, G., & Valtonen, M. 1990, ApJ, 350, 89

Cowie, L. L., & Songaila, A. 1977, Nature, 266, 501

Dressler, A. 1980, ApJ, 236, 351

Dressler, A., et al. 1997, ApJ, 490, 577

Drinkwater, M. J., Jones, J. B., Gregg, M. D., Phillipps, S. 2000, PASA, 17, 227

Dubinski, J. 1998, ApJ, 502, 141

Dubinski, J., Mihos, J. C., & Hernquist, L. 1999, ApJ, 526, 607

Fischer, P., et al. 2000, AJ, 120, 1198

Gallagher, J. S., III, & Ostriker, J. P. 1972, ApJ, 77, 288

Geha, M., Guhathakurta, P., & van der Marel, R. P. 2002, AJ, 124, 3073

Gerhard, O. 1981, MNRAS, 197, 179

Ghigna, S., Moore, B., Governato, F., Lake, G., Quinn, T., & Stadel, J. 1998, MNRAS, 300, 146

Gnedin, O. Y. 2003a, ApJ, 582, 141

Gnedin, O. Y. 2003b, ApJ, 589, 752

Grebel, E. K. 2001, ApS&SS, 277, 231

Gunn, J. E., & Gott, J. R. 1972, ApJ, 176, 1

Hausman, M. A., & Ostriker, J. P. 1978, ApJ, 224, 320

Holmberg, E. 1941, ApJ, 94, 385

Hoyle, F. 1951, in Problems of Cosmical Aerodynamics, ed. J. M. Burgers & H. C. van de Hulst (Dayton: Central Air Documents Office)

Kauffmann, G. 1995, MNRAS, 274, 153

Kormendy, J. 1989, ApJ, 342, L63

Mayer, L., Governato, F., Colpi, M., Moore, B., Quinn, T., Wadsley, J., Stadel, J., & Lake, G. 2001, ApJ, 559, 754

Merritt, D. 1983, ApJ, 264, 24

Mihos, C. 1999, Ap&SS, 266, 195

Moore, B., Katz, N., & Lake, G. 1996a, ApJ, 457, 455

Moore, B., Katz, N., Lake, G., Dressler, A., & Oemler, A. 1996b, Nature, 379, 613

Moore, B., Lake, G., & Katz, N. 1998, ApJ, 495, 139

Naab, T., Burkert, A., & Hernquist, L. 1999, ApJ, 523, L133

Norman, C. A., & Silk, J. 1979, ApJ, 233, L1

Nulsen, P. E. J. 1982, MNRAS, 198, 1007

Ostriker, J. P., & Tremaine, S. D. 1975, ApJ, 202, L113

Quilis, V., Moore, B., & Bower, R. 2000, Science, 288, 1617

Richstone, D. O. 1976, ApJ, 204, 642

Sarazin, C. L. 1988, X-ray Emission from Clusters of Galaxies (Cambridge: Cambridge Univ. Press)

Smail, I., Dressler, A., Couch, W. J., Ellis, R. S., Oemler, A., Butcher, H., & Sharples, R. 1997, ApJS, 110, 213

Solanes, J. M., & Salvador-Sole, E. 1992, ApJ, 395, 91

Solanes, J. M., Salvador-Sole, E., & Sanroma, M. 1989, AJ, 98, 798

Spitzer, L., Jr., & Baade, W. 1951, ApJ, 113, 413

Summers, F. J., Davis, M., & Evrard, A. E. 1995, ApJ, 454, 1

Toomre, A. 1977, in The Evolution of Galaxies and Stellar Populations, ed. B. M. Tinsley & R. B. Larson (New Haven: Yale Univ. Observatory), 401

Valluri, M. 1993, ApJ, 408, 57

Valluri, M., & Jog, C. J. 1991, ApJ, 374, 103

van Dokkum, P. G., Franx, M., Kelson, D. D., Illingworth, G. D., Fisher, D., & Fabricant, D. 1998, ApJ, 500, 714

Vollmer, B., Balkowski, C., & Cayatte, V 2002, Ap&SS, 281, 359

Vollmer, B., Marcelin, M., Amram, P., Balkowski, C., Cayatte, V., & Garrido, O. 2000, A&A, 364, 532

White, S. D. M. 1976, MNRAS, 177, 717

19

Interaction of galaxies with the intracluster medium

JACQUELINE H. VAN GORKOM
Department of Astronomy, Columbia University

Abstract

Although speculations of an interaction between galaxies and the intracluster medium date back more than 30 years, the impact and importance of a possible interaction have long remained elusive. In recent years the situation has completely changed. A wealth of data and detailed hydrodynamical simulations have appeared that show the effects of interactions. Single-dish observations show that cluster galaxies are deficient in their neutral hydrogen content out to two Abell radii. The deficient galaxies tend to be on radial orbits. Detailed imaging of the neutral hydrogen distribution in individual galaxies in two nearby clusters show a remarkable trend of H I extent with location in the cluster. These trends can be reproduced in simulations of ram pressure stripping by the intracluster medium using SPH and full three-dimensional hydrodynamic codes. Detailed imaging studies of individual galaxies have found a number of galaxies with undisturbed stellar disks, truncated gas disks that are much smaller than the stellar disks, asymmetric extraplanar gas in the center, and enhanced central star formation. These phenomena have all been predicted by hydrodynamical simulations. For the first time detailed observations of gas morphology and kinematics are used to constrain simulations. Simple models of ram pressure stripping are consistent with the data for some galaxies, while for other galaxies more than one mechanism must be at work. Optical imaging and spectroscopic surveys show that small H I disks go together with truncated star-forming disks, that hydrogen deficiency correlates with suppressed star formation rates, and that the spatial extent of H I deficiency in clusters is matched, or even surpassed, by the extent of reduced star formation rates.

Recent volume-limited imaging surveys of clusters in the local Universe show that most gas-rich galaxies are located in smaller groups and subclumps, which have yet to fall into the clusters. These groups form an ideal environment for interactions and mergers to occur, and we see much evidence for interactions between gas-rich galaxies.

19.1 Introduction

It has long been known that in the local Universe the mix of morphological types differs in different galactic environments, with ellipticals and S0s dominating in the densest clusters and spirals dominating the field population (Hubble & Humason 1931). This so-called density-morphology relation has been quantified by Oemler (1974) and Dressler (1980), and is found to extend over 5 orders of magnitude in space density (Postman & Geller 1984). Whether this relation arises at formation (nature) or is caused by density-driven evolutionary effects (nurture) remains a matter of debate. More recent studies of clusters of galaxies at intermediate redshifts show that both the morphological mix and the star formation rate strongly evolve with redshift (Dressler et al. 1997; Poggianti et al. 1999; Fasano et al. 2000). In particular the fraction of S0s goes down and the spiral fraction and star formation rate go up with increasing redshift. There are many physical mechanisms at work in clusters or during the growth of clusters that could affect the star formation rate and possibly transform spiral galaxies into S0s. In this review I will limit myself to the role that the hot intracluster medium (ICM) may play.

The first suggestion that an interaction between the ICM and disk galaxies may affect the evolution of these galaxies was made immediately after the first detection of an ICM in clusters (Gursky et al. 1971). In a seminal paper on "the infall of matter into clusters," Gunn & Gott (1972) discuss what might happen if there is any intergalactic gas left after the clusters have collapsed. The interstellar material in a galaxy would feel the ram pressure of the ICM as it moves through the cluster. A simple estimate of the effect assumes that the outer disk gas gets stripped off when the local restoring force in the disk is smaller than the ram pressure. Thus, disks gets stripped up to the so-called stripping radius, where the forces balance. They estimate that for a galaxy moving at the typical velocity of 1700 km s^{-1} through the Coma cluster the interstellar medium (ISM) would be stripped in one pass. This would explain why so few normal spirals are seen in nearby clusters. In particular it would explain the existence of so many gas-poor, non-star-forming disk galaxies, first noticed by Spitzer & Baade (1951) and later dubbed "anemics" by van den Bergh (1976).

Ram pressure stripping is but one way in which the ICM may affect the ISM. The effects of viscosity, thermal conduction, and turbulence on the flow of hot gas past a galaxy were considered by Nulsen (1982), who concluded that turbulent viscous stripping will be an important mechanism for gas loss from cluster galaxies. While the above-mentioned mechanisms would work to remove gas from galaxies and thus slow down their evolution, an alternative possibility is that an interaction with the ICM compresses the ISM and leads to ram pressure-induced star formation (Dressler & Gunn 1983; Gavazzi et al. 1995).

On the observational side there has long been evidence that spiral galaxies in clusters have less neutral atomic hydrogen than galaxies of the same morphological type in the field (for a review, see Haynes, Giovanelli, & Chincarini 1984). The CO content, however, does not seem to depend on environment (Stark et al. 1986; Kenney & Young 1989). Both single-dish observations and synthesis-imaging results of the Virgo cluster show that the H I disks of galaxies in projection close to the cluster center are much smaller than the H I disks of galaxies in the outer parts (Giovanelli & Haynes 1983; Warmels 1988a,b,c; Cayatte et al. 1990, 1994). All of these phenomena could easily be interpreted in terms of ram pressure stripping. Dressler (1986) made this even more plausible by pointing out that the gas-deficient galaxies seem statistically to be mostly on radial orbits, which would carry them into the dense environment of the cluster core. However, nature turned out to be more complicated than that. In a comprehensive analysis of H I data on six nearby clusters, Magri et al. (1988) conclude that the data cannot be used to distinguish between inbred and evolutionary gas-deficiency mechanisms, or among different environmental effects. Although H I deficiency varies with projected radius from the cluster center, with the most H I-poor objects close to the cluster centers, no correlation is found between deficiency and relative radial velocity, as would be expected from ram pressure stripping.

In more recent years a number of developments have taken place. First, there was a flurry of activity on the theoretical front, and for the first time detailed numerical simulations on the effects of ram pressure stripping appeared. Since then, both improved statistics on H I deficiency and detailed multiwavelength observations of cluster galaxies undergoing trauma appeared. More recently, detailed comparisons have been made between individual systems and numerical simulations. Finally, synthesis imaging of neutral hydrogen no longer needs to be limited to a few selected systems in nearby clusters, and results of volume-limited surveys of entire clusters at redshifts between $z = 0$ and $z = 0.2$ have started to appear in the literature. In this review I will first discuss what we have learned about the statistical properties of the H I content of cluster galaxies. Then I will review some of the recent numerical work that has been done and compare these with observational results. After that I will discuss what we have learned from imaging surveys, and in conclusion I will discuss the importance of ICM interaction for galaxy evolution.

19.2 The Statistics of H I Deficiency

The most comprehensive survey on H I content in cluster galaxies to date is the work by Solanes et al. (2001). These authors compiled H I data on 1900 spiral galaxies in 18 nearby clusters. The data are mostly obtained with the Arecibo telescope, a single pixel telescope, and give information about the total amount of neutral hydrogen within the Arecibo beam ($3'$) centered on optically selected galaxies. Galaxies are earmarked as belonging to a cluster when they fall within a projected distance of 5 Abell radii (R_A), i.e. within $7.5\,h^{-1}$ Mpc, from the cluster center and have a radial velocity that is less than 3 times the average velocity dispersion from the cluster mean. Only clusters are included for which there are good H I data for at least 10 galaxies within $1\,R_A$ of the cluster center. H I deficiency is calculated according to the recipe of Haynes & Giovanelli (1984): $\log_{10}(M_{H\,I}$ observed$/M_{H\,I}$ expected), where the expected H I mass is derived from a sample of isolated spirals of the same morphological type and optical diameter. To get significant statistics the cluster sample is then divided into two groups, a deficient cluster sample and a non-deficient cluster sample. The deficient sample contains all clusters for which the H I deficiency distribution over galaxies is significantly different within $1\,R_A$ from that of the galaxies outside $1\,R_A$. One of the most remarkable results of the analysis of that database is shown in Figure 19.1. It shows the H I-deficiency fraction, i.e. the fraction of galaxies with a deficiency greater than 0.3, in bins of projected radius from the cluster center for the superposition of all the H I-deficient clusters. Galaxies with a deficiency greater than 0.3 are galaxies that are deficient in neutral hydrogen by a factor of 2 or more as compared to isolated galaxies of the same morphological type and size in the field. The percentage of H I-deficient spirals increases monotonically going inward. What is surprising is that this monotonic rise starts as far out as $2\,R_A$. This suggests that the effect of the cluster environment can be felt out to 2 Abell radii, far beyond the reaches of the dense ICM.

No correlation is found for the fraction of H I-deficient spirals (i.e. the number of spirals with an H I deficiency DEF ≥ 0.30 within $1\,R_A$ of the cluster center compared to all galaxies of that type found in that region) with global cluster properties, such as X-ray luminosity, X-ray temperature, and radial velocity dispersion. As pointed out in their paper, this could be a selection bias. If stripped spirals would lose all their gas they may be transformed into S0s and they would be left out from the statistics. It is somewhat plausible that this is indeed the case since the fraction of spirals is clearly anti-correlated with X-ray luminosity. The most important result of the paper, apart from the extent of the occurrence of deficient galaxies, is the correlation between deficiency and orbital parameters. This is shown in Fig-

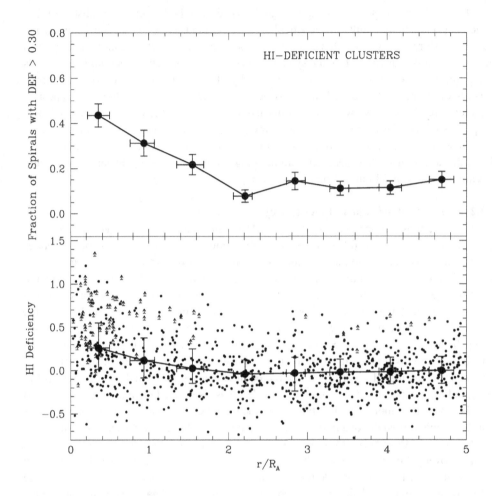

Fig. 19.1. *Top*: H I-deficiency fraction, in bins of projected radius from the cluster center, for the superposition of all the H I-deficient clusters. *Bottom*: H I deficiency versus projected radius from the cluster center. Small dots show the radial variation of H I deficiency for individual galaxies, while the arrows identify non-detections plotted at their estimated lower limit. Large dots are the medians of the binned number distribution. (Taken from Solanes et al. 2001.)

ure 19.2. It shows, for the composite deficient cluster, the radial run of the line-of-sight velocity dispersion for the most deficient spirals, the non-deficient spirals, for all spirals, and for ellipticals and lenticulars. If galaxies are on radial orbits the measured velocity dispersion should decrease at large distances from the cluster center. Although all spirals show a decrease in velocity dispersion at large distances, this effect is by far the most pronounced in the H I-deficient spirals. The ellipticals and S0s have a constant velocity dispersion with radius. This confirms the result by Dressler (1986). These results suggest that deficiency is most pronounced when galaxies go through the dense cluster center at high velocities and, as such, support the idea that ram pressure stripping causes the deficiency.

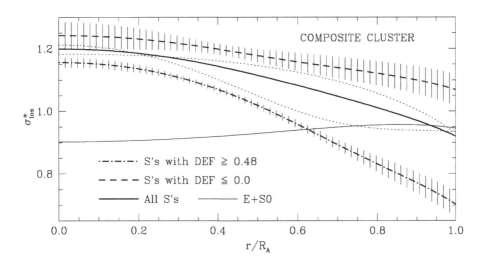

Fig. 19.2. Radial run of normalized line-of-sight velocity dispersion for the composite H I-deficient cluster. (Taken from Solanes et al. 2001.)

19.3 Simulations

In the early 70s, several papers appeared considering the effect of the ICM on galaxies in clusters, including ram pressure stripping (Gunn & Gott 1972), evaporation (Cowie & Songaila 1977), and turbulent viscosity and evaporation (Nulsen 1982). Not much work was done, however, to connect theory with observations. The first paper that specifically looks at observational characteristics is that by Stevens, Acreman, & Ponman (1999). This paper focuses on the impact of the ICM on an elliptical galaxy with a hot ISM and calculates the observational signatures of this in the hot ICM, predicting bow shocks, wakes, and tails. Some, but still precious few, examples of structures that could be interpreted like this exist in the X-ray literature (Stevens et al. 1999). Observationally, there is much more evidence for the impact of the ICM on the cool ISM of disk galaxies. From a view point of galaxy evolution this is also the more important question. In clusters star formation rates are known to evolve rapidly between intermediate redshifts and the local Universe (Balogh et al. 1999; Poggianti et al. 1999). A mechanism is required to bring star formation almost completely to a halt, and a major issue is whether ram pressure stripping could do this to disk galaxies. The original analytical estimates of Gunn & Gott (1972) predict that gas gets stripped from a galaxy up to a stripping radius, within which the restoring force from the disk exceeds the ram pressure. The first numerical simulations (Abadi, Moore, & Bower 1999), using a 3-dimensional SPH/*N*-body simulation to study ram pressure stripping of gas from spiral galaxies orbiting in clusters, confirm that gas in disk galaxies gets stripped up to the stripping radius estimated by Gunn & Gott (1972). At small radii the potential provided by the bulge component contributes considerably. They estimate that a galaxy passing through the center of Coma would have its gaseous disk truncated to ∼4 kpc, losing about 80% of its gas. However, the process is in general not efficient enough to account for the rapid and widespread truncation of star formation observed in cluster galaxies. Quilis, Moore, & Bower (2000) use a finite-difference code to achieve higher resolution in order to be able to include complex turbulent and viscous stripping at the interface of cold and hot gaseous components, as well as the formation of bow shocks in the ICM ahead of the galaxy. From only a few selected runs on galaxies with holes in the central gas distribution, they reverse the conclusion of Abadi et al. (1999) and state that ICM-ISM interaction could explain the morphology of S0 galaxies and the rapid truncation of star formation implied by spectroscopic observations. The main difference with the Abadi et al. result is the use of a complex multi-phase structure of the ISM. They show that the presence of holes and bubbles in the diffuse H I can greatly enhance the stripping efficiency. As the ICM streams through the holes in the ISM it ablates the edges and prevents stripped gas from falling back. Schulz & Struck (2001), in a comprehensive study using SPH, an adaptive mesh HYDRA code, and including radiative cooling, confirm that low-column density gas is promptly removed from the disk. They also find that the onset of the ICM wind has a profound effect on the gas in the disk that does not get stripped. The remnant disk is compressed and slightly displaced relative to the halo center. This can trigger gravitational instability, angular momentum gets transported outward, and the disk compresses further, forming a ring. This makes the inner disk resistant to further stripping, but presumably susceptible to global starbursts. These various simulations appear to more or less agree on the effects of the ISM. All of the above work modeled the ICM as a constant wind. Vollmer et al. (2001) took a different approach. Using an *N*-body/sticky-particle code, they simulate galaxies in radial orbits through the gravitational potential of the Virgo cluster. The galaxies thus experience a time-variable ram pressure, and maximal damage to their gaseous disks only becomes apparent well after closest approach to the Virgo cluster center. Thus, if we see galaxies with truncated H I disks or distorted velocity fields, they are likely to be on their way out from the center. They also find that a considerable part of the stripped total gas mass remains bound to the galaxy and falls back onto the galactic disk after the stripping event, possibly causing a central starburst. The results of Schulz & Struck (2001) and Vollmer et al. (2001) are the first to produce simultaneously stripping in the outer parts and a mechanism to enhance star formation in the inner parts. This may help use up any remaining gas in the central regions, and it could possibly do some secular bulge building. There is observational evidence for stripped H I disks with enhanced central H I surface densities (Cayatte et al. 1994), and there are several lines of evidence that the most recent episode of star formation in cluster galaxies occurred in the central parts (e.g., Rose et al. 2001).

19.4   Comparison of Simulations with H I Imaging

The wealth of single-dish data on the H I content of selected spirals in the cluster environment has shown beyond doubt that H I deficiency occurs among cluster spirals. These data are less suitable to study the mechanisms that remove the gas. Projection effects along the line of sight and uncertainties about the orbital history of individual galaxies complicate matters. H I imaging has so far provided far less statistics, but in the imaging data individual galaxies can be selected that appear to have distortions in their H I morphology or kinematics that are unique to the cluster environment. A prime example is the occurrence of tiny H I disks in Virgo (Warmels 1988a,b; Cayatte et al. 1990). The size of the gaseous disks is considerably smaller than the optical disk. The effect is most pronounced close to the cluster center and gently decreases at increasing distance from the center. In addition to Virgo, this

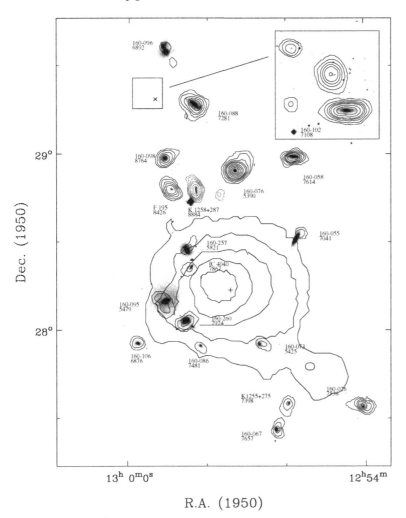

Dec. (1950)

29°

28°

13^h 0^{m}0^s 12^{h}54^m

R.A. (1950)

Fig. 19.3. Composite of individual H I images of Coma spirals observed with the VLA. Galaxies are shown at their proper position and they are magnified by a factor of 7. The H I images (contours) are overlaid on DSS optical images (greyscale). The large-scale contours sketch the X-ray emission as observed by Vikhlinin, Forman, & Jones (1997). (Taken from Bravo-Alfaro et al. 2000.)

has now also been seen in the Coma cluster (Bravo-Alfaro et al. 2000, 2001). Figure 19.3 shows an overlay of the total H I emission (contours) on an Digital Sky Survey (DSS) optical image in greyscale. Each galaxy is located at its proper position in Coma, but the images are blown up by a factor of 7. The thick contours are the X-ray emission as observed with *ROSAT*. The first thing to note is that the H I disks seen in projection on the X-ray emission are in general smaller compared to the optical image than for galaxies far from the center of Coma. An example in case is the galaxy CGCG 160-095 (NGC 4921) east of the center, where H I is only seen in one half of the disk. This must be caused by a mechanism that only affects the gas, and ram pressure stripping is a good candidate. Figure 19.4 shows the

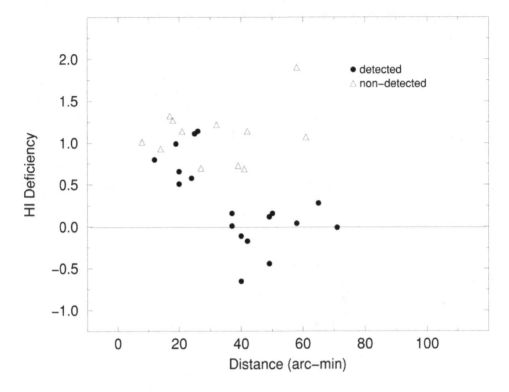

Fig. 19.4. Distribution of H I-deficiency parameter as a function of the projected distance from the center of Coma. Filled circles correspond to H I-detected galaxies and triangles to the lower limits of the deficiency parameter for the galaxies that are not detected in H I. (Taken from Bravo-Alfaro et al. 2000.)

H I deficiency versus projected distance from the center. Its interpretation is already more complicated. Though none of the galaxies projected on to the X-ray emission has a normal H I content (deficiency 0.0), there are several non-detections out to large projected radii. Possibly these galaxies have already gone through the center.

Imaging of Virgo and Coma indicates that H I disks that are smaller than the optical ones may be generic to cluster galaxies. Numerical simulations show that this is what one expects from ram pressure stripping. Abadi et al. (1999) show the results of a simulation with a constant ram pressure typical of the Virgo cluster ICM and galaxies moving with relative velocities of 1000 km s^{-1}, for galaxies of different size. The dependence of stripping radius on disk scale length (plotted in their Fig. 3) is consistent with the result found by Cayatte et al. (1994). Even more impressive is the result of Vollmer et al. (2001) shown in Figure 19.5. The simulation specifically models galaxies on radial orbits through the Virgo potential. The figure plots H I deficiency versus H I to optical diameter. Both the model data (stars) and observed values (Cayatte et al. 1994) are shown. The solid line corresponds to a model where only the outer parts of the disk get stripped, and the constant central H I radial surface density remains unchanged in the stripping process.

Vollmer, in a series of papers (Vollmer et al. 1999, 2000, 2001; Vollmer 2003), tries to reproduce observed gas distribution and kinematics of selected Virgo spirals with his *N*-

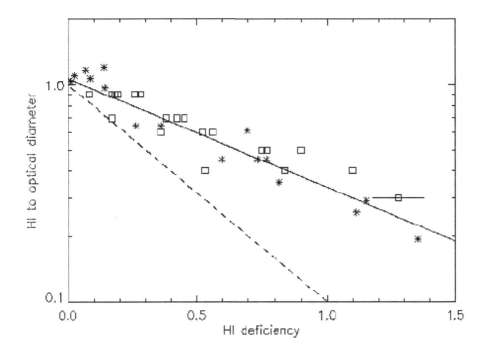

Fig. 19.5. Normalized H I to optical diameter as a function of the H I deficiency for the Virgo cluster. *Squares*: observed values (Cayatte et al. 1994); *stars*: model values. The solid line assumes that the H I surface density has the same value before and after the stripping event. (Taken from Vollmer et al. 2001.)

body/sticky particle simulations. The galaxies are selected based on their H I morphology, and all show truncated H I disks. This is the first time ever that both gas distribution and kinematics are put to test in comparison with models. In several cases, the simulations can reproduce the observed signatures in the gas, and all of these galaxies are found to be on their way out of the cluster, having passed through the dense center. In one case (NGC 4654), Vollmer (2003) shows that both ram pressure stripping and a gravitational interaction must be at play.

One of the most interesting things that has been found recently is a number of galaxies with truncated H I disks, normal or enhanced star formation in the central regions, and some extraplanar gas on one side of the galaxy, while the stellar disks are completely undisturbed. These may be the best candidates for galaxies that are currently undergoing an ICM-ISM interaction. A prime example is NGC 4522, studied by Kenney and collaborators in great detail. Kenney & Koopmann (1999) first pointed out that NGC 4522 is one of the best candidates for ICM-ISM stripping in action. Figures 19.6–19.8 summarize the main characteristics of this galaxy. NGC 4522 is located within a subclump of the Virgo cluster centered on M49. A *ROSAT* map (Fig. 19.6) shows weak extended X-ray emission at the projected location of NGC 4522. All the known peculiarities of NGC 4522 are associated with gas, dust, and H II regions, not with the older stars. The H I (Fig. 19.7) is spatially coincident

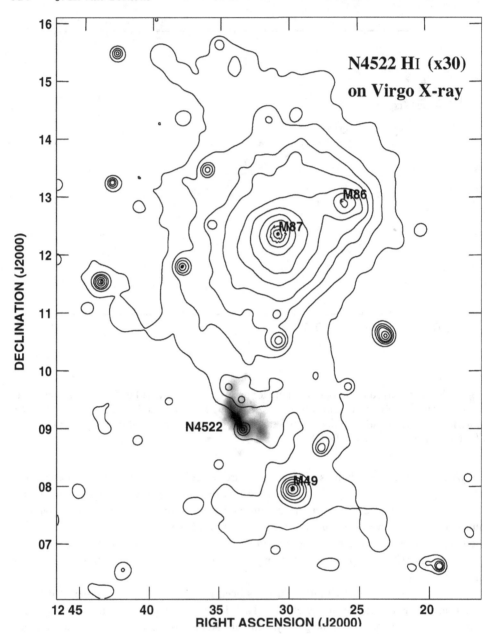

Fig. 19.6. H I greyscale image of NGC 4522, scaled up in size by a factor of 30, superposed on a *ROSAT* X-ray image of the Virgo cluster from Böhringer et al. (1994). The image indicates the locations of the giant ellipticals M87, M86 and M49, which are associated with sub-clusters. (Taken from Kenney et al. 2004.)

with the undisturbed stellar disk in the central 3 kpc (0.4 R_{25}) of the galaxy. At 0.4 R_{25} the H I truncates abruptly and is only seen above the plane to the SW. About half of the total H I appears to be extraplanar, extending to ~3 kpc above the plane (Kenney, van Gorkom,

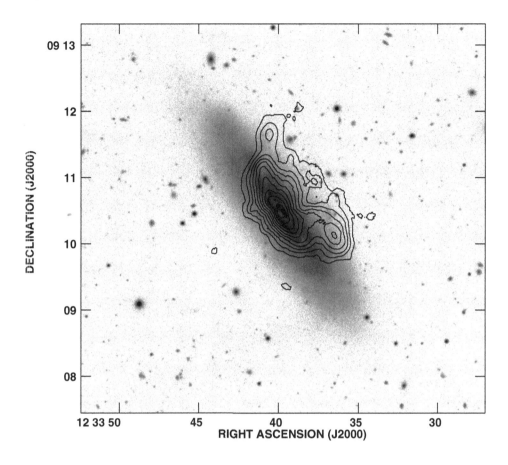

Fig. 19.7. H I contours overlaid on an *R*-band greyscale image from the WIYN telescope (Kenney & Koopmann 1999). Note the undisturbed outer stellar disk. (Taken from Kenney et al. 2004.)

& Vollmer 2004). There is striking similarity between the spatial distribution of the Hα emission and the H I emission (Fig. 19.8). The Hα emission from the disk is confined to the inner 3 kpc as well, and extraplanar Hα filaments (10% of the Hα emission) emerge from the outer edge of the Hα disk (Kenney & Koopmann 1999). Note that there are H II regions associated with each of the two major extraplanar H I peaks, and that those in the SW are much more luminous. These data strongly suggest an ICM-ISM interaction. The undisturbed stellar disk rules out a gravitational interaction. The truncated H I disk suggests ram pressure stripping is at work. The extraplanar gas has almost certainly been swept out of the disk. A detailed simulation by Vollmer et al. (2000) suggests that the gas is falling

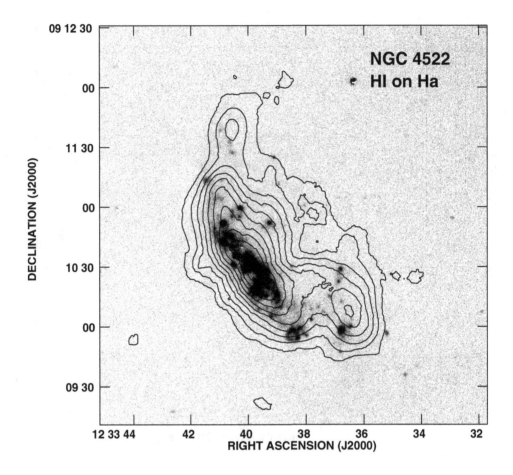

Fig. 19.8. H I contours overlaid on an Hα greyscale image from Kenney & Koopmann (1999). Note that there are H II regions associated with each of the two major extraplanar H I peaks. (Taken from Kenney et al. 2004.)

back after a stripping event, but this is inconsistent with the observed kinematics (Kenney et al. 2004). More likely, the gas is still on its way out due to ram pressure stripping (Kenney & Koopmann 1999). The combined characteristics such as a stripped H I disk, only central Hα emission in the disk, and extraplanar gas on only one side of the galaxy make this such a convincing candidate for an ICM-ISM interaction. Evidence for enhanced central star formation in some other examples makes this even more interesting. A central starburst would help use up any remaining gas, and a morphological transformation would be in place.

19.5 Surveys and the Importance of Interactions with the ICM

So far I have only presented evidence, based on selected cases, that interaction with the ICM occurs. The small gaseous disks in the optically selected samples of Virgo and Coma are almost certainly due to ram pressure stripping. A growing number of spiral galaxies are found with an unusual morphology in H I, Hα, and radio continuum, such as the long-known examples in A1367 (Gavazzi et al. 1995), in Virgo galaxies such as NGC 4522, NGC 4388 (Veilleux et al. 1999), NGC 4569, and NGC 4438 (B. Vollmer, in preparation), and in Coma (Bravo-Alfaro et al. 2001; Beijersbergen 2003; Gregg, Holden, & West 2004). These are prime candidates for ongoing ram pressure stripping.

How important are these ICM-ISM interactions for the evolution of galaxies in clusters? An important first step to address this question is the imaging study by Koopmann & Kenney (1998, 2004) of 55 Virgo cluster spirals in Hα and *R* band. They find that the total massive star formation rates in Virgo cluster spirals have been reduced by factors up to 2.5 in the median compared to isolated spirals. The reduction in total star formation is caused primarily by truncation of the star-forming disks (seen in 52% of the spirals). Some of these have undisturbed stellar disks and are likely the product of ICM-ISM stripping, but others have disturbed stellar disks, and are likely the product of tidal interactions or minor mergers, possibly in addition to ICM-ISM stripping. Some evidence is found for enhanced star formation rates due to low-velocity tidal interactions and possibly accretion of H I gas. A strong correlation is found between H I deficiency and normalized Hα flux. The authors conclude that the survey provides strong evidence that ICM-ISM interactions play a significant role in the evolution of most Virgo spirals by stripping gas from their outer disks.

So far I have only discussed H I results obtained on individual galaxies that were selected because they were interesting or, at best, because they were in some optical flux-limited sample. To see how the gas content and morphology depends on cluster environment, optically unbiased studies need to be done. Ideally one should probe the entire volume of clusters, including the low-density outskirts, to get some idea of the gas content and star formation properties as function of local or global density.

The first volume-limited H I survey of a cluster was done of the Hydra cluster (McMahon 1993; van Gorkom 1996). Dickey (1997) surveyed two clusters in the rich group of clusters in the Hercules cluster. These surveys already show that there is a great variety in the neutral hydrogen properties of clusters. Hydra shows barely any evidence for hydrogen deficiency, despite the fact that it is very similar in its global properties to the Virgo cluster. The most likely explanation of these results is that Hydra is in fact a superposition of at least three groups along the line of sight, seen in projection close to each other. The most striking result of the Hercules survey (Dickey 1997) is the spatial variation of H I properties within the clusters. Galaxies in A2147 and the southwest of A2151 show strong H I deficiency, while galaxies in the northeast of A2151 are gas rich. It is perhaps one of the most convincing demonstrations of environmental impact on galaxy properties. The X-ray luminous clusters have strong H I deficiency, whereas the parts of the clusters that have no detectable ICM have an abundance of gas-rich galaxies.

A more systematic survey of five nearby clusters (Abell 85, 754, 496, 2192, 2670) is currently being done at the VLA (Poggianti & van Gorkom 2001; van Gorkom et al. 2003). Each cluster is completely covered out to $2\,R_A$, thus covering the dense inner parts and the low-density outer parts, and the entire optical velocity range is probed. The most striking result is that in all clusters the H I detections are highly clustered both spatially and in

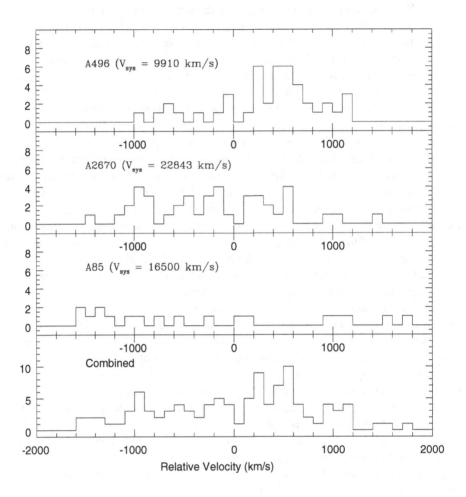

Fig. 19.9. The velocity distribution of the H I-detected galaxies in Abell 496, 2670 and 85 with respect to the mean velocity of the cluster. At the bottom is shown the combined distribution for the three clusters. Note the very non-Gaussian distribution of the velocities. In Figure 19.11 an example is shown of spatial and velocity clustering of H I-detected galaxies.

velocity. Figure 19.9 shows the velocity distribution of the H I detections in three of the clusters. Although the velocity distribution of the optically cataloged galaxies in each of the clusters is Gaussian, the velocity distribution of the gas-rich galaxies is far from Gaussian. Figure 19.10 shows the total H I image of Abell 2670. Contours represent the integrated H I emission for individual galaxies. At first glance the image looks like the images of the Virgo and Coma cluster, with small H I disks close to the center and large H I disks further out. But this image now shows the H I emission from all galaxies in the cluster with H I masses $\geq 2 \times 10^8 M_\odot$. Figure 19.11 shows an overlay of a group of galaxies to the NW on the DSS. The internal velocity dispersion of this group is only a few$\times 100$ km s^{-1}. These galaxies are very gas-rich, and obviously conditions for interactions and merging are ideal.

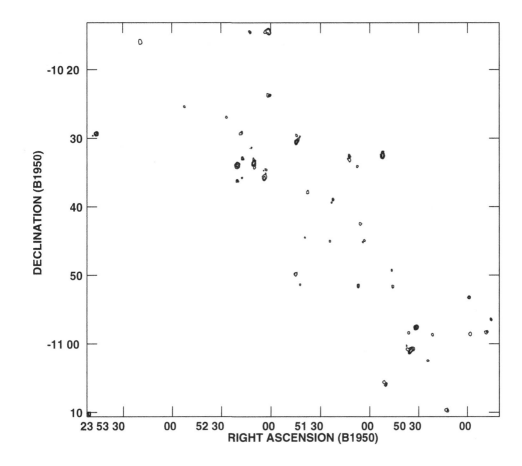

Fig. 19.10. H I emission of the cluster Abell 2670 at a redshift of $z = 0.08$. The image is centered on the center of the cluster, and the H I detections are spread over a region of $5 \times 2h^{-1}$ Mpc. All galaxies with an H I mass $\geq 2 \times 10^8 M_\odot$ in the velocity range of the cluster are shown.

Several galaxies do, in fact, show evidence for distorted H I. These results indicate that gas-rich disk galaxies that make it into the center of a cluster are likely to be seriously affected by interaction with the ICM. It is likely that a significant fraction of disk galaxies falling into clusters are located in low-velocity dispersion, loose groups. The interactions in these groups, before the actual infall, may be more damaging to the morphology than any ICM interaction thereafter. The most dramatic example (Fig. 19.12) of that is the H I image of a number of S0 galaxies in the outskirts of the Ursa Major cluster by Verheijen & Zwaan (2001). Optically one would not have guessed that anything dramatic is about to happen to these galaxies. The H I shows that strong interactions are already taking place.

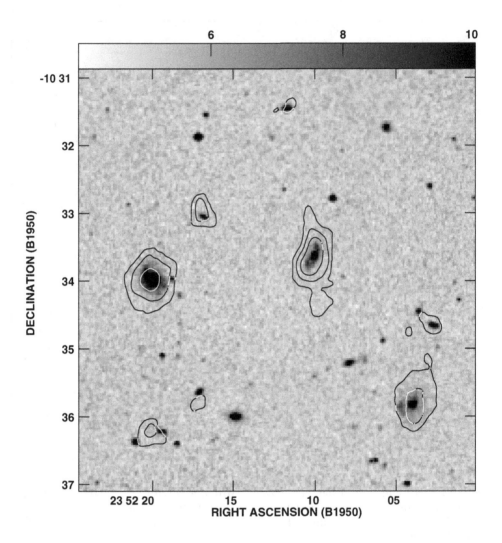

Fig. 19.11. An overlay of the total H I emission (contours) on a DSS optical image (greyscale) of a group of galaxies in the NE part of the A2670 cluster. The internal velocity dispersion of this group is only a few$\times 100$ km s^{-1}. These are very gas-rich galaxies. Note the distorted H I contours indicative of interactions.

19.6   Concluding Remarks

We can now begin to answer the question posed in the introduction: are interactions with the ICM important for the evolution of disk galaxies. The answer is a definite yes for disk galaxies in cluster environments. We see individual galaxies that show all the signs of an ongoing interaction, signs that are predicted by detailed hydrodynamical simulations. We

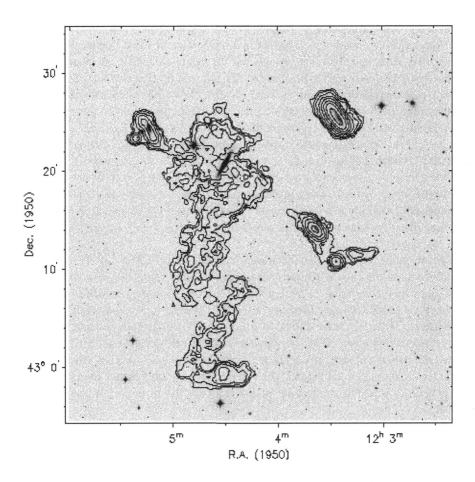

Fig. 19.12. Recently discovered H I filaments near the S0 galaxy NGC 4111 in the Ursa Major cluster. The S0 galaxy is located in a small group, and the H I morphology indicates that tidal interaction between the galaxies is taking place. (Taken from Verheijen & Zwaan 2001.)

see trends in galaxy properties, such as the truncated gaseous disks in the center of Virgo and Coma, and truncated star formation disks in Virgo (Koopmann & Kenney 2004), which can be reproduced in simulations of ICM interactions with the cold ISM in disk galaxies. These interactions should produce a population of early-type, non-star-forming disk galaxies with a range of bulge-to-disk ratios, as was predicted by Dressler & Gunn (1983) and as has been found in Virgo (Koopmann & Kenney 2004). However, other effects also play a role in clusters. Galaxies are found that experience both stripping and gravitational interactions (Vollmer 2003; Koopmann & Kenney 2004), and examples of gravitationally induced star formation have also been found (Rose et al. 2001; Sakai et al. 2002; Koopmann & Kenney 2004). However, the dominant environmental effect on cluster disk galaxies is a reduction of the star formation rate, which goes hand-in-hand with hydrogen deficiency, and for most galaxies this is due to ram pressure stripping (Koopmann & Kenney 2004).

On larger scales, Solanes et al. (2001) found evidence that the H I deficiency goes out as far as $2\,R_A$. Although this is surprisingly far, the fact that the deficient galaxies at large distances from the cluster tend to be on radial orbits makes it plausible that the cause of the deficiency is ram pressure stripping as well. The extent of the H I deficiency fits in nicely with the results of Balogh et al. (1998), who find that the star formation rate, as measured by the [O II] equivalent width, is depressed in clusters out to $2\,R_{200}$ ($\sim 2\,R_A$), and more recently the analyses of the 2dF and the SDSS survey results (Lewis et al. 2002; Gómez et al. 2003; Nichol 2004), which indicate that star formation rates begin to drop between 1 and 2 virial radii. These results are somewhat at odds with the H I imaging results, where evidence is found that the groups in the outskirts of clusters are very gas rich. Many examples of ongoing interactions are found in these locations. In a simple scenario the interactions would bring gas to large distances from the galaxies, which could then easily be stripped as the galaxies fall into the denser ICM.

It is an intriguing possibility that the impact and reach of the ICM is closely related to the dynamical state of the cluster. Cluster-(sub)cluster merging can give rise to bulk motions, shocks, and temperature structure within the ICM. In merging clusters, observational evidence has been found for large velocities in the ICM (Dupke & Bregman 2001), enhanced star formation (Miller & Owen 2003; Miller 2004), and distortion of radio sources by the ICM motions (Bliton et al. 1998). If stripping would mostly depend on the motions of the ICM and the dynamical state of the cluster, it would more easily explain why the effects are seen far out into the infall region. The radial orbits measured for the more distant H I-deficient galaxies would then reflect the infall direction of the most recent accretion event in the cluster.

Acknowledgements. I am grateful to the organizers of this conference for inviting me to give this review. I thank Jose Solanes, Marc Verheijen, Hector Bravo-Alfaro, Dwarakanath, Jeff Kenney, Bianca Poggianti, Raja Guhathakurta, Ann Zabludoff, David Schiminovich, Monica Valluri, Eric Wilcots, and Bernd Vollmer for help with figures and many discussions on these topics. I thank the Kapteyn Institute, where part of this work was done, for their hospitality. This research was supported by an NSF grant to Columbia University and a NWO bezoekers beurs to Groningen.

References

Abadi, M. G., Moore, B., & Bower, R. G. 1999, MNRAS, 308, 947
Balogh, M. L., Morris, S. L., Yee, H. K. C., Carlberg, R. G., & Ellingson, E. 1999, ApJ, 527, 54
Balogh, M. L., Schade, D., Morris, S. L., Yee, H. K. C., Carlberg, R. G., & Ellingson, E. 1998, ApJ, 504, L75
Beijersbergen, M. 2003, Ph.D. Thesis, Univ. of Groningen
Bliton, M., Rizza, E., Burns, J. O., Owen, F. N., & Ledlow, M. J. 1998, MNRAS, 301, 328
Böhringer, H., Briel, U. G., Schwarz, R. A., Voges, W., Hartner, G., & Trumper, J. 1994, Nature, 368, 828
Bravo-Alfaro, H., Cayatte, V., van Gorkom, J. H., & Balkowski, C. 2000, AJ, 119, 580
——. 2001, A&A, 379, 347
Cayatte, V., Kotanyi, C., Balkowski, C., & van Gorkom, J. H. 1994, AJ, 107, 1003
Cayatte, V., van Gorkom, J. H., Balkowski, C., & Kotanyi, C. 1990, AJ, 100, 604
Cowie, L. L., & Songaila, A. 1977, Nature, 266, 501
Dickey, J. M. 1997, AJ, 113, 1939
Dressler, A. 1980, ApJ, 236, 351
——. 1986, ApJ, 301, 35

Dressler, A., et al. 1997, ApJ, 490, 577

Dressler, A., & Gunn, J. E. 1983, ApJ, 270, 7

Dupke, R. A., & Bregman, J. N. 2001, ApJ, 562, 226

Fasano, G., Poggianti, B. M., Couch, W. J., Bettoni, D., Kjærgaard, P., & Moles, M. 2000, ApJ, 542, 673

Gavazzi, G., Contursi, A., Carrasco, L., Boselli, A., Kennicutt, R., Scodeggio, M., & Jaffe, W. 1995, A&A, 304, 325

Giovanelli, R., & Haynes, M. P. 1983, AJ, 88, 881

Gómez, P. L., et al. 2003, ApJ, 584, 210

Gregg, M. D., Holden, B. P., & West, M. J. 2004, in Galaxy Evolution in Groups and Clusters, ed. C. Lobo, M. Serote-Roos, & A. Biviano (Dordrecht: Kluwer), in press (astro-ph/0301459)

Gunn, J. E., & Gott, J. R. 1972, ApJ, 176, 1

Gursky, H., Kellogg, E., Murray, S., Leong, C., Tananbaum, H., & Giacconi, R. 1971, ApJ, 167, L81

Haynes, M. P., & Giovanelli, R. 1984, AJ, 89, 758

Haynes, M. P., Giovanelli, R., & Chincarini, G. D. L. 1984, ARA&A, 22, 445

Hubble. E., & Humason, M. L. 1931, ApJ, 74, 43

Kenney, J. D. P., & Koopmann, R. A. 1999, AJ, 117, 181

Kenney, J. D. P., van Gorkom, J. H., & Vollmer, B. 2004, AJ, submitted

Kenney, J. D. P., & Young, J. S. 1989, ApJ, 344, 171

Koopmann, R. A., & Kenney, J. D. P. 1998, ApJ, 497, L75

——. 2004, ApJ, submitted (astro-ph/0209547)

Lewis, I., et al. 2002, MNRAS, 334, 673

Magri, C., Haynes, M. P., Forman, W., Jones, C., & Giovanelli, R. 1988, ApJ, 333, 136

McMahon, P. M. 1993, Ph.D. Thesis, Columbia Univ.

Miller, N. A. 2004, in Carnegie Observatories Astrophysics Series, Vol. 3: Clusters of Galaxies: Probes of Cosmological Structure and Galaxy Evolution, ed. J. S. Mulchaey, A. Dressler, & A. Oemler (Pasadena: Carnegie Observatories, http://www.ociw.edu/ociw/symposia/series/symposium3/proceedings.html)

Miller, N. A., & Owen, F. N. 2003, AJ, 125, 2427

Nichol, R. C. 2004, in Carnegie Observatories Astrophysics Series, Vol. 3: Clusters of Galaxies: Probes of Cosmological Structure and Galaxy Evolution, ed. J. S. Mulchaey, A. Dressler, & A. Oemler (Cambridge: Cambridge Univ. Press), in press

Nulsen P. 1982, MNRAS, 198, 1007

Oemler, A., Jr. 1974, ApJ, 194, 10

Poggianti, B. M., Smail, I., Dressler, A., Couch, W. J., Barger, A. J., Butcher, H., Ellis, R. S., & Oemler, A. 1999, ApJ, 518, 576

Poggianti, B. M., & van Gorkom, J. H. 2001, in Gas and Galaxy Evolution, ed. J. E. Hibbard, M. Rupen, & J. H. van Gorkom (San Francisco: ASP), 599

Postman, M., & Geller, M. J. 1984, ApJ, 281, 95

Quilis, Q., Moore, B., & Bower, R. G. 2000, Science, 288, 1617

Rose, J. A., Gaba, A. E., Caldwell, N., & Chaboyer, B. 2001, AJ, 121, 793

Sakai, S., Kennicutt, R. C., van der Hulst, J. M., & Moss, C. 2002, ApJ, 578, 842

Schulz, S., & Struck, C. 2001, MNRAS, 328, 185

Solanes, J. M., Manrique, A., García-Gómez, C., González-Casado, G., Giovanelli, R., & Haynes, M. P. 2001, ApJ, 548, 97

Spitzer, L., Jr., & Baade, W. 1951, ApJ, 113, 413

Stark, A. A., Knapp, G. R., Bally, J., Wilson, R. W., Penzias, A. A.,& Rowe, H. E. 1986, ApJ, 310, 660

Stevens, I. R., Acreman, D. M., & Ponman, T. 1999, MNRAS, 310, 663

van den Bergh, S. 1976, ApJ, 206, 883

van Gorkom, J. H. 1996 in the Minnesota Lectures on Extragalactic Neutral Hydrogen, ed. E. D. Skillman (San Francisco: ASP), 293

van Gorkom, J. H. et al. 2003, http://www.aoc.nrao.edu/vla/html/vlahome/largeprop/

Veilleux, S., Bland-Hawthorn, J., Cecil, G., Tully, R. B., & Miller, S. T. 1999, ApJ, 520, 111

Verheijen, M. A. W., & Zwaan, M. 2001, in Gas and Galaxy Evolution, ed. J. E. Hibbard, M. Rupen, & J. H. van Gorkom (San Francisco: ASP), 867

Vikhlinin, A., Forman, W., & Jones, C. 1997, ApJ, 474, 7

Vollmer, B. 2003, A&A, 398, 525

Vollmer, B., Cayatte, V., Balkowski, C., Boselli, A., & Duschl, W. J. 1999, A&A, 349, 411

Vollmer, B., Cayatte, V., Balkowski, C., & Duschl, W. J. 2001, ApJ, 561, 708

Vollmer, B., Marcelin, M., Amram, P., Balkowski, C., Cayatte, V., & Garrido, O. 2000, A&A, 364, 532
Warmels, R. H. 1988a, A&AS, 72, 19
——. 1988b, A&AS, 72, 57
——. 1988c, A&AS, 72, 427

20

The difference between clusters and groups: a journey from cluster cores to their outskirts and beyond

RICHARD G. BOWER and MICHAEL L. BALOGH
Institute for Computational Cosmology, University of Durham, UK

Abstract

In this review, we take the reader on a journey. We start by looking at the properties of galaxies in the cores of rich clusters. We have focused on the overall picture: star formation in clusters is strongly suppressed relative to field galaxies at the same redshift. We will argue that the increasing activity and blue populations of clusters with redshift results from a greater level of activity in field galaxies rather than a change in the transformation imposed by the cluster environment. With this in mind, we travel out from the cluster, focusing first on the properties of galaxies in the outskirts of clusters and then on galaxies in isolated groups. At low redshift, we are able to efficiently probe these environments using the Sloan Digital Sky Survey and 2dF redshift surveys. These allow an accurate comparison of galaxy star formation rates in different regions. The current results show a strong suppression of star formation above a critical threshold in local density. The threshold seems similar regardless of the overall mass of the system. At low redshift at least, only galaxies in close, isolated pairs have their star formation rate boosted above the global average. At higher redshift, work on constructing homogeneous catalogs of galaxies in groups and in the infall regions of clusters is still at an early stage. In the final section, we draw these strands together, summarizing what we can deduce about the mechanisms that transform star-forming field galaxies into their quiescent cluster counterparts. We discuss what we can learn about the impact of environment on the global star formation history of the Universe.

20.1 Introduction

Let us start with an outline of this review. We will begin by looking at galaxies in the cores of clusters. We have been observing clusters for many years. Some milestones are the papers on the morphological differences between cluster galaxies and the general field (Hubble & Humason 1931), the discovery of a global morphology-density relation (Oemler 1974; Dressler 1980), and the realization of the importance of the color-magnitude relation (Sandage & Visvanathan 1978). We will attempt to summarize what we have learned from looking at clusters since this time. In particular, recent observations now span a wide range of redshift, allowing us to look directly at how the galaxy populations evolve.

In the second section, we will investigate how galaxy star formation rates vary with radius and local density. In particular, we will focus on the recent results from the 2dF galaxy redshift survey. The aim here is to understand how galaxy properties are influenced by their environment. As we will discuss, it seems that the group environment is critical to the evolution of galaxies, creating a distinctive threshold.

In the third section, we will review some of the ideas about how this can all be put to-gether, and how we can hope to use the environmental studies that many groups are under-taking to build a better understanding of the evolution of the Universe.

Throughout, this paper will focus on galaxy star formation rates as the measure of galaxy properties, and will leave aside the whole issue of galaxy morphology for other reviewers to deal with. Clearly, the two issues are related since galaxy morphology partly reflects the strength of H II regions in the galaxy disk (Sandage 1961), but the two factors are not uniquely linked. Morphology and star formation may be influenced differently by different environments (Dressler et al. 1997; Poggianti et al. 1999; McIntosh, Rix, & Caldwell 2004). We will also skirt around the important issue of E+A galaxies (Couch & Sharples 1987; Dressler & Gunn 1992; Barger et al. 1996; Balogh et al. 1999; Poggianti et al. 1999) and star formation that is obscured from view in the optical (Poggianti & Wu 1999; Smail et al. 1999; Duc et al. 2002; Miller & Owen 2002). These are discussed in detail in Poggianti (2004). Wherever possible we will use $H\alpha$ as the star formation indicator (Kennicutt 1992), but as we probe to higher redshift, we are forced to use [O II] $\lambda3727$ unless we shift our strategy to infrared spectrographs.

We will also stick to talking about bright galaxies, by which we mean galaxies brighter than 1 mag fainter than L_*. It would need another complete review if we were to compare the properties of dwarf galaxies over the same range of environments. A good place to start would be Drinkwater et al. (2001), or the many presentations on cluster dwarfs at this Symposium. By the same token, we will avoid discussion of the evolution of the galaxy lu-minosity function (Barger et al. 1998; De Propris et al. 1999); this is summarized in Rudnick et al. (2004).

To avoid confusion, it is worth laying out exactly what we mean by the terms "cluster" and "group." We will use the term cluster to mean a virialized halo with mass greater than $10^{14} M_\odot$ and the term group to mean a halo more massive than about $10^{13} M_\odot$ (but less than $10^{14} M_\odot$). If an isolated L_* galaxy has a halo mass of order $10^{12} M_\odot$ (Evans & Wilkinson 2000; Guzik & Seljak 2002; Sakamoto, Chiba, & Beers 2003), then our definition of a group contains more than five L_* galaxies at the present day. At higher redshift, the conversion between mass and galaxy numbers is more complicated since it depends on whether L_* evolves or not. If we stick to a definition in terms of mass, then at least everything is clear from a theoretical perspective, and we can make quite definite predictions about the numbers of such halos, their clustering as a function of redshift (Press & Schechter 1974; Jenkins et al. 2001; Sheth, Mo, & Tormen 2001), and how mass accumulates from smaller halos into large clusters (Bond et al. 1991; Bower 1991; Lacey & Cole 1993; Mo & White 2002).

20.2   Clusters of Galaxies

At the outset, its worth reminding ourselves of why we study galaxy evolution in clusters. One popular reason is that the cluster is a good laboratory in which to study galaxy evolution. Another is that it is "easy"—when we observe the galaxy spectra, we know that most objects will be in this dense environment and that our observations will be highly efficient. The same reason allows us to recognize clusters out to very high redshifts and thus to extend our studies to a very long baseline. But we should remember that clusters do have a significant drawback: they are rare objects. For the standard ΛCDM cosmology ($\Omega_m = 0.3$, $\Omega_\Lambda = 0.7$, $h = 0.7$, $\sigma_8 = 0.9$), the space density of $> 10^{14} M_\odot$ halos is $7 \times 10^{-5} h^{-3} \mathrm{Mpc}^{-3}$.

Even though such clusters contain $\sim 100\,L_*$ galaxies, less than 10% of the cosmic galaxy population is found in such objects.

There is an emerging consensus that suggests that the stellar populations of galaxies in cluster cores are generally old, with most of the stars formed at $z > 2$. Most of these galaxies also have early-type morphology. It is possible to derive remarkably tight constraints from looking at colors (Bower, Lucey, & Ellis 1992; Bower, Kodama, & Terlevich 1998; Gladders et al. 1998; van Dokkum et al. 1998), at the Mg-σ relation (Gúzman et al. 1992), or at the scatter in the fundamental plane (Jørgensen et al. 1999; Fritz et al. 2004). These results rely on the argument that recent star formation would lead to excessive scatter in these tight relations, unless it was in some way coordinated, or the color variations due to age were cancelled out by variations in metal abundance (Faber et al. 1999; Ferreras, Charlot, & Silk 1999). Line-index measurements generally suggest very old populations (Jørgensen 1999; Poggianti et al. 2001), but these relations tend to show somewhat more scatter. This has been interpreted as evidence for the cancellation effects in broad-band colors.

To improve the evidence, one can compare clusters at high redshift. For example, if we concentrate on the color-magnitude relation, we would expect the narrow relation seen in local clusters to break down as we approach the epoch when star formation was prevalent. In fact, we have discovered that the color-magnitude relation is well established in high-redshift clusters (Ellis et al. 1997; van Dokkum et al. 1998), and that the line-index correlations, fundamental plane (Kelson et al. 2001), and Tully-Fisher relation measurements (e.g., Metavier 2004; Ziegler et al. 2004; but see Milvang-Jensen et al. 2003) also show little increase in scatter compared to local clusters. So far, tight relations have been identified in clusters out to $z = 1.27$ (van Dokkum et al. 2000; van Dokkum & Stanford 2003; Barrientos et al. 2004). The tight relation does eventually seem to break down, and we are not aware of any strong color-magnitude relation that has been identified in "proto-clusters" at $z > 2$.

There is a bias here, however, that should be clearly recognized . Although we are discovering that clusters at high redshifts seem also to contain old galaxies, this does not mean that all galaxies in local clusters must have these old populations. A large fraction of galaxies that are bound into local clusters would have been isolated "field" galaxies at $z \approx 1$. An even stronger bias of this type has been termed "progenitor bias" by van Dokkum & Franx (2001). They point out that if only a subset of the cluster populations is studied (for example only the galaxies with early-type morphology), then it is quite easy to arrive at a biased view. To get the full picture, one needs to study the galaxy population of the cluster as a whole.

An interesting strategy is therefore to simply measure the star formation rate in clusters at different epochs. The general consensus seems to be that there is little star formation (relative to field galaxies at the same redshift) in virialized cluster cores below $z = 1.5$. For example, Couch et al.'s (2001) survey of the AC114 cluster found that star formation was suppressed by an order of magnitude compared to the field. Similar levels of suppression are seen in poor clusters (Balogh et al. 2002). While these studies find some exciting objects (see Finn & Zaritsky 2004 for further examples), the general trend is for the star formation rate to be strongly suppressed relative to the field at the same redshift. infrared measurements (Duc et al. 2002) and radio measurements (Morrison & Owen 2003; Miller 2004) have generally come to similar conclusions. The E+A galaxies (Dressler & Gunn 1992) or post-starburst galaxies (Couch & Sharples 1987) are a puzzling exception. The large numbers found by the MORPHs group (Dressler et al. 1997) suggest that there was strong star formation activity in the recent past in many galaxies (but see Balogh et al. 1999). A possible explanation is that these galaxies have only recently arrived in the cluster from much lower-density environments. Indeed, field studies at low redshift have shown this type of object to be more common in low-density regions than in clusters (Zabludoff et al. 1996; Goto et al. 2003; Quintero et al. 2004). Therefore, the greater numbers of E+A galaxies found in high-redshift clusters may result from the greater star formation activity of galaxies outside clusters—this idea gains strong support from Tran et al.'s (2004) observations presented at the Symposium.

The next step is to compare the star formation rates in clusters cores at different redshifts. Work is only just starting on this using emission-line strengths (e.g., Ellingson et al. 2001), since it is essential to control systematic uncertainties, such as the aperture through which the star formation rate is measured. However, extensive comparisons have been made on the basis of colors, starting with Butcher & Oemler (1978, 1984) and Couch & Newell (1984). These papers showed a startling increase in the numbers of blue galaxies in $z > 0.2$ galaxy clusters compared to the present day. These results have been confirmed by more recent studies (e.g., Rakos & Schombert 1995; Margoniner et al. 2001), although the effect of the magnitude limit and cluster selection play at least as important a role as the redshift (Fairley et al. 2002).

There are two issues that complicate the comparison of the galaxies in cluster cores, however. Firstly, we must be careful how we select galaxies that are to be compared. Most of the blue galaxies lie close to the photometric completeness limit. These galaxies will fade by up to 1 mag if star formation is turned off, and thus they are not directly comparable to the red-galaxy population selected at the same magnitude limit (Smail et al. 1998; Kodama & Bower 2001). Secondly, we are observing galaxy clusters in projection. There is little doubt that the field galaxy population at intermediate redshift is much bluer than in the local Universe (Lilly et al. 1995; Madau, Pozzetti, & Dickinson 1998); thus, although a small level of contamination by field galaxies has little influence on the overall color distribution, the same contamination will have a much bigger impact on the distribution at intermediate redshift. This problem is only partially eliminated if a complete sample of galaxy redshifts is available since the velocity dispersion of the cluster makes it impossible to distinguish cluster members from "near-field" galaxies that are close enough to the cluster to be indistinguishable in redshift space (Allington-Smith et al. 1993; Balogh et al. 1999; Ellingson et al. 2001). This idea is reinforced by experiments with numerical simulations. Galaxies can be associated with dark matter particles, and then "observed" to measure the extent to which radial information is lost. Diaferio et al. (2001) found that a contamination of 10% can easily occur; furthermore, since most of the contaminating galaxies are blue (and in these models most genuine cluster galaxies are red), the fraction of blue galaxies can then be boosted by 50%. Despite this, Ellingson et al. (2001) conclude that the rate at which clusters are being built up must also be higher in the past in order for this explanation to work. Kauffmann (1995) shows that there is good theoretical justification for this.

It will be interesting to see if the evolution in the colors of the cluster population are consistent with the evolution in the emission-line strengths. We might expect to see a difference because of the different time scales probed by colors and by emission lines. For example, if galaxies that fall into the cluster have their star formation quickly suppressed, they will remain blue (in the Butcher-Oemler sense) for a significant period after the line emission subsides (Ellingson et al. 2001). Combining these factors, it seems quite possible to accom-

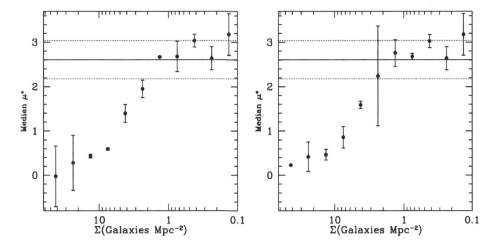

Fig. 20.1. The median star formation rate as a function of local density from galaxies around clusters in the 2dF survey (based on Lewis et al. 2002). The *left* panel shows the star formation rate of all galaxies in the sample; the *right* panel shows the effect of removing galaxies within 2 virial radii of the cluster center. Error bars are jackknife estimates. The relations are amazingly similar, showing that the local density is more important than the overall mass of the group or cluster.

modate both weak evolution in emission-line strength and more rapid evolution in the colors of cluster galaxies.

20.3 The Other Axis: Density

20.3.1 The Cluster Outskirts

So far we have been discussing the properties of galaxies within the cores of clusters, but the dependence on density (or, nearly equivalently, cluster-centric radius) provides another axis over which to study galaxy properties. We have seen that star formation is strongly suppressed in the cores of rich clusters—but at what radius do the galaxies become more like the field? We should also realize that it might be better to compare galaxy properties with their local densities (Dressler 1980; Kodama et al. 2001), as the large-scale structure surrounding clusters may have the dominant impact on galaxy evolution.

One of the first steps at studying galaxies in the transition zone around clusters were made with the CNOC2 survey (Balogh et al. 1999). They showed that there was a strong radial dependence in the star formation rate, but that the star formation rate had not yet reached the field value even at $r \approx r_{vir}$. The Sloan Digital Sky Survey (SDSS) and 2dF galaxy redshift survey surveys have allowed us to make a huge leap forward in this respect. In the local Universe, we are able to map galaxy star formation rates, using the complete redshift information to eliminate contamination by interlopers. In this section we will concentrate on what we have learned from the 2dF survey (Lewis et al. 2002), but the results from the SDSS give very consistent answers (Gómez et al. 2003). Figure 20.1 shows the median star formation rate as a function of local density. What is remarkable in this plot is that there is quite a sharp transition between galaxies with field-like star formation rates at $\Sigma < 1\,\mathrm{Mpc}^{-2}$

and galaxies with low star formation rates comparable to cluster cores ($\Sigma > 7\,\mathrm{Mpc}^{-2}$). The switch is complete over a range of less than 7 in density.

The density at which the transition occurs corresponds to the density at the virial radius. If star formation is plotted against radius, the transition is considerably smeared out, but does occur at around the cluster virial radius—well outside the core region on which a lot of previous work has been focused. The 2dF galaxy redshift survey sample is sufficiently large that we can remove the cluster completely from this diagram. By only plotting galaxies more than 2 virial radii from the cluster centers, we concentrate on the filaments of infalling material. The correlation with local density is shown in Figure 20.1. Amazingly, the relation hardly changes compared to the complete cluster diagram.

This is a great success: we have identified the region where galaxy transformation occurs! It is in the infalling filaments (consisting of chains of groups) where galaxies seem to change from star-forming, field-like galaxies to passive, cluster-like objects. Of course, it is tempting to associate the transformation in star formation rate with a transformation from late- to early-type morphology. Unfortunately, this test cannot be undertaken with the available 2dF data, but we can expect clearer results from SDSS.

What happens at higher redshift? In fact, the first claim of a sharp transition in galaxy properties was made by Kodama et al. (2001) for the distant cluster A851 at $z = 0.41$ (top panel in Fig. 20.2). Kodama et al. (2001) used photometric redshifts to eliminate foreground objects, and thus to reduce contamination of the cluster members to a level that allowed the color distribution to be studied in the outer parts of the cluster. Their results show an amazing transition in color. Direct comparison with the local clusters is difficult, however, as the magnitude limits are very different (Kodama et al.'s photometric data reach much fainter than the local spectroscopic samples), but Gómez et al. (2003) concluded that the threshold seen by Kodama et al. (2001) was at a significantly higher local density. Perhaps dwarf galaxies are more robust to this environmental transformation; we are not going to attempt to cover this issue.

A number of researchers are now engaged in spectroscopic programs to study the transformation threshold in higher-redshift systems. The results of Treu et al. (2003) are perhaps the most advanced. They also have the advantage of panoramic WFPC2 imaging that will allow them to compare the transformation of galaxy morphology (see Treu 2004).

The highest redshifts that can be studied require a combination of photometric preselection of objects for spectroscopy. Nakata et al. (2004) have used the photometric technique to map the large-scale structure around the Lynx cluster at $z = 1.27$ (lower panel in Fig. 20.2), and similar techniques are described by Demarco et al. (2004). These groups identify several candidate filaments; spectroscopy of these regions is now underway.

20.3.2   *Galaxy Groups*

Returning to the local Universe, it is interesting to see if we can probe the properties of galaxies in groups directly. A lot of work has been carried out looking at small samples of groups selected from the CfA redshift survey (Geller & Huchra 1983; Moore, Frenk, & White 1993), from the Hickson compact group catalog (Hickson, Kindl, & Auman 1989), and also from X-ray surveys (Henry et al. 1995; Mulchaey et al. 2003).

In the era of the 2dF and SDSS redshift surveys, we can construct robust catalogs

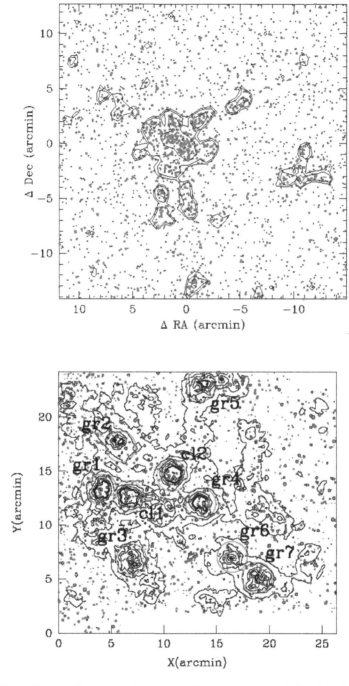

Fig. 20.2. *Top*: Contour lines pinpoint the transition zone around the rich cluster A851 at $z = 0.41$ (from Kodama et al. 2001). *Bottom*: $z = 1.27$ groups tracing the large-scale structure surrounding the Lynx clusters (based on Nakata et al. 2004). These figures illustrate how photometric methods can be used to reduce contamination rates in the outskirts of distant clusters so that the galaxy population can be studied there.

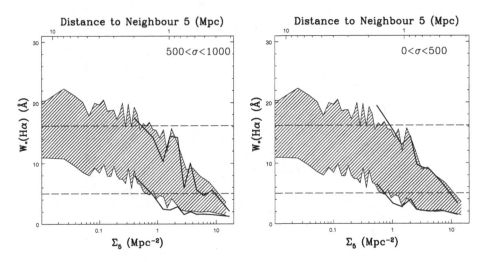

Fig. 20.3. The star formation rate as a function of density, comparing groups of galaxies with clusters. The upper and lower horizontal dashed lines show the 75th percentile and the median of the equivalent widths. The hashed region shows the relation for the complete sample, while the solid line shows the relation for systems with $500 \, \text{km s}^{-1} < \sigma < 1000 \, \text{km s}^{-1}$ (*left*) and $\sigma < 500 \, \text{km s}^{-1}$ (*right*). The dependence on local density is identical irrespective of the velocity dispersion of the whole system. (Figure based on Balogh et al. 2004.)

containing thousands of groups (Eke et al. 2004). It is interesting to compare the star formation rate in the groups as a function of local density, with the relation found in clusters. The relation for the 2dF survey is shown in Figure 20.3 (Balogh et al. 2004). The panels show the effect of selecting systems on the basis of their velocity dispersion. There is actually very little difference between the trends. The galaxies in dense regions suffer the same suppression of their star formation rate, regardless of the system's total mass. It is also possible to show that the groups in the infall regions of clusters show the same pattern as isolated groups. We have to conclude that the suppression of star formation is very much a local process. This is an important clue to distinguish between the different transformation mechanisms.

Interestingly, in the local Universe, there is little evidence for the environment producing a rise in the star formation rate above the field value. The only exception to this appears to be the close, low-velocity encounters of isolated galaxies (Barton, Geller, & Kenyon 2000; Lambas et al. 2004). Figure 20.4 shows the star formation rate as a function of separation for systems of different total velocity dispersion. A spike in the median star formation rate appears only in the smallest bin of the first panel. It will be interesting to study this trend within groups and clusters (Balogh et al. 2004).

One of the next goals is to extend studies of groups to higher redshifts. The first steps in this direction were made by Allington-Smith et al. (1993). They used radio galaxies to pick out galaxy groups at redshifts up to 0.5. By stacking photometric catalogs, they showed that the galaxy populations of rich groups ($N_{0.5}^{-19} > 30$)[*] became increasingly blue with redshift, while poorer groups contained similar populations of blue galaxies at all redshifts. A survey

---

[*] Group richness defined as the number of galaxies with $M_V \leq -19$ mag within a 0.5 Mpc radius of the radio galaxy ($H_0 = 50 \, \text{km s}^{-1} \, \text{Mpc}^{-1}$ and $q_0 = 0$ assumed).

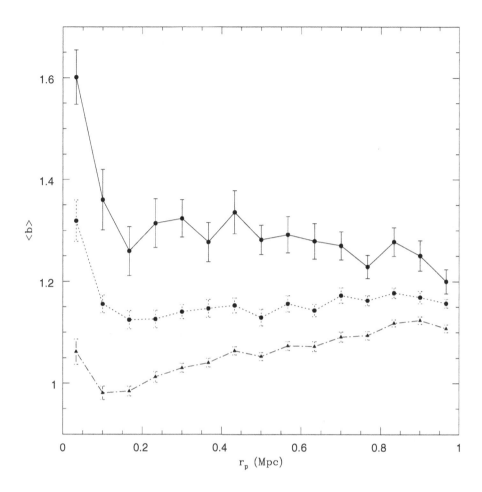

Fig. 20.4. The star formation rate of galaxy pairs (*b*) as a function of separation. The different line styles distinguish the spectral type of the first galaxy of the pair, with the sequence of dot-dash, dotted, and solid lines showing the effect of restricting the sample to more and more active central objects. A strong enhancement is only seen when the separation is less than 100 kpc. (Figure based on Lambas et al. 2004.)

of redshift-space selected groups at intermediate redshift became possible with the CNOC2 redshift survey. Carlberg et al. (2001) report a statistical sample of 160 groups out to redshift 0.4. On the Magellan telescopes, we have been following up the systems at $z > 0.3$ in order to determine the complete membership and measure total star formation rates. Figure 20.5 shows the membership of a sample group. The initial results are exciting—star formation in many galaxies are more comparable to the surrounding field values. If these results are confirmed as we derive more redshifts and improve the group completeness, it represents a very interesting change from the properties seen in the 2dF groups. At higher redshifts, a tantalizing glimpse of the properties of a few groups can be obtained from the Caltech redshift survey (Cohen et al. 2000).

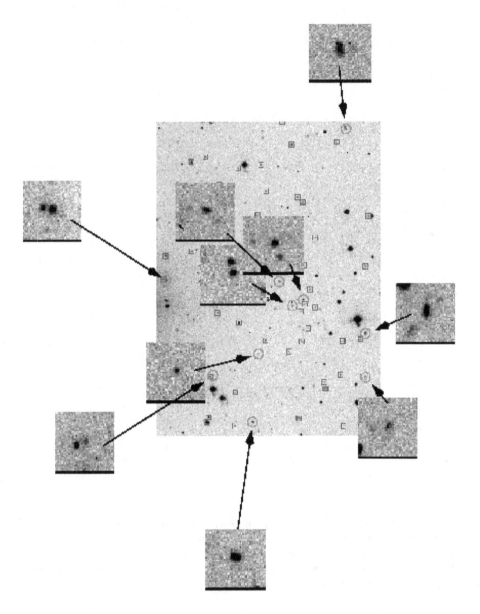

Fig. 20.5. A sample group from the CNOC2/Magellan group survey. This poor group, containing 10 members, is at $z = 0.393$.

20.4   What Does It All Mean?

20.4.1   *The Mechanisms Driving Galaxy Evolution*

The mechanisms that have been proposed to drive galaxy evolution in dense environments can be broadly separated into three categories.

- *Ram pressure stripping.* Galaxies traveling through a dense intracluster medium suffer a strong ram pressure effect that sweeps cold gas out of the stellar disk (Gunn & Gott 1972; Abadi, Moore, & Bower 1999; Quilis, Moore, & Bower 2000). The issue with this mechanism is whether it can be effective outside dense, rich cluster cores where the galaxy velocities are very high and the intracluster medium is very dense. Quilis et al. (2000) found that incorporating holes in the galaxy H I distribution made galaxies easier to strip (Fig. 20.6), but it still required clusters more massive than the Virgo cluster to have a great effect.

- *Collisions and harassment.* Collisions or close encounters between galaxies can have a strong effect on their star formation rates. The tidal forces generated tend to funnel gas toward the galaxy center (Barnes & Hernquist 1991; Barnes 2002; Mihos 2004). It is likely that this will fuel a starburst, ejecting a large fraction of the material (Martin 1999). Gas in the outer parts of the disk, on the other hand, will be drawn out of the galaxy by the encounter. Although individual collisions are expected to be most effective in groups because the velocity of the encounter is similar to the orbital time scale within the galaxy, Moore et al. (1996) showed that the cumulative effect of many weak encounters can also be important in clusters of galaxies.

- *Strangulation.* Current theories of galaxy formation suggest that isolated galaxies continuously draw a supply of fresh gas from a hot, diffuse reservoir in their halo (Larson, Tinsley, & Caldwell 1980; Cole et al. 2000). Although the reservoir is too cool and diffuse to be easily detected (Benson et al. 2000; Fang, Sembach, & Canizares 2003), this idea is supported by the observation that 90% of the baryonic content of clusters is in the from of a hot, diffuse intracluster medium. The baryon reservoir in galaxy halos is entirely analogous. When an isolated galaxy becomes part of a group, it may loose its preferential location at the center of the halo and thus be unable to draw further on the baryon reservoir. Without a mechanism for resupplying the material that is consumed in star formation and feedback, the galaxies' star formation rate will decline. The exact rate depends on the star formation law that is used (Schmidt 1959; Kennicutt 1989) and on whether feedback is strong enough to drive an outflow from the disk.

Semi-analytic models (e.g., Cole et al. 2000) generally incorporate only the third of these mechanisms. The observational data strongly suggest that the ram pressure stripping scenario cannot be important for the majority of galaxies. As we have seen, the suppression of star formation seems to occur well outside of the clusters and is equally effective in low-velocity groups, which do not possess a sufficiently dense intracluster medium. Distinguishing between the remaining two scenarios is rather harder, since they have similar dependence on environment. Indeed, they may both play a role. The key difference is the time scale on which they operate: collisions are expected to produce changes in galaxy properties on short time scales ($\sim 100\,\mathrm{Myr}$), while the changes due to strangulation are much more gradual ($> 1\,\mathrm{Gyr}$). The time scale for harassment is less well defined; while the individual encounters may induce short-lived bursts of star formation, the overall effect may accumulate over several Gyr. The radial gradients that we observe appear to prefer long time scales and,

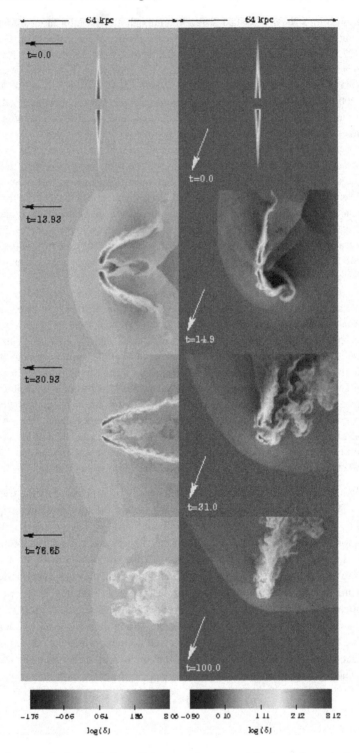

Fig. 20.6. A numerical simulation of the ram pressure stripping of a spiral galaxy. The left and right panels compare the effect when the galaxy is face-on and almost edge-on. (Reprinted with permission from Quilis, V., Moore, B., & Bower, R. 2000, Science, 288, 1617. Copyright 2000 AAAS.)

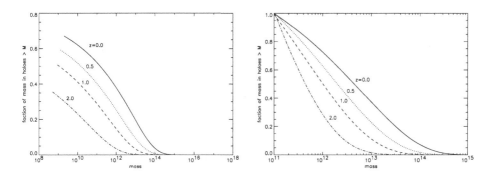

Fig. 20.7. The fraction of mass in the Universe bound into objects more massive than the mass scale on the horizontal axis. The *left* panel shows the curves with their overall normalization. Note that most of the mass is actually in subgalactic halos even in the $z = 0$ Universe. The *right* panel shows the curves renormalized at $10^{11} M_\odot$. Between $z = 1$ and the present day, there is a dramatic increase in the fraction of the renormalized mass contained in groups. The issue is to determine whether the growth of structure is responsible for the decline of the cosmic star formation rate.

hence, a mechanism like strangulation or harassment (Balogh, Navarro, & Morris 2000). To make further progress in this area, we need to compile detailed observations of galaxies that are caught in the transition phase. In particular, morphological measurements will provide another important distinction (e.g., McIntosh et al. 2004).

20.4.2 The Star Formation History of the Universe

In addition to its own intrinsic interest, one reason for studying the impact of the environment on galaxy evolution is to understand the down-turn in the cosmic star formation rate. Studies of the global star formation rate show a decline of a factor 3–10 since the peak of star formation at $z = 1$–2 (Lilly et al. 1995; Madau et al. 1998; Glazebrook et al. 1999; Wilson et al. 2002). We can simplify the possible explanations into two alternative hypotheses: (1) the down-turn is caused by galaxies running out of a supply of material for star formation, or (2) it is driven by the growth of the mass structure of the Universe. In the first scenario, the down-turn is intrinsic to the galaxy population; in the second, it is caused by the changing environments of galaxies. Of course, the truth probably lies somewhere in between. In popular "semi-analytic" models (for example, Kauffmann et al. 1999; Somerville & Primack 1999; Cole et al. 2000), the decline occurs because galaxies that are not at the centers of their halo potential cannot accrete fresh gas, because this supply is being switched off by the growth of the halo mass.

It is useful to consider a toy model to investigate whether the growth of structure in scenario (2) is sufficiently rapid to be viable. Defining groups and clusters by their mass (as discussed in the Introduction), we can plot the fraction of the mass in "groups" and "clusters" as a function of redshift. Figure 20.7 shows that this fraction is always small, but that the fraction of mass in groups changes by a factor of 3–5 since the peak of cosmic star formation history. Of course, this model is highly simplified, so one should only treat it as an illustration of the idea.

By observing groups and clusters of galaxies at different redshifts, we can hope to com-

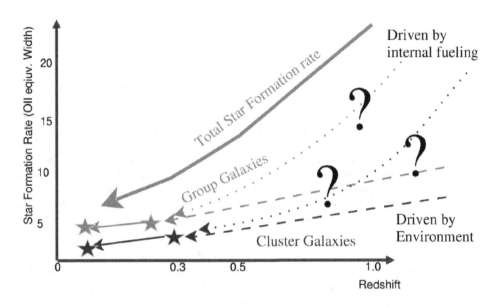

Fig. 20.8. An illustration of how we might expect the cosmic star formation history to vary between galaxies in different environments. As we look back in time, do galaxies in clusters and groups mirror the trend in the total star formation rate, or is star formation in galaxies in dense environments suppressed to similarly low levels at all redshifts? The plot hints at how we might interpret these different environment-dependent histories. Note that the curves drawn in this figure refer to the star formation rates of galaxies observed to be in groups and clusters at the redshift shown. The history of an individual galaxy (which cannot be observed, but may be extracted from simulations) will jump from one track to another as it is accreted from the field into groups and filaments and finally into the cluster core.

bine the data from different environments to make an "environmental Madau plot" where we break the contribution to the star formation rate down into its contributions from different environments. This is a task that is becoming more easily within our grasp. Some suggestions for how the plot might look are shown in Figure 20.8. An extension to the concept is to treat galaxy formation as an inverse problem. We have good models for how the dark matter halos of galaxies evolve and combine, so we can connect together galaxies in groups at $z = 1$ with galaxies in clusters at $z = 0$. By combining our observations of galaxies in different environments at different redshifts with these numerical models, we can solve for the star formation histories of galaxies along this trajectory.

20.4.3 A Closing Thought

Throughout this review we have taken it for granted that the reason why galaxies in clusters end up looking different from galaxies in the field is causally related to their present environment—that there is a definite moment of transformation when field galaxies are transformed into passive cluster/group-like objects. But it is worth pausing to consider whether this is necessarily true. We wonder if it is still possible to believe that the cluster galaxies initially form differently (e.g., in a much more rapid collapse) and then just happen to end up in clusters (i.e., the "nature" scenario). In this case there is no causal connection. For example, if we ask the question "where are the Lyman-break galaxies now?", numerical simulations have shown that these galaxies are now preferentially located in cluster cores (Governato et al. 1998). This would seem to support the "nature" conjecture. However, we do not believe that the converse is necessarily true. For example, if one selects all the galaxies in a present-day cluster and ask in what environment (halo mass) they were located at $z = 2$, one will find that there is a huge spread in the distribution of environments, and that it is not straightforward to distinguish this histogram from the corresponding histogram for galaxies identified in the present-day field. Given that the correspondence between the present-day environment and its environment at $z = 2$ is relatively weak, we find it difficult to see how galaxies can "predict" their present-day environment at the time they are being formed. Clearly, this argument needs to be placed on a stronger footing by combining simulations and semi-analytic techniques to accurately define the environment of galaxies at each epoch (Diaferio et al. 2001; Springel et al. 2001; Helly et al. 2003).

Acknowledgments. RGB thanks the Leverhulme Foundation for its support, and MLB thanks PPARC for its support. This work made extensive use of the NASA's Astrophysics Data System online bibliographic database, and the ArXiv/Astro-ph electronic preprint server.

References

Abadi, M. G., Moore, B., & Bower, R. G. 1999, MNRAS, 308, 947

Allington-Smith, J. A., Ellis, R. S., Zirbel, E. L., & Oemler, A. 1993, ApJ, 404, 52

Balogh, M. L., et al. 2004, in preparation

Balogh, M. L., Bower, R. G., Smail, I., Ziegler, B. L., Davies, R. L., Gaztelu, A., & Fritz, A. 2002, MNRAS, 337, 256

Balogh, M. L., Morris, S. L., Yee, H. K. C., Carlberg, R. G., & Ellingson, E. 1999, ApJ, 527, 54

Balogh, M. L., Navarro, J. F., & Morris, S. L. 2000, ApJ, 540, 113

Barger, A. J., et al. 1998, ApJ, 501, 522

Barger, A. J., Aragón-Salamanca, A., Ellis, R. S., Couch, W. J., Smail, I., & Sharples, R. M. 1996, MNRAS, 279, 1

Barnes, J. E. 2002, MNRAS, 333, 481

Barnes, J. E., & Hernquist, L. E. 1991, ApJ, 370, L65

Barrientos, L., Manterola, M. C., Gladders, M. D., Yee, H. K. C., Infante, L., Hall, P., & Ellingson, E. 2004, in Carnegie Observatories Astrophysics Series, Vol. 3: Clusters of Galaxies: Probes of Cosmological Structure and Galaxy Evolution, ed. J. S. Mulchaey, A. Dressler, & A. Oemler (Pasadena: Carnegie Observatories, http://www.ociw.edu/ociw/symposia/series/symposium3/proceedings.html)

Barton, E. J., Geller, M. J., & Kenyon, S. J., 2000, ApJ, 530, 660

Benson, A. J., Bower, R. G., Frenk, C. S., & White, S. D. M. 2000, MNRAS, 314, 557

Bond, J. R., Kaiser, N., Efstathiou, G., & Cole, S. 1991, ApJ, 379, 440

Bower, R. G. 1991, MNRAS, 248, 332

Bower, R. G., Kodama, T., & Terlevich, A. 1998, MNRAS, 299, 1193

Bower, R. G., Lucey, J. R., & Ellis, R. S. 1992, MNRAS, 254, 601

Butcher, H., & Oemler, A., Jr. 1978, ApJ, 219, 18

——. 1984, ApJ, 285, 426

Carlberg, R. G., Yee, H. K. C., Morris, S. L., Lin, H., Hall, P. B., Patton, D. R., Sawicki, M., & Shepherd, C. W. 2001, ApJ, 563, 736

Cohen, J. G., Hogg, D. W., Blandford, R. D., Cowie, L. L., Hu, E., Songalia, A., Shopbell, P., & Richberg, K. 2000, ApJ, 538, 29

Cole, S., Lacey, C. G., Baugh, C. M., & Frenk, C. S. 2000, MNRAS, 319, 168

Couch, W. J., Balogh, M. L., Bower, R. G., Smail, I., Glazebrook, K., & Taylor, M. 2001, ApJ, 549, 820

Couch, W. J., & Newell, E. B. 1984, ApJS, 56, 143

Couch, W. J., & Sharples, R. M. 1987, MNRAS, 229, 423

Demarco, R., Rosati, P., Lidman, C., Nonino, M., Mainieri, V., Stanford, A., Holden, B., & Eisenhardt, P. 2004, in Carnegie Observatories Astrophysics Series, Vol. 3: Clusters of Galaxies: Probes of Cosmological Structure and Galaxy Evolution, ed. J. S. Mulchaey, A. Dressler, & A. Oemler (Pasadena: Carnegie Observatories, http://www.ociw.edu/ociw/symposia/series/symposium3/proceedings.html)

De Propris, R., Stanford, S. A., Eisenhardt, P. R., Dickinson, M., & Elston, R. 1999, ApJ, 118, 719

Diaferio, A., Kauffmann, G., Balogh, M. L., White, S. D. M., Schade, D., & Ellingson, E. 2001, MNRAS, 323, 999

Dressler, A. 1980, ApJS, 42, 565

Dressler, A., et al. 1997, ApJ, 490, 577

Dressler, A., & Gunn, J. E. 1992, ApJS, 78, 1

Drinkwater, M. J., Gregg, M. D., Holman, B. A., & Brown, M. J. I. 2001, MNRAS, 326, 1076

Duc, P.-A., et al. 2002, A&A, 382, 60

Eke, V., et al. 2004, in preparation

Ellingson, E., Lin, H., Yee, H. K. C., & Carlberg, R. G., 2001, ApJ, 547, 609

Ellis, R. S., Smail, I., Dressler, A., Couch, W. J., Oemler, A., Jr., Butcher, H., & Sharples, R. M. 1997, ApJ, 483, 582

Fairley, B. W., Jones, L. R., Wake, D. A., Collins, C. A., Burke, D. J., Nichol, R. C., & Romer, A. K. 2002, MNRAS, 330, 755

Fang, T., Sembach, K. R., & Canizares, C. R. 2003, ApJ, 586, L49

Faber, S. M., Trager, S. C., González, J. J., & Worthey, G. 1999, Ap&SS, 267, 273

Ferreras, I., Charlot, S. & Silk, J. 1999, ApJ, 521, 81

Finn, R. A., & Zaritsky, D. 2004, in Carnegie Observatories Astrophysics Series, Vol. 3: Clusters of Galaxies: Probes of Cosmological Structure and Galaxy Evolution, ed. J. S. Mulchaey, A. Dressler, & A. Oemler (Pasadena: Carnegie Observatories, http://www.ociw.edu/ociw/symposia/series/symposium3/proceedings.html)

Fritz, A., Ziegler, B. L., Bower, R. G., Smail, I., & Davies, R. L. 2004, in Carnegie Observatories Astrophysics Series, Vol. 3: Clusters of Galaxies: Probes of Cosmological Structure and Galaxy Evolution, ed. J. S. Mulchaey, A. Dressler, & A. Oemler (Pasadena: Carnegie Observatories, http://www.ociw.edu/ociw/symposia/series/symposium3/proceedings.html)

Geller, M. J., & Huchra, J. P. 1983, ApJS, 52, 61

Gladders, M. D., Lopez-Cruz, O., Yee, H. K. C., & Kodama, T. 1998, ApJ, 501, 571

Glazebrook, K., Blake, C., Economou, F., Lilly, S., & Colless, M. 1999, MNRAS, 306, 843

Gómez, P. L., et al. 2003, ApJ, 584, 210

Goto, T., et al. 2003, PASJ, 55, 771

Governato, F., Baugh, C. M., Frenk, C. S., Cole, S., Lacey, C. G., Quinn, T., & Stadel, J. 1998, Nature, 392, 359

Gunn, J. E., & Gott, J. R. 1972, ApJ, 176, 1

Guzik, J., & Seljak, U. 2002, MNRAS, 335, 311

Gúzman, R., Lucey, J. R., Carter, D., & Terlevich, R. J. 1992, MNRAS, 257, 187

Helly, J. C., Cole, S., Frenk, C. S., Baugh, C. M., Benson, A., & Lacey, C. 2003, MNRAS, 338, 903

Henry, J. P., et al. 1995, ApJ, 449, 422

Hickson, P., Kindl, E., & Auman, J. R. 1989, ApJS, 70, 687

Hubble E., & Humason, M. L. 1931, ApJ, 74, 43

Jenkins, A., Frenk, C. S., White, S. D. M., Colberg, J. M., Cole, S., Evrard, A. E., Couchman, H. M. P., & Yoshida, N. 2001, MNRAS, 321, 372

Jørgensen, I. 1999, MNRAS, 306, 607

Jørgensen, I., Franx, M., Hjorth, J., & van Dokkum, P. G. 1999, MNRAS, 308, 833

Kauffmann, G. 1995, MNRAS, 274, 153

Kauffmann, G., Colberg, J. M., Diaferio, A., & White, S. D. M. 1999, MNRAS, 303, 188

Kelson, D. D., Illingworth, G. D., Franx, M., & van Dokkum, P. G. 2001, ApJ, 552, L17

Kennicutt, R. C. 1989, ApJ, 344, 685

——. 1992, ApJ, 388, 310

Kodama, T., & Bower, R. G. 2001, MNRAS, 321, 18

Kodama, T., Smail, I., Nakata, F., Okamura, S., & Bower, R. G. 2001, ApJ, 562, 9

Lacey, C., & Cole, S. 1993, MNRAS, 262, 627

Lambas, D. G., Tissera, P. B., Sol Alonso, M., & Coldwell, G. 2004, MNRAS, submitted (astro-ph/0212222)

Larson, R. B., Tinsley, B. M., & Caldwell, C. N. 1980, ApJ, 237, 692

Lewis, I. J., et al. 2002, MNRAS, 334, 673

Lilly, S. J., Tresse, L., Hammer, F., Crampton, D., & Le Févre, O. 1995, ApJ, 455, 108

Madau, P., Pozzetti, L., & Dickinson, M. 1998, ApJ, 498, 106

Margoniner, V. E., de Carvalho, R. R., Gal, R. R., & Djorgovski, S. G. 2001, ApJ, 548, L143

Martin, C. L. 1999, ApJ, 513, 156

McIntosh, D. H., Rix, H.-W., & Caldwell, N. 2004, ApJ, submitted (astro-ph/0212427)

Metavier, A. J. 2004, in Carnegie Observatories Astrophysics Series, Vol. 3: Clusters of Galaxies: Probes of
    Cosmological Structure and Galaxy Evolution, ed. J. S. Mulchaey, A. Dressler, & A. Oemler (Pasadena:
    Carnegie Observatories, http://www.ociw.edu/ociw/symposia/series/symposium3/proceedings.html)

Mihos, J. C. 2004, in Carnegie Observatories Astrophysics Series, Vol. 3: Clusters of Galaxies: Probes of
    Cosmological Structure and Galaxy Evolution, ed. J. S. Mulchaey, A. Dressler, & A. Oemler (Cambridge:
    Cambridge Univ. Press), in press

Miller, N. A. 2004, in Carnegie Observatories Astrophysics Series, Vol. 3: Clusters of Galaxies: Probes of
    Cosmological Structure and Galaxy Evolution, ed. J. S. Mulchaey, A. Dressler, & A. Oemler (Pasadena:
    Carnegie Observatories, http://www.ociw.edu/ociw/symposia/series/symposium3/proceedings.html)

Miller, N. A., & Owen, F. N. 2002, AJ, 124, 2453

Milvang-Jensen, B., Aragón-Salamanca, A., Hau, G., Jørgensen, I., & Hjorth, J. 2003, MNRAS, 339, 1

Mo, H. J., & White, S. D. M. 2002, MNRAS, 336, 112

Moore, B., Frenk, C. S., & White, S. D. M. 1993, MNRAS, 261, 827

Moore, B., Katz, N., Lake, G., Dressler, A., & Oemler, A. 1996, Nature, 379, 613

Morrison, G. E., & Owen, F. N. 2003, AJ, 125, 506

Mulchaey, J. S., Davis, D. S., Mushotzky, R. F., & Burstein, D. 2003, ApJS, 145, 39

Nakata, F., et al. 2004, in preparation

Oemler, A., Jr. 1974, ApJ, 194, 10

Poggianti, B. M. 2004, in Carnegie Observatories Astrophysics Series, Vol. 3: Clusters of Galaxies: Probes of
    Cosmological Structure and Galaxy Evolution, ed. J. S. Mulchaey, A. Dressler, & A. Oemler (Cambridge:
    Cambridge Univ. Press), in press

Poggianti, B. M., et al. 2001, ApJ, 563, 118

Poggianti, B. M., Smail, I., Dressler, A., Couch, W. J., Barger, A. J., Butcher, H., Ellis, R. S., & Oemler, A., Jr.
    1999, ApJ, 518, 576

Poggianti, B. M., & Wu, H. 2000, ApJ, 529, 157

Press, W. H., & Schechter, P. 1974, ApJ, 187, 425

Quilis, V., Moore, B., & Bower, R. G. 2000, Science, 288, 1617

Quintero, A. D., et al. 2004, ApJ, submitted (astro-ph/0307074)

Rakos, K. D., & Schombert, J. M. 1995, ApJ, 439, 47

Rudnick, G. H., De Lucia, G., White, S. D. M., & Pelló, R. 2004, in Carnegie Observatories Astrophysics Series,
    Vol. 3: Clusters of Galaxies: Probes of Cosmological Structure and Galaxy Evolution, ed. J. S. Mulchaey, A.
    Dressler, & A. Oemler (Pasadena: Carnegie Observatories,
    http://www.ociw.edu/ociw/symposia/series/symposium3/proceedings.html)

Sakamoto, T., Chiba, M., & Beers, T. C. 2003, A&A, 397, 899

Sandage A., 1961, in The Hubble Atlas of Galaxies (Washington, DC: Carnegie Int. of Washington)

Sandage A., & Visvanathan, N. 1978, ApJ, 223, 707

Schmidt, M. 1959, ApJ, 129, 243

Sheth, R. K., Mo, H. J., & Tormen, G. 2001, MNRAS, 323, 1

Smail, I., Edge, A. C., Ellis, R. S., & Blandford, R. D. 1998, MNRAS, 293, 124

Smail, I., Morrison, G., Gray, M. E., Owen, F. N., Ivison, R. J., Kneib, J. P., & Ellis, R. S. 1999, ApJ, 525, 609

Somerville, R. S., & Primack, J. R. 1999, MNRAS, 310, 1087

Springel, V., White, S. D. M., Tormen, G., & Kauffmann, G. 2001, MNRAS, 328, 726

Tran, K., Franx, M., Illingworth, G., & van Dokkum, P. 2004, in Carnegie Observatories Astrophysics Series, Vol.
    3: Clusters of Galaxies: Probes of Cosmological Structure and Galaxy Evolution, ed. J. S. Mulchaey, A.

Dressler, & A. Oemler (Pasadena: Carnegie Observatories,
   http://www.ociw.edu/ociw/symposia/series/symposium3/proceedings.html)
Treu, T. 2004, in Carnegie Observatories Astrophysics Series, Vol. 3: Clusters of Galaxies: Probes of
   Cosmological Structure and Galaxy Evolution, ed. J. S. Mulchaey, A. Dressler, & A. Oemler (Cambridge:
   Cambridge Univ. Press), in press
Treu, T., Ellis, R. S., Kneib, J.-P., Dressler, A., Smail, I., Czoske, O., Oemler, A., & Natarajan P. 2003, ApJ, 591,
   53
van Dokkum, P. G., & Franx, M. 2001, ApJ, 553, 90
van Dokkum, P. G., Franx, M., Fabricant, D., Illingworth, G. D., & Kelson, D. D. 2000, ApJ, 541, 95
van Dokkum, P. G., Franx, M., Kelson, D. D., Illingworth, G. D., Fischer, D., & Fabricant, D. 1998, ApJ, 500, 714
van Dokkum, P. G., & Stanford, S. A. 2003, ApJ, 585, 78
Wilson, G., Cowie, L. L., Barger, A. J., & Burke, D. J. 2002, AJ, 124, 1258
Zabludoff, A. I., Zaritsky, D., Lin, H., Tucker, D., Hashimoto, Y., Shectman, S. A., Oemler, A., & Kirshner, R. P.
   1996, ApJ, 466, 104
Ziegler, B., Böhm, A., Jager, K., Fritz, A., & Heidt J. 2004, in Carnegie Observatories Astrophysics Series, Vol. 3:
   Clusters of Galaxies: Probes of Cosmological Structure and Galaxy Evolution, ed. J. S. Mulchaey, A. Dressler,
   & A. Oemler (Pasadena: Carnegie Observatories,
   http://www.ociw.edu/ociw/symposia/series/symposium3/proceedings.html)

21

Galaxy groups at intermediate redshift and the mechanisms of galaxy evolution

RAY G. CARLBERG
Department of Astronomy, University of Toronto

Abstract

The CNOC2 intermediate-redshift spectroscopic survey is the first covering the redshift regime $z \approx 0.5$ from which galaxy groups can be identified in a manner nearly identical to the low-redshift kinematic groups. The CNOC2 groups contain about 20% of $L > 0.2L_*$ galaxies. The group galaxy properties are, on average, correlated with the group's velocity dispersion and the position of the galaxies within the group. The galaxies in groups with line-of-sight velocity dispersion (σ_1) above 150 km s^{-1} show a luminosity function and color evolution similar to those of the field galaxy population. The gradient of mean galaxy color in $\sigma_1 > 150$ km s^{-1} groups becomes redder toward the center, indicating a decline in star formation activity. No statistically secure color gradient is present in lower-velocity dispersion groups. The volume rate of merging of group galaxies is estimated using Monte Carlo simulations of the future evolution of each group. Merging is significant, but not population altering, at 2.0% per Gyr per group galaxy at $z \simeq 0.4$. On average, galaxies in groups are dragged inward a factor of 2 in radius over 5 Gyr, with the details depending on radial location and velocity dispersion of the group. The dominant trend from redshift 0.6 to the current epoch is the overall decline of star formation to low redshift, which is common to both group and field (non-group) galaxies. The physical driver of group galaxy evolution is most likely to be gravity, predominantly tidal stripping of halo gas and satellites, with some residual galaxy merging. Given that the pairwise velocity dispersion indicates that the average field galaxy is in a group with $\sigma_1 \simeq 200$ km s^{-1}, this suggests that evolution of field galaxies is driven by the same gravitational processes characterized in our multi-member groups.

21.1   Introduction

Galaxy groups are a much more typical environment for a galaxy to find itself than the relatively rare environment of a rich cluster. The average group in the field, with a line-of-sight velocity dispersion $\sigma_1 \approx 200$ km s^{-1} (e.g., Ramella, Geller, & Huchra 1989; Tucker et al. 2000), is the source of much of the galaxy two-point correlation function and the pairwise velocity dispersion in the field (e.g., Davis & Peebles 1983). At the low-velocity dispersion end of the group range, ~ 100 km s^{-1}, groups of galaxies overlap with the internal velocity dispersions of individual galaxies. At low velocity dispersion and in the dense environment of a group's dark halo, dynamical friction is overwhelming, leading to merging of galaxies (Barnes 1990; Hickson 1997). In the field, the evolution out to $z \approx 0.5$ is dramatic in the blue galaxies and modest in the red galaxies (e.g., Lin et al. 1999). Evolutionary effects have not been directly examined using a sample that covers a significant range in redshift. Here we first examine the statistics of group evolution, then search for some probable causes.

At low redshift there have been extensive studies of groups and their galaxy contents. The groups have properties, such as numbers, clustering, and mass-to-light ratio, consistent with their being located in dark matter halos (Gott & Turner 1976; Ramella, Geller, & Huchra 1989, 1990; Girardi, Boschin, & da Costa 2000; Girardi & Giuricin 2000). The "light traces mass" aspect is not particularly surprising in 1000 km s^{-1} clusters because dynamical friction on galaxies is slow (Merritt 1985). However, since the dynamical friction time scale varies as σ_1^3, it is much more important in $\sim$200 km s^{-1} groups, as borne out by extensive numerical simulations (for instance, Barnes 1990). At the same time, the effects of ram pressure are proportional to $\rho_{gas}\sigma_1^2$, with ρ_{gas} showing a similar (scaled) density profile with σ_1. Hence, the relative importance of friction relative to gas pressure at a given fraction of the virial radius varies as σ_1^5. Therefore, frictional infall and merging should dominate over ram pressure gas removal in groups. The countervailing gravitational process is tidal stripping, which acts to disrupt infalling smaller groups and remove satellite galaxies. For an approximately universal density profile, tidal effects, which are proportional to the local mass density, are not dependent on group velocity dispersion.

There is substantial observational evidence at low redshift that tides and dynamical friction processes are playing an important role in groups (Toomre 1977; Hickson 1982, 1987). A telling discovery is the "ghost group" (Mulchaey & Zabludoff 1999), in which an isolated elliptical is found within the extended X-ray emission profile that is characteristic of a group of galaxies, rather than of an isolated elliptical. The inference is that the galaxies have merged together to make the elliptical, leaving behind the extended dark matter halo and the X-ray gas. Over the 5 Gyr look-back time of the CNOC2 survey, about half (based on a roughly constant process) of such current-epoch "ghost groups" should be present as small groups, possibly in the process of merging. The rate of merging within the intermediate-redshift groups can be quantified using the observational data and straightforward orbital integrations.

In this paper we first examine the statistical properties of the galaxy population in groups at intermediate redshift. Second, we undertake a Monte Carlo dynamical analysis of the orbital evolution of the group galaxies over a 5 Gyr time span. Both evolutionary studies require a sample selected from a high velocity precision redshift survey; otherwise, projection effects will overwhelm the sample with interlopers. The VIRMOS (Le Fèvre et al. 2001) and DEEP2 (Davis et al. 2001) surveys are now underway and will generate large, new samples, but at the present time the only available survey is the CNOC2 sample (Carlberg et al. 2000; Yee et al. 2000). We note, in passing, that under the assumption of pure luminosity evolution the CNOC2 analysis (Lin et al. 1999) predicts that galaxies like the LMC and SMC will statistically brighten above $M_R = -18.5$ mag at $z \approx 0.5$. Accordingly, the Milky Way at $z \approx 0.5$ would, in principle, be a member of our group catalog, underscoring that these small, intermediate-redshift groups have a substantial connection to the current-epoch field.

21.2   Properties of the CNOC2 Groups

The CNOC2 redshift survey extends from redshift 0.15 to 0.7, with statistically complete coverage of all galaxy spectral types over the redshift range 0.15 to 0.55. The groups are identified in redshift space using a technique designed to incorporate all galaxies

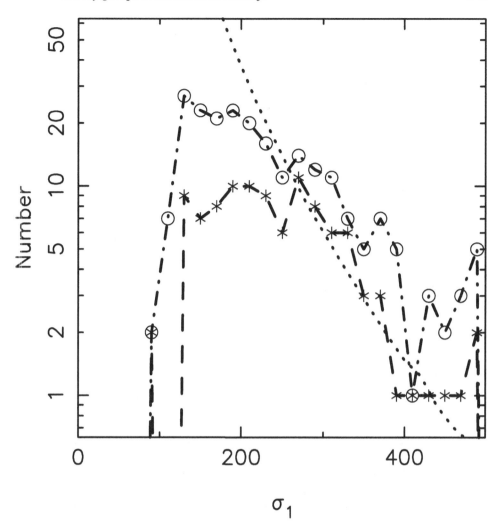

Fig. 21.1. The $n(\sigma_1)$ relation for our standard approach to identifying groups used in our analysis. Both three (upper, solid line) and four member groups (lower, dashed line) are shown. The dotted line is the Press-Schechter prediction. The expected fraction of redshift-space interloper galaxies is only about 15%.

within a virialized halo. The algorithm (Carlberg et al. 2001a) is a redshift-space analog to the algorithm used in N-body simulations to identify virialized dark matter halos by the criterion that the mean interior density exceed 200 times the critical density. Groups are identified as having three or more members within 1.5 times (for $\Omega_m \simeq 0.3$) the radius corresponding to a mean interior density of 200 times the critical density.

Kinematic identification of groups using separation on the sky and along the line of sight using velocities will always be subject to interlopers in velocity space. That is, at velocity separation Δv, it is impossible to tell the difference between two galaxies at the same distance with a relative peculiar velocity Δv and two galaxies moving with the Hubble flow

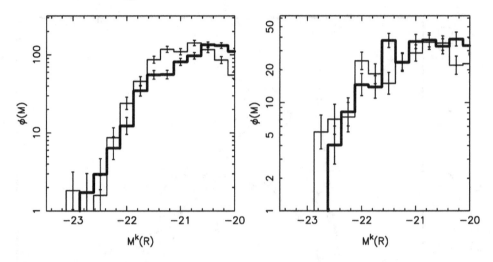

Fig. 21.2. The luminosity functions of red galaxies, $(B-R)_0 > 1.25$ mag, in the field (*left*) and in groups (*right*), for galaxies at a mean redshift of 0.27 (thick line) and 0.43 (thin line). No significant evolution is present.

separated along the line of sight by a distance $\Delta r_z = \Delta v / H(z)$, where $H(z)$ is the Hubble constant at their redshift. The probability distribution of Δr_z is evaluated using the two-point correlation function with $r_0 \simeq 4h^{-1}$ Mpc (co-moving). The result is that any two galaxies grouped on the sky with separation r_p have a probability of about 50% that $\Delta r_z \leq r_p$. This does not mean that only 50% of the kinematic groups are physical groups. As the number of kinematic group members goes up, so does the probability that the group is a physical group. For two kinematic members it is about 1/2; for three kinematic members the chances that it is at least a physical pair is about 2/3; for four kinematic members the probability of it being a physical group of three is about 3/4, and for it being at least a physical pair is close to 100%, where we use the fact that the three-point correlation function is proportional to the square of the two-point correlation function. In Figure 21.1 we show the groups with at least three and at least four kinematic members over the same range of velocity dispersion. Beyond $\sigma_1 \simeq 150$ km s^{-1}, three-member groups do not dominate the groups, and groups with four and more members comprise $\sim 80\%$ of the $\sigma_1 > 200$ km s^{-1} groups, consistent with the analysis of group reality given above. We also note that the numbers as a function of σ_1 are in good correspondence with the Press-Schechter (1974) predictions. The membership status of any particular member of any size group will always have an element of uncertainty that cannot be removed using a redshift survey alone.

21.3   Group Galaxy Evolution

Group galaxies show evolution trends similar, but not entirely identical, to those of field galaxies. We divide galaxies by color at a k-corrected $(B-R)_0 = 1.25$ mag. The red galaxies have essentially identical luminosity function evolution in the group and the field, as shown in Figure 21.2. The modest evolution of the red galaxies is well known to be consistent with passive evolution alone.

The evolution of the blue galaxy luminosity function in groups (Fig. 21.3) has an additional twist relative to the red galaxies. The luminosity function of galaxies in high-velocity

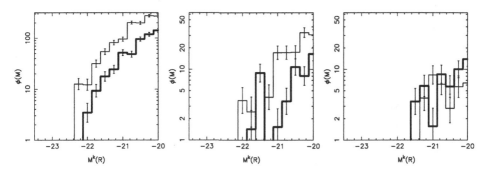

Fig. 21.3. The luminosity functions of blue galaxies, $(B-R)_0 < 1.25$ mag, in the field (*left*) and in groups above (*middle*) and below (*right*) $\sigma_1 = 150$ km s^{-1} at a mean redshift of 0.27 (thick line) and 0.43 (thin line).

dispersion groups evolves in a manner similar to the field. In both groups and the field the evolution is very strong, approximately half a magnitude in luminosity over a redshift interval of only $\Delta z = 0.16$. On the other hand, the low-velocity dispersion groups show effectively no luminosity function evolution.

The similarity of evolution in the field and groups has two possible causes. One possible explanation is that both sets of galaxies share the same cosmic clock and are genuinely evolving in parallel due to a universal driver common to both. A physical cause would be the expansion of the Universe and the decline in density of gas and galaxies. The other possible explanation is that the conditions in groups cause group galaxies to evolve, and that since all galaxies are statistically known to be in a virialized group of some sort, on the basis of the correlation function and extending the group definition to include low-luminosity galaxies, then galaxy evolution *is* group galaxy evolution.

21.4   Radial Color Gradients of Groups

Groups show a significant radial color gradient, with galaxies being statistically redder the closer they are to the center, as shown in Figure 21.4. The color gradient is naturally interpreted as a variation in star formation rate. The star formation gradient interpretation is supported directly with unpublished $U-I$ colors and [O II] line-strength measurements. A similar result holds at low redshift (Hashimoto et al. 1998).

The existence of a color gradient in groups indicates that the group galaxy properties are correlated with their group environment. A key question is what is the physical origin of the group color gradient. There is a reasonably clear idea of how groups build up with time. Any dark halo is, on average, subject to ongoing infall from surrounding field galaxies. Infall causes the object to increase in mass and size, with the new galaxies being added at the periphery. Therefore, as one moves from the center outward through the group the galaxies become, on average, progressively more recent additions to the group. In the case of small groups having only three or four members the "time of entry" gradient is substantially blurred by orbital motions. In spite of that, the relatively red colors of the central galaxies have two possible explanations: either they had some internal clock that indicated at formation that they would become red, or they were average field galaxies that the environment of the group forced to become redder.

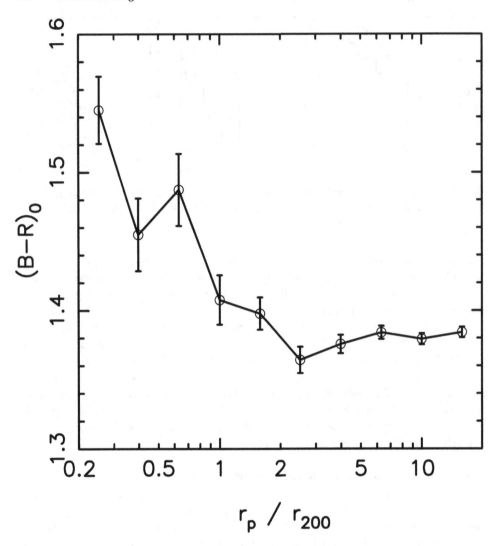

Fig. 21.4. The gradient of color across the groups with a velocity dispersion greater than 150 km s^{-1}. Mergers are expected to be rare in this subsample, although the galaxies are drawn inward by dynamical friction by about a factor of 2 in their orbits over the 5 Gyr time interval between today and the sample's redshift. This net inward flow demands some transformation of the galaxies, not purely an initial distribution.

21.5   Merging and the Inward Flow of Group Galaxies

The importance of merging of group galaxies is estimated using Monte Carlo integrations of each galaxy in its group. The median line-of-sight velocity dispersion of the CNOC2 groups, 230 km s^{-1}, is close to the velocity dispersion above which dynamical friction in a cosmological dark halo is no longer able to drag any galaxies down to the center in a Hubble time. The calculations use the measured (x, y, v_z) of every galaxy and distribute the galaxies over (z, v_x, v_y) using the measured statistical distribution in r_p and line-of-sight velocity dispersion σ, assuming an isotropic velocity ellipsoid. The dark halo is modeled

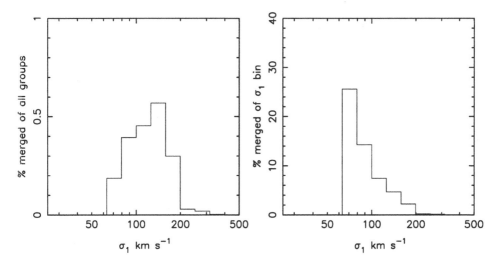

Fig. 21.5. (*Left*) The fraction of the whole sample that is merged in each bin of velocity dispersion in 1 Gyr. The integrated rate is 2.0% per Gyr per group galaxy. (*Right*) The percentage of galaxies in groups of σ_1 merged per Gyr.

as an isothermal sphere. Any orbit that comes within 0.03 units of the center is assumed to merge into a central galaxy. Some of the resulting starting points have orbits unbound from the cluster; these were removed from the analysis.

The left panel of Figure 21.5 shows the fraction of all group galaxies that merge in 1 Gyr as a function of the group velocity dispersion. The rate is calculated on a galaxy-by-galaxy basis using the observed positions and velocities of the group galaxies. At a fixed velocity dispersion, the merger rate is proportional to σ_1^{-3}, as expected, as shown in the right panel of Figure 21.5, where the fractions are normalized to the numbers in the bins.

The group catalog is constructed using all galaxies down to about $0.2L_*$ and does not include any low-mass galaxies. Given the relatively limited range of masses, the mean mass of galaxies merging is essentially equal to the mean mass of all galaxies in the group sample. The merger rate integrated over all groups is a mean rate of mass increase of 2.0% per Gyr. Although this rate is nontrivial, it is not sufficient to lead to gross changes in the galaxy population between redshift of 0.5 and the current epoch. The merging will lead to a small population of recent major-merger galaxies at the current epoch, 2.0% per Gyr, for 5 Gyr, equal to 21% of galaxies in groups or 2% of all galaxies. The merger rate is low enough that it presents no serious threat to the existence of the group population. At higher redshift the mass density within the virialized groups rises in proportion to $H^2(z) \sim (1+z)^{2-3}$, leading to a rise in the importance of merging with redshift. The density effect accounts for a rise in the group merger rate of roughly $\sqrt{\rho} \sim (t_0/t) \sim (1+z)^{1-1.5}$.

Dynamical friction causes the orbits of the group galaxies to spiral inward, sometimes fairly quickly. In the presence of a radial inflow a color gradient will be reduced, possibly to no gradient at all, as a result of blue galaxies moving inward. If the gradient remains present, then the group environment is acting to suppress star formation as blue, star-forming galaxies enter the group.

The same calculations that predict the merger rate also follow the individual galaxy orbits.

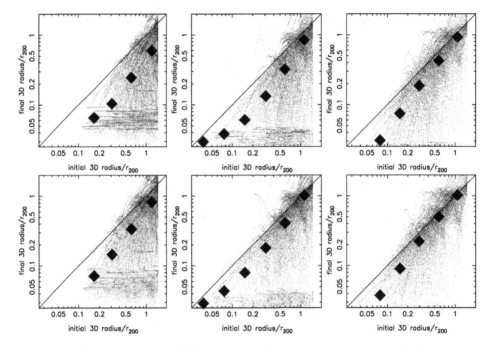

Fig. 21.6. The three-dimensional radial distance from the center after 5 Gyr or at time of central merger plotted against the initial projected radius for galaxies in groups with $\sigma_1 < 150$ km s^{-1} (*left*), 150 km s^{-1} $< \sigma_1 < 250$ km s^{-1} (*middle*) and $\sigma_1 > 250$ km s^{-1} (*right*). The top row uses $M/L = 100$, and the bottom uses $M/L = 30$. The diamonds show the mean final radius for the galaxies at that range of initial radius.

We display the final orbital radius as a function of initial orbital radius in Figure 21.6. Also shown is the average final radius as a function of initial radius. Within our sample of group galaxies as a whole, the average galaxy moves inward by about a factor of 2 in radius. The figure shows that the details depend on group velocity dispersion and initial galaxy radius, with a fairly weak dependence on the assumed total M/L for the galaxies plus their remnant dark matter halos. The expected trends of an increased inward drift with lower velocity dispersion and smaller initial orbital radius appear, although it is diluted by the projection effects that spread galaxies out of the groups along the line of sight and give some galaxies relatively large tangential velocities.

On the basis of these integrations we have found that there is substantial radial inspiral of group galaxies. The radial inflow in systems with $\sigma_1 < 150$ km s^{-1} is likely to be so rapid that it can explain the absence of any secure color gradient in these groups, and possibly an excess of strongly star-forming galaxies at the center (Hashimoto & Oemler 2000; Carlberg et al. 2001b). The color gradient is present in groups at 150 km s^{-1} $< \sigma_1 < 250$ km s^{-1}, where the net radial inflow is about a factor of 2 in radius. A reasonable interpretation is that the presence of galaxies redder than the field toward the centers of groups is largely due to the influence of the group environment. One important force in groups is simply the tidal effect of the group, which can gravitationally strip off halo gas and smaller, gas-rich companion galaxies that infalling galaxies bring with them as they fall into the group (or equally, cluster).

21.6   Consequences of Galaxy Evolution in Groups

At intermediate redshift, group galaxies evolve parallel to galaxies in the field. Given that groups comprise about 21% of the field, if we include galaxies with luminosities greater than $0.2L_*$, and would rise if we include lower-luminosity galaxies, at least some of the cause of the evolution of galaxies in the field is that they are in a group environment and have their star formation rates reduced as they move toward the centers of groups.

Low-velocity dispersion groups are where mergers occur. Low-velocity groups are under-represented relative to the Press-Schechter prediction in our sample. At higher redshift the importance of mergers will be much greater because dark halo densities rise, and it is likely that groups are the primary source of major mergers.

At much earlier epochs, say redshifts beyond about 2, we can speculate that the role of groups will change. At low redshift their dominant role appears to be to suppress star forma-tion activity. At higher redshifts there will be relatively fewer groups with $\sigma_1 > 150$ km s^{-1}, and the mean mass density will rise in proportion to the local Hubble constant, $\rho_{200} \propto H^2(z)$. Hence, dynamical friction and merging are bound to be much more important than at low redshift. Furthermore, it is likely that the galaxies are much more gas rich. Consequently one expects that there will be increased merger activity, almost certainly with increased induced star formation. The task of extending a redshift survey with $\sim 10^5$ galaxies to redshifts in the range 2 to 5 is an observational challenge required to develop an understanding of the links between structure in the dark matter and the formation of galaxies.

21.7   Conclusions

The groups at intermediate redshift are the galaxy population that creates much of the clustering signal measured in the two-point correlation function and pairwise velocities. These groups have the following properties.

- The luminosity function of both blue and red galaxies in groups with velocity dispersion $\sigma_1 > 150$ km s^{-1} evolves like field galaxies.
- The $\sigma_1 > 150$ km s^{-1} groups have internal color gradient, becoming redder toward the center.
- The rate of mass increase due to merging of group galaxies is 2.0% per Gyr.
- On average, dynamical friction will drag group galaxies inward about a factor of 2 in 5 Gyr.
- We speculate that the gravitational forces that cause a decline in star formation in group galaxies as they move inward is also a significant driver of the evolution of the field as a whole, on the basis that the average galaxy is in a ~ 200 km s^{-1} group.

The CNOC2 groups will statistically evolve over an average of about 5 Gyr into most of the group galaxies seen today. Based on the merger rate in groups, the mean mass of galaxies will increase only about 2% due to group merging; approximately half of those will become single galaxies. On the other hand, the tendency of galaxies to become redder as they move toward the group center will affect virtually all group galaxies and is at least partially responsible for the decline in star formation between redshift 1 and 0.

References

Barnes, J. E. 1985, MNRAS, 215, 517
———. 1990, Nature, 344, 379

Carlberg, R. G., Yee, H. K. C., Morris, S. L., Lin, H. Hall, P. B., Patton, D., Sawicki, M., & Shepherd, C. W. 2000, ApJ, 542, 57

——. 2001a, ApJ, 552, 427

——. 2001b, ApJ, 563, 736

Davis, M., Newman, J. A., Faber, S. M., & Phillips, A. C. 2001, in Deep Fields, ed. S. Cristiani, A. Renzini, & R. E. Williams (Berlin: Springer), 241

Davis, M., & Peebles, P. J. E. 1983, ApJ, 267, 465

Girardi, M., Boschin, W., & da Costa, L. N. 2000, A&A, 353, 57

Girardi, M., & Giuricin, G. 2000, ApJ, 540, 45

Gott, J. R., III, & Turner, E. L. 1976, ApJ, 209, 1

Hashimoto, Y., & Oemler, A., Jr. 2000, ApJ, 530, 652

Hashimoto, Y., Oemler, A., Lin, H., & Tucker, D. L. 1998, ApJ, 499, 589

Hickson, P. 1982, ApJ, 255, 382

——. 1997, ARA&A, 35, 357

Le Fèvre, O., et al. 2001, in Deep Fields, ed. S. Cristiani, A. Renzini, & R. E. Williams (Berlin: Springer), 236

Lin, H., Yee., H. K. C., Carlberg, R. G., Morris, S. L., Sawicki, M., Patton, D., Wirth, G., & Shepherd, C. W. 1999, ApJ, 518, 533

Merritt, D. 1985, ApJ, 289, 18

Mulchaey, J. S., & Zabludoff, A. I. 1999, ApJ, 514, 133

Press, W. H., & Schechter, P. 1974, ApJ, 187, 425

Ramella, M., Geller, M. J., & Huchra, J. P. 1989, ApJ, 344, 57

——. 1990, ApJ, 353, 51

Toomre, A. 1977, in Evolution of Galaxies and Stellar Populations, ed. B. M. Tinsley & R. B. Larson (Yale Observatory: New Haven), 401

Tucker, D. L., et al. 2000, ApJS, 130, 237

Yee, H. K. C., et al. 2000, ApJS, 129, 475

22

The intragroup medium

JOHN S. MULCHAEY

The Observatories of the Carnegie Institution of Washington

Abstract

X-ray surveys over the last decade have shown that approximately half of all nearby groups of galaxies contain spatially extended X-ray emission. This X-ray emission is produced by a low-density gas component referred to as the intragroup medium. All groups with an X-ray emitting intragroup medium contain at least one early-type galaxy. In most cases, the X-ray emission is centered on an elliptical galaxy. The presence of a hot gas halo indicates that many groups are real, physical systems and not chance superpositions. Assuming the hot gas is in hydrostatic equilibrium, the temperature of the gas can be used to estimate the mass of individual groups. Such estimates indicate that X-ray groups are massive systems dominated by dark matter. The intragroup medium also carries an imprint of the physical processes that have occurred in these systems. There is now considerable evidence that groups depart from the simplest predictions of cold dark matter models. In particular, nongravitational processes have been very important in groups. While there has been considerable effort by the theoretical community in the last few years to understand these processes, recent observations suggest most of the proposed models have important flaws.

22.1   Introduction

Groups of galaxies constitute the most common galaxy associations, containing at least 50% of all galaxies at the present day (Geller & Huchra 1983; Tully 1987; Nolthenius & White 1987). Groups are, therefore, an important laboratory for studying the processes associated with galaxy formation and evolution. However, we know surprisingly little about these systems compared to rich clusters. This is because group studies are hampered by small number statistics—a typical poor group contains only a few bright galaxies. For this reason, the dynamical properties of any individual group are always rather uncertain. In fact, many cataloged groups may not be real physical systems at all, but rather chance superpositions or large-scale structure filaments viewed edge-on (Frederic 1995; Hernquist, Katz, & Weinberg 1995; Ramella, Pisani, & Geller 1997).

The discovery that many groups are X-ray sources has provided considerable new insight into these important systems. *ROSAT* observations indicate that the X-ray emission in groups is often extended on scales of hundreds of kiloparsecs. X-ray spectroscopy suggests the emission mechanism is most likely a combination of thermal bremsstrahlung and line emission. This interpretation requires that the entire volume of groups be filled with a hot, low-density gas. This diffuse, extended gas component is referred to as the intragroup medium by analogy with the intracluster medium found in rich clusters of galaxies (see Mushotzky 2004).

The mere presence of a hot gas halo indicates that many groups are real, physical systems. This fact has been further demonstrated by optical studies that show that X-ray groups contain a large number of galaxies (at least ~20–50 members down to few magnitudes fainter than M_*; Zabludoff & Mulchaey 1998). Like rich clusters, these groups also contain a high fraction of early-type galaxies and follow a morphology-density relation (Helsdon & Ponman 2003a). Thus, in some ways, X-ray groups can be thought of as "mini-clusters." However, some important physical differences exist between these systems. Many of the mechanisms that have been proposed to drive galaxy evolution in clusters (such as ram-pressure stripping and galaxy harassment) are not expected to be important in groups. Galaxy-galaxy interactions, however, are likely more frequent in groups than in rich clusters, given the lower velocity dispersions of groups. The lower masses of groups also means that nongravitational processes such as heating by supernovae and active galactic nuclei (AGNs) likely have a bigger impact on the X-ray emitting gas than in rich clusters. In fact, the potential wells of groups are small enough that they cannot be considered "closed box" systems like clusters: strong galactic winds may be able to drive baryons out of these systems (Renzini et al. 1993; Renzini 1997; Davis, Mulchaey, & Mushotzky 1999).

Although there were clues from the *Einstein* Observatory that some groups are X-ray sources (Biermann, Kronberg, & Madore 1982; Biermann & Kronberg 1983; Bahcall, Harris, & Rood 1984), detailed studies of the intragroup medium were not possible prior to the *ROSAT* mission. The low internal background, large field of view, and good sensitivity to soft X-rays made the *ROSAT* PSPC detectors ideal instruments for studying nearby groups. This meeting marks the 10-year anniversary of the first *ROSAT* papers on poor groups (Mulchaey et al. 1993; Ponman & Bertram 1993). While most of what we know about the intragroup medium is still based on *ROSAT* data, this situation is rapidly changing. I expect that in a few years the current generation of X-ray telescopes (*Chandra* and *XMM-Newton*) will play a dominant role in our understanding of the hot gas in groups.

22.2   *ROSAT* Studies

It was clear from the earliest surveys with *ROSAT* that not all groups contain an X-ray emitting intragroup medium (Ebeling, Voges, & Böhringer 1994; Pildis, Bregman, & Evrard 1995; Saracco & Ciliegi 1995; Mulchaey et al. 1996a; Ponman et al. 1996). The fraction of groups that contain extended X-ray emission has been estimated from several studies. Mahdavi et al. (2000) used the *ROSAT* All-Sky Survey to search for X-ray emission from a sample of 260 groups taken from the combined Center for Astrophysics and Southern Sky Redshift Surveys. After accounting for selection effects, they estimate that ~40% of the groups are extended X-ray sources. Deeper exposures taken from the "pointed mode" phase of the *ROSAT* mission (during this phase *ROSAT* pointed at specific targets in the sky for exposure times typically 10 to 50 times longer than those of the All-Sky Survey) are consistent with the Mahdavi et al. (2000) estimate (Helsdon & Ponman 2000a; Mulchaey et al. 2003). Thus, approximately half of all nearby groups contain an X-ray emitting intragroup medium.

Many studies have looked for trends between the presence of X-ray gas and the global properties of the group. There appears to be very few, if any, trends between properties such as compactness and optical luminosity and the X-ray properties of groups (Mulchaey et al.

1996a; Ponman et al. 1996; Helsdon & Ponman 2000b). However, the presence of extended X-ray emission in groups is strongly linked to the presence of early-type galaxies. In fact, all of the groups detected by *ROSAT* contain at least one early-type galaxy (Mulchaey et al. 2003). This result has implications for our own Local Group, which by the present definition would be defined as a spiral-only group and thus would not be expected to be X-ray detected. (The spiral fraction of a group is generally determined from only the most luminous members of each group, and at the distance of a typical object in the *ROSAT* surveys, the Local Group would only include the Milky Way, M31, and M33).

The failure to detected spiral-only groups with *ROSAT* is not surprising when one considers the expected gas temperature and density for these systems. Based on their velocity dispersions, the virial temperatures of spiral-only groups should be lower than those of their early-type dominated counterparts (Mulchaey et al. 1996b). For a system like the Local Group, the gas temperature is likely 2–3 million degrees Kelvin (e.g., Maloney & Bland-Hawthorn 1999). Such gas would be very difficult to detect with *ROSAT*. However, a "warm" intragroup medium might produce detectable features in the far-UV or X-ray spectra of background quasars (Mulchaey et al. 1996b; Perna & Loeb 1998; Hellsten, Gnedin, & Miralda-Escudé 1998; Fang & Canizares 2000). While many high-ionization absorption features have now been detected (e.g., Fang et al. 2002; Mathur, Weinberg, & Chen 2003; Sembach et al. 2003), associating these lines with a warm intragroup medium has proved difficult. The most promising case to date was the possible detection with *Chandra* of an O VIII Lyα absorption feature at the recessional velocity of a small spiral-rich group along the sight line toward PKS 2155−304 (Fang et al. 2002). However, subsequent observations of the same target with *XMM-Newton* do not detect the O VIII feature (Rasmussen, Kahn, & Paerels 2003).

22.3   Spatial Properties of the Intragroup Medium

The morphology of the X-ray emission can provide important clues into the nature of the hot gas. Most of the groups studied with *ROSAT* have round, symmetrical morphologies (e.g., NGC 2563 group; Fig. 22.1). These morphologies are consistent with the groups being relaxed systems. In almost all cases, the X-ray emission is peaked on a luminous elliptical galaxy (Mulchaey & Zabludoff 1998; Helsdon & Ponman 2000a). The position of the central galaxy is also indistinguishable from the center of the group potential, as defined by the mean velocity and projected spatial centroid of the group galaxies (Zabludoff & Mulchaey 1998). Therefore, the brightest elliptical galaxy lies near the dynamical center of the group. In addition, the properties of the central galaxy are closely related to the group potential. For example, Mulchaey & Zabludoff (1998) note a trend for the position angle of the optical light in the central galaxy to align with the position angle of the large-scale X-ray emission. The X-ray luminosities of the central galaxies are also more strongly correlated with the X-ray luminosity of the group than with the optical light of the central galaxy (Helsdon & Ponman 2003b). These observations suggest that the formation and evolution of the central galaxy is strongly linked to the overall group potential.

A few nearby groups have bimodal X-ray morphologies (e.g., NGC 4065; Fig. 22.1). In all these cases, the second X-ray peak is also centered on a luminous early-type galaxy. These objects may be two groups in the process of merging. Based on *ROSAT* surveys, such groups are relatively rare in the nearby Universe. Although the statistics are still poor, bimodal X-ray groups may be more common at higher redshifts (Bauer et al. 2002).

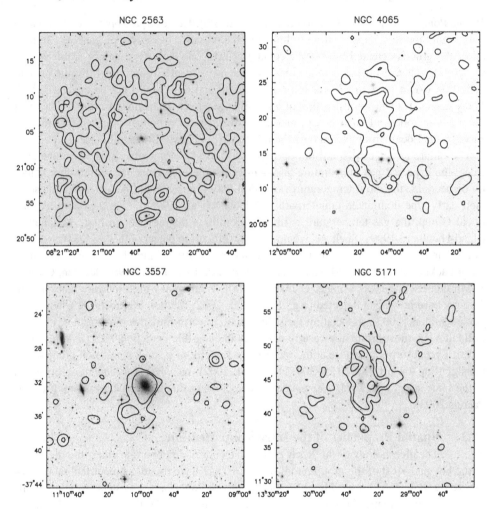

Fig. 22.1. Contour maps of diffuse X-ray emission of four groups measured with *ROSAT* overlaid on optical images. (Adapted from Mulchaey et al. 2003.)

For a few groups, the X-ray emission is not peaked on an individual galaxy (e.g., the NGC 5171 group, Fig. 22.1; see Osmond & Ponman 2004 for a detailed study of this system). In many of these systems, the X-ray emission may not be related to the global group potential at all. For example, much of the X-ray emission in HCG 16 appears to be associated with star formation and nuclear activity (Dos Santos & Mamon 1999; Turner et al. 2001; Belsole al. 2003). In HCG 92, the morphology of the X-ray emission matches the morphology of the radio continuum and optical emission-line maps, suggesting the X-ray emission is due to a large-scale shock (Pietsch et al. 1997; Sulentic et al. 2001). Regardless of the exact origin of the gas, the clumpy X-ray morphologies suggest the gas is likely not virialized in these cases.

Recent *Chandra* and *XMM-Newton* observations have revealed "holes" in the surface brightness distributions of some groups (Vrtilek et al. 2000; Buote et al. 2003). These

X-ray cavities have also been observed in many clusters with *Chandra* (Fabian et al. 2000; McNamara et al. 2000; Blanton et al. 2001; David et al. 2001; McNamara et al. 2001; Heinz et al. 2002; Sun et al. 2003) and are believed to be produced by an interaction with radio lobes from a central AGN (Churazov et al. 2001; Nulsen et al. 2002; Soker, Blanton, & Sarazin 2002). Neither of the two groups that have been studied in detail (HCG 62 and NGC 5044) show evidence for a central AGN or for radio lobes in existing radio maps. One possibility is that these cavities were created during an earlier epoch when the central galaxies were active and the high-energy, radio-emitting electrons have been lost. In addition to cavities, the NGC 5044 group also displays sharp edge-like features (David et al. 1994; Buote et al. 2003) that have been interpreted as "cold fronts," discontinuities between regions of different temperature and density, in clusters (Markevitch, Vikhlinin, & Mazzotta 2001). It is difficult to determine how common such features are in groups at the present time because so few objects with deep *Chandra* or *XMM-Newton* exposures have been analyzed. This situation should change dramatically in the next few years.

Traditionally, a hydrostatic isothermal model has been used to describe the surface brightness profiles of rich clusters (e.g., Jones & Forman 1984). Most authors have also adopted this model for galaxy groups. With King's (1962) analytic approximation to the isothermal sphere, the X-ray surface brightness at a projected radius r is given by:

$$S(r) = S_0(1 + (r/r_c)^2)^{-3\beta+0.5}, \tag{22.1}$$

where r_c is the core radius of the gas distribution. This model is often referred to as the standard β model in the literature. The parameter β is the ratio of the specific energy in galaxies to the specific energy in the hot gas:

$$\beta \equiv \mu m_p \sigma^2 / kT_{gas}, \tag{22.2}$$

where μ is the mean molecular weight, m_p is the mass of the proton, σ is the one-dimensional velocity dispersion, and T_{gas} is the temperature of the intragroup medium. In general, the standard β model provides a good fit to the surface brightness profiles of rich clusters, with a typical β value of about 2/3 (e.g., Arnaud & Evrard 1999; Mohr, Mathiesen, & Evrard 1999).

Although the hydrostatic isothermal model has almost universally been used for groups, in most cases it provides a poor fit to the data. In general, the central regions of groups exhibit an excess of emission above the extrapolation of the β model to small radii. Mulchaey & Zabludoff (1998) suggested that the *ROSAT* profiles of groups could be adequately fit using two separate β models: one associated with the central galaxy (the "central" component) and one associated with the large-scale group emission (the "extended" component). Subsequent work by Helsdon & Ponman (2000a) found a strong correlation between the β value of the "extended" component and gas temperature, with cooler systems having lower β values. This trend has recently been verified by Sanderson et al. (2003) using a combination of *ROSAT* and *ASCA* data. Helsdon & Ponman (2000a) found a mean β value of ~ 0.46 for the "extended" component in groups, which is significantly lower than the typical value for a rich cluster. Higher resolution *Chandra* and *XMM-Newton* data are also consistent with groups having lower β values (Helsdon, Ponman, & Mulchaey 2004; Mushotzky et al. 2004). In the absence of nongravitational heating and/or cooling, the standard cold dark matter (CDM) model predicts that the shape of the gas density profile should be similar for

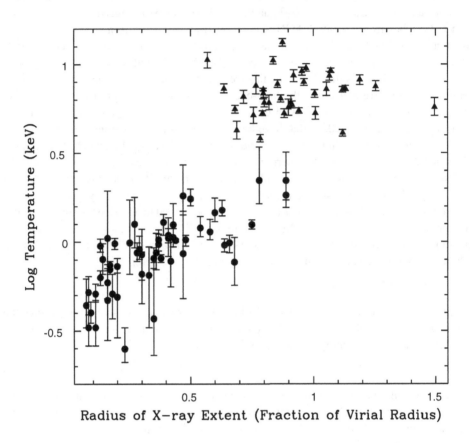

Fig. 22.2. Total radius of X-ray extent, plotted as a fraction of the virial radius, of each system versus the logarithm of the temperature for a sample of groups (circles) and rich clusters (triangles). The groups points are taken from Mulchaey et al. (2003) and the cluster points are from White (2000).

groups and clusters. Thus, the trend for β values to increase with temperature (i.e., mass) suggests that nongravitational processes have been important in groups (see § 22.6).

However, there is some concern that the measured β values may not reflect the true values of the gas density profiles in groups. Simulations of clusters indicate that the β value derived from a surface brightness profile depends strongly on the range of radii used in the fit (Navarro, Frenk, & White 1995; Bartelmann & Steinmetz 1996; Lewis et al. 2000). Indeed, Vikhlinin, Forman, & Jones (1999) find evidence for a slight steepening of the gas density slope with radius in rich clusters. In general, the X-ray emission in groups is detected out to a small fraction of the virial radius (Helsdon & Ponman 2000a; Mulchaey 2000). In fact, there is a strong trend for hotter systems to be detected to a larger fraction of the virial radius than cooler systems (Mulchaey 2000; Fig. 22.2). This is important because it suggests that a larger fraction of the gas mass, and thus X-ray luminosity, is detected in the hotter systems. Because groups tend to be detected to a smaller fraction of the virial radius than rich clusters, the derived β values for groups may also be biased low (Mulchaey 2000). Sanderson et al. (2003) have examined this effect by fitting the *ROSAT* data for two clusters over a range of radii and conclude that it is unlikely to introduce a large systematic relative bias between groups and clusters. While this is somewhat reassuring, more detailed studies of the surface brightness profiles of groups using *Chandra* and *XMM-Newton* will probably be required to put this issue to rest.

22.4   Spectral Properties of the Intragroup Medium

Although the spectral resolution of the *ROSAT* PSPC detectors was very crude ($\Delta E / E \approx 0.4$ at 1 keV), these instruments were used to derive important information about the hot gas. In particular, the *ROSAT* observations provided temperature measurements for a large number of nearby groups. The derived gas temperatures for groups range from about 0.3 keV to 2 keV (Mulchaey et al. 1996a, 2003; Ponman et al. 1996; Helsdon & Ponman 2000a). Systems with temperatures higher than ~ 2 keV would generally be considered clusters, while groups cooler than 0.3 keV would be difficult to detect with *ROSAT*. Subsequent work with the *ASCA* telescope suggests that for systems with a temperature less than about 1.5–2 keV, the *ROSAT* data provide an accurate measurement of the temperature. For hotter systems, the *ROSAT* data tend to underestimate the true gas temperature (Hwang et al. 1999; Horner 2001; Mulchaey et al. 2003).

Recent work with *Chandra* suggests that *ROSAT* and *ASCA* spectra may be significantly contaminated by unresolved point source emission in some cases. For example, Helsdon et al. (2004) find that up to 30%–40% of the *ROSAT* flux in two nearby groups is due to points sources that can be identified with the higher resolution *Chandra* data. In one of the two groups, the point source contamination appears to have had a large impact on the temperature measured by *ROSAT*. Both of the groups studied by Helsdon et al. (2004) have low X-ray luminosities. It is likely that point source contamination is less important for the more luminous X-ray groups that have dominated most *ROSAT* studies. Still, it will be important to study many more groups with *Chandra* to determine the importance of point sources contamination on the *ROSAT* and *ASCA* measurements.

Radial temperature profiles have now been measured for a large number of groups with *ROSAT*, *ASCA*, *Chandra*, and *XMM-Newton* (Ponman & Bertram 1993; David et al. 1994; Doe et al. 1995; Davis et al. 1996; Trinchieri, Fabbiano, & Kim 1997; Mulchaey & Zabludoff 1998; Buote 2000; Helsdon & Ponman 2000a; Buote et al. 2003; Sanderson et al. 2003; Helsdon et al. 2004; Mushotzky et al. 2004). Most authors have found that the *ROSAT* and *ASCA* data are consistent with the gas being largely isothermal outside of the very centers of groups. However, *Chandra* observations suggest there is considerable temperature structure in the gas that might have been missed in the low-resolution data (Fig. 22.3). Although very few *Chandra* group results have been published to date, a quick look at groups in the *Chandra* archive suggests the profile shapes in Figure 22.3 are typical: the temperature rises rapidly from the center of the group to a maximum before declining significantly at larger radii.

The cool cores observed in the *Chandra* data are one of the standard signatures of cooling flows (Fabian 1994). However, as discussed in Donahue & Voit (2004), recent

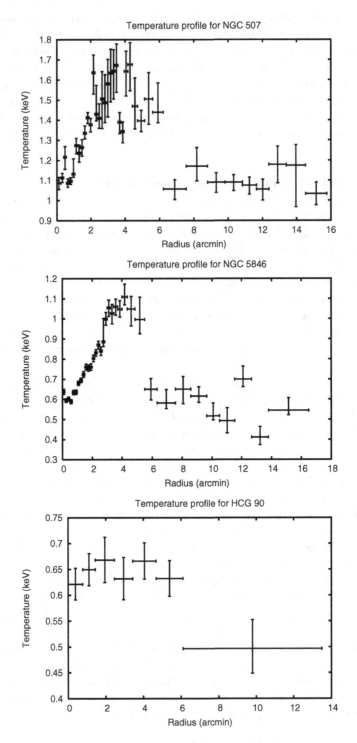

Fig. 22.3. Temperature profiles for three nearby groups derived from archival *Chandra* ACIS data.

high-resolution spectroscopy of the centers of cooling flow clusters has failed to find the emission lines expected from the cooling gas. Although only a couple of objects have been looked at in detail so far, groups seem to exhibit the same problem (Xu et al. 2002; Tamura et al. 2003). Thus, whatever mechanism is invoked to remove the cool gas must be in operation over the full range of masses from poor groups to the richest clusters. The discovery of X-ray cavities in groups and clusters (see § 22.3) suggests that AGN heating may play a role in removing the cooling gas (but see Brighenti & Mathews 2003 for potential problems with applying the cluster models to groups and individual galaxies).

In addition to measuring gas temperatures, *ROSAT* PSPC and *ASCA* observations of groups have been used to estimate the metal content of the intragroup medium. For gas at temperatures of ~ 1 keV ($\sim 10^7$ K), the most important emission-line features in the X-ray spectra include the K-shell ($n = 1$) transitions of carbon through sulfur and the L-shell ($n = 2$) transitions of silicon through iron. Particularly important is the Fe L-shell complex in the spectral range ~ 0.7–2.0 keV (Liedahl, Osterheld, & Goldstein 1995). The wealth of line features in the soft X-ray band potentially provides powerful diagnostics of the physical conditions of the gas, including the excitation mechanism and the elemental abundance (Mewe 1991; Liedahl et al. 1995).

Unfortunately, the great potential to study the chemical composition of the gas has been largely diminished by the poor spectral resolution of the X-ray instruments flown to date. While many authors quote metallicity measurements based on *ROSAT* data, it is universally agreed that such measurements are highly suspect given the poor spectral resolution of the PSPC detectors. It is generally believed that the metallicity measurements from *ASCA* are more reliable, but different authors have derived vastly different results using the same data. Most authors have derived sub-solar abundances from *ASCA* data (Fukazawa et al. 1998; Davis et al. 1999; Hwang et al. 1999; Finoguenov, David, & Ponman 2000; Horner 2001). However, Buote (2000) derived approximately solar metallicity for a sample of 12 groups. The differences between metallicities derived by Buote and other authors are primarily due to different assumptions about the spectral model. Most authors have assumed a single-temperature model to derive metallicities, while Buote adopts a two-temperature model. Recent work by Buote and collaborators using *XMM-Newton* and *Chandra* (Buote 2002; Buote et al. 2003) favors a two-temperature medium in at least two groups. In the case of the NGC 5044 group, Buote et al. (2003) find that the cooler temperature component ($T \approx 0.7$ keV) has a temperature consistent with the kinetic temperature of the stars in the central galaxy, while the hotter component (($T \approx 1.4$ keV) is consistent with the virial temperature of the group.

The Buote (2000) results have proved to be very controversial, and some authors have dismissed the idea of a multi-phase intragroup medium. To resolve this issue, high-spectral resolution observations are needed. It is now possible to obtain such spectra for a few groups with *XMM-Newton*'s Reflection Grating Spectrometers (RGS). The RGS produce spectra integrated over a region $\sim 1'$–$2'$ in diameter. Unfortunately, it is only feasible to study the bright cores of groups with this instrument. Xu et al. (2002) have studied the center of the NGC 4636 group and find no evidence for the emission-line features expected in a strongly multi-phase gas. This would appear to be inconsistent with the Buote (2000) model. However, Tamura et al. (2002) find evidence for two temperature components in the RGS data for the NGC 5044 group. Thus, a strong conclusion cannot be reached at this time. To truly test the Buote (2000) results, it will be important to obtain spatially resolved, high-spectral resolution observations outside of the bright central cores of groups. This type of observation will be possible with the X-ray microcalorimeter on the *Astro E-2* mission (scheduled to be launched in 2005). Regardless of the final outcome, the work of Buote (2000) is an important reminder that the properties derived from X-ray spectroscopy are very sensitive to the choice of input model.

22.5   Mass and Baryon Fraction Estimates

The X-ray gas in groups can be used to estimate the masses of individual systems using the hydrostatic equilibrium method originally developed for rich clusters (e.g., Fabricant, Lecar, & Gorenstein 1980; Fabricant, Rybicki, & Gorenstein 1984; Cowie, Henriksen, & Mushotzky 1987). In addition to the assumption that the gas is in rough hydrostatic equilibrium, it is also assumed that the gas temperature is a direct measure of the potential depth. However, there is now considerable evidence that nongravitational processes are important for groups of galaxies (see § 22.6). How these processes have affected the measured temperatures is currently not clear. Semi-analytic "preheating" models suggest that for systems with global temperatures greater than about 0.8 keV, the measured temperature is a fair tracer of the potential depth. From their models that include both heating and cooling, Voit et al. (2002) find that these processes tend to increase the temperature of the gas near the system's center. However, these processes also decrease the gas density near the core, shifting the bulk of the system's luminosity to larger radii, where the gas temperatures are smaller. The net effect is that the luminosity-weighted temperatures of groups are not substantially changed by nongravitational processes in the Voit et al. (2002) models. Still, it is worth bearing in mind that these processes may be having some effect on the mass estimates.

With the further assumption of spherical symmetry, the mass interior to radius r is given by (Fabricant et al. 1984):

$$M_{\text{total}}(<r) = \frac{kT_{\text{gas}}(r)}{G\mu m_{\text{p}}} \left[\frac{d\log\rho}{d\log r} + \frac{d\log T}{d\log r} \right] r, \tag{22.3}$$

where k is Boltzmann's constant, $T_{\text{gas}}(r)$ is the gas temperature at radius r, G is the gravitational constant, μ is the mean molecular weight, m_{p} is the mass of the proton, and ρ is the gas density. In principle, all of the unknowns in this equation can be calculated from the X-ray data. Typically, the gas temperature is measured directly from the X-ray spectrum, and the gas density profile is determined by fitting the standard β model to the surface brightness profile. With *ROSAT* data, it is often necessary to further assume that the gas is isothermal (i.e., $\frac{d\log T}{d\log r} = 0$). The isothermal assumption appears to result in an error in the mass of no more than about 10% (e.g., David et al. 1994; Davis et al. 1996).

The hydrostatic mass has generally only been calculated for the most luminous groups, all of which have temperatures around ~ 1 keV. It is therefore not surprising that the total masses of all of these systems tend to cluster around a single value ($\sim 10^{13}h_{100}^{-1}M_{\odot}$ out to approximately 1/3 of the virial radius; Mulchaey et al. 1993, 1996a; Ponman & Bertram 1993; David et al. 1994; Henry et al. 1995; Pildis et al. 1995). Based on the *ROSAT* mass estimates, X-ray groups make a substantial contribution to the total mass density of the Universe (Henry et al. 1995; Mulchaey et al. 1996a).

Adding up the baryons in galaxies and intragroup gas and comparing to the total mass, one finds that the known baryonic components typically account for only 10%–20% of the total mass (Mulchaey et al. 1993, 1996a; Ponman & Bertram 1993; David et al. 1994; David,

Jones, & Forman 1995; Davis et al. 1995; Doe et al. 1995; Pildis et al. 1995; Pedersen, Yoshii, & Sommer-Larsen 1997). Thus, X-ray groups are dark matter dominated. The gas mass fraction is considerably lower in groups than in clusters at the same fixed fraction of the virial radius (Sanderson et al. 2003). In general, it is not possible to accurately estimate the total gas mass in groups because the X-ray emission is detected to less than half of the virial radius (Helsdon & Ponman 2000a; Mulchaey 2000). If the observed gas density profiles extend out to the virial radius, than the intragroup medium comprises a significant reservoir of baryons in the local Universe (e.g., Fukugita, Hogan, & Peebles 1998).

22.6 Evidence for "Additional" Physics

Cold dark matter models have been remarkably successful at predicting the formation of structure in the Universe. In the standard model, the shape of dark matter potentials is expected to be the same over a wide range of mass (Navarro, Frenk, & White 1997). In the absence of radiative cooling and supernova/AGN heating, the gas density profiles are also expected to be similar over a range in mass. This leads to the expectation that clusters and groups of galaxies should follow similar scaling relationships between properties such as X-ray luminosity (L_X) and temperature (T). For example, assuming the emission is dominated by bremsstrahlung, one expects $L_X \propto T^2$. It has been known for over a decade, however, that the observed relationship between L_X and T is considerably steeper for rich clusters ($L_X \propto T^3$; Edge & Stewart 1991; White, Jones, & Forman 1997; Markevitch 1998).

Considerable effort has gone into understanding the nature of the $L_X - T$ relationship in the group regime. Based on *ROSAT* observations, Helsdon & Ponman (2000a) find that the $L_X - T$ relationship is considerably steeper for groups than for clusters ($L_X \propto T^{4.9}$). However, Horner (2001) finds that while there is more scatter in the relationship at low X-ray luminosities (i.e., groups), a single relationship with $L_X \propto T^3$ holds over a factor of 10^4 in X-ray luminosity. Some caution is warranted when comparing groups and clusters because the X-ray luminosities of groups are generally measured to smaller fraction of the virial radius than rich clusters (Fig. 22.2; Helsdon & Ponman 2000a; Mulchaey 2000). Thus, in general, a smaller fraction of the total gas mass and X-ray luminosity is detected in groups. A detailed study of groups and clusters taking this aperture effect into account would be valuable.

Many authors have explored the failure of the simplest CDM models to predict the slope of the observed $L_X - T$ relationship in groups and clusters. It is generally believed that the inclusion of physics other than gravity can explain the discrepancy. This additional physics may also explain the entropy ($S \propto T n^{-2/3}$) of the gas in groups and clusters. In the standard CDM model, most of the entropy in the X-ray gas is predicted to be from the accretion shock of the gas falling into the potential (Eke, Navarro, & Frenk 1998). This leads to the prediction that the entropy of the gas should increase with the mean temperature of the system. The observed entropy in groups is actually much higher than that expected from the standard model (Ponman, Cannon, & Navarro 1999; Lloyd-Davies, Ponman, & Cannon 2000; Finoguenov et al. 2002). It has been suggested by Ponman and collaborators that there is an "entropy floor": a lower limit to the entropy that a collapsed system can have.

A large amount of theoretical work has been invested in recent years to understand the excess entropy and the shape of the $L_X - T$ relationship. The proposed explanations fall into one of three main classes: (1) "preheating" models, where the intergalactic medium was heated prior to the formation of groups and clusters (Evrard & Henry 1991; Kaiser 1991;

Cavaliere, Menci, & Tozzi 1998; Balogh, Babul, & Patton 1999; Valageas & Silk 1999; Tozzi & Norman 2001; Babul et al. 2002; Dos Santos & Doré 2002; Nath & Roychowdhury 2002), (2) internal heating models, where the gas was heated inside the bound system by supernovae or AGNs (Bower 1997; Loewenstein 2000; Voit & Bryan 2001), or (3) cooling models, where the low-entropy gas was removed from the system, producing an effect similar to heating (Knight & Ponman 1997; Bryan 2000; Pearce et al. 2000; Davé, Katz, & Weinberg 2002; Muanwong et al. 2002; Wu & Xue 2002). While other mechanisms are possible (e.g., magnetic pressure, cosmic ray pressure), these have not been studied in detail.

Generally, all three classes of models can match the shape of the $L_X - T$ relationship and produce the excess entropy in groups, but the required parameters introduce problems in each case. For example, the preheating scenarios require an uncomfortably large amount of energy to be injected into the intergalactic medium at high redshift (e.g., Babul et al. 2002). Internal heating models utilizing supernovae require nearly all of the available energy to heat the gas, while AGN heating models need to be very finely tuned to match the observations. Pure cooling models overpredict the observed mass of stars in groups and clusters. Many of these problems can probably be overcome by considering more realistic scenarios with both heating and cooling (Borgani et al. 2002; Voit et al. 2002).

The shape of the entropy profile in groups provides a much stronger discriminant of the various models than the $L_X - T$ relation. In particular, preheating models predict a flat entropy profile out to radii where shocks begin to dominate (Babul et al. 2002; Voit et al. 2002). For low-mass systems like groups, the entropy profile should be isentropic over much of the observed emission. In contrast, the internal heating and cooling models tend to produce entropy profiles that are rising from the very centers of groups (although the exact shape of the profiles depends somewhat on the history of the system and the amount of cooling or heating; Brighenti & Mathews 2001; Borgani et al. 2002).

Several recent studies indicate that galaxy groups do not have large isentropic cores, but rather profiles that rise with increasing radius (Fig. 22.4; Ponman, Sanderson, & Finoguenov 2003; Pratt & Arnaud 2003; Mushotzky et al. 2004). These observations are inconsistent with current preheating models. While the observed profiles may be consistent with internal heating and cooling models, to zeroth order these models may be inconsistent with other observations. For example, although Mushotzky et al. (2004) find a wide range in entropy values at a given fraction of the virial radius in their small sample, the total stellar masses of these groups are remarkably uniform. This would appear to be inconsistent with models where cooling dominates because one would expect a correlation between entropy and stellar mass (assuming that the low-entropy gas that drops out forms stars). If supernova heating was the dominant source of the excess entropy, one might naturally predict a correlation between gas metallicity and entropy. No such trend is observed (Finoguenov et al. 2002; Mushotzky et al. 2004). At the moment, the number of systems that have detailed entropy profiles derived from *XMM-Newton* and *Chandra* data is very small. Further observations will prove useful to see if these trends (or lack thereof) still hold when larger samples are available. Regardless, more complex models will probably be required to understand the history of the hot gas in groups (e.g., Ponman et al. 2003).

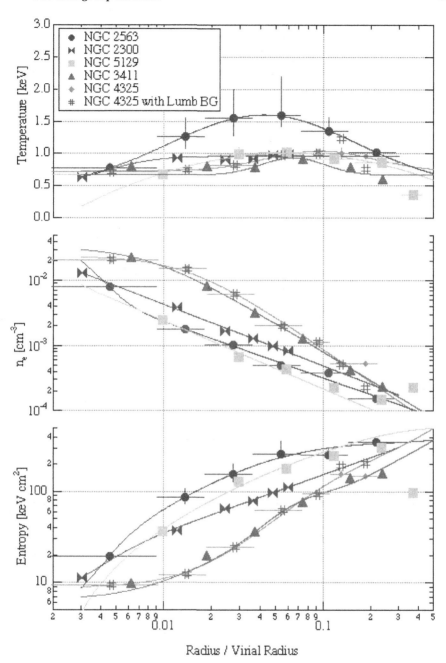

Fig. 22.4. Temperature, density and entropy profiles for five X-ray groups derived from *XMM-Newton* data (Mushotzky et al., in preparation). Note that none of the entropy profiles are consistent with the existence of isentropic cores as predicted by current preheating models.

22.7   The Intragroup Medium in The Local Group

There has been a lot of interest in the last few years in the intragroup medium in the Local Group. As noted in § 22.2, spiral-only groups like the Local Group are not detected in X-ray surveys. However, the intragroup medium in these groups is expected to produce detectable absorption features in the far-UV or X-ray spectra of background quasars. While a clear detection of a warm intragroup medium in other spiral-only groups has yet to be made, the case for the detection of warm gas in the Local Group is now considerably stronger. The grating spectrometers on *Chandra* and *XMM-Newton* have recently been used to detect O VII, O VIII, and Ne IX absorption features at $z \approx 0$ in the spectra of several background AGNs (Kaspi et al. 2002; Nicastro et al. 2002; Fang, Sembach, & Canizares 2003; Rasmussen et al. 2003). Assuming collisional equilibrium, Rasmussen et al. (2003) estimate that the electron temperature of the absorbing gas is in the range of $(2-5) \times 10^6$ K. They also derive a lower limit for the length scale of the gas of 140 kpc. Based on these properties, they conclude that the hot gas is associated with the Local Group's intragroup medium. Nicastro et al. (2002) have studied the bright blazar PKS 2155–304, and in this case the X-ray features are associated with a known O VI UV absorber detected by *Far Ultraviolet Spectroscopic Explorer (FUSE)*. If a single temperature is assumed for both the UV and X-ray features, the physical properties of the absorber are also consistent with the expectations for the Local Group intragroup medium. However, the single-temperature assumption may not be a good one. Heckman et al. (2002) have considered a radiative cooling model that contains gas over a broad range of temperatures. In this model, the scale length of the gas can be much smaller than in the isothermal models, allowing the gas to be associated with the disk or halo of the Milky Way (see also Fang et al. 2003 for a nice discussion of the high-ionization absorbers along the line of sight to 3C 273). Thus, the origin of these X-ray features is still not clear.

At the temperatures expected for a warm intragroup medium, many of the most prominent absorption features fall in the far-UV region. The strongest of these lines is expected to be the O VI doublet at $\lambda\lambda 1031.93, 1037.62$ Å. A few of the O VI systems discovered with *HST* have properties that may be consistent with intervening groups (Bergeron et al. 1994; Savage, Tripp, & Lu 1998). The number of known O VI systems has increased dramatically in the last few years with the launch of *FUSE*. Particularly interesting is the discovery of a class of high-velocity O VI clouds (Sembach et al. 2000, 2003). Nicastro et al. (2003) have examined the velocity distribution of these objects and find that the amplitude of the average velocity vector decreases significantly as one moves from the Local Standard of Rest to the Local Group rest frame. This suggests that these clouds may originate in a warm intragroup medium distributed throughout the Local Group. However, other explanations are possible. In particular, it seems likely that at least some of the O VI high-velocity clouds originate in the Milky Way (Fang et al. 2003; Sembach et al. 2003; Sternberg 2003).

Several other observations imply the presence of warm gas in the Local Group. Wang & McCray (1993) found evidence in the soft X-ray background for a thermal component with temperature $\sim$0.2 keV, which could be due to a warm intragroup medium in the Local Group (see, however, Sidher, Sumner, & Quenby 1999, who argue that the X-ray halo of the Galaxy dominates). Maloney & Bland-Hawthorn (1999) have shown that the ionizing flux produced by warm intragroup gas in the Local Group may be sufficient to induce detectable Hα emission from the outer parts of H I disks. Indeed, an encounter between intragroup gas and the Magellanic Stream may be responsible for the strong Hα emission detected by

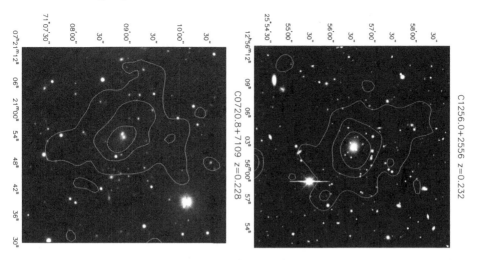

Fig. 22.5. X-ray contours derived from *XMM-Newton* data overlaid on optical images of two moderate-redshift groups (Mulchaey et al. 2004). These *XMM-Newton* data have enough counts to allow a temperature and metallicity measurement.

Weiner & Williams (1996; the Magellanic Stream has also been detected in O VI, Sembach et al. 2000, 2003). Future searches for Hα emission near the outer parts of H I disks in the Local Group could set interesting limits on intragroup gas in our own system.

22.8   Concluding Remarks

While X-ray studies of rich clusters have been underway for several decades now, the study of the hot gas in groups is still a relatively new field. We have learned a tremendous amount from the *ROSAT* and *ASCA* missions. A large number of groups have already been observed with *Chandra* and *XMM-Newton*, and I anticipate many new results in the coming years. X-ray studies of rich clusters have now been extended to redshifts beyond $z \approx 1$. So far, X-ray group studies have been restricted to the nearby Universe. This situation is rapidly changing as groups are now being studied out to moderate redshifts with *XMM-Newton* and *Chandra* (Fig. 22.5; Grant et al. 2004; Mulchaey et al. 2004). Therefore, it should be possible to directly study the evolution of the intragroup medium out to $z \approx 0.5$ in the near future.

Acknowledgements. I would like to thank David Davis, Stephen Helsdon, Lori Lubin, Richard Mushotzky, Trevor Ponman, Ken Sembach, and Ann Zabludoff for useful discussions during the preparation of this work. I would also like to thank Stephen Helsdon for providing Figure 22.3 and Mark Donikian for help with the other figures. Finally, I would like to acknowledge Luis Ho for the incredible amount of effort he has spent organizing Carnegie's centennial symposia and overseeing the publication of these volumes.

References

Arnaud, M., & Evrard, A. E. 1999, MNRAS, 305, 631
Babul, A., Balogh, M. L., Lewis, G. F., & Poole, G. B. 2002, MNRAS, 330, 329
Bahcall, N. A., Harris, D. E., & Rood, H. J. 1984, ApJ, 284, L29
Balogh, M. L., Babul, A., & Patton, D. R. 1999, MNRAS, 307, 463

Bartelmann, M., & Steinmetz, M. 1996, MNRAS, 283, 431

Bauer, F. E., et al. 2002, AJ, 123, 1163

Belsole, E., Sauvageot, J.-L., Ponman, T. J., & Bourdin, H. 2003, A&A, 398, 1

Bergeron, J., et al. 1994, 436, 33

Biermann, P., & Kronberg, P. P. 1983, ApJ, 268, L69

Biermann, P., Kronberg, P. P., & Madore, B. F. 1982, ApJ, 256, L37

Blanton, E. L., Sarazin, C. L., McNamara, B. R., & Wise, M. W. 2001, ApJ, 558, 15

Borgani, S., Governato, F., Wadsley, J., Menci, N., Tozzi, P., Quinn, T., Stadel, J., & Lake, G. 2002, MNRAS, 336, 409

Bower, R. 1997, MNRAS, 288, 355

Brighenti, F., & Mathews, W. G. 2001, ApJ, 553, 103

——. 2003, ApJ, 587, 580

Bryan, G. L. 2000, ApJ, 544, L1

Buote, D. A. 2000, ApJ, 539, 172

——. 2002, ApJ, 574, L135

Buote, D. A., Lewis, A. D., Brighenti, F., & Mathews, W. G. 2003, ApJ, 594, 741

Cavaliere, A., Menci, N., & Tozzi, P. 1998, ApJ, 501, 493

Churazov, E., Brüggen, M., Kaiser, C. R., Böhringer, H., & Forman, W. 2001, ApJ, 554, 261

Cowie, L. L., Henriksen, M., & Mushotzky, R. F. 1987, ApJ, 317, 593

Davé, R., Katz, N., & Weinberg, D. H. 2002, ApJ, 579, 23

David, L. P., Jones, C., & Forman, W. 1995, ApJ, 445, 578

David, L. P., Jones, C., Forman, W., & Daines, S. 1994, ApJ, 428, 544

David, L. P., Nulsen, P. E. J., McNamara, B. R., Forman, W., Jones, C., Ponman, T., Robertson, B., & Wise, M. 2001, ApJ, 557, 546

Davis, D. S., Mulchaey, J. S., & Mushotzky, R. F. 1999, ApJ, 511, 34

Davis, D. S., Mulchaey, J. S., Mushotzky, R. F., & Burstein, D. 1996, ApJ, 460, 601

Doe, S. M., Ledlow, M. J., Burns, J. O., & White, R. A. 1995, AJ, 110, 46

Donahue, M. E., & Voit, G. M. 2004, in Carnegie Observatories Astrophysics Series, Vol. 3: Clusters of Galaxies: Probes of Cosmological Structure and Galaxy Evolution, ed. J. S. Mulchaey, A. Dressler, & A. Oemler (Cambridge: Cambridge Univ. Press), in press

Dos Santos, S., & Doré, O. 2002, A&A, 383, 450

Dos Santos, S., & Mamon, G. A. 1999, A&A, 352, 1

Ebeling, H., Voges, W., & Böhringer, H. 1994, ApJ, 436, 44

Edge, A. C., & Stewart, G. C. 1991, MNRAS, 252, 428

Eke, V. R., Navarro, J. F., & Frenk, C. S. 1998, ApJ, 503, 569

Evrard, A. E., & Henry, J. P. 1991, ApJ, 383, 95

Fabian, A. C. 1994, ARA&A, 32, 277

Fabian, A. C., et al. 2000, MNRAS, 318, 65

Fabricant, D., Lecar, M., & Gorenstein, P. 1980, ApJ, 241, 552

Fabricant, D., Rybicki, G., & Gorenstein, P. 1984, ApJ, 286, 186

Fang, T., & Canizares, C. R. 2000, ApJ, 539, 532

Fang, T., Marshall, H. L., Lee, J. C., Davis, D. S., & Canizares, C. R. 2002, ApJ, 565, 86

Fang, T., Sembach, K. R., & Canizares, C. R. 2003, ApJ, 586, L49

Finoguenov, A., David, L. P., & Ponman, T. J. 2000, ApJ, 544, 188

Finoguenov, A., Jones, C., Böhringer, H., & Ponman, T. J. 2002, ApJ, 578, 74

Frederic, J. J. 1995, ApJS, 97, 259

Fukazawa, Y., Makishima, K., Tamura, T., Ezawa, H., Xu, H., Ikebe, Y., Kikuchi, K., & Ohashi, T. 1998, PASJ, 50, 187

Fukugita, M., Hogan, C. J., & Peebles, P. J. E. 1998, ApJ, 503, 518

Geller, M. J., & Huchra, J. P. 1983, ApJS, 52, 61

Grant, C. E., Bautz, M. W., Chartas, G., & Garmire, G. P. 2004, ApJ, submitted (astro-ph/0305137)

Heckman, T. M., Norman, C. A., Stickland, D. K., & Sembach, K. R. 2002, ApJ, 577, 691

Heinz, S., Choi, Y.-Y., Reynolds, C. S., & Begelman, M. C. 2002, ApJ, 569, 79

Hellsten, U., Gnedin, N. Y., & Miralda-Escudé J. 1998, ApJ, 509, 56

Helsdon, S. F., & Ponman, T. J. 2000a, MNRAS, 315, 356

——. 2000b, MNRAS, 319, 933

——. 2003a, MNRAS, 339, 29

————. 2003b, MNRAS, 340, 485

Helsdon, S. F., Ponman, T. J., & Mulchaey, J. S. 2004, ApJ, submitted

Henry, J. P., et al. 1995, ApJ, 449, 422

Hernquist, L., Katz, N., & Weinberg, D. H. 1995, ApJ, 442, 57

Horner, D. J. 2001, Ph.D. Thesis, Univ. Maryland

Hwang, U., Mushotzky, R. F., Burns, J. O., Fukazawa, Y., & White, R. A. 1999, ApJ, 516, 604

Jones, C., & Forman, W. 1984, ApJ, 276, 38

Kaiser, N. 1991, ApJ, 383, 104

Kaspi, S., et al. 2002, ApJ, 574, 643

King, I. R. 1962, AJ, 67, 471

Knight, P. A., & Ponman, T. J. 1997, MNRAS, 289, 955

Lewis, G. F., Babul, A., Katz, N., Quinn, T., Hernquist, L., & Weinberg, D. H. 2000, ApJ, 536, 623

Liedahl, D. A., Osterheld, A. L., & Goldstein, W. H. 1995, ApJ, 438, L115

Lloyd-Davies, E. J., Ponman, T. J., & Cannon, D. B. 2000, MNRAS, 315, 689

Loewenstein, M. 2000, ApJ, 532, 17

Mahdavi, A., Böhringer, H., Geller, M. J., & Ramella, M. 2000, ApJ, 534, 114

Maloney, P. R., & Bland-Hawthorn, J. 1999, ApJ, 522, L81 (erratum ApJ, 550, L231)

Markevitch, M. 1998, ApJ, 504, 27

Markevitch, M., Vikhlinin, A., & Mazzotta, P. 2001, ApJ, 562, L153

Mathur, S., Weinberg, D., & Chen, X. 2003, ApJ, 582, 82

McNamara, B. R., et al. 2000, ApJ, 534, 135

————. 2001, ApJ, 562, 149

Mewe, R. 1991, A&AR, 3, 127

Mohr, J. J., Mathiesen, B., & Evrard, A. E. 1999, ApJ, 517, 627

Muanwong, O., Thomas, P. A., Kay, S. T., & Pearce, F. R. 2002, MNRAS, 336, 527

Mulchaey, J. S. 2000, ARA&A, 38, 289

Mulchaey, J. S., Davis, D. S., Mushotzky, R. F., & Burstein, D. 1993, ApJ, 404, L9

————. 1996a, ApJ, 456, 80

————. 2003, ApJS, 145, 39

Mulchaey, J. S., Lubin, L. M., Helsdon, S. F., & Rosati, P. 2004, in preparation

Mulchaey, J. S., Mushotzky, R. F., Burstein, D., & Davis, D. S. 1996b, ApJ, 456, L5

Mulchaey, J. S., & Zabludoff, A. I. 1998, ApJ, 496, 73

Mushotzky, R. F. 2004, in Carnegie Observatories Astrophysics Series, Vol. 3: Clusters of Galaxies: Probes of Cosmological Structure and Galaxy Evolution, ed. J. S. Mulchaey, A. Dressler, & A. Oemler (Cambridge: Cambridge Univ. Press), in press

Mushotzky, R. F., Figueroa-Feliciano, E., Loewenstein, M., & Snowden, S. L. 2004, ApJ, in press

Nath, B. B., & Roychowdhury, S. 2002, MNRAS, 333, 145

Navarro, J. F., Frenk, C. S., & White, S. D. M. 1995, ApJ, 275, 720

————. 1997, ApJ, 490, 493

Nicastro, F., et al. 2002, ApJ, 573, 157

————. 2003, Nature, 421, 719

Nolthenius, R., & White, S. D. M. 1987, MNRAS, 225, 505

Nulsen, P. E. J., David, L. P., McNamara, B. R., Jones, C., Forman, W. R., & Wise, M. 2002, ApJ, 568, 163

Osmond, J. P., & Ponman, T. J. 2004, in Carnegie Observatories Astrophysics Series, Vol. 3: Clusters of Galaxies: Probes of Cosmological Structure and Galaxy Evolution, ed. J. S. Mulchaey, A. Dressler, & A. Oemler (Pasadena: Carnegie Observatories, http://www.ociw.edu/ociw/symposia/series/symposium3/proceedings.html)

Pearce, F. R., Thomas, P. A., Couchman, H. M. P., & Edge, A. C. 2000, MNRAS, 317, 1029

Pedersen, K., Yoshii, Y., & Sommer-Larsen, J. 1997, ApJ, 485, L17

Perna, P., & Loeb, A. 1998, ApJ, 503, L135

Pietsch, W., Trinchieri, G., Arp, H., & Sulentic, J. W. 1997, A&A, 322, 89

Pildis, R. A., Bregman, J. N., & Evrard, A. E. 1995, ApJ, 443, 514

Ponman, T. J., & Bertram 1993, Nature, 363, 51

Ponman, T. J., Bourner, P. D. J., Ebeling, H., & Böhringer, H. 1996, MNRAS, 283, 690

Ponman, T. J., Cannon, D. B., & Navarro, J. F. 1999, Nature, 397, 135

Ponman, T. J., Sanderson, A. J. R., & Finoguenov, A. 2003, MNRAS, 343, 331

Pratt, G. W., & Arnaud, M. 2003, A&A, 408, 1

Ramella, M., Pisani, A., & Geller, M. J. 1997, AJ, 113, 483

Rasmussen, A., Kahn, S. M., & Paerels, F. 2003, in The IGM/Galaxy Connection: The Distribution of Baryons at z = 0, ed. J. L. Rosenberg & M. E. Putman (Dordrecht: Kluwer), 109

Renzini, A. 1997, ApJ, 488, 35

Renzini, A., Ciotti, L., D'Ercole, A., & Pellegrini, S. 1993, ApJ, 419, 52

Sanderson, A. J. R., Ponman, T. J., Finoguenov, A., Lloyd-Davies, E. J., & Markevitch, M. 2003, MNRAS, 340, 989

Saracco, P., & Ciliegi, P. 1995, A&A, 301, 348

Savage, B. D., Tripp, T. M., & Lu, L. 1998, AJ, 115, 436

Sembach, K. R., et al. 2000, ApJ, 538, L31

——. 2003, ApJS, 146, 165

Sidher, S. D., Sumner, T. J., & Quenby, J. J. 1999, A&A, 344, 333

Soker, N., Blanton, E. L., & Sarazin, C. L. 2002, ApJ, 573, 533

Sternberg, A. 2003, Nature, 421, 708

Sulentic, J. W., Rosado, M., Dultzin-Hacyan, D., Verdes-Montenegro, L., Trinchieri, G., Xu, C. & Pietsch, W. 2001, AJ, 122, 2993

Sun, M., Jones, C., Murray, S. S., Allen, S. W., Fabian, A. C., & Edge, A. C. 2003, ApJ, 587, 619

Tamura, T., Kaastra, J. S., Makishima, K., & Takahashi, I. 2003, A&A, 399, 497

Tozzi, P., & Norman, C. 2001, ApJ, 546, 63

Trinchieri, G., Fabbiano, G., & Kim, D.-W. 1997, A&A, 318, 361

Tully, R. B. 1987, ApJ, 321, 280

Turner, M. J. L., et al. 2001, A&A, 365, 110

Valageas, P., & Silk, J. 1999, A&A, 350, 725

Vikhlinin, A., Forman, W. & Jones, C. 1999, ApJ, 525, 47

Voit, G. M., & Bryan, G. L. 2001, Nature, 414, 425

Voit, G. M., Bryan, G. L. Balogh, M. L., & Bower, R. G. 2002, ApJ, 576, 601

Vrtilek, J. M., David, L. P., Grego, L., Jerius, D., Jones, C., Forman, W., Donnelly, R. H., & Ponman, T. J. 2000, in Constructing the Universe with Clusters of Galaxies, ed. F. Durret and D. Gerbal (Paris: IAP, http://www.iap.fr/Conferences/Colloque/coll2000/)

Wang, Q. D., & McCray, R. 1993, ApJ, 409, L37

Weiner, B. J., & Williams, T. B. 1996, AJ, 111, 1156

White, D. A. 2000, MNRAS, 312, 663

White, D. A., Jones, C., & Forman, W. 1997, MNRAS, 292, 419

Wu, X.-P., & Xue, Y.-J. 2002, ApJ, 572, 19

Xu, H., et al. 2002, ApJ, 579, 600

Zabludoff, A. I., & Mulchaey, J. S. 1998, ApJ, 496, 39

23

Symposium summary

JEREMIAH P. OSTRIKER

Princeton University, USA and University of Cambridge, UK

23.1   Introduction

I will refrain from taking the normal path followed by lecturers asked to summarize conferences, who tend to emphasize three things: the physical principles behind the science, recent dramatic observational results, and, finally, a summary of their own work. The first two have been well presented by the individual lecturers, and the last is available to the interested reader on ADS. Thus, references to my own and others' papers will be kept to a minimum. Instead, I will focus on the questions (rather than the answers): those asked and those not very much addressed.

23.2   Important Results Addressed by Speakers

23.2.1   *Theory on Halos*

The talks by S. White, A. Evrard, B. Moore, and others amply demonstrated that, within the context of relatively standard, hierarchical dark matter dominated schemes for understanding cosmic structure formation, the calculation of certain "halo" properties is well in hand. Dark matter simulations of a specific cosmological model by different practitioners give essentially identical results—if only the gross features are analyzed. Specifically, if one takes some agreed-on density threshold to define a halo (e.g., such that the mean density within structures defined by such a contour level is 200 times the critical density of the Universe at the epoch in question) and then terms the structures so isolated "individual halos," and then one defines the distribution $dP = N_h(M, z)dV\, dM$, which gives the probability that a halo of mass $M \rightarrow M + dM$ will be in dV at epoch z—this function is well understood. Good approximations abound (Press-Schechter, Sheth-Tormen), accurate N-body simulations can easily be made, and different codes/investigators will agree on the results at the 10%–20% level. But there is great uncertainty on other more detailed, closely related subjects: substructure within halos and inner halo slope are two such uncertainties. Are there well-defined subhalos within each halo, and, if so, how can they best be characterized and what is their spatial and mass distribution? This is an important question since it is primarily subhalos, not halos, that would be the homes of observable galaxies. Secondly, does the inner slope α (proportional to $r^{-\alpha}$) have a value close to 1.0, as advocated by Navarro, Frenk, & White (1996), or close to 1.5, as advocated by Moore et al. (1998), or does it have a fair dispersion, with the mean value dependent on the slope of the power spectrum (or effectively on the halo mass), as advocated by Subramanian, Cen, & Ostriker (2000) and Ricotti (2003)? The jury is out on this very important question.

23.2.2 Halos Equal Galaxies?

Second, even if we did understand the properties of the halos completely, how good would be the extremely plausible theoretical assumption—common to all the semi-analytical methods for galaxy formation—that there is a unique correspondence between halo properties and galaxy properties (cf. White and Kauffmann, this conference). Within individual observed systems, the mass distributions of the stellar and halo components are largely non-coextensive; stars and gas dominate the inner parts of galaxies and dark matter halos (if they are present) only dominate in the outer parts, where there are relatively few observational tracers. Insofar as direct numerical simulations have studied this issue, the evidence is mixed and uncertain (Cen & Ostriker 2000: note Fig. 1, galaxies without halos; but also see Yoshida et al. 2002 and Hernquist & Springel 2003: star formation rates from semi-analytical methods and direct numerical simulations agree), so at this point, we would have to say that we simply do not know if the fundamental assumption of semi-analytical methods is valid, or is a pretty good approximation, or is inadequate.

23.3 EROs

One important byproduct of a naive application of this assumption definitely appears to be wrong. While it is certainly true that low-mass dark matter halos form early and high-mass dark matter halos form late, the same does not appear to be true for galaxies. The extremely red objects ("EROs") found by R. Ellis and co-workers (this conference), A. Renzini and co-workers (this conference), and others are clearly giant ellipticals in a fairly complete state of formation seen at moderate to high redshift in numbers approaching those of local giant ellipticals; they are "old" at redshift 2–3. But, low-mass local systems, from dwarf irregulars to some dwarf spheroidals, seem to have had ongoing star formation and, in some cases, seem astrophysically young (i.e., dominated by gas and young stars). Does this invalidate the hierarchical structure formation scenario? I think not. But it does invalidate the simpler versions of semi-analytical methods. In the hydro simulations (e.g., Nagamine et al. 2001, Fig. 3), low-mass galaxies do start forming first, but in the high-density regions, they rapidly merge to make massive systems; and then further infall and star formation ceases (Butcher-Oemler and Dressler-Gunn effects), since they find themselves surrounded by hot gas, which is unable to cool or accrete onto the formed galaxies. But in lower-density regions the dwarf systems survive and continue to accrete gas with ongoing intermittent and inefficient bursts of star formation. Thus, the surviving examples of massive systems are old, and the surviving examples of low-mass systems are relatively young. But, clearly, much more observational and theoretical work will need to be done to test if this proposal is correct or if the existence of EROs does pose a challenge to the ΛCDM paradigm.

23.3.1 Feedback

M. Norman and co-workers (this conference), L. Hernquist and co-workers, and others have wrestled with how to include the "feedback" from star and galaxy formation onto the subsequent star and galaxy formation. One aspect of this problem is well understood and, if not solved, at least on its way to solution. The output of ionizing (and hence heating) radiation from stars, accreting black holes, supernova remnants, etc., is vitally important in reheating and reionizing the Universe and provides, globally, the thermodynamic context for galaxy formation. This is already included in the hydrodynamic codes. Locally, where we should make appropriate allowance for individual sources and sinks of radiation, we know what to do, but are only gradually implementing the detailed radiative transfer required to do it right (cf., for example, Cen 2002; Razoumov et al. 2002). Observations provide strong guidance here, since we can actually see in the UV the relevant radiation from galaxies and can thus make sure that our theoretical modeling is accurate. But perhaps equally important is the mechanical energy input into the intergalactic medium from supernovae (collectively galactic winds and "superwinds") and jets from AGNs. We observe (cf. Heckman and co-workers; e.g., Cid Fernandes et al. 2001) the outpouring of mechanical energy from nearby starburst galaxies and Steidel and Adelberger (this conference) have found that this is generically present in the Lyman-break galaxies at redshifts $z = 2 - 3$. How does one model this physical process, and is it important in affecting further galaxy formation? The answer, simply put, is that we do not presently know how to model it and we believe that it is important. The difficulty is that most of the relevant codes do not include magnetic fields or cosmic rays, nor do they allow for the consequent multiphase medium induced when supernova energy is injected into such a medium (cf. Cox 1979; McKee & Ostriker 1988). In such a medium, where most of the gaseous mass is contained in "clouds" or "filaments" occupying a small fraction of the volume, blast waves will propagate efficiently through the low-density component and can emerge from a galaxy with only moderate losses. But accurate computation with current generation high-resolution codes (e.g., Gnedin & Bertschinger 1996), which do not allow for the high-energy component and consequently do not generate multiphase media, simply find that the supernova blasts are radiatively quenched. One solution to the problem (Yoshida, Sugiyama, & Hernquist 2003) is to put a plausible multiphase medium into the codes "by hand," while another (Cen & Ostriker 2004) is to bypass the difficult problem of propagation of blasts within a galaxy by injecting the energy in the volume surrounding the galaxy and normalizing the energy output by observations of starburst galaxies and Lyman-break galaxies. It is not known now if either of these approaches will be satisfactory.

23.3.2 *Clusters Gas Cooling Problem*
Many authors (e.g., R. Mushotzky, M. Norman, and others, all at this conference) have stressed that we see gas in clusters of galaxies that has a radiative cooling time shorter than the Hubble time, but the gas does not seem to be cooling and "dropping out" as stars. How is one to understand this? This is a real and important point, but it is necessary to stress that this is also true for almost all other gas in the Universe. It is true for the general intergalactic medium (not just the cluster gas); it is also true for the general interstellar medium (where star formation on a cooling timescale is less than one part per thousand); and it is true for the gas in the Sun! But, in all these other venues, we know the explanation; there exists some heating process, either steady or intermittent, which balances—in a time averaged sense—the inevitable (and observed) radiative cooling. We remain uncertain what that process is in the clusters of galaxies, with the best guess to date (cf. Binney and co-workers; e.g., Binney & Tabor 1995) being mechanical energy input from AGNs. This problem (and its solution) is probably related to the problem that the cluster temperature-luminosity relation is not as expected theoretically, a mystery pointed out by several lecturers at this conference.

23.3.3   *Cluster Star Formation Rates*

A. Renzini, A. Dressler, and others at this conference have stressed that the star formation rate is not highest in the highest-density regions. A simple application of "bias" would say that galaxy density monotonically increases with dark matter density (in the simplest picture increases linearly), and so, dividing by the Hubble time, we should find the highest rates of star formation in the richest clusters. But, the contrary is correct: star formation rates are unusually low in such regions! Why? Over the years, physical collisions, ram pressure stripping, and tidal shocks have been proposed as mechanisms to explain this phenomenon. All must occur and must contribute to the solution, but surely the root cause is that the escape velocity from typical galaxies is less than the rms velocity of galaxies in clusters, or (equivalently) the speed of sound in the cluster gas. Thus, energetically, there is simply no incentive for gas to settle in or to stay in galaxy potential wells for galaxies in rich clusters. To date, semi-analytical models, which tend to treat the "halo" as the unit that fixes the properties of the hypothetically embedded galaxy, have not easily included the interaction between a forming galaxy and the surrounding intergalactic medium. But the hydro methods (cf. Blanton et al. 2000, Fig. 12) naturally reproduce the Butcher-Oemler and Dressler-Gunn effects.

23.4   Observations, Phenomenology, and Data Interpretation

There appears to be a universal agreement (cf. W. Couch, C. Miller, R. Nichol, and others at this conference) concerning the star formation rate at the other extreme—regions of low stellar density. At low redshift, the star formation rate decreases rapidly as the stellar density decreases. We know of this as an empirical result, valid if we average over regions with a smoothing length measured in kiloparsecs or megaparsecs. We do not understand the cause of this phenomenon, and it would be most helpful if it could be quantified better by the observers. The relation cannot be true at all epochs or otherwise galaxy formation could never have commenced (since at early time $\rho_* \to 0$ everywhere)! I suspect that the critical variable is, in fact, gas density, not star density, and that at the current epoch most regions of low stellar density also have a low density of cold gas, but that at earlier epochs, galaxy formation was initiated in regions of low stellar density, which had a high density of dark matter and of gas. In general, both observations and theory indicate that if there are regions within which the gas density dominates over both the stellar and dark matter densities, and both cooling and Jeans criteria are satisfied, then star formation is prompt.

23.4.1   *High-density Regions*

J. Huchra, R. Mushotzky, R. Davies, B. Poggianti, A. Renzini, L. Jones, T. Treu, M. Franx, D. Kelson, A. Dressler, and others at this conference agree that, in high-density regions (e.g., clusters of galaxies), galaxies are brighter, older, and perhaps somewhat more metal rich than in lower-density regions. In fact, if the giant ellipticals formed at redshifts $z \gtrsim 2.5$, then the circumambient gas must have been enriched in metals even before the clusters formed. It is an open question if these important facts, which are universally agreed on by observers, are consistent with the current standard $\Omega_m = 0.3$ ΛCDM models. My own guess is that they will force us to consider lower-density models (e.g., $\Omega_m = 0.2$), within which galaxy formation commences earlier than in $\Omega_m = 0.3$ models.

23.4.2   X-ray Cluster Evolution

The closely related finding (cf. P. Rosati, B. Holden, and others at this conference) that the X-ray clusters have evolved relatively little from the present back to $z \approx 1$ provides an interesting and important puzzle. This is likely to be another strong indicator that we live in a low-density Universe (cf. N. Bahcall and co-workers).

23.4.3   Internal X-ray Cluster Structure

R. Mushotzky, M. Donahue, A. Renzini, and others at this conference have stressed that in the "X-ray concentrated" clusters, the central cores are higher in density, lower in temperature, and higher in metallicity than the other parts $[T_c \approx (1/3)\langle T \rangle]$. But these are not cooling flow clusters, since the absence of still lower-temperature gas shows that the expected cooling flows are not occurring. This presents, as noted earlier, a set of serious theoretical puzzles. What keeps the gas in such regions from fragmenting and from collapsing further to higher densities and lower temperatures?

23.5   Methodology and Technology

23.5.1   Weak and Strong Gravitational Lenses

Several authors (e.g., J. Cohen, I. Smail, M. Postman, and others at this conference) have stressed the potential of techniques to measure the distortion or multiplication of galactic images due to the effects of intervening gravitational inhomogeneities. These techniques, still in their infancy, have enormously powerful potential in that they measure directly the gravitational field perturbations along the line of sight, and are thus unaffected by our insecurity in understanding the complicated relations between mass and light distributions. Preliminary results obtained are roughly consistent with those obtained by other methods. For weak lensing, improvements will be made as one is better able to accurately correct for instrumental effects in measuring low-level light distortions; for strong lensing, we await complete optical surveys with well-defined selection criteria (e.g., similar to the radio CLASS survey). Since strong lensing is dependent on the very small fraction (of order 1%–3%) of matter in the highest-density regions (on the Gaussian tail of the cosmic spectrum of fluctuations), it is very sensitive to cosmic model variations.

23.5.2   Sunyaev-Zel'dovich Effects

Similarly, M. Birkinshaw, K. Romer, and others at this conference have stressed the exciting potential of future Sunyaev-Zel'dovich surveys. This well-understood physical effect can now be used to study large numbers of clusters containing hot gas. It is linear in the integrated surface pressure, and detectability is independent of distance for resolved sources. Interpretation will require high-precision theoretical modeling.

23.5.3   Multi-band, Large-area, Deep Photometry (with Spectroscopic Follow-up)

Work by C. Steidel and co-workers (reported at this conference) and by other groups has already given us more information on galaxy formation at $z \approx 2-3$ than any other method. These techniques are still under development, but already they have provided us with vital information on the formation of galactic bulges, on the QSO-galaxy relation, on the prevalence of galactic wind feedback, etc. With additional large telescopes having large-area detectors coming online, this approach holds great promise.

23.6   Important Results Based on Clusters Not Addressed

23.6.1   *Cosmology*

One can guess that the primary reason that using clusters as a tool to study cosmology was not addressed specifically at this conference is that there was another Carnegie Symposium on cosmology. But it is worth stressing that both the classical cosmological parameters (Ω_m, H_0, q_0) and the newer ones (Ω_Λ, Ω_b, σ_8) are measurable using well-defined cluster samples. As an important example, Peebles and, separately, N. Bahcall have used clusters to estimate Ω_m. Essentially, one uses the virial theorem (in a manner similar to that pioneered by F. Zwicky) to estimate the total mass within a fiducial radius in a rich cluster, and observes the light from the same region. Then, the so derived mass-to-light ratio can be multiplied by the light density of the Universe to provide an estimate of the global matter density. The result is $\Omega_m = 0.12 - 0.20$, somewhat below the estimates obtained by other modern techniques, but not obviously in error. Resolving this discrepancy is an important task. The Hubble constant can be estimated (for a given cosmological model) by combining the distance independent Sunyaev-Zel'dovich measurements with X-ray fluxes, which are inversely proportional to the square of the luminosity distance. When large samples are available (to average over projection effects), this should provide an important independent method for estimating H_0. The number density of rich clusters is a strong function of $\sigma_8 \Omega_m^{0.6}$ and therefore provides a significant constraint on the matter density perturbations at scales smaller than those directly measured by cosmic background radiation probes. But the evolution of the cluster number density is more sensitive to (Ω_m, Ω_Λ), and thus the combination of observations at a range of redshifts can significantly constrain the possible cosmological models (cf. Bahcall & Bode 2003). The lecturers S. White, M. Birkinshaw, M. Postman, and others at this conference did note the important cosmological implications of cluster measurements, but it is worth stressing that in terms of cosmic scale studied and cosmic epoch, cluster measurements are complementary to cosmic background radiation probes and thus help to remove or reduce the serious degeneracies still present even after analysis of exquisitely accurate cosmic background radiation data (cf. Bridle et al. 2003).

23.7   Discussion

I will not attempt to summarize the lively discussion section that concluded the meeting. Rather, I would prefer to emphasize my agreement with the mood of optimism. While these are among the largest and longest studied cosmological objects, the potential they provide for helping us unravel nature's secrets remains enormous. The current combination of new and much more powerful observational techniques with a well-defined theoretical paradigm for the growth of structure and, lastly, increasingly powerful computational simulations to link observations and theory will surely lead to a very rapid expansion of our understanding. I would guess that the next half century will be at least as productive as the half century that followed Zwicky's pioneering work.

References

Bahcall, N. A., & Bode, P. 2003, ApJ, 588, L1
Binney, J., & Tabor, G. 1995, MNRAS, 276, 663
Blanton, M., Cen, R., Ostriker, J. P., & Strauss, M. A. 2000, ApJ, 531, 1
Bridle, S. L., Lahav, O., Ostriker, J. P., & Steinhardt, P. J. 2003, Science, 299, 1532
Cen, R. 2002, ApJS, 141, 211
Cen, R., & Ostriker, J. P. 2000, ApJ, 538, 83

———. 2004, in preparation

Cid Fernandes, R., Jr., Heckman, T. M., Schmitt, H. R., Golzález Delgado, R. M., & Storchi-Bergmann, T. 2001, ApJ, 558, 81

Cox, D. P. 1979, ApJ, 234, 863

Gnedin, N., & Bertschinger, E. 1996, ApJ, 470, 115

Hernquist, L., & Springel, V. 2003, MNRAS, 341, 1253

McKee, C. F., & Ostriker, J. P. 1988, Rev. Mod. Phys., 60, 1

Moore, B., Governato, F., Quinn, T., Stadel, J., & Lake, G. 1998, ApJ, 499, 5

Nagamine, K., Fukugita, M., Cen, R., & Ostriker, J. P. 2001, MNRAS, 327, L10

Navarro, J. F., Frenk, C. S., & White, S. D. M. 1996, ApJ, 462, 563

Razoumov, A., Norman, M. L., Abel, T., & Scott, D. 2002, ApJ, 572, 695

Ricotti, M. 2003, MNRAS, 344, 1237

Subramanian, K., Cen, R., & Ostriker, J. P. 2000, ApJ, 538, 528

Yoshida, N., Stoehr, F., Springel, V., & White, S. D. M. 2002, MNRAS, 335, 762

Yoshida, N., Sugiyama, N., & Hernquist, L. 2003, MNRAS, 344, 481

Credits

The following figures in this volume were reproduced with permission from the original author and publisher.

Figure 1.2: Jenkins, A. et al. 2001, MNRAS, 321, 372, "The Mass Function of Dark Matter Haloes." Reproduced with permission from Blackwell Publishing.

Figure 1.3: Reed, D., et al. 2003, MNRAS, 346, 565, "Evolution of the Mass Function of Dark Matter Haloes." Reproduced with permission from Blackwell Publishing.

Figure 2.1: Bennett, C. L., et al. 2003, ApJS, 148, 1, "First-Year Wilkinson Microwave Anisotropy Probe (WMAP) Observations: Preliminary Maps and Basic Results." Reproduced with permission from the American Astronomical Society.

Figure 2.8: Gómez, P. L., et al. 2003, ApJ, 584, 210, "Galaxy Star-Formation as a Function of Environment in the Early Data Release of the SDSS." Reproduced with permission from the American Astronomical Society.

Figure 2.10: Goto, T. et al. 2003, PASJ, 55, 757, "The Environment of Passive Spiral Galaxies in the SDSS." Reproduced with permission from The Astronomical Society of Japan.

Figure 3.2–3.6: De Propris, R., et al. 2003, MNRAS, 342, 725, "The 2dF Galaxy Redshift Survey: The Luminosity Function of Cluster Galaxies." Reproduced with permission from Blackwell Publishing.

Figure 3.7–3.8: Lewis, I. J., et al. 2002, MNRAS, 334, 673, "The 2dF Galaxy Redshift Survey: The Environmental Dependence of Galaxy Star Formation Rates near Clusters." Reproduced with permission from Blackwell Publishing.

Figure 5.2: Vikhlinin, A., et al. 2002, ApJ, 578, L107, "Evolution of the Cluster X-ray Scaling Relations since $z > 0.4$." Holden, B. P., et al. 2002, AJ, 124, 33, "Moderate-Temperature Clusters of Galaxies from the RDCS and the High-Redshift Luminosity-Temperature Relation." Reproduced with permission from the American Astronomical Society.

Figure 5.4: Rosati, P., Borgani, S., & Norman, C. 2002, ARA&A, 40, 539, "The Evolution of X-ray Clusters of Galaxies." Reproduced with permission from the *Annual Reviews of Astronomy and Astrophysics*, Volume 40©2002 by Annual Reviews, www.annualreviews.org.

Figure 5.5: Allen, S. W., Schmidt, R. W., & Fabian, A. C. 2002, MNRAS, 334, L11, "Cosmological Constraints from the X-ray Gas Mass Fraction in Relaxed Lensing Clusters Observed with Chandra." Reproduced with permission from Blackwell Publishing.

Figure 7.4: Smith, G. P., et al. 2001, ApJ, 552, 493, "A Hubble Space Telescope Lensing Survey of X-ray Luminous Galaxy Clusters. I. A383." Reproduced with permission from the American Astronomical Society.

Figure 7.5: Allen, S. W. 1998, MNRAS, 296, 392, "Resolving the Discrepancy between X-ray and Gravitational Lensing Mass Measurements for Clusters of Galaxies." Reproduced with permission from Blackwell Publishing.

Figure 7.7: Natarajan, P., Kneib, J.-P., & Smail, I. 2002, ApJ, 580, L11, "Evidence for Tidal Stripping of Dark Matter Halos in Massive Cluster Lenses." Reproduced with permission from the American Astronomical Society.

Figure 9.2: Peterson, J. R., et al. 2003, ApJ, 590, 207, "High-resolution X-ray Spectroscopic Constraints on Cooling-flow Models for Clusters of Galaxies." Reproduced with permission from the American Astronomical Society.

Figure 9.4: Oegerle, W. R., et al. 2001, ApJ, 560, 187, "FUSE Observations of Cooling-flow Gas in the Galaxy Clusters A1795 and A2597." Reproduced with permission from the American Astronomical Society.

Figure 9.5: Balogh, M. L., et al. 2001, MNRAS, 326, 1228, "Revisiting the Cosmic Cooling Crisis." Reproduced with permission from Blackwell Publishing.

Figure 9.6: Fabian, A. C., Voigt, L. M., & Morris, R. G. 2002, MNRAS, 335, L71, "On Conduction, Cooling Flows and Galaxy Formation." Reproduced with permission from Blackwell Publishing.

Figure 10.2: Molnar, S. M., & Birkinshaw, M. 2000, ApJ, 537, 542, "Contributions to the Power Spectrum of Cosmic Microwave Background from Fluctuations Caused by Clusters of Galaxies." Reproduced with permission from the American Astronomical Society.

Figure 10.3: Worrall, D., & Birkinshaw, M. 2004, MNRAS, in press (astro-ph/0301123), "The Temperature and Distribution of Gas in CL 0016+16 Measured with XMM-Newton." Reproduced with permission from Blackwell Publishing.

Figure 11.4: Koopmans, L. V. E., & Treu, T. 2003, ApJ, 583, 606, "The Structure and Dynamics of Luminous and Dark Matter in the Early-type Lens Galaxy of 0047–281 at $z = 0.485$." Reproduced with permission from the American Astronomical Society.

Figure 12.1: Bower, R. G., Lucey, J. R., & Ellis, R. S. 1992, MNRAS, 254, 601, "Precision Photometry of Early-type Galaxies in the Coma and Virgo Clusters: A Test of the Universality of the Colour-magnitude Relation. II. Analysis." Reproduced with permission from Blackwell Publishing.

Figure 12.2: van Dokkum, P. G., & Stanford, S. A. 2003, ApJ, 585, 78, "The Fundamental

Plane at $z = 1.27$: First Calibration of the Mass Scale of Red Galaxies at $z > 1$." Reproduced with permission from the American Astronomical Society.

Figure 12.3: van Dokkum, P. G., et al. 2001, ApJ, 552, L101, "The Galaxy Population of Cluster RXJ0848+4453 at $z = 1.27$." Reproduced with permission from the American Astronomical Society.

Figure 12.5: Franx, M., et al. 2003, ApJ, 587, L79, "A Significant Population of Red, Near-infrared-selected High-redshift Galaxies." Reproduced with permission from the American Astronomical Society.

Figure 14.1: Durrell, P. R., et al. 2002, ApJ, 570, 119, "Intracluster Red Giant Stars in the Virgo Cluster." Reproduced with permission from the American Astronomical Society.

Figure 14.2: Terlevich, A. I., et al. 1999, MNRAS, 310, 445, "Colour-magnitude Relations and Spectral Line Strengths in the Coma Cluster." Reproduced with permission from Blackwell Publishing.

Figure 14.3: Pimbblet, K. A., et al. 2002, MNRÁS, 331, 333, "The Las Campanas/AAT Rich Cluster Survey II: The Environmental Dependance of Galaxy Colours in Clusters at $z \approx 0.1$." Reproduced with permission from Blackwell Publishing.

Figure 14.4: Lewis, I. J., et al. 2002, MNRAS, 334, 673, "The 2dF Galaxy Redshift Survey: The Environmental Dependence of Galaxy Star Formation Rates near Clusters." Reproduced with permission from Blackwell Publishing.

Figure 14.7: Davies, R. L., et al. 2001, ApJ, 548, L33, "Galaxy Mapping with the SAURON Integral-field Spectrograph: The Star Formation History of NGC 4365." Reproduced with permission from the American Astronomical Society.

Figure 15.2: Couch, W. J., & Sharples, R. M. 1987, MNRAS, 229, 423, "A Spectroscopic Study of Three Rich Galaxy Clusters at $z = 0.31$." Reproduced with permission from Blackwell Publishing.

Figure 15.3, 15.5: Kodama, T., & Bower, R. G. 2001, MNRAS, 321, 18, "Reconstructing the History of Star Formation in Rich Cluster Cores." Reproduced with permission from Blackwell Publishing.

Figure 15.4: Smail, I., et al. 2001, MNRAS, 323, 839, "A Photometric Study of the Ages and Metallicities of Early-type Galaxies in A2218." Reproduced with permission from Blackwell Publishing.

Figure 16.2: De Grandi, S., & Molendi, S., 2001, ApJ, 551, 153, "Metallicity Gradients in X-ray Clusters of Galaxies." Reproduced with permission from the American Astronomical Society.

Figure 17.1: Gnedin, O. Y. 2003, ApJ, 582, 141, "Tidal Effects in Clusters of Galaxies." Reproduced with permission from the American Astronomical Society.

Figure 17.6: Feldmeier, J. J., et al. 2002, ApJ, 575, 779, "Deep CCD Surface Photometry of

Galaxy Clusters I: Methods and Initial Studies of Intracluster Starlight." Reproduced with permission from the American Astronomical Society.

Figure 19.1–19.2: Solanes, J. M., et al. 2001, ApJ, 548, 97, " The H I Content of Spirals. II. Gas Deficiency in Cluster Galaxies." Reproduced with permission from the American Astronomical Society.

Figure 19.3–19.4: Bravo-Alfaro, H., et al. 2000, AJ, 119, 580, "VLA H I Imaging of the Brightest Spiral Galaxies in Coma." Reproduced with permission from the American Astronomical Society.

Figure 19.5: Vollmer, B., et al. 2001, ApJ, 561, 708, "Ram Pressure Stripping and Galaxy Orbits: The Case of the Virgo Cluster." Reproduced with permission from the American Astronomical Society.

Figure 19.12: Verheijen, M. A. W., & Zwaan, M. A. 2001, in Gas and Galaxy Evolution, ed. Hibbard, Rupen, & van Gorkom (San Francisco: ASP), 867. Reproduced with permission from The Astronomical Society of the Pacific.

Figure 20.2: Kodama, T., et al. 2001, ApJ, 562, L9, "The Transformation of Galaxies within the Large-scale Structure around a $z = 0.41$ Cluster." Reproduced with permission from the American Astronomical Society.

Figure 20.6: Reprinted with permission from Quilis, V., Moore, B., & Bower, R. 2000, Science, 288, 1617, "Gone with the Wind: The Origin of S0 Galaxies in Clusters." Copyright 2000 AAAS.

Figure 22.1: Mulchaey, J. S., et al. 2003, ApJS, 145, 39, "An X-ray Atlas of Groups of Galaxies." Reproduced with permission from the American Astronomical Society.